Cell Biology of the Eye

This is a volume in
CELL BIOLOGY
A series of monographs

Editors: D. E. Buetow, I. L. Cameron, G. M. Padilla, and A. M. Zimmerman

A complete list of the books in this series appears at the end of the volume.

Cell Biology of the Eye

Edited by

David S. McDevitt

Department of Animal Biology
University of Pennsylvania
The School of Veterinary Medicine
Philadelphia, Pennsylvania

1982

ACADEMIC PRESS
A Subsidiary of Harcourt Brace Jovanovich, Publishers
New York London
Paris San Diego San Francisco São Paulo Sydney Tokyo Toronto

ACADEMIC PRESS, INC.
111 Fifth Avenue, New York, New York 10003

United Kingdom Edition published by
ACADEMIC PRESS, INC. (LONDON) LTD.
24/28 Oval Road, London NW1 7DX

Library of Congress Cataloging in Publication Data
Main entry under title:

Cell biology of the eye.

(Cell biology)
Includes bibliographies and index.
1. Eye. 2. Cytology. I. McDevitt, David S.
II. Series. [DNLM: 1. Eye--Cytology. 2. Eye--Physiology. WW 101 C393]
QP475.C44 1982 599.01'823 82-8880
ISBN 0-12-483180-X AACR2

PRINTED IN THE UNITED STATES OF AMERICA

82 83 84 85 9 8 7 6 5 4 3 2 1

Contents

3. Growth Factors: Effect on Corneal Tissue

D. GOSPODAROWICZ AND L. GIGUÈRE

4. Ontogeny and Localization of the Crystallins in Eye Lens Development and Regeneration

DAVID S. McDEVITT AND SAMIR K. BRAHMA

5. Transdifferentiation of Lens Cells and Its Regulation

TUNEO YAMADA

6. Biological-Physical Basis of Lens Transparency

FREDERICK A. BETTELHEIM AND ERNEST L. SIEW

7. Control of Cell Division in the Ocular Lens, Retina, and Vitreous Humor

JOHN R. REDDAN

8. Retinoids in Ocular Tissues: Binding Proteins, Transport, and Mechanism of Action

GERALD J. CHADER

9. Chromatic Organization of the Retina

ROBERT E. MARC

List of Contributors

Numbers in parentheses indicate the pages on which the authors' contributions begin.

FREDERICK A. BETTELHEIM (243), Chemistry Department, Adelphi University, Garden City, New York 11530

SAMIR K. BRAHMA (143), Department of Medical Anatomy and Embryology, The State University, Janskerkhof 3A, 3512 BK Utrecht, The Netherlands

GERALD J. CHADER (377), National Eye Institute, National Institutes of Health, Bethesda, Maryland 20205

L. GIGUÈRE (97), Cancer Research Institute and the Departments of Medicine and Ophthalmology, University of California Medical Center, San Francisco, California 94143

D. GOSPODAROWICZ (97), Cancer Research Institute and the Departments of Medicine and Ophthalmology, University of California Medical Center, San Francisco, California 94143

GERALD W. HART (1), The Johns Hopkins University School of Medicine, Baltimore, Maryland 21218

ROBERT E. MARC (435), Sensory Sciences Center and Department of Ophthalmology, The University of Texas Health Science Center, Houston, Texas 77025

DAVID S. McDEVITT (143), Department of Animal Biology, University of Pennsylvania, The School of Veterinary Medicine, Philadelphia, Pennsylvania 19104

WILLIAM W. MORGAN (533), Department of Anatomy, The University of Texas, Health Science Center at San Antonio, San Antonio, Texas 78284

DAVID S. PAPERMASTER (475), Department of Pathology, V. A. Medical Center, West Haven, Connecticut 06516, and Yale Medical School, New Haven, Connecticut 06510

JOHN R. REDDAN (299), Department of Biological Sciences, Oakland University, Rochester, Michigan 48063

MICHAEL V. RILEY (53), Department of Biological Sciences, Oakland University, Rochester, Michigan 48063

BARBARA G. SCHNEIDER (475), Department of Pathology, V. A. Medical Center, West Haven, Connecticut 06516, and Yale Medical School, New Haven, Connecticut 06510

ERNEST L. SIEW (243), Chemistry Department, Adelphi University, Garden City, New York 11530

TUNEO YAMADA (193), Swiss Institute for Experimental Cancer Research, CH-1066 Épalinges, Switzerland

Preface

This treatise includes contributions that demonstrate the importance of the eye tissues in elucidating questions at the forefront of cell biological research today. The eye has proved exceedingly valuable in investigations on active transport, growth factors, receptors, and differentiation—all topics of immediate interest and likely to remain so, and all discussed in this volume. We believe a treatment of such diversity within the common conceptual framework of the eye to be unique.

This mixture of fact and speculation should be of use to graduate and professional students, teachers, researchers, and clinicians (none of whom are mutually exclusive!), and to all in the vision field. The chronology of the subject matter presented as it pertains to cornea, lens, and retina adheres to this classic order to some extent. We chose not to make this division strict in the interest of "flow" as well as the stress of the three tissues' interdependence on one another to permit the wonder of vision. This book will serve as a reference text and as a source for ideas for future basic and applied eye research.

David S. McDevitt

1

Corneal Proteoglycans

GERALD W. HART

I. INTRODUCTION

A. Scope and Purpose

The highly specialized extracellular matrix of the corneal stroma contains ordered orthogonal arrays of type I collagen fibrils (Trelstad and Coulombre, 1971) and a unique population of proteoglycans (Meyer *et al.*, 1953). Next to those from cartilage, proteoglycans from the vertebrate cornea have been among the most intensively investigated, both in terms of their structure–function relationships and with respect to their biosynthesis. The biochemistry and biology of the extracellular matrix of the vertebrate cornea have been rather exten-

Cell Biology of the Eye

ISBN 0-12-483180-X

sively studied for several reasons: (1) The extracellular matrix composition of the cornea is unique, is comparatively simple, and plays an important role in the development and maintenance of corneal transparency (Meyer *et al.*, 1953; Davidson and Meyer, 1954; Conrad, 1970a,b; Berman, 1970; Castellani *et al.*, 1970; Toole and Trelstad, 1971; Anseth, 1972; Conrad and Dorfman, 1974; Hart, 1976; Klintworth, 1977b; Hart 1978a,b). (2) At nearly all stages of their development, corneas can easily be obtained totally free of other tissues in large amounts for biochemical studies. (3) The morphological details of both developing and adult corneas are well characterized, allowing the correlation of biochemical and morphological phenomena (Hay and Revel, 1969). (4) The degree of success of corneal development or maintenance of normal corneal morphogenesis (i.e., transparency) and the role of the extracellular matrix can easily be studied or quantitated (Coulombre and Coulombre, 1964; Coulombre, 1965). (5) The contributions to the extracellular matrix, and the interactions among each of the three corneal cellular layers, can readily be studied since each cellular layer can be easily isolated, grown apart *in vitro,* and utilized in reconstitution experiments (Conrad, 1970b; Dodson and Hay, 1971; Meier and Hay, 1973; Conrad and Dorfman, 1974; Hart, 1978b).

The development and maintenance of corneal transparency are a result of the concerted action of the three major corneal cell types. The corneal epithelium establishes the primary acellular stroma, which serves as a template for the extracellular matrix of the adult stroma, and it also provides a protective barrier in addition to contributing a unique array of its own macromolecules to the extracellular matrix of the cornea (Hay and Revel, 1969; Hart, 1978b).

The stromal keratocytes synthesize the bulk of the collagen, glycoproteins, and proteoglycans of the adult cornea, and these highly specialized cells synthesize the major proteoglycans of the cornea, which are unique to this tissue (Conrad and Dorfman, 1974; Hart 1978b). The corneal endothelium actively controls corneal hydration (see Chapter 2 by Riley, this volume), in addition to playing an important role in stromal cell invasion (Toole and Trelstad, 1971) and supplying its own contribution of macromolecules to the extracellular matrix (Hart, 1978b). The three cell types also influence each other's biosynthetic activities (Wortman, 1960; Anseth, 1971; Meier and Hay, 1973, 1974a,b), and the achievement of a functional transparent cornea can be regarded as the product of their interactions.

This chapter will focus on the biochemistry and biology of proteoglycans and glycosaminoglycans (the polysaccharide component of proteoglycans) of the vertebrate cornea. Proteoglycans from other tissues will be discussed to provide background and for comparisons of the biochemistry and functions of proteoglycans in the cornea. Numerous recent and excellent reviews on the general aspects of proteoglycan structure (Muir and Hardingham, 1975; Rodén, 1980), biosynthesis (Schachter and Rodén, 1973; Rodén and Schwartz, 1975; Lindahl *et al.*, 1977; Schachter 1978), physical properties (Lindahl and Höök, 1978; Chakrabarti and Park, 1980), and biological functions exist (Balazs, 1970; Slavkin, 1972; Slavkin and Greulich, 1975; Comper and Laurent, 1978; Jaques, 1980). In addition, there are excellent reviews on the morphological aspects of corneal development, structure, and disease (Hay and Revel, 1969; Klintworth, 1977b). Prior to discussing the proteoglycans of the cornea, it will be valuable to mention briefly current notions concerning the overall biological roles of these polyanionic protein-polysaccharides in order to provide a perspective for the discussion of their functions in the cornea.

B. Overview of Biological Aspects of Proteoglycans

The interest of cell and developmental biologists in proteoglycans is on the upswing, since evidence is mounting for their role as mediators of cell-cell and cell-matrix interactions, intercellular ion and protein flux, and morphogenesis. Proteoglycans represent some of the most complex molecules in nature. These very large macromolecules (MW 4×10^3–8×10^6, depending upon the source and methods of isolation) generally consist of a protein core to which are covalently attached various polyanionic polysaccharide chains (glycosaminoglycans; for review, see Muir and Hardingham, 1975; Rodén, 1980). The best three-dimensional structural analogy of individual proteoglycan molecules is that of a bottle brush with the glycosaminoglycans as bristles protruding from the protein core (Mathews and Lozaityte, 1958; Sajdera and Hascall, 1969; Phelps, 1975; Rosenberg *et al.*, 1975; Hardingham *et al.*, 1976).

Relatively little is known about the number of types or the structure of the protein cores of proteoglycans. Recently, a low-molecular-weight chain of chondroitin sulfate was found to be covalently attached to human thyroglobulin (Spiro, 1977), and low-sulfated or nonsulfated keratan sulfate-like polysaccharides have been found to be major components in human red blood cell membranes (Järnfelt *et al.*, 1978). A very recent, detailed study on isolated membrane glycoproteins of normal and Rous sarcoma-transformed baby hamster kidney cells indicates that several membrane glycoproteins are hybrids, in that they contain small glycosaminoglycan side chains in addition to N-glycosidically and/or O-glycosidically linked oligosaccharide side chains covalently attached to the same protein core (Baker *et al.,* 1980). These observations suggest the exciting possibility that glycosaminoglycan side chains might be commonly present on secretory and membrane proteins which are not considered classical proteoglycans.

Our lack of detailed knowledge of the chemistry and biology of intact proteoglycans arises mostly from the technical problems associated with their large size, composite structure, and extreme heterogeneity (Rodén, *et al.,* 1972; Kimura *et al.,* 1978; Rodén, 1980). Proteoglycans are difficult to extract from tissues without extensive degradation or denaturation [requiring extensive mechanical disruption, a denaturing solvent, or very high salt (Section II)]. Their large size and very nonideal behavior make physical studies very difficult (Muir and Hardingham, 1975), and they are very polydisperse (a 10-fold variation in the size of proteoglycans of similar composition is common) (Phelps, 1975; Heinegård, 1977). Proteoglycans within any preparation not only contain variations in the spacing, number, and type of glycosaminoglycans attached, but also the individual types of glycosaminoglycans vary in chain length, degree of sulfation, and minor sugar components within themselves (Brimacombe and Webber, 1964; Rosenberg, *et al.,* 1975). It is precisely this heterogeneity and potential information content, taken together with their extracellular location, that originally interested cell and developmental biologists in these molecules.

The information content of glycosaminoglycans and proteoglycans is largely determined by the spatial relationships among hydroxyl, sulfate, carboxyl, and acetyl side groups along the polysaccharide backbone. The uniqueness of each individual polysaccharide is therefore determined by the types of monosaccharide residues, position and configuration of glycosidic bonds, location and spacing of substituents, conformation of individual monosaccharides, and overall shape and size of the chain(s). Specific glycosyltransferases determine many of these properties, and others are the result of postpolymerization modifications (Rodén, 1970; Rodén and Schwartz, 1975; Lindahl *et al*., 1977; Rodén, 1980).

Since proteoglycans and glycosaminoglycans are major constitutents of the intercellular spaces, along with collagens and elastins, the interactions of these polyanionic polyelectrolytes with water, ions, hormonal substances, other cell surface glycoproteins, and proteins of the intercellular matrix account for many of the physical and biological properties of tissues (Slavkin and Greulich, 1975; Comper and Laurent, 1978; Chakrabarti and Park, 1980).

It is clear from numerous studies that proteoglycans and collagens play a regulatory role in cell and tissue interactions (Nevo and Dorfman, 1972; Kosher *et al.*, 1973; Fitton-Jackson, 1975; Kosher and Church, 1975; Meier and Hay, 1975; Lash and Vasan, 1977; Hart, 1978b). Each stage of morphogenesis and differentiation of a tissue is not only characterized by distinct cell types but also by a unique extracellular matrix (for review, see Manasek, 1973; Slavkin and Greulich, 1975; Lash and Burger, 1977).

Proteoglycans interact strongly with collagens and regulate fibrillogenesis of these proteins (Toole and Lowther, 1968; Gelman and Blackwell, 1974; Greenwald *et al.*, 1975; Oegema *et al.*, 1975). In fact, it appears that all glycosaminoglycans, except perhaps hyaluronic acids and keratan sulfates, bind tightly to collagen at physiological pH and ionic strength. Proteoglycans retard the nucleation phase of collagen fibrillogenesis and visibly alter the organization of the fibril when viewed by electron microscopy (Oegema *et al.*, 1975). Altered collagen fibrillogenesis in the development of the organized multicellular hierarchy is obviously of great importance.

Proteoglycans and glycosaminoglycans have been demonstrated to influence directly both quantitatively and qualitatively the gene expression of cells (for review, Slavkin and Greulich, 1975). The presence of chondroitin sulfate proteoglycans in the culture medium greatly stimulates proteoglycan synthesis by embryonic chondrocytes (Nevo and Dorfman, 1972) and by chondrocytes from adult cartilage (Wiebkin and Muir, 1973). Even very small amounts of exogenously added hyaluronic acids greatly inhibit proteoglycan biosynthesis by both adult (Wiebkin and Muir, 1973) and embryonic chondrocytes (Solurish *et al.*, 1974). On the other hand, hyaluronic acids have no effect on chondroblasts, dermal fibroblasts, or synovial cells in these studies. The observed effects were specific for hyaluronic acids and involved binding of the molecules to the cell surface (Wiebkin *et al.*, 1975). More recent studies indicate that the inhibition of proteoglycan biosynthesis by hyaluronic acids results from inhibition of the biosynthesis of the core protein, not from a reduction in glycosyltransferase activity (Handley and Lowther, 1976). Similar studies on cultured corneal epithelium indicate that heparan sulfates and chondroitin sulfates stimulate glycosaminoglycan biosynthesis by this tissue (Meier and Hay, 1974a). Hyaluronic acids had no effect on corneal epithelium in these studies. In addition, it was found that keratan sulfate I inhibited the synthesis of heparan sulfates by corneal epithelium without affecting the biosynthesis of other glycosaminoglycans (Meier and Hay, 1974a). It has been suggested that glycosaminoglycans affect protein biosynthesis at both the transcriptional and translational levels. Glycosaminoglycans loosen chromatin structure (Arnold *et al.*, 1972; Saiga and Kinoshita, 1976) and increase the template activity of isolated chromatin toward DNA polymerases (Kraemer and Coffey, 1970). Nuclei from many cells appear to contain glycosaminoglycans (Kinoshita, 1974; Bhavanandan and Davidson, 1975; Fromme *et al.*, 1976).

Glycosaminoglycans are also components of plasma, and their interactions with plasma proteins have been extensively studied. The interactions between heparins and coagulation factors have been well studied for obvious reasons of clinical importance (Fujikawa *et al.*, 1973; Gentry and Alexander, 1973; Anderson *et al.*, 1975; Chisell and Catapano, 1979; Jaques, 1980). Many studies on glycosaminoglycans associated with vascular endothelium have examined the possible roles of these glycoconjugates in atherosclerosis (Zilversmit, 1973; Buonassisi and Root, 1975; Rosenberg, 1977; Olivecrona *et al.*, 1977). Both low-density lipoproteins (LDLs) and very low-density lipoproteins (VLDLs) bind glycosaminoglycans (Bernfeld, 1966; Nakashima *et al.*, 1975). LDL is the major cholesterol-carrying lipoprotein which delivers its cholesterol by a receptor-mediated process (Brown *et al.*, 1975). The binding of LDL to the receptor at the cell surface is reversed by heparin or dermatan sulfate (Goldstein *et al.*, 1976). It appears that extracellular glycosaminoglycans might mediate the binding of LDL to the cell surface. In fact, cell surface heparan sulfate-related molecules might actually be components of the LDL receptor. In an analogous manner, it is quite likely that glycosaminoglycans mediate a large number of receptor–ligand interactions at the cell surface.

Cells contain a wide variety of glycosaminoglycans associated with their cell surfaces—especially heparan sulfates (Kraemer, 1971; Kraemer and Smith, 1973; Conrad and Hart, 1975; Kleinman *et al.*, 1975; Roblin *et al.*, 1975). Evidence has been presented that these cell surface glycosaminoglycans mediate cell–cell communication (Roblin *et al.*, 1975), shield surface receptors (Kraemer and Smith, 1973), and regulate cell growth (Ohnishi *et al.*, 1975). Sulfated glycosaminoglycans associated with the cell surfaces of 3T3 cells are markedly decreased after transformation (Goggins *et al.*, 1972; Roblin *et al.*, 1975; Dunham and Hynes, 1978). The addition of exogenous sulfated glycosaminoglycans to cells in culture results in a significant alteration in their mitotic rate (Lippman, 1968). Hyaluronate biosynthesis is increased during cell division and in some transformed cells (Satoh *et al.*, 1973). Hyaluronate produced by cells in culture aggregates lymphoma cells (Pessac and Defendi, 1972; Wasteson *et al.*, 1973), while normal lymphocytes and lymphosarcomas are not aggregated by hyaluronic acids. Hyaluronic acids have also been implicated in cellular adhesion that appears to require their binding to the cell surface (Culp, 1976).

In elegant studies on hyaluronate metabolism in the early development of several tissues, Toole and co-workers established close correlations between hyaluronate synthesis and cell migration, and between hyaluronate removal and cell differentiation (Toole and Trelstad, 1971; Toole and Gross, 1971; Toole, 1972, 1976). Other investigators have also described the presence of hyaluronates at the time and at the site of mesenchymal cell migrations during development. Pratt *et al.* (1975) have demonstrated a corrleation with the production of hyaluronate and the initiation of neural crest migration. Migration of cushion cells from the endocardium into cardiac jelly takes place in the presence of high concentrations of hyaluronates (Manasek *et al.*, 1973).

Sulfated glycosaminogycans are also synthesized early in development, and their increased biosynthesis has been correlated with early cytokinesis (Kinoshita, 1969; Kosher and Searls, 1973) and cell migrations during development (Sugiyama, 1972; Kosher and Searls, 1973; Karp and Solurish, 1974; Kinoshita and Yoshii, 1979). Sulfate deprivation dramatically inhibits the biosynthesis of sulfated glycosaminoglycans and sulfated glycoproteins by sea urchin embryos and concomitantly causes them to undergo aberrant development (Immers and Runnstrom, 1965; Sugiyama, 1972; Karp and Solurish, 1974). Interestingly, sulfate deprivation, and the resulting blockage in the synthesis of sulfated glycoconjugates, blocks the development of the embryos at gastrulation and particularly affects the migration of primary mesenchyme cells (Karp and Solurish, 1974). Very interesting studies involving the microinjection of proteoglycans into developing sea urchins demonstrate that *p*-nitrophenyl-β-xylosides specifically block development at gastrulation and that the blockage is prevented by coinjection of sea urchin proteoglycans isolated from developmental stages at or later than gastrulation but not of proteoglycans from earlier stages of development (Kinoshita and Yoshii, 1979).

In epitheliomesenchymal organs, such as thymus, salivary gland, lung, and pancreas, normal morphogenesis is strictly dependent upon the interaction between the two cellular components (Bernfield *et al.*, 1973). Both glycosaminoglycans and collagens appear to play major roles in the normal development and morphogenesis of these organs.

The interactions between proteoglycans and collagens have led to the suggestion that self-assembly of these extracellular components plays a functional role in the development of the epithelial organ form (Trelstad *et al.*, 1974). Bernfield and co-workers have suggested that the collagen- and glycosaminoglycan-containing basal lamina imposes morphological stability on the epithelium, while the mesenchyme is involved in the changes in morphology that occur (Banjeree *et al.*, 1977; Cohn *et al.*, 1977). Glycosaminoglycans of epithelial basal lamina are synthesized by the epithelium itself (Dodson and Hay, 1971; Trelstad *et al.*, 1974).

II. COMPOSITION AND STRUCTURE OF CORNEAL PROTEOGLYCANS

A. General Features of Proteoglycans

1. *Model Proteoglycans of Cartilage*

Proteoglycans of bovine nasal cartilage serve as the prototype of proteoglycan structure (for a recent review, see Rodén, 1980). Although cartilage proteoglycans are undoubtedly highly specialized and differ in many respects from proteoglycans of other tissues, they serve as the model because our knowledge of their structures and higher-order complexes *in vivo* far exceeds that for proteoglycans of other tissues.

When proteoglycans of cartilage are extracted under dissociative conditions (Sajdera and Hascall, 1969; e.g. 4 *M* guanidine • HCl or 3 *M* $MgCl_2$) and the dissociative solute is dialyzed away, the proteoglycan monomers associate specifically with two other components of the extracellular matrix, hyaluronic acid and "link" proteins, to form highly ordered complex aggregates (Gregory, 1973; Heinegård and Hascall, 1974a; Hardingham and Muir, 1974; Kimura *et al.*, 1978, 1979; Tang *et al.*, 1979; Faltz *et al.*, 1979; Baker and Caterson, 1979; Kimura *et al.*, 1980). The polypeptide core of cartilage proteoglycan monomers has a molecular weight of approximately 200,000 with about 100 chondroitin sulfate chains and 40–60 keratan sulfate chains attached (Hascall and Riolo, 1972; Muir and Hardingham, 1975). One end of the protein core (about 40%) is devoid of glycosaminoglycans and contains the link protein- and hyaluronic acid-binding sites (Heinegård and Hascall, 1974a; Faltz *et al.*, 1979). The chondroitin sulfate side chains are preferentially localized at the opposite end, and the keratan sulfate side chains are clustered between the hyaluronic acid-binding region and the chondroitin sulfate-binding region (Heinegård and Axelsson, 1977).

Recent data which analyzed the assembly of newly synthesized proteoglycans and link proteins in cultures of swarm rat chondrosarcoma cells suggest that the aggregate structures of cartilage proteoglycans contain one link protein per monomer and that the active binding sites for monomer and link protein to hyaluronate in the aggregates are located within a length of less than 30 monosaccharide residues (Kimura *et al.*, 1980).

Cartilage proteoglycan monomers, in addition to containing chondroitin sulfate and keratan sulfate covalently linked side chains, also contain several different types of shorter oligosaccharide side chains (DeLuca *et al.*, 1980; Lohmander *et al.*, 1980).

2. Glycosaminoglycans

At our present level of understanding, the properties of proteoglycans are largely determined by the number and types of glycosaminoglycan side chains attached (Mathews, 1967; Rees, 1975; Comper and Laurent, 1978; Chakrabarti and Park, 1980). In general, eight operationally distinct families of glycosaminoglycans are found in vertebrate tissues. However, several of these families are closely interrelated biosynthetically or structurally. Table I summarizes their properties. Each family is primarily classified on the basis of protein linkage and the repeating disaccharide unit [hexosamine and hexuronic acid (galactose in keratan sulfates)] they contain. However, individual members of each family vary considerably in structure as described earlier.

a. Hyaluronic Acids. Hyaluronic acids are the only completely nonsulfated glycosaminoglycans and are widely distributed throughout nature in both prokaryotes and eukaryotes (Mathews, 1975). As shown in Table I, hyaluronic acids are often over 10 times larger than other types of glycosaminoglycans and have as a backbone the repeating disaccharide structure {[(1→4)-β-D-glucuronosyl-(1→3)-β-D-*N*-acetylglucosaminyl]$_n$}. Also, unlike other glycosaminoglycans, hyaluronic acids do not occur as multichain components of proteoglycans but rather as long single-chain molecules (Laurent, 1970). Whether or not the long single chains of hyaluronic acids are covalently linked to protein remains to be conclusively demonstrated. However, there is evidence suggesting that in eukaryotes hyaluronic acids probably are covalently bound to protein (Scher and Hamerman,

TABLE I

Properties of Glycosaminoglycans[a]

Family	Repeating disaccharide (RD)	Tissues[b]	Linkage to protein	Sulfate M/RD[f]	Intrinsic viscosity	MW ($\times 10^3$)	Elution, Dowex-1 (Cl^-) NaCl (*M*)	Carbazole orcinol[c]
Hyaluronic acids	[(1 → 4)-β-D-glucuronosyl-(1 → 3)-β-D-*N*-acetylglucosaminyl]$_n$	Vitreous humor, skin, umbilical cord, aorta, synovial fluid	?	0	0.2–48	4–8	0.5	1.5–2.0
Chondroitin 4-sulfates	[(1 → 4)-β-D-glucuronosyl-(1 → 3)-β-D-*N*-acetylgalactosaminyl]$_n$	Cartilage, bone, skin, notochord, cornea	Ser(Thr)-Xyl-Gal-Gal	0.1–1.0	0.2–1.0	5–50	1.5	2.0
Chondroitin 6-sulfates		Cartilage, skin, umbilical cord, skin, aorta	Same	0.2–1.3	0.2–1.8	5–50	1.5	2.0
Dermatan sulfates	Same as chondroitin 4-sulfates but with varying amounts of L-iduronate	Skin, tendon, aorta	Same	1.0–2.0	0.5–1.0	15–40	1.7	0.25–0.35
Keratan sulfates I	[(1 → 3)-β-D-galactosyl-(1 → 4)-β-D-*N*-acetylglucosaminyl]$_n$	Cornea	Man-Man-Asp-GlcNAc	0.9–1.7	0.2–0.5	4–19	3.0	—
Keratan sulfates II		Cartilage, nucleus pulposus, annulus	GalNAc-Ser (Thr)	1.1–1.8	0.2–0.5	8–12	2.0	—
Heparan sulfates	[(1 → 4)-β-D-glucuronosyl-(1 → 4)-β-D-*N*-acetylglucosaminyl]$_n$	Aorta, beef lung, liver amyloid	Same as chondroitin sulfates	0.1–2.8	—	15–40	1.3	3.0–4.0
Heparins		Lung, liver, skin, (mast cells)	Usually not linked to protein	2–3	0.1–0.25	7–16	2.0	3.0–4.0

	Susceptibility to degradation									
		Hyaluronidases				Chondroitinases				
Family	β-Elimination[d]	Leech	Testicular	Streptococcal	Streptomyces	AC	ABC	HNO_2	KSase[g]	Remarks
Hyaluronic acids	+	+	+	+	+	±	±	−	−	
Chondroitin 4-sulfates	+	−	+	±[e]	−	+	+	−	−	Disaccharides have been found which contain sulfate both at the 4- and 6-positions
Chondroitin 6-sulfates	+	−	+	±[e]	−	+	+	−	−	
Dermatan sulfates	+	−	±	−	−	−	+	−	−	Also contain D-glucuronic acid
Keratan sulfates I	−	−	−	−	−	−	−	−	+	Also contain D-Gal, sialic acids, and L-fucose and are often very branched
Keratan sulfates II	+	−	−	−	−	−	−	−	±	
Heparan sulfates	+	−	−	−	−	−	−	+	−	Contain variable amounts of *N*-sulfate, *O*-sulfate, and L-iduronic acid residues
Heparins	+	−	−	−	−	−	−	+	−	

[a] Summarized from several references; see text for details and references.
[b] Richest sources.
[c] Colorimetric assays for the uronic acid ratio are a measure of L-iduronic acid.
[d] Susceptibility of protein–carbohydrate linkage to cleavage.
[e] Will degrade under sulfated chondroitin sulfates.
[f] Mole ratio of sulfate per repeating disaccharide.
[g] Keratan sulfate endo-β-galactosidase.

1972). Hyaluronic acids have recently been demonstrated to have an α-helical structure in solution (Rees, 1975).

b. Chondroitin Sulfates. Chondroitin sulfates (Table I) are also widely distributed and have a disaccharide unit {[(1→4)-β-D-glucuronosyl-(1→3)-β-D-*N*-acetylgalactosaminyl]$_n$} with a structure similar to that of hyaluronic acids, except that they contain *N*-acetyl-D-galactosamine instead of *N*-acetyl-D-glucosamine and, in addition, may contain ester sulfate at either carbon 4 or carbon 6 of the *N*-acetylgalactosamine moiety. Recent evidence suggests that many chondroitin sulfates exist as hybrid copolymers of both 4-sulfated and 6-sulfated disaccharides on the same chain (Seno, 1974; Seno *et al.*, 1975; Faltynek and Silbert, 1978).

c. Dermatan Sulfates. Dermatan sulfates (Table I; chondroitin sulfate B; Meyer *et al.*, 1956) represent an isomeric form of chondroitin 4-sulfates in which many of the D-glucuronic acid moieties are converted to L-iduronic acid residues by C-5 epimerization at the macromolecular level (Malström *et al.*, 1975; Lindahl, 1976). In addition, dermatan sulfates contain an ester sulfate on many of their iduronate residues (Suzuki, 1960; Fransson *et al.*, 1974). Again, considerable structural variability has been found. In fact, the D-glucuronic acid content of various dermatan sulfates can range from 90% to almost negligible.

d. Heparins and Heparan Sulfates. Heparins and heparan sulfates (Table I; for reviews, see Lindahl *et al.*, 1977; Lindahl and Höök, 1978; Rodén and Horowitz, 1978; Jaques, 1980) both contain *N*-acetyl-D-glucosamine and uronic acid in their disaccharide structure {[(1→4)-β-D-glucuronosyl-(1→4)-α-D-*N*-acetylglucosaminyl]$_n$}, but a large proportion of D-glucosamine residues are N-sulfated instead of N-acetylated (Lindahl, 1970). Also, much of the uronic acid is epimerized to L-iduronic acid (up to 70–90% in heparins) (Lindahl and Axelsson, 1971). The reactions involved in the deacetylation of *N*-acetylglucosamine residues followed by subsequent sulfation of the free amino groups are unique to the formation of heparin-like polysaccharides (Lindahl *et al.*, 1973; Höök *et al.*, 1975; Silbert *et al.*, 1975; Jansson *et al.*, 1975; Lindahl *et al.*, 1976). Heparan sulfates contain a greater proportion of *N*-acetyl-D-glucosamine, less *N*-sulfate, a much lower degree of sulfation (both *O*- and *N*-sulfate), and fewer L-iduronic acid moieties than heparins (Cifonelli and King, 1970, 1972, 1973; Linker and Hovingh, 1973; Taylor *et al.*, 1973). In general, low-sulfated, D-glucuronic acid-rich polysaccharides of this family are classified as heparan sulfates, and highly sulfated, L-iduronic acid-rich polymers are classified as heparins. However, intermediate species appear to exist.

e. **Keratan Sulfates.** Keratan sulfates (Table I) are the most complex types of glycosaminoglycans and have been divided into two families based mostly on their protein linkages (Meyer *et al.*, 1953; Seno *et al.*, 1965; Mathews and Cifonelli, 1965). Both types have similar repeating disaccharide {[(1→3)-β-D-galactosyl-(1→4)-β-D-*N*-acetylglucosaminyl]$_n$} backbone structures. Since they lack uronic acid, and also since they are often branched, contain L-fucose, sialic acids, and *N*-acetyl-D-galactosamine, they are more similar in structure to the glycosyl moieties of classical glycoproteins (Gottschalk, 1972; Lennarz, 1980) than are other glycosaminoglycans. However, their large size, high degree of sulfation, location within tissues (in cartilage, keratan sulfates and chondroitin sulfates are found on the same protein cores; Hascall and Riolo, 1972; Section II,A,1), and underlying repeating disaccharide structures justify including keratan sulfates with glycosaminoglycans in any classification scheme (Marshall, 1972). Keratan sulfates type I (corneal type) are linked to protein by an alkali-stable amide linkage involving asparagine and *N*-acetyl-D-glucosamine (Baker *et al.*, 1975; Brekle and Mersmann, 1980). Keratan sulfates type II (skeletal type) are more highly branched, contain more sialic acid, and are linked to protein by an alkali-labile 0-glycosidic linkage between *N*-acetyl-D-galactosamine and serine (Seno *et al.*, 1965; Cifonelli *et al.*, 1967; Rodén, 1970).

B. Corneal Proteoglycans

1. *Isolation and Characterization of Intact Proteoglycans*

The methodology for the isolation and study of corneal proteoglycans is largely a result of procedures adapted from Hascall's and Sajdera's work on cartilage (1969, 1970). However, it is now evident that the proteoglycans of the vertebrate cornea are very different both compositionally and structurally than those of cartilage (Table II).

Berman (1970) extracted bovine corneal proteoglycans using 0.1 *M* K_2CO_3 and obtained 85% yield. After removal of the nonacidic components by ion exchange, proteolytic fragments of the proteoglycans were analyzed by $(NH_4)_2SO_4$ precipitation. He found that a 70% NH_4Cl supernatant contained mostly proteochondroitin sulfates (PChS), which were separable into four fractions by ion exchange, and that

TABLE II

Proteoglycans of Cornea and Cartilage[a]

Cornea	Cartilage
KS I and chondroitin-dermatan sulfates on a *different* polypeptide core	KS II and CS on *same* polypeptide core
Protein core: CS, MW 70,000-90,000; KS, MW 32,000	Protein core, MW 200,000
CS proteoglycan, 70% protein; KS proteoglycan, 74% protein	Proteoglycan, 10% protein
Proteoglycan monomers: CS, MW 100,000-150,000; KS, MW 40,000-70,000	Proteoglycan monomers, MW 2.5×10^6
KS I proteoglycan contains *only* one to two KS chains (MW 7000-15,000) and several asparagine linked oligosaccharides	Proteoglycan monomer contains 110 KS II and O-linked oligosaccharides
CS proteoglycan contains *only* one or two CS chains (MW 55,000) and several "mucin"-type saccharide chains	85 CS chains, MW 25,000

[a] See text for details and references. KS, keratan sulfate; CS, chondroitin sulfate.

each had marked differences in sulfate and neutral sugar composition. On the other hand, the 70% $(NH_4)_2SO_4$ pellet contained mostly acidic glycoproteins and proteokeratan sulfate (PKS) which yielded seven fractions upon ion-exchange chromatography. The later eluting (more highly charged) PKSs were free of galactosamine and uronic acid even though they contained one-sixth of the total PKS.

Muthiah *et al.* (1974) compared the efficiency of 3*M* $MgCl_2$ and 4 *M* guanidinium chloride in the extraction of bovine corneal proteoglycans and were able to extract 70% of the total uronic acids and 100% of the neutral sugar using 4 *M* guanidinium chloride, while only 43% of the total uronic acid and 23% of the neutral sugar were extracted with 3 *M* $MgCl_2$. Sequential extraction with 4 *M* followed by 8 *M* guandinium chloride and 2 *M* $CaCl_2$ for 48 hr each solubilized 90% of the total uronic acid and 100% of the total neutral sugar. No significant differences were found in the CsCl density gradient patterns between calf and adult corneal proteoglycans. However, analysis of the glycosaminoglycans by Dowex-1 suggested that chondroitin sulfates from calf were more highly sulfated than those from adult corneas, whereas keratin sulfates from adult corneas appeared to be more highly sulfated than those from calf corneas.

Antonopoulous *et al.* (1974) modified existing procedures for cartilage proteoglycans and systematically developed procedures for the isolation of proteoglycans from noncartilaginous tissues such as cornea, sclera, and aorta. Since most tissues contain much smaller proportions of proteoglycans than cartilage and a larger proportion of nonproteoglycans, and since their proteoglycans are of lower buoyant density than those of cartilage, CsCl density gradients, so widely used in cartilage studies, do not work well in purifying proteoglycans from other tissues. Antonopoulous *et al.* (1974) tested 1-5 *M* guanidinium chloride and obtained maximal extraction (88% of hexosamine) with 3 or 4 *M*. Guanidinium chloride then was exchanged for 7 *M* urea, and the extract was chromatographed on DEAE-cellulose equilibrated with 7 *M* urea. On these columns, significant amounts of glycoproteins eluted with urea or 0.15 *M* NaCl-urea but, interestingly, a substantial amount (69%) of the glycopeptides eluted with the proteoglycan fraction in 2 *M* NaCl-urea, unlike the results of similar studies on other tissues. The protein content of corneal proteoglycans was found to be larger (40% dry wt) than that of other tissue proteoglycans, and the amino acid content was unusually low in hydroxyamino acids. The asparagine content was much higher than that of other proteoglycans, consistent with the linkage of keratan sulfates to protein. Also, the buoyant densities of corneal proteoglycans were found to be lower than those of proteoglycans from other tissues.

Axelsson and Heinegård (1975) modified the methods of Antonopoulous *et al.* (1974) and extended investigations of proteoglycans from bovine corneal stroma. They demonstrated that in calf cornea about 5-8% of the total glycosaminoglycans was dermatan sulfate, and they confirmed that corneal proteoglycans had a higher protein content and lower buoyant densities than proteoglycans from aorta, sclera, and cartilage. Using ethanol precipitation in the presence of guanidinium chloride, Axelsson and Heinegård (1975) fractionated corneal proteoglycans into four fractions with different ratios of keratan sulfates to galactosaminoglycans: The 50, 60, and 70% ethanol pellets (designated 50P, 60P, and 70P) contained 33, 33, and 24%, respectively, of the total hexosamine and had similar hexosamine/protein ratios but different amino acid compositions. The 50P fraction was found to be rich in dermatan-chondroitin sulfate proteoglycans but still contained large proportions of keratan sulfate-like proteoglycans. The 60P fraction had an intermediate amino acid composition with respect to 50P and 70P, galactose was the major neutral sugar, and the glucosamine/galactosamine ratio was 54:45 (w/w). The 70P fraction contained leucine, asparagine, and glutamic acid as the predominant amino acids, galactose and mannose accounted for 96% of the neutral sugar, and less than 1% of the total hexosamine was galactosamine. Although Stuhlsatz *et al.* (1971) had suggested the existence of a pure keratan sulfate proteoglycan from their metabolic labeling studies, Axelsson and Heinegard (1975) were the first to isolate such a unique intact proteoglycan directly.

Recently, Axelsson and Heinegård (1978) have characterized the 70P fraction containing the keratan sulfate proteoglycan from bovine corneal stroma and found it to contain about one-third of the proteoglycans of the bovine cornea, a large proportion of keratan sulfate, small amounts of oligosaccharides, but no significant galactosaminoglycans. The keratan sulfate proteoglycans in fraction 70P could be fractionated into two major components by gel filtration chromatography on 4% agarose (designated 70P-A and 70P-B in order of elution). The lower-molecular-weight fraction, 70P-B, predominates, is much smaller in size than cartilage proteoglycan monomers, and aggregates at a low pH or after the reduction of disulfide bonds. However, interestingly, aggregation requires the presence of components from fraction 70P-A, and most of the 70P-A species can be dissociated into monomers similar in size to 70P-B by concentrations of sodium dodecyl sulfate–guanidinium chloride. Ultracentrifugation studies have confirmed that the 70P corneal keratan sulfate proteoglycans consist of a major population of monomers which aggregate under certain conditions and a small proportion of aggregates which do not dissociate readily and are necessary for aggregation of the monomers. The monomer molecular weight and Stokes radius were found to be 72,000 and 12 nm, respectively. Therefore, the keratan sulfate proteoglycan monomer is only 3–7% the size of the cartilage proteoglycan monomer. In addition, it was found that each keratan sulfate proteoglycan monomer contained two to three intrachain disulfide bridges which appeared to be important in maintaining its three-dimensional structure. Also, a considerable fraction of the hexosamine in the 70P keratan sulfate proteoglycan fraction was found in neutral or acidic oligosaccharide chains, and the evidence suggests that, in addition to keratan sulfate, several oligosaccharide chains of the glycoprotein type are also covalently attached to the protein core. These oligosaccharides were found to be compositionally similar to a mixture of corneal glycosaminoglycan linkage regions (Moczar *et al.*, 1969).

Recent similar studies on bovine corneal proteoglycan interactions (Speziale *et al.,* 1978) confirm the types of aggregation phenomena demonstrated for corneal proteoglycan by Axelsson and Heinegård (1975, 1978).

The information to date on the bovine corneal keratan sulfate proteoglycan can be summarized as follows: Monomers have a molecular weight of approximately 72,000 and Stokes radii of 7–12 nm; by weight they are 45% protein, 30% keratan sulfate, 10–12% oligosaccharides, and the remainder counterions. The molecular weight of the core protein is approximately 32,000. The total molecular weight of the keratan sulfate chains is 22,000, and the total molecular weight of the oligosaccharide chains is 8000. Monomers contain two to three intrachain disulfide bonds and only one polypeptide core. Based upon the average observed molecular weight for isolated keratan sulfate chains (Laurent and Anseth, 1961; Greiling and Stuhlsatz, 1966), each proteoglycan monomer, on the average, contains one, two, or three keratan sulfate chains but could contain several smaller chains. The oligosaccharide chains have a molecular weight of about 700–1000, and therefore each proteoglycan monomer must contain about 10–12 oligosaccharides (Axelsson and Heinegård, 1978).

Using procedures similar to those of Axelsson and Heinegård (1975, 1978), Hassell *et al.* (1980) have utilized metabolic labeling of short-term organ cultures and direct chemical analyses to study the corneal proteoglycans of the rhesus monkey. The utility of this approach was demonstrated by the fact that the proteoglycans labeled in organ culture were similar or identical to those chemically detected in the stroma, although the absolute proportions of those synthesized *in vitro* differed from those accumulated in the stroma *in vivo*. The studies indicated that there were two distinct proteoglycans present in intact rhesus monkey corneal stroma and synthesized by corneal explants. The larger proteoglycan species was shown to be a chondroitin-dermatan sulfate proteoglycan by its high *N*-acetylgalactosamine content and sensitivity to chondroitinase ABC. The chondroitin-dermatan sulfate proteoglycan contained about 70% protein and an average buoyant density of 1.34 g/ml. Papain or β-elimination released chondroitin-dermatan sulfate chains of molecular weight 55,000 and smaller oligosaccharides of molecular weight 2500 similar to oligosaccharide chains of mucin-type glycoproteins. The quantitative data suggest that rhesus monkey corneal chondroitin-dermatan sulfate proteoglycan consists of a protein core with only one, or at most two, chondroitin-dermatan sulfate side chains and several glycoprotein side chains. The smaller keratan sulfate proteoglycan from rhesus monkey corneal stroma consists of 74% protein and has a buoyant density of 1.32 g/ml. The keratan sulfate proteoglycan monomers have a molecular weight of about 70,000–100,000. The core-bound saccharides consist of keratan sulfate I chains of molecular weight 7000–15,000 and smaller mannose-labeled oligosaccharides of molecular weight 2500 which are similar to the oligosaccharide chains of asparagine-linked glycoproteins. Quantitative data indicate that the keratan sulfate proteoglycan consists of a protein core with one or two keratan sulfate chains and a few glycoprotein-type oligosaccharides. Although the corneal proteoglycan monomers are smaller than those from cartilage, the keratan sulfate side chains are nearly the same size and the chondroitin sulfate chains are nearly twice as large as those from cartilage.

2. *Structural Studies on Corneal Glycosaminoglycans*

As early as 1894 (Mörner, 1894) "mucinlike" substances were known to exist in corneas. Polysaccharides containing D-glucosamine, glucuronic acids, a sulfate ester, and acetyl groups were isolated from bovine corneas and named mucoitin sulfuric acid or hyaluronosuluric acid in several early studies (Levene and López-Suárez, 1918; Meyer and Chaffee, 1940; Sato, 1952). Woodin (1952) isolated an electrophoretically "homogeneous," sulfated fraction which contained D-galactose, D-glucosamine, and D-galactosamine. Meyer *et al.* (1953) used ethanol precipitation of the calcium salts of the corneal polysaccharides and identified three distinct species: chondroitin sulfate, a uronic acid and hexosamine-containing polymer compositionally similar to hyaluronic acid (later shown to be chondroitin by Davidson and Meyer, 1954), and a polymer composed of *N*-acetylglucosamine, galactose, and sulfate in approximately equimolar proportions, which they named keratosulfate (keratan sulfate I).

Rosen *et al.* (1960) demonstrated that corneal keratan sulfates contained small amounts of methylpentose, that they were cleavable by blood group-cleaving enzymes, and that the desulfated polysaccharide had cross-reactivity with anti-blood group sera. Lloyd *et al.* (1960) showed that pure fractions of corneal keratan sulfate could be obtained by DEAE chromatography of proteolytic digests.

Hirano *et al.* (1961) studied the structure of bovine corneal keratan sulfate by methylation analysis of sulfated and desulfated polysaccharide and established that the *O*-β-D-galactopyranosyl-(1→4)-(2-acetamido-2-deoxy-D-glucose 6-sulfate) linked (1→3) to the D-galactose of the next disaccharide was the repeating disaccharide sequence of the polysaccharide backbone. Anseth and Laurent (1961) developed a method based upon ECTEOLA-cellulose chromatography for fractionating corneal glycosaminoglycans and, in conjunction with ethanol precipitation, were able to separate glucosaminoglycans from galactosaminoglycans completely.

Seno *et al.* (1965) performed detailed analyses of keratan sulfates from various cartilages and bovine corneas and found that skeletal keratan sulfates (type II) had amino acids at their reducing termini different from those of corneal keratan sulfates (type I). Based upon amino acid analysis and the sensitivities of the carbohydrate protein linkages to dilute alkali, they were the first to demonstrate that skeletal keratan sulfates were *O*-glycosidically linked to serine or threonine via alkali-sensitive bonds and that corneal keratan sulfates were linked via *N*-glycosylamine alkali-stable linkages. In addition, the presence of sialic acid and methylpentose (fucose) in varying and nonstoichiometric amounts was confirmed for all the keratan sulfates examined. Meyer and Anderson (1965) demonstrated that the chondroitin sulfates of adult bovine cornea were sulfated at the 4-position of *N*-acetylgalactosamine.

Mathews and Cifonelli (1965) performed a detailed comparative study on keratan sulfates from several species of cornea and cartilage. Corneal keratan sulfates from fish, bird, or mammal contained equimolar amounts of galactose and glucosamine and had an average sulfate/hexosamine molar ratio of 1.0–1.3. The major amino acid after hydrolysis was aspartic acid, and quantitative data, based on a number average molecular weight of 10,000 for the keratan sulfate chains, suggested that only asparagine was involved in the protein–carbohydrate linkage.

Cartilage keratan sulfate was shown to differ from corneal keratan sulfate in that it had a higher degree of sulfation, a higher amount of *N*-acetylgalactosamine, insufficient aspartate to account for protein linkage, and higher amounts of residual threonine, glutamate, and proline at the reducing termini of the carbohydrate. Cifonelli *et al.* (1967) examined the distribution of the sulfate content of various keratan sulfate fractions isolated from adult and embryonic mammalian tissues using the Dowex-1 fractionation procedures developed by Schiller *et al.* (1961). Fractions of keratan sulfates with sulfate/hexosamine ratios ranging from 0.5 to 1.6 were isolated. Bovine cornea and nasal septum cartilage were found to contain higher proportions of low-sulfated keratan sulfates than other tissues examined.

Bhavanandan and Meyer (1967) reinvestigated the structure of bovine corneal keratan sulfates by improved permethylation analysis and found that on the average about 74% of the *N*-acetylglucosamine residues were sulfated at position 6. Furthermore, at least 40% of the galactosyl residues were also substituted (primarily sulfated) at position 6. It was also observed that the polysaccharide contained about a 25% molar excess of galactosyl over hexoaminyl residues, and the data also suggest that these excess residues occur at branch points. Later work by Bhavanandan and Meyer (1968) indicated that both corneal and skeletal keratan sulfates contained mannose, however, recent data suggest that the mannose associated with skeletal keratan sulfate resides in a copurifying oligosaccharide chain (Rodén, 1980).

Greiling *et al.* (1970) isolated β-*N*-aspartyl-*N*-acetylglucosamine resulting from partial acid hydrolysis of corneal PKS and confirmed the structure with authentic 2-acetamide-1,2-dideoxy-β-D-glucose. Based on compositional data of PKS peptides and the isolated linkage fragment, the following structure of the linkage region was proposed:

$$[\text{Gal—}\underset{\substack{|\\ SO_4}}{\text{GluNAc}}]_N\text{—Man—Man—GlcNAc—}\overset{|}{\underset{|}{\text{Asn}}}$$

Arnott *et al.* (1974) studied x-ray diffraction patterns from stretched films of bovine corneal keratan sulfates and found that the molecules were twofold helices with an axial rise per disaccharide residue of 0.945. They proposed a model of keratan sulfate as an extended polysaccharide chain fringed with charged sulfate side groups, which is similar to earlier models for chondroitin sulfates. Choi and Meyer (1975) performed a detailed comparative structural study on various keratan sulfates and concluded that bovine corneal keratan sulfates contained two mannose residues per chain, each of which was linked by three different substituents. It was also observed that fucose and *N*-acetylneuraminic acid were nonreducing terminal residues and that the sialic acid was linked to the 3-position of the galactosyl residues. They further confirmed that the sulfate esters were all located at the 6-position of *N*-acetylglucosamine and/or galactose residues.

Baker *et al.* (1975) investigated the linkage region of corneal keratan sulfate after exhaustive proteolysis to remove the bulk of the peptide portion of the proteoglycan. Amino acid analysis of acid hydrolyzates indicated that 59% of the total amino acids were aspartic acids. The linkage fragment 2-acetamido-1-(L-β-aspartamido)-1,2-dideoxy-β-D-glucose (Asn-GlcNAc) was isolated and rigorously characterized, thus confirming previous linkage studies. Brekle and Mersmann (1980) have recently used hydrazinolysis of papain-digested bovine corneal keratan sulfate in conjunction with sodium borotritide reduction, nitrous acid deamination, and subsequent stabilization of deamination products by cyanhydrin reaction to isolate the linkage fragment. Importantly, under the hydrazinolysis conditions used, sulfate groups were unaffected and there was no evidence of degradation of the sugar chain, but the amino acids at the reducing end appeared to be completely removed. The resulting free reducing termini were labeled by sodium borotritide reduction. The data, in general, support the structure of the linkage region proposed by Choi and Meyer (1975) from their methylation analyses.

In summary, the carbohydrate chain of the corneal keratan sulfate proteoglycan is linked to protein via an N-glycosidic bond between asparagine and *N*-acetylglucosamine, the linkage region oligosaccharide contains several mannosyl residues, and the repeating, relatively unbranched polysaccharide backbone contains about 8–30 (1→3)-D-galactopyranosyl-(1→4)-*N*-acetyl-β-D-glucosaminyl disaccharide units sulfated at varying degrees on C-6 of the *N*-acetylglucosamine and/or galactose.

III. BIOSYNTHESIS OF CORNEAL PROTEOGLYCANS AND GLYCOSAMINOGLYCANS

A. Developmental Biosynthesis Studies

Smelser and Ozanics (1956, 1957, 1960) studied the biosynthesis of sulfated polysaccharides in day-14 to -16 fetal rabbit corneas by autoradiography and demonstrated the incorporation of $^{35}SO_4^{2-}$ immediately after the appearance of the first stromal cells. Metachromatic staining was first detected on day 17 and, based upon the susceptibility of both the radioactivity and the metachromatic staining to testicular hyaluronidase, which degrades chondroitin sulfates but not keratan sulfates, it was suggested that keratan sulfates appeared late in development.

Anseth (1961b) used his recently developed procedures for the quantitation and fractionation of glycosaminoglycans by ECTEOLA-cellulose chromatography (Anseth, 1961a) to examine the glycosaminoglycans of developing corneal stroma of chickens and cattle. He observed an increase in the total amount of glycosaminoglycans per dry weight of corneal tissue after day 14 of development in the chick, and the increase was accounted for by the appearance of new species of polysaccharides requiring higher (2.0 *M*) ionic strength for elution from ECTEOLA-cellulose. However, whether the polysaccharides were larger, more highly sulfated, or both was not determined. In the chick, on day 14 of development the eyelids first begin to cover the cornea completely, the stromal fibroblast number reaches a maximum, and there is a rapid onset of dehydration, a decrease in stromal thickness, and a rapid increase in the transparency of the cornea (Fig. 1). Evidence suggests that many of these events may be under thyroid control and triggered by elevated thyroxin levels (Coulombre and Coulombre, 1964; for review, see Hay and Revel, 1969). An increase in the total amount of glycosaminoglycans per dry weight of tissue from early embryonic until postnatal life was also observed for bovine cornea (Anseth, 1961b). Interestingly, very little

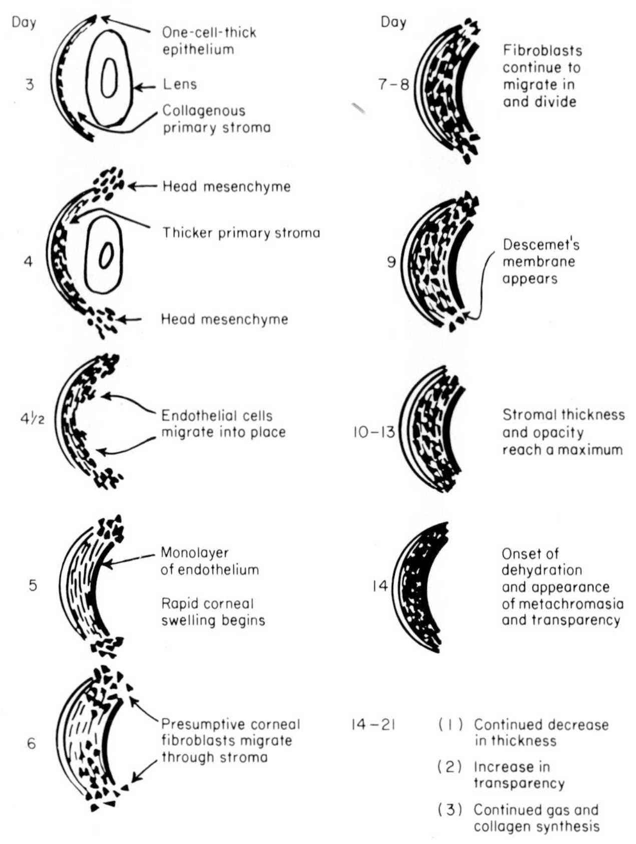

Fig. 1. Development of the chick cornea. See text and Hay and Revel (1969) for details.

low-sulfated chondroitin sulfate was evident in embryonic bovine cornea as compared to adult bovine cornea (Anseth and Laurent, 1961; Laurent and Anseth, 1961; Anseth, 1971).

Conrad (1970a) labeled embryonic chick corneas *in ovo* or *in vitro* with a radioactive precursor, $^{35}SO_4^{2-}$ or [^{14}C]glucosamine, and analyzed the distribution of radioactivity in the various glycosaminoglycans by the cetylpyridinium chloride–cellulose (CPC–cellulose) chromatographic procedure of Svegcar and Robertson (1967). The incorporation of ^{35}S radioactivity into CPC-precipitable material was demonstrated to occur as early as day 5 of development in the chick, well before invasion of the primary acellular stroma by fibroblasts. Elution profiles of radioactivity from the CPC–cellulose chromatography suggested that a complete array of corneal-type glycosaminoglycans were being synthesized as early as day 9. These studies also indicated that the $^{35}S/^{3}H$ ratio incorporated into CPC-precipitable material increased steadily from day 9, reaching a peak at day 14 of development. Since corneal keratan sulfates are to a large and variable extent not precipitated by CPC, these data clearly suggested an increase in the relative degree of sulfation of the corneal chondroitin sulfates during these later developmental stages. These observations of an apparent increase in the relative sulfation of corneal chondroitin sulfates from day 9 to day 14 were further supported by the findings of parallel increases in the proportions of [^{14}C]glucosamine-labeled glycosaminoglycans requiring a higher ionic strength for elution from CPC–cellulose columns.

Similarly, Conrad (1970b) examined glycosaminoglycan biosynthesis in mass cultures and clones of stromal fibroblasts and observed a correlation between cell density and the degree of sulfation of the glycosaminoglycans synthesized. This latter observation is of potential significance when related to the increased degree of sulfation of stromal glycosaminoglycans which appears to occur at about day 14 of development when corneal dehydration and compaction takes place (Fig. 1). An important contribution of these studies was the demonstration by direct clonal analysis that every stromal fibroblast synthesized an array of glycosaminoglycans and collagen, thus ruling out the alternative possibility in which one fibroblast is committed to synthesize primarily collagen and another predominantly glycosaminoglycans. Similar developmental studies were also performed by Praus and Brettschneider (1971) who confirmed the disappearance of an undersulfated species of glycosaminoglycan as development proceeded in the embryonic chick cornea.

Toole and Trelstad (1971) isotopically labeled chick corneas *in ovo* and *in vitro* and found that early in the development of the cornea hyaluronic acids were a major component.

Moczar and Moczar (1972) extracted embryonic and postnatal bovine corneal stroma with buffered 1 *M* $CaCl_2$ and examined the saccharides present in the soluble and insoluble fractions. They compared the age-dependent distribution and types of saccharides of the three major macromolecular stromal constituents: collagens, proteoglycans, and glycoproteins. In older corneas they observed an increase in the hexosamine/uronic acid ratio, suggesting an age-related increase in the relative proportion of keratan sulfates in the insoluble fractions.

Meier and Hay (1973) investigated the synthesis of sulfated glycosaminoglycans by isolated 4- to 6-day embryonic chick corneal epithelia and found that the glycosaminoglycans synthesized were entirely accountable for as chondroitin sulfates and heparan sulfates.

Importantly, the isolated epithelia lost substantial amounts of glycosaminoglycan to the labeling medium when grown on Millipore filters, but lost substantially less glycosaminoglycan to the medium when grown on a frozen killed lens capsule or stroma. Also of significance is their observation that the relative amounts of heparan sulfates synthesized by isolated chick corneal epithelia decreased from 50% at day 4 to 12% by day 12. A similar decline in the proportion of heparan sulfates was observed with time *in vitro*.

Conrad and Dorfman (1974) examined the effects of *in vitro* culture on the synthesis of cornea-specific glycosaminoglycans by day-14 embryonic chick corneas and isolated corneal cell types and observed that *in vitro* cultured intact corneas or corneal fibroblasts rapidly appeared to lose the ability to synthesize cornea-type keratan sulfate but did not lose the capacity to synthesize other glycosaminoglycans. Also, when elution of chondroitinase ABC-treated glycosaminoglycans from Dowex-1 (Cl^-) columns was used as an assay, their data suggested that the stromal fibroblasts were the cells most likely involved in the bulk of the synthesis of keratan sulfate-like material *in vivo*. Unfortunately, because of the limitations of the available technology at the time of these studies, the synthesis of undersulfated keratan sulfate by the other two corneal cell types was not excluded.

Meier and Hay (1974a) extended their earlier observations on the effects of collagen matrices on the biosynthesis of glycosaminoglycans by isolated chick corneal epithelia. Epithelia grown *in vitro* on frozen killed lens capsule synthesized approximately 2½ times as much glycosaminoglycan as when grown on Millipore filters or other noncollagenous substrates. Similar levels of enhanced glycosaminoglycan biosynthesis were also observed when epithelia were grown on chemically pure chondrosarcoma collagen. The observation was confirmed using a variety of collagen substrates. In subsequent studies, Meier and Hay (1974b) also demonstrated that the addition of 200 μg/ml of chondroitin sulfates or heparins to the culture medium of isolated corneal epithelia stimulated their own synthesis of chondroitin sulfates and heparan sulfates by about twofold. Collagen biosynthesis appeared unaffected by the addition of exogenous glycosaminoglycans to the cultures. Interestingly, chondroitin sulfate proteoglycans had a similar effect, but dermatan sulfates or hyaluronic acids had no effect, and keratan sulfates appeared to depress glycosaminoglycan biosynthesis. These observations could be extremely relevant with respect to the interactions among the three corneal cell types, especially in regard to our current knowledge of the kinds of glycosaminoglycans each is known to synthesize (see later).

Conrad and Hart (1975) studied the biosynthesis of heparan sulfate by embryonic tissues and primary fibroblast populations of the chick including those of cornea. The proportion of [^{35}S]glycosaminoglycans found to be in heparan sulfates was observed to increase from 12% in freshly isolated intact corneas labeled *in vitro* for 10 hr to 34% in corneas that had been cultured for 3 days and then also labeled for 10 hr. Both freshly isolated and 14-day cultures of isolated corneal fibroblasts incorporated $^{35}SO_4$ and [^{3}H]glucosamine into glycosaminoglycans of which about 38% were heparan sulfates. Interestingly, as observed earlier by Kraemer (1971) for other cell types, nearly all these heparan sulfates were preferentially cell-associated, in contrast to the other glycosaminoglycans which were preferentially secreted into the medium.

It is evident from the preceding discussion that an inherent major difficulty in many of the earlier biosynthetic studies on corneal glycosaminoglycans was the unambiguous identification of keratan sulfates. This problem was, in part, due to the close compositional similarity between keratan sulfates and glycoproteins, their variable degrees of sulfation, and their anomalous behavior in quaternary ammonium detergents such as CPC. The discovery, isolation, and characterization, by Nakazawa and Suzuki (1975), of an enzyme, keratan sulfate-β-endogalactosidase, which specifically degrades keratan sulfates, allowed for the first time an accurate, quantitative, sensitive approach for the unequivocal determination of the small amounts of keratan sulfate in developing tissues. The excellent initial studies on the specificity of keratan sulfate-β-endogalactosidase were later confirmed using highly purified reference standard glycosaminoglycans and chromatography on Sephadex G-50 as a very sensitive assay of degradation (Hart, 1976). Heparins, heparan sulfates, hyaluronic acids, dermatan sulfates, chondroitin 6-sulfates, and chondroitin 4-sulfates are unaffected by the enzyme, whereas keratan sulfates type II (skeletal) are degraded at rates about 10 times slower than those of keratan sulfates type I (corneal).

Hart (1976) labeled 5- to 20-day embryonic chick corneas with [^{3}H] glucosamine and $^{35}SO_4^{2-}$ and studied the biosynthesis of individual corneal glycosaminoglycans, paying particular attention to the corneal keratan sulfates. The data demonstrate that, contrary to earlier suggestions, keratan sulfate biosynthesis in the cornea begins at the time of the first invasion of the acellular stroma by fibroblasts between days 5 and 6, and at least 8 days prior to the onset of transparency which occurs on day 14 (Figs. 1 and 2). Significantly, the increasing rate of keratan sulfate biosynthesis, as development proceeds, levels off on day 14 coincident with the achievement of maximal stromal cell number (Coleman *et al.*, 1965). The proportion of ^{3}H and ^{35}S radioactivity found in corneal keratan sulfates was seen to reach maximal levels as early as day 9 (Fig. 2). Probably the most significant observation of these studies is that the keratan sulfates synthesized on and after day 14 of development appeared to be substantially more sulfated than those synthesized earlier in development (Fig. 3). The appearance of more highly sulfated species of corneal keratan sulfates exactly coincides with the onset of corneal transparency. Similarly, it was observed that the ratio of 4-sulfated to 6-sulfated chondroitin sulfates increased sharply during corneal development and that this ratio also peaked on day 14 of development (Fig. 4). It is clear that these developmental changes in glycosaminoglycan biosynthesis undoubtedly reflect the important role these macromolecules play in corneal morphogenesis and in the achievement of corneal transparency.

Conrad *et al.* (1977) analyzed the glycosaminoglycans synthesized by

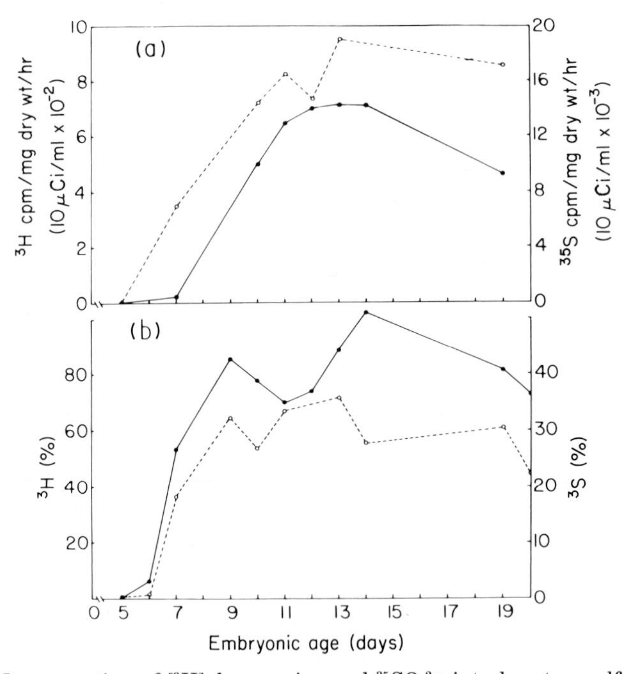

Fig. 2. Incorporation of [^{3}H]glucosamine and $^{35}SO_4^{2-}$ into keratan sulfates during corneal development in the chick. (a) Rate of incorporation of radioactivity into keratan sulfates per dry weight of cornea. (b) Percentage of radioactivity in total glycosaminoglycans that are keratan sulfates. (From Hart, 1976, with permission.)

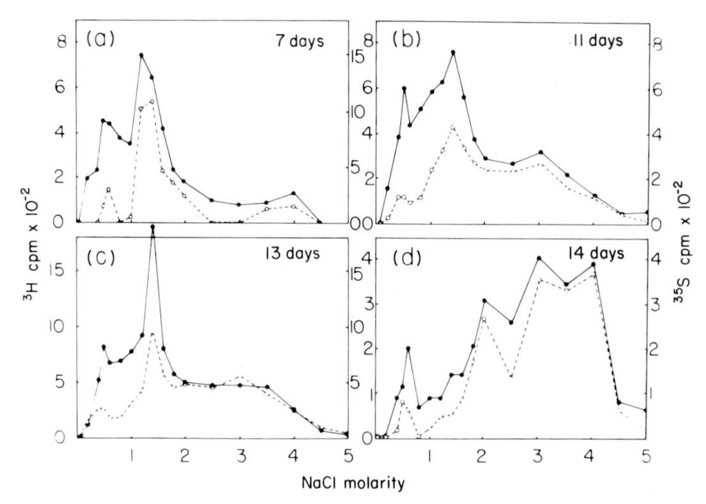

Fig. 3. Ion-exchange chromatography of keratan sulfates isolated during chick corneal development. Isolated chick corneal keratan sulfates were examined for relative degrees of sulfation by chromatography on Dowex-1 (Cl^-). (a–c) Typical profiles obtained for keratan sulfates from embryonic corneas younger than day 14. (d) Typical results obtained for keratan sulfates from corneas older than day 14. (From Hart, 1976, with permission.)

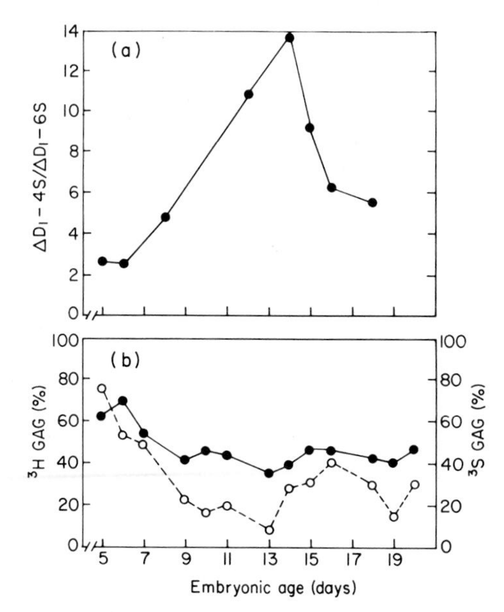

Fig. 4. Incorporation of [^{3}H]glucosamine and $^{35}SO_4^{2-}$ into chondroitin sulfates during corneal development in the chick. (a) ΔDi-4S/ΔDi-6S ratio during chick corneal development. (b) percentage of radioactivity in total glycosaminoglycans that is chondroitin sulfates. (From Hart, 1976, with permission.)

purified populations of fibroblasts from the back skin, heart ventricle, and cornea of day-14 chick embryos. These extraordinarily careful and quantitative studies have provided the most accurate measure available to date of the amounts and types of glycosaminoglycans synthesized by long-term cultures of fibroblasts. Both glycosaminoglycans secreted into the medium and those associated with the cell layer were determined independently enough times for statistical evaluation of the accuracy (P values generally less than 0.001) of the measurement. Glycosaminoglycans found in the medium of corneal fibroblasts consisted of 12% heparan sulfates, 48% hyaluronates, 4% undersulfated chondroitin sulfates, 9% chondroitin 6-sulfate, 9% chondroitin 4-sulfate, 5% dermatan sulfate, 5% keratan sulfate-like material, and 8% unidentified species. Cell-associated glycosaminoglycans consisted of 30% heparan sulfates, 19% hyaluronic acids, 8% undersulfated chondroitin sulfates, 13% chondroitin 4-sulfate, 13% chondroitin 6-sulfate, 2% dermatan sulfate, 3% keratan sulfate, and 12% unidentified species. These studies also demonstrated with statistical significance that the arrays of glycosaminoglycans synthesized by purified corneal fibroblasts were clearly different from those of purified fibroblasts from heart and skin from the same animals that were grown under identical conditions of culture.

In later studies, the types of glycosaminoglycans synthesized and the amounts of each type of glycosaminoglycan contributed to the whole cornea by each of the freshly isolated three corneal cell types were determined (Hart, 1978b). The data from these experiments clearly established that nearly all corneal keratan sulfates were derived from stromal fibroblasts. Corneal heparan sulfates appeared to be predominantly derived from the epithelium, especially at later developmental ages. Corneal hyaluronic acids were largely derived from the endothelium. Of substantial interest was the observation that only the corneal epithelium synthesized a large proportion of sulfated glycoproteins and a high-molecular-weight sulfated polysaccharide or large glycopeptide which was resistant to treatments which degrade all known types of glycosaminoglycans. Thus it appears that each corneal cell type contributes a unique array of distinct classes of glycosaminoglycans and other glycoconjugates to the extracellular matrix of the cornea.

B. Biosynthesis Studies on Adult Corneas

Wortman (1960) found a decrease in the uptake of $^{35}SO_4^{2-}$ by adult corneal stroma after removal of the epithelium and endothelium, with the bovine corneal stroma affected less than that from rabbit. Anseth (1971) was the first to examine the effect of epithelial removal on the incorporation of $^{35}SO_4^{2-}$ into individual fractions of glycosaminoglycans. He found that the rate of incorporation of ^{35}S into glycosaminoglycans was about two- to fourfold less without the epithelium, and most importantly that the specific activity of ^{35}S per hexosamine was also substantially reduced in both the keratan sulfate and chondroitin sulfate fractions. These results suggest an involvement of epithelium in maintaining the normal levels of sulfation of glycosaminoglycans in the intact cornea.

Handley and Phelps (1972) labeled adult bovine corneal stroma with [^{14}C]glucose and fractionated the glycosaminoglycans by CPC-cellulose chromatography. The major labeled glycosaminoglycans consisted of 71% keratan sulfates, 17% chondroitin 4-sulfates, and 4% chondroitin 6-sulfates. Their examination of the specific activities and pool sizes of radioactive sugar nucleotides (glycosaminoglycan precursors) suggested that the ratios of UDP-hexose and UDP-*N*-acetylhexosamine epimers in the stroma corresponded to a close approximation to the steady-state equilibrium of the respective epimerases involved in sugar nucleotide interconversions. The data also indicated that the relatively high levels of UDP-xylose (0.072 m*M*) in the stroma would inhibit by 90% the activity of the UDP-glucose dehydrogenase responsible for the conversion of UDP-glucose to UDP-glucuronic acid. This latter observation is important in view of the relative proportions of uronic acid containing glycosaminoglycans compared to keratan sulfates in the cornea. Perhaps the most significant observation from these biosynthetic analyses was the estimated turnover rate for the glycosaminoglycans of adult cornea. Keratan sulfates were observed to turn over every 723 hr, whereas chondroitin 6-sulfates turned over every 251 hr, which suggests that these glycosaminoglycans probably turn over much more slowly in the adult than in the fetal or neonatal cornea.

Dahl *et al.* (1974) established primary cultures of adult rabbit stromal cells in order to study regulation of the synthesis of corneal glycosaminoglycans. The glycosaminoglycans produced in their cultures consisted of no more than 10% keratan sulfates, 20% chondroitin sulfates, and 60–70% hyaluronic acids. They could detect no significant variation with the cell growth cycle, however, the degree of sulfation increased with the age of the cultures from about 1 sulfate per 10 hexosamines to about 1 sulfate per 3 hexosamines in the oldest cultures.

Gnädinger and Schwager-Hübner (1975) isolated adult rabbit or bovine corneal epithelium and stromal fibroblasts, labeled them with $^{35}SO_4$ or [^{3}H]glucosamine, and fractionated glycosaminoglycans by CPC-cellulose chromatography. From their fractionation patterns it was suggested that the isolated epithelium produced glycosaminoglycans of a lower apparent degree of sulfation than those produced by the stromal fibroblasts. In similar studies, Schwager-Hübner and Gnädinger examined long-term (21-day) cultures of the three corneal cell types of the adult rabbit and demonstrated that under their culture conditions the arrays of glycosaminoglycans synthesized by the epithelium and endothelium were similar and that those synthesized by the fibroblasts of the stroma were distinct.

Yue *et al.* (1976) labeled confluent monolayer cultures of adult rabbit corneal endothelial and stromal cells with $^{35}SO_4^{2-}$ and [^{3}H] glucosamine and characterized the glycosaminoglycans by DEAE-cellulose chromatography in conjunction with specific degrading enzymes. Both endothelial and stromal cultures synthesized hyaluronic acids as the major products. The patterns of glycosaminoglycans were similar to those of other long-term cultures and were reminiscent of those of regenerating or developing tissue.

Klintworth and Smith (1976) performed a comparative study of extracellular sulfated glycosaminoglycans synthesized by rabbit corneal fibroblasts in confluent cultures and by corneal and scleral explants. The glycosaminoglycans synthesized by corneal explants differed considerably from those obtained from confluent corneal fibroblast cultures and scleral explants. The study supports Conrad's and Dorfman's observation that cultured corneal explants or stromal fibroblasts rapidly lose the ability to synthesize keratan sulfate. The study also confirmed the observation that removal of the corneal epithelium and endothelium reduced the biosynthetic capacity of the stromal cells by 50%.

Yue *et al.* (1978) repeated their earlier study (Yue *et al.*, 1976) using long-term cultures of human cells instead of rabbit cells and found that both the endothelial cells and stromal cells synthesized a variety of glycosaminoglycans. Of the sulfated species, chondroitin 6-sulfates were the major products, and chondroitin 4-sulfates, dermatan sulfates, and heparan sulfates were present in smaller amounts. Keratan sulfates were synthesized in very low amounts, or not at all. Both cell types also synthesized hyaluronic acids and, interestingly, endothelial cells from a 1-day-old infant still synthesized more hyaluronic acids than those from older corneas.

It is very clear from the numerous studies on cultured corneal cells that, although they may provide a useful model for certain purposes, their biosynthetic capacities in culture do not reflect the same capacities *in vivo*. In fact, the vast majority of the data suggest that cultured corneal cell types synthesize glycosaminoglycan arrays typical of injured or regenerating tissues.

Recent studies have examined the biosynthesis of intact corneal proteoglycans in organ cultures of intact cornea or in confluent cell cultures of stromal fibroblasts from adult rabbits (Dahl and Cöster, 1978). Proteoglycans were extracted by guanidinium-HCl and purified by DEAE-cellulose chromatography and gel filtration. Stromal cell cultures accumulated polysaccharides linearly for over 48 hr, and about 90% of the glycosaminoglycans were secreted into the medium. The material in the medium consisted mostly of hyaluronic acids and a dermatan sulfate proteoglycans. Negligible keratan sulfate proteoglycan was synthesized by stromal cell cultures. Interestingly, even though short-term organ cultures synthesized the chondroitin sulfate and keratan sulfate proteoglycans typical of cornea, they still synthesized mainly heparan sulfate and dermatan sulfate proteoglycans under the conditions of this study, indicating that the change in biosynthetic patterns from the *in vivo* state to that commonly observed *in vitro* may be even more rapid than previously thought. As mentioned earlier (section II,B), Hassell *et al.* (1980) have performed similar biosynthetic studies using short-term organ cultures of rhesus monkey corneas. Their careful studies clearly demonstrate the utility of isotopic labeling in *short-term* organ culture for studying the biosynthesis and distribution of intact corneal proteoglycans.

C. Cell-Free Biosynthesis Studies on Cornea

1. Sulfation of Glycosaminoglycans

Wortman (1961) examined the sulfotransferase activities in homogenates of adult bovine corneal epithelia and demonstrated the existence of both phenol sulfotransferase and glycosaminoglycan sulfotransferase activities. Phenol sulfotransferase and 3′-phosphoadenosine 5′-phosphate (PAP) were used as a feeder system to generate 3-phosphoadenosine 5′-phosphosulfate (PAPS), the sulfate donor utilized by glycosaminoglycan sulfotransferases (Fig. 5). The cell-free sulfation of endogenous chondroitin sulfates was demonstrated, and the enzymatic properties of the system were investigated.

Pasternak *et al.* (1963) investigated the activation of sulfate in homogenates from corneal epithelia and stromata of normal and

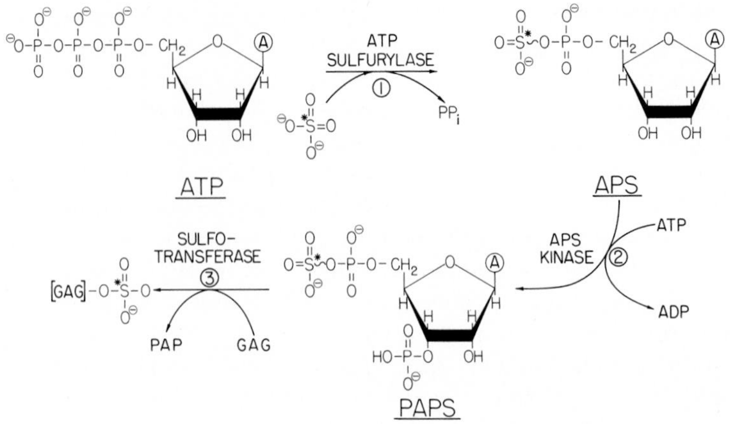

Fig. 5. Reactions involved in glycosaminoglycan sulfation. Reactions 1 and 2 represent the sulfate activation system. GAG, Glycosaminoglycan; PAP, adenosine 3′,5′-diphosphate; A, adenine.

vitamin A-deficient calves, rabbits, guinea pigs, and rats. Earlier studies on extracts from colonic mucosa had suggested that sulfate activation was reduced in vitamin A deficiency (Varandani *et al.*, 1960). The enzymes ATP:sulfate adenylyltransferase (EC 2.7.7.4, ATP-sulfurylase) and ATP:-adenylsulfate 3′-phosphotransferase (EC 2.7.1.25, APS-kinase), which catalyze the activation of sulfate, were demonstrated in several parts of the eye, including the corneal epithelium and stroma (Fig. 5). It was further found that vitamin A deficiency did not lead to decreased sulfate activation. In later studies, Pasternak and Pirie (1964) reexamined the effects of vitamin A deficiency on sulfate activation and suggested that perhaps the reduced sulfate activation observed in earlier studies with vitamin A-deficient animals was due to infections superimposed on the vitamin deficiency rather than the deficiency itself.

Wortman and Locke (1969) have presented data suggesting that the incorporation of $^{35}SO_4^{2-}$ into corneas is cation-dependent and involves a Na^+,K^+-activated, ATPase-linked transport system. In their studies, inhibition of the Na^+,K^+-ATPase by ouabain and chlorpromazine led to a decrease in sulfate incorporation rather than an increase as was anticipated because of the increased availability of ATP. Since sulfation in cell-free systems is independent of a sodium–potassium requirement, the observed inhibition of sulfation caused by these drugs suggested that the Na^+,K^+-ATPase was involved in sulfate transport in the cornea.

Since data from earlier studies (Conrad, 1970a; Hart, 1976) indicated that the apparent relative degree of sulfation of corneal keratan sulfate increased dramatically on and after day 14 of development in the chick, coincident with the onset of transparency, and since the location and number of sulfates on corneal chondroitin sulfates showed an analogous change at this pivotal developmental period, Hart (1978b) initiated an investigation to determine the regulation of glycosaminoglycan sulfation in the developing cornea. Conditions for the assay of embryonic corneal glycosaminoglycan sulfotransferases, with defined exogenous substrates as acceptors, were established, and the relative specific activities of the individual glycosaminoglycan sulfotransferases were determined throughout corneal development. The *in ovo* and *in vitro* effects of serum factors and thyroid hormones on corneal glycosaminoglycan sulfotransferases were also examined. These studies represented the first time that glycosaminoglycan sulfotransferases had been examined directly throughout the development of a tissue. Surprisingly, the specific activities of all the glycosaminoglycan sulfotransferases did not change significantly throughout corneal development. In view of the dramatic increases in sulfation that occur during corneal development, it was suggested that sulfation of corneal glycosaminoglycans might be regulated by the levels of the sulfate donor, PAPS, rather than by the specific activities of sulfotransferases. Conrad and Woo (1980) tested this hypothesis by assaying the ability of cell-free homogenates to synthesize PAPS and 3′-adenosine phosphosulfate (APS) from ATP and $^{35}SO_4^{2-}$ throughout corneal development of the chick cornea. The data from these studies indicated that the equilibrium concentrations of [^{35}S]APS did not change significantly during corneal development, but that the ability of the corneal homogenate to synthesize [^{35}S]PAPS increased by at least 2½-fold between days 8 and 16, coincident with the onset of transparency and increased sulfation of the glycosaminoglycans. These data support the notion that the degree of sulfation of corneal glycosaminoglycans is controlled by the availability of PAPS rather than by changes in the specific activities of glycosaminoglycan sulfotransferases.

2. *Polysaccharide Chain—Pathways and Regulation*

As mentioned above, although there have been numerous studies on the pathways of biosynthesis of glycosaminoglycans in cartilage (for review, see Rodén and Schwartz, 1975), there have been very few such studies on corneas directly. It is probably safe to assume that the general mechanisms of glycosaminoglycan biosynthesis studied in other tissues also apply in the cornea. However, there have been few studies directly concerning corneal glycosaminoglycan biosynthesis either in cell-free systems or concerned with elucidation of the pathways or mechanisms of regulation.

Balduini *et al.* (1970) examined the role of UDP-xylose in the regulation of corneal glycosaminoglycan biosynthesis using adult bovine corneas. They observed a decrease in the rate of biosynthesis of chondroitin sulfate and an increase in the biosynthesis of keratan sulfate when bovine corneas were inoculated with UDP-xylose. Based on their data, they suggested that the exogenous UDP-xylose inhibited the formation of UDP-glucuronic acid by UDP-glucose dehydrogenase, causing an accumulation of UDP-glucose and thus an increase in UDP-galactose and keratan sulfates.

Since virtually nothing was known concerning the details of corneal keratan sulfate biosynthesis and since, as described earlier, this glycosaminoglycan is linked to protein via an N-glycosidic linkage between asparagine and *N*-acetylglucosamine, exactly in the same manner as the large class of asparagine-linked membrane and secretory glycoproteins, Hart and Lennarz (1978) decided to determine if the biosynthetic pathway for the protein linkage region of corneal keratan sulfate was the same as that utilized by asparagine-linked glycoproteins. The effects of the antibiotic tunicamycin on the biosynthesis of all types of glycosaminoglycans by the chick cornea were investigated. Tunicamycin specifically inhibits the formation of *N*-acetylglucosaminylpyrophosphorylpolyisoprenol, the first lipid-saccharide intermediate in the lipid-linked pathway for biosynthesis of asparagine-linked glycoproteins (for review, see Waechter and Lennarz, 1976). The data from these studies strongly indicate that corneal keratan sulfates are synthesized via the lipid-linked glycosylation pathway and that the other corneal glycosaminoglycans are synthesized one sugar at a time directly from sugar nucleotides, as has been demonstrated for cartilage glycosaminoglycans (Fig. 6). It thus appears likely that the mannose- and *N*-acetylglucosamine-containing core oligosaccharide that serves to link the sulfated, repeating disaccharide unit of keratan sulfate to protein is assembled on a lipid carrier and subsequently transferred *en bloc* to the protein, as is the case in the biosynthesis of the more conventional asparagine-linked glycoproteins (Hart *et al.*, 1979). These clear distinctions in the biosynthetic pathways of corneal keratan sulfates and those of other glycosaminoglycans take on greater significance in light of observations of the rapid loss of corneal synthetic capacities for keratan sulfate but not for other glycosaminoglycans when the organs are cultured *in vitro* (Section III,A).

In more recent cell-free studies on corneal keratan sulfate biosyn-

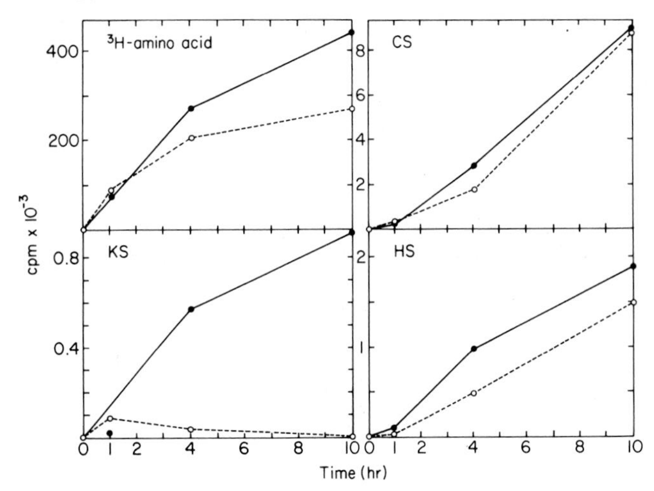

Fig. 6. Time course of the incorporation of radioactivity into proteins and glycosaminoglycans in the presence or absence of tunicamycin. (a) Incorporation of ^{3}H-labeled amino acids into protein. (b) Incorporation of [^{3}H]glucosamine into chondroitin sulfates. (c) Incorporation of [^{3}H]glucosamine into keratan sulfates. (d) Incorporation of [^{3}H]glucosamine into heparan sulfates. ○, With tunicamycin; ●, without tunicamycin; CS, chondroitin sulfates; KS, keratan sulfates; HS, heparan sulfates. (From Hart and Lennarz, 1978, with permission.)

thesis, Christner *et al.* (1979) purified and characterized a soluble galactosyltransferase from bovine corneas which was involved in synthesis of the *O*-β-galactopyranosyl-(1→4)-*N*-acetylglucosamine repeating disaccharide of corneal keratan sulfate. The enzyme was purified 22,000-fold, was shown to transfer galactose from UDP-galactose to several acceptors including bovine corneal agalactokeratan, and was shown by many properties to be similar or identical to the galactosyltransferase of bovine milk.

IV. FUNCTIONAL ASPECTS

A. Hydration and Transparency

The role of glycosaminoglycans in corneal hydration has received considerable study because of the importance of hydration to corneal transparency (Maurice, 1957). Isolated corneal stroma swells drastically when immersed in aqueous solutions and rapidly become opaque. Early studies using dilute alkali (Heringa *et al.*, 1940) extraction demonstrated a correlation between loss of total sulfur and greatly reduced water uptake by the stroma and suggested that the reduction in water uptake was due to the extraction of sulfated polysaccharides.

Graubert (1948) observed a similar reduction in corneal swelling after prolonged extraction with 5% $CaCl_2$ at a pH greater than 4.0 and suggested that this swelling was originally due to glycosaminoglycans. Based on electron microscopic studies on cornea stromal swelling, Francois *et al.* (1954) concluded that the volume increase in the interfibrillar substance was responsible for the swelling of the stroma and that the diameter of the individual collagen fibrils did not differ from those in normal cornea. Maurice (1957) performed refraction studies on collagen fibrils of nearly dried and swollen corneas and found that the hydration of the fibrils was unchanged despite the fact that the cornea swelled to three times its normal thickness.

Hedbys (1961) used dried bovine corneal stromata to examine the role of stromal glycosaminoglycans in the swelling properties of the tissue. The swelling was found to be little affected by temperatures lower than those causing protein denaturation and was also found to be dependent on the concentration of electrolytes in the swelling medium but not by nonionic compounds such as glucose. Hedby also demonstrated a close correlation between the inhibition of swelling by CPC and the solubility properties of CPC–glycosaminoglycan complexes, again suggesting the involvement of these polysaccharides in the swelling properties of the tissue. Enzymatic digestion of stroma with testicular hyaluronidase, which degrades chondroitin sulfates and hyaluronic acids, also inhibited swelling of the stromata. It has been concluded that the swelling of the cornea and the concomitant loss of its transparency are determined by such factors as the type and concentration of the polysaccharides, the degree of dissociation of their acidic groups, the type of counterion, the types and concentrations of other ions in the media, and the dielectric constant of the solvent.

Although the nonswelling properties of elasmobranch stroma can, in part, be explained by the existence of sutural fibers which traverse the stroma in the anterior–posterior direction (Smelser, 1962), their unusual swelling properties have also been attributed to the unusually highly sulfated galactosaminoglycans comprising their stroma (Praus and Goldman, 1970). These highly sulfated chondroitin sulfate-like glycosaminoglycans comprising the elasmobranch corneal stroma appear to contain sulfate on their uronic acid moiety as well as on C-6 of their *N*-acetylgalactosamine moiety (Suzuki, 1960).

Hodson (1971) proposed the notion that the corneal stroma swells because of the Donnan potential associated with the acidic groups of glycosaminoglycans and discounts any masking of the glycosaminoglycan acidic groups by bound cations. Bettelheim and Plessy (1975) studied the hydration behavior of isolated proteoglycans of bovine cornea and investigated the differences in such behavior between intact proteoglycans and their saccharide side chains, the glycosaminoglycans. In their studies the macromolecule–water interaction was followed by determining the water vapor sorption and subsequent desorption from macromolecules which had been dried *in vacuo*. From their studies, they concluded that the protein core of proteoglycans participated in water uptake only to a small extent, and that chondroitin 4-sulfates created a tight network via Ca^{2+} bridges, hydrogen bonding, etc., resulting in limited water uptake but high retention in the polysaccharide network of the proteoglycan. Keratan sulfates, on the other hand, were thought to create a more open polysaccharide network, allowing a large swelling capacity but very little retentive power. Of interest in this regard are the observations of Bettelheim and Goetz (1976), which demonstrated in bovine cornea that there were relatively more chondroitin 4-sulfate chain-bearing proteoglycans in the anterior part of the cornea than in the posterior part. Since, as mentioned above, proteoglycans bearing predominantly keratan sulfate side chains absorb two to three times as much water as proteoglycans with a high chondroitin 4-sulfate content (Bettelheim and Plessy, 1975) and, since they also transfer the water easily, as measured by desorption, their positioning in the cornea would be expected to facilitate movement of water into the cornea from the aqueous humor, thus assisting the active pump of the endothelium. Having a high concentration of chondroitin 4-sulfate containing proteoglycans beneath the epithelium would likewise prevent water loss through dehydration (Bettelheim and Goetz, 1976).

B. Proteoglycans and Collagen Fibrillogenesis

As alluded to earlier, many studies have indicated that collagen fibrillogenesis is regulated by proteoglycans found in the extracellular matrix (Lowther and O'Toole, 1967; Meier and Hay, 1973, 1974a,b; Gelman and Blackwell, 1974; Trelstad, Hayashi, and Toole, 1974; Slavkin and Gruelich, 1975). In an elegant study, Borcherding *et al.* (1975) compared the type and organization of collagen fibers, as determined by electron microscopy, with the composition and types of proteoglycans present at discrete regions from midcornea to sclera in corneal–scleral tissue of humans. In their studies, a direct relationship between the degree of fiber organization and uniformity and the concentrations of various glycosaminoglycans was found. They concluded that the precisely ordered spacing of collagen fibers in the cornea was determined by the specific molecular constraints imposed by the conformation of keratan sulfate proteoglycans. They further suggested that the loss of uniformity and organization of the collagen fibers at the cornea–sclera interface was due to the absence of keratan sulfates and the presence of other glycosaminoglycans such as dermatan sulfates.

C. Proteoglycans of Traumatized Corneas

As discussed earlier, the biosynthetic patterns of the cornea change dramatically upon injury or removal from the eye, especially with regard to a decrease in keratan sulfate and an increase in heparan or dermatan sulfate biosynthesis. These changes are uniformally accompanied by a loss of transparency (Aurell, 1954; Maurice, 1957; Anseth, 1972). Dunnington and Smelser (1958) observed a decrease in the metachromatic staining at the edge of perforating corneal incisions in rabbits 24–48 hr after surgery. Similar studies using autoradiography demonstrated rapid incorporation of [^{35}S] sulfate at the edges of perforating wounds in the rabbit cornea during the healing process (Dohlman, 1957; Dunnington and Smelser, 1958). Anseth (1961a) examined glycosaminoglycans during the healing process of perforating corneal wounds in rabbits in a more quantitative manner. He found that the concentration of glycosaminoglycans in the wound area decreased rapidly during the healing process and that keratan sulfates of high molecular weight and chondroitin sulfates of low sulfate content were greatly reduced. His data suggested that galactosaminoglycans with a high sulfate content preferentially accumulated in the wounded area. It was later suggested that the principal polysaccharide of corneal scar tissue was chondroitin sulfate (Anseth and Laurent, 1961).

Anseth (1969) examined the effects of transient stromal edema (brought about by careful removal of the endothelium from rabbit corneas) on the glycosaminoglycans within the stroma. Not surprisingly, he observed a considerable loss of glycosaminoglycans. Both keratan sulfates and chondroitin 4-sulfates were reduced by 50%, and after healing the levels returned to normal. In later studies, Anseth and Fransson (1969) induced corneal scars by removal of the corneal endothelium and Descemet's membrane and determined the glycosaminoglycans which accumulated in the wounded area. Dermatan sulfate was found to be a major component of the scar tissue. Similar studies on nontransparent, unsuccessful corneal grafts also indicated that dermatan sulfate was a major component of the cloudy, unsuccessful grafted corneas (Anseth, 1969). Other studies have also confirmed these basic observations with respect to glycosaminoglycans in rabbit corneal scar formation (Praus and Dohlman, 1969; Cintron *et al.,* 1973).

D. Proteoglycans of Diseased Corneas

1. *Mucopolysaccharidoses*

A detailed discussion of these genetic diseases is beyond the scope of this article, however, corneal opacification is a clinical feature of several of these diseases which involves defective catabolism of glycosaminoglycans ("mucopolysaccharide" is the old term for glycosaminoglycan) (for review, see McKusick, 1972; Dorfman *et al.,* 1972; Neufeld *et al.,* 1975; McKusick *et al.,* 1978). Among the major mucopolysaccharidoses which involve corneal opacification are type I-H, Hurler's syndrome (α-L-iduronidase deficiency); type I-S, Scheie's syndrome (α-L-iduronidase deficiency); the Maroteaux–Lamy syndrome (*N*-acetylgalactosamine sulfatase deficiency); and Morquio's syndrome (keratan sulfate catabolic deficiency). Surprisingly, Hunter's syndrome, type II mucopolysaccharidosis (iduronate sulfatase deficiency) does not involve corneal opacification. The accumulation of the glycosaminoglycans in the cornea and the resulting opacification are reminiscent of dermatan sulfate accumulation in opaque corneal scar tissue. In any case, these genetic defects clearly indicate the importance of normal corneal glycosaminoglycan homostasis in maintaining the transparency and function of the cornea.

2. *Macular Corneal Dystrophy*

Macular corneal dystrophy is a genetic disease characterized by certain abnormalities of the corneal stroma and gradually increasing corneal opacity. The onset of corneal opacification is usually first detected at puberty as a diffuse clouding which becomes progressively worse until total visual impairment occurs, usually at about age 50 (Francois, 1966; Klintworth, 1977b). Klintworth and Vogel (1964) suggested that the disease was a metabolic disorder restricted to the cornea, which was similar to the mucopolysaccharidoses in that there was both intra- and extracellular accumulation of glycosaminoglycan-like material. In support of this contention, accumulations of glycosaminoglycan-like material have been identified histochemically in numerous studies on the disease (Jones and Zimmerman, 1959, 1961; Klintworth and Vogel, 1964; Morgan, 1966; Teng, 1966; Garner,

1969; Ghosh and McCulloch, 1973; Tremblay and Dubé, 1973; Francois and Victoria-Troncoso, 1965). In such studies, the persistence of staining even after treatment with testicular hyaluronidase or chondroitinase ABC led to the conclusion that the genetic defect involved the metabolism of corneal-type keratan sulfate (Garner, 1969; Klintworth and Vogel, 1964). The staining properties of the accumulations seen in the disease have also led to the suggestion that the material is more like a glycoprotein than a proteoglycan (Graf *et al.,* 1974). Klintworth and Smith (1977) studied glycosaminoglycan biosynthesis and degradation in cultured corneal fibroblasts and corneal explants from patients with macular corneal dystrophy. They compared the properties of their respective glycosaminoglycans with those from corneas of normal individuals and patients with mucopolysaccharidoses and found an impaired synthesis of keratan sulfate in the macular dystrophic corneas. Their data suggests that macular corneal dystrophy was not due to abnormal catabolism of glycosaminoglycans but rather to altered synthesis of keratan sulfate and perhaps chondroitin 6-sulfate.

Recently, Hassell *et al.* (1980) studied intact proteoglycan biosynthesis in short-term organ cultures of macular dystrophic corneas and found that, even though normal corneas synthesized both of the typical chondroitin sulfate and keratan sulfate proteoglycans of cornea, macular corneas synthesized a normal chondroitin sulfate proteoglycan but did not synthesize either keratan sulfate or a mature keratan sulfate proteoglycan. Macular corneas, however, synthesized a novel glycoprotein with unusually large oligosaccharide chains, which was not observed in normal corneas. Based on recent evidence that the biosynthesis of corneal keratan sulfate proceeds via the dolichol-saccharide pathway through a glycoprotein intermediate (Hart and Lennarz, 1978; Hassall *et al.*, 1980; see also Fig. 6), it has been suggested that macular corneal dystrophy results from an enzymatic defect in the processing of such an intermediate to yield the mature keratan sulfate proteoglycan. Besides being of great importance in our understanding of macular corneal dystrophy, these findings are of major significance in unraveling the complexities of the regulation of glycoprotein biosynthesis and the processing of oligosaccharide chains in general.

3. *Other Abnormalities*

Abnormalities in corneal glycosaminoglycans have been observed in viral keratitis (Anseth, 1969) and keratoconus (Praus and Goldman, 1971; Yue *et al.,* 1979). Anseth (1969) examined glycosaminoglycans in macular corneas and corneas from patients with keratitis, lattice degeneration, and keratoconus and found a decrease in the glycosaminoglycan content in the first three diseases. Higher levels of dermatan sulfates were also found in keratitis and lattice degeneration. Yue *et al.* (1976) studied the biosynthesis of glycosaminoglycans in cultures of corneal fibroblasts from normal and keratoconus corneas. It was found that both synthesized similar amounts of glycosaminoglycans, except that a higher proportion of those synthesized by keratoconus corneas were secreted into the labeling medium. It was found upon closer examination that the synthesis of cell surface-associated heparan sulfate was markedly reduced in keratoconus corneal cells.

REFERENCES

Anderson, L., Borg, H., and Miller-Anderson, M. (1975). Purification and characterization of human factor IX. *Thromb. Res.* **7,** 451-459.

Anseth, A. (1961a). Studies on corneal polysaccharide. III. Topographic and comparative biochemistry. *Exp. Eye Res.* **1,** 106-115.

Anseth, A. (1961b). Glycosaminoglycans in the developing corneal stroma. *Exp. Eye Res.* **1,** 116-121.

Anseth, A. (1969). Studies on corneal polysaccharides. V. Changes in corneal glycosaminoglycans in transient stromal edema. *Exp. Eye Res.* **8,** 297-301.

Anseth, A. (1971). Influence of corneal epithelium on the incorporation of $^{35}SO_4$ into stromal glycosaminoglycans. *Exp. Eye Res.* **11,** 251-254.

Anseth, A. (1972). Polysaccharide chemistry in corneal opacification. *Isr. J. Med. Sci.* **8,** 1543-1545.

Anseth, A., and Fransson, L. A. (1969). Studies on corneal polysaccharides. VI. Isolation of dermatan sulfate from corneal scar tissue. *Exp. Eye Res.* **8,** 302-309.

Anseth, A., and Laurent, T. C. (1961). Studies on corneal polysaccharides. I. Separation. *Exp. Eye Res.* **1,** 25-38.

Antonopoulous, C. A., Axelsson, I., Heinegård, D., and Gardell, S. (1974). Extraction and purification of proteoglycans from various types of connective tissue. *Biochim. Biophys. Acta* **338,** 108-119.

Arnold, E., Yawn, D., Brown, D., Wylie, R., and Coffey, A. (1972). Structural alteration in isolated rat liver nuclei after removal of template restriction by polyanions. *J. Cell Biol.* **53,** 737-757.

Arnott, S., Guss, J., Hukins, D., Dea, I., and Rees, D. (1974). Conformation of keratan sulphate. *J. Mol. Biol.* **88,** 175-184.

Aurell, G. (1954). Healing processes in the cornea with special regard to structure and metachromasia. *Acta Ophthalmol.* **32,** 307-313.

Axelsson, I., and Heinegård, E. (1975). Fractionation of proteoglycans from bovine corneal stroma. *Biochem. J.* **145,** 491-500.

Axelsson, I., and Heinegård, D. (1978). Characterization of the keratan sulphate proteoglycans from bovine corneal stroma. *Biochem. J.* **169,** 517-530.

Baker, J. R., and Caterson, B. (1979). The isolation and characterization of the link proteins from proteoglycan aggregates of bovine nasal cartilage. *J. Biol. Chem.* **254,** 2387-2393.

Baker, J. R., Cifonelli, J. A., and Rodén, L. (1975). The linkage of corneal keratan sulfate to protein. *Connect. Tissue Res.* **3**, 149-156.

Baker, J. R., Blithe, D., Buck, C., and Warren, L. (1980). Glycosaminoglycans and other carbohydrate groups bound to proteins of control and transformed cells. *J. Biol. Chem.* **255**, 8719-8728.

Balazs, E. A. (1970). "Chemistry and Molecular Biology of the Intercellular Matrix." Academic Press, New York.

Balduini, C., Brovelli, A., and Castellani, A. A. (1970). Biosynthesis of glycosaminoglycans in bovine cornea: The effect of uridine diphosphate xylose. *Biochem. J.* **120**, 719-723.

Banjeree, S., Cohn, R., and Bernfield, M. (1977). Basal lamina of embryonic salivary epithelia. *J. Cell Biol.* **73**, 445-473.

Berman, E. R. (1970). Proteoglycans of bovine corneal stroma. *In* "Chemistry and Molecular Biology in the Intercellular Matrix" (E. A. Balazs, ed.), Vol. 2, pp. 879-886. Academic Press, New York.

Bernfeld, P. (1966). Interaction of polyanions with blood components. *In* "The Amino Sugars" (E. A. Balazs and R. W. Jeanloz, eds.), Vol. 2B, p. 252. Academic Press, New York.

Bernfield, M. R., Cohn, R. H., and Banerjee, S. D. (1973). Glycosaminoglycans and epithelial organ formation. *Am. Zool.* **13**, 1067-1083.

Bettelheim, F. A., and Goetz, D. (1976). Distribution of hexosamines in bovine cornea. *Invest. Ophthalmol.* **15**, 301-304.

Bettelheim, F. A., and Plessy, B. (1975). The hydration of proteoglycans of bovine cornea. *Biochim. Biophys. Acta* **381**, 203-214.

Bhavanandan, V., and Davidson, E. (1975). Mucopolysaccharides associated with nuclei of cultured mammalian cells. *Proc. Natl. Acad. Sci. U.S.A.* **72**, 2032-2036.

Bhavanandan, V., and Meyer, K. (1967). Studies on keratosulfates: Methylation and partial acid hydrolysis of bovine corneal keratosulfate. *J. Biol. Chem.* **242**, 4352-4359.

Bhavanandan, V., and Meyer, K. (1968). Studies on keratosulfates: Methylation desulfation and acid hydrolysis studies on old human rib case cartilage keratosulfate. *J. Biol. Chem.* **243**, 1052-1059.

Borcherding, M., Blacik, L., Sittig, R., Bizzell, J., Breen, M., and Weinstein, H. G. (1975). Proteoglycans and collagen fibre organization in human corneoscleral tissue. *Exp. Eye Res.* **21**, 59-70.

Brekle, A., and Mersmann, G. (1980). The carbohydrate-protein binding region in keratan sulfate from bovine cornea. I. Isolation and partial characterization. *Hoppe-Seyler's Z. Physiol. Chem.* **361**, 31-39.

Brimacombe, J. S., and Webber, J. M. (1964). Mucopolysaccharides. *Biochim. Biophys. Acta Library* **6**.

Brown, M., Dana, S., and Goldstein, J. (1975). Cholesterol ester formation in cultured human fibroblasts. *J. Biol. Chem.* **250**, 4025-4027.

Buonassisi, V., and Root, M. (1975). Enzymatic degradation of heparin-related mucopolysaccharides from the surface of endothelial cell cultures. *Biochim. Biophys. Acta* **385**, 1-10.

Castellani, A. A., Balduini, C., and Brovelli, A. (1970). Regulatory role of UDP-xylose on the biosynthesis of glycosaminoglycans in cornea. *In* "Chemistry and Molecular Biology of the Intercellular Matrix" (E. A. Balazs, ed.), Vol. 2, pp. 921-928. Academic Press, New York.

Chakrabarti, B., and Park, J. (1980). Glycosaminoglycans: Structure and interaction. *CRC Crit. Rev. Biochem.* **8**, 225-313.

Chisell, G., and Catapano, A. (1979). Glycosaminoglycans and lipoprotein metabolism: An overview. *Pharmacol. Res. Commun.* **11,** 571-583.

Choi, H. U., and Meyer, K. (1975). The structure of keratan sulphates from various sources. *Biochem. J.* **151,** 543-553.

Christner, J. E., Distler, J. J., and Jourdian, G. W. (1979). Biosynthesis of keratan sulfate: Purification and properties of a galactosyltransferase from bovine cornea. *Arch. Biochem. Biophys.* **192,** 548-558.

Cifonelli, J. A., and King, J. (1970). Variable content of 2-acetamido-2-deoxy-D-glucose and neutral sugars in mammalian heparins. *Carbohyd. Res.* **12,** 391-417.

Cifonelli, J. A., and King, J. (1972). The distribution of 2-acetamido-2-deoxy-D-glucose residues in mammalian heparins. *Carbohyd. Res.* **21,** 173-186.

Cifonelli, J. A., and King, J. (1973). Structural studies on heparins with unusually high *N*-acetylglucosamine contents. *Biochim. Biophys. Acta* **320,** 331-340.

Cifonelli, J. A., Saunders, A., and Gross, J. I. (1967). Keratan sulfate fractions from bovine and human tissues. *Carbohyd. Res.* **3,** 478-485.

Cintron, C., Schneider, H., and Kublin, C. (1973). Corneal scar formation. *Exp. Eye Res.* **17,** 251-259.

Cohn, R., Banerjee, S., and Bernfield, M. (1977). Basal lamina of embryonic salivary epithelia: Nature of glycosaminoglycans and organization of extracellular matrix. *J. Cell Biol.* **73,** 464-478.

Coleman, J. R., Herrman, H., and Bess, B. (1965). Biosynthesis of collagen and non-collagen protein during development of the chick cornea. *J. Cell Biol.* **25,** 69-78.

Comper, W. D., and Laurent, T. C. (1978). Physiological function of connective tissue polysaccharides. *Physiol. Rev.* **58,** Jan. 1978.

Conrad, G. W. (1970a). Collagen and mucopolysaccharide biosynthesis in the developing chick cornea. *Dev. Biol.* **21,** 292-317.

Conrad, G. W. (1970b). Collagen and mucopolysaccharide biosynthesis in mass cultures and clones of chick corneal fibroblasts *in vitro*. *Dev. Biol.* **21,** 611-635.

Conrad, G. W., and Dorfman, A. (1974). Synthesis of sulfated mucopolysaccharides by chick corneal fibroblasts *in vitro*. *Exp. Eye Res.* **81,** 421-433.

Conrad, G. W., and Hart, G. W. (1975). Heparan sulfate biosynthesis by embryonic tissues and primary fibroblast populations. *Dev. Biol.* **44,** 253-269.

Conrad, G. W., and Woo, M. L. (1980). Synthesis of 3′-phosphoadenosine-5′-phosphosulfate (PAPS) increases during corneal development. *J. Biol. Chem.* **255,** 3086-3091.

Conrad, G. W., Hamilton, C., and Haynes, E. (1977). Differences in glycosaminoglycans synthesized by fibroblast-like cells from chick cornea, heart and skin, *J. Biol. Chem.* **252,** 6861-6870.

Coulombre, A. J. (1965). Problems in corneal morphogenesis. *Adv. Morphog.* **4,** 81-109.

Coulombre, A. J., and Coulombre, J. L. (1964). Corneal development. III. The role of the thyroid in dehydration and the development of transparency. *Exp. Eye Res.* **3,** 105-114.

Culp, L. (1976). Electrophoretic analysis of substrate-attached proteins from normal and virus-transformed cells. *Biochemistry* **15,** 4094-4104.

Dahl, I.-M. S., and Cöster, L. (1978). Proteoglycan biosynthesis in cultures of corneas and corneal stroma cells from adult rabbits. *Exp. Eye Res.* **26,** 175-190.

Dahl, I.-M. S., Johnsen, W., Anseth, A., and Prydz, H. (1974). The synthesis of glycosaminoglycans by corneal stroma cells in culture. *Exp. Cell Res.* **88,** 193-197.

Davidson, E. A., and Meyer, K. (1954). Chondroitin, a new mucopolysaccharide. *J. Biol. Chem.* **211,** 604-611.

DeLuca, S., Lohmander, S., Nilsson, B., Hascall, V., and Caplan, A. (1980). Proteoglycans from chick limb bud chondrocyte cultures. *J. Biol. Chem.* **255**, 6077-6083.

Dodson, J. W., and Hay, E. D. (1971). Secretion of collagens stroma by isolated epithelium grown *in vitro*. *Exp. Cell Res.* **65**, 215-220.

Dohlman, C. H. (1957). Incorporation of radioactive sulfate into the rabbit eye. *Acta Ophthalmol.* **35**, 115-119.

Dorfman, A., Matalon, R., Cifonelli, J. A., Thompson, J., and Dawson, G. (1972). The degradation of acid mucopolysaccharides and the mucopolysaccharidoses. *In* "Sphingolipids, Sphingiolipidoses and Allied Disorders" (B. W. Volk and S. M. Aronson, eds.), p. 195. Plenum, New York.

Dunham, J., and Hynes, R. (1978). Differences in the sulfated macromolecules synthesized by normal and transformed hamster fibroblasts. *Biochim. Biophys. Acta* **506**, 242-255.

Dunnington, J. H., and Smelser, G. K. (1958). Incorporation of S-35 in healing wounds in normal and devitalized cornea. *AMA Arch. Ophthalmol.* **60**, 116-121.

Faltynek, C. R., and Silbert, J. E. (1978). Copolymers of chondroitin 4-sulfate and chondroitin 6-sulfate in chick embryo epiphyses and other cartilage. *J. Biol. Chem.* **253**, 7646-7649.

Faltz, L., Reddi, A., Hascall, G., Martin, D., Pita, J., and Hascall, V. (1979). Characteristics of proteoglycans extracted from the swarm rat chondrosacoma with associative solvents. *J. Biol. Chem.* **254**, 1375-1380.

Fitton-Jackson, S. (1975). The influence of tissue interactions and extracellular macromolecules on the control of phenotypic expression and synthetic capacity of bone and cartilage. *In* "Extracellular Matrix Influences on Gene Expression" (H. Slavkin and R. Greulich, eds.), pp. 489-495. Academic Press, New York.

Francois, J. (1966). Heredo-familial corneal dystrophies. *Trans. Ophthalmol. Soc. U.K.* **86**, 367-416.

Francois, J., and Victoria-Troncoso, V. (1965). Physiopathologie du kératocyte dans les diptrophies grillagée et tachetée de la cornée. *Arch. Ophthalmol. (Paris)* **35**, 49-55.

Francois, J., *et al.* (1954). L'ultrastructure des tissus oculaires au microscope ēlectronique; ētude de la cornēe et de la sclērotique. *Ophthalmologica* **127**, 74-85.

Fransson, L.-A., Coster, L., Havsmark, B., Malmström, A., and Sjöberg, I. (1974). The copolymeric structure of pig skin dermatan sulphate: Isolation and characterization of L-idurono-sulphate-containing oligosaccharides from copolymeric chains. *Biochem. J.* **143**, 379-389.

Fromme, H., Buddecke, E., von Figura, K., and Kresse, H. (1976). Localization of sulfated glycosaminoglycans within cell nuclei by high-resolution autoradiography. *Exp. Cell Res.* **102**, 445-449.

Fujikawa, K., Thompson, A., Legaz, M., Meyer, R., and Davie, E. (1973). Isolation and characterization of bovine factor IX. *Biochemistry* **12**, 4938-4945.

Garner, A. (1969). Amyloidosis of the cornea. *Br. J. Ophthalmol.* **53**, 73-81.

Gelman, R. A., and Blackwell, J. (1974). Interactions between mucopolysaccharides and cationic polypeptides in aqueous solution: Hyaluronic acid, heparitin sulphate and karatan sulphate. *Biopolymers* **13**, 139-156.

Gentry, P. W., and Alexander, B. (1973). Specific coagulation factor adsorption to insoluble heparin. *Biochem. Biophys. Res. Commun.* **50**, 500-509.

Ghosh, M., and McCulloch, C. (1973). Macular corneal dystrophy. *Can. J. Ophthalmol.* **8**, 515-521.

Gnädinger, M. C., and Schwager-Hübner, M. E. (1975). Biosynthesis of glycosaminoglycans by mammalian corneal epithelium and fibroblasts *in vitro*. I. Isolation and fractionation differences of GAG from two cell types. *Albrecht von Graefes Arch.*

Klin. Ophthalmol. **196,** 21-30.

Goggins, J., Johnson, G., and Paston, I. (1972). The effect of dibutyryl cyclic ademosine monophosphate on synthesis of sulfated acid mucopolysaccharides by transformed fibroblasts. *J. Biol. Chem.* **147,** 5759-5764.

Goldstein, J., Basu, S., Brunschede, G., and Brown, M. (1976). Release of low density lipoprotein from its cell surface receptor by sulfated glycosaminoglycans. *Cell* **7,** 85-95.

Gottschalk, A. (1972) "Glycoproteins: Their Composition, Structure and Function." Elsevier, Amsterdam.

Graf, B., Pouliquen, Y., Frouin, M. A., Faure, J.-P., and Offret, G. (1974). Cytochemical study of macular dystrophy of the cornea (Groenouw II): An ultrastructural study. *Exp. Eye Res.* **18,** 163-169.

Graubert, D. (1948). Dissertation, Paris, Imprimerie Doris, Amsterdam.

Greenwald, R. A., Schwartz, C. E., and Cantor, J. O. (1975). Interaction of cartilage proteoglycans with collagen substituted agarous gels. *Biochem. J.* **145,** 601-605.

Gregory, J. D. (1973). Multiple aggregation factors in cartilage proteoglycan. *Biochem. J.* **133,** 383-386.

Greiling, H., and Stuhlsatz, H. W. (1966). Struktur und Stoffwechsel von Glykosaminoglykan-proteinen. I. Die Keratansulfat-peptide der Finder-cornea. *Hoppe-Seyler's Z. Physiol. Chem.* **345,** 236-245.

Greiling, H., Stuhlsatz, H. W., and Kisters, R. (1970). Structure and metabolism of proteokeratan sulfate. *In* "Chemistry and Molecular Biology of the Intercellular Matrix" (E. A. Balazs, ed.), pp. 873-877. Academic Press, New York.

Handley, C. J., and Lowther, D. A. (1976). Inhibition of proteoglycan biosynthesis by hyaluronic acid in chondroitins in cell culture. *Biochim. Biophys. Acta* **444,** 69-74.

Handley, C. J., and Phelps, C. F. (1972). The biosynthesis *in vitro* of keratan sulfate in bovine cornea. *Biochem. J.* **128,** 205-213.

Hardingham, T. E., and Muir, H. (1974). Hyaluronic acid in cartilage and proteoglycan aggregation. *Biochem. J.* **139,** 565-581.

Hardingham, T., Ewins, R., and Muir, H. (1976). Cartilage proteoglycans. *Biochem. J.* **157,** 127-143.

Hart, G. W. (1976). Biosynthesis of glycosaminoglycans during corneal development. *J. Biol. Chem.* **251,** 6513-6521.

Hart, G. W. (1978a). Glycosaminoglycan sulfotransferases of the developing chick cornea. *J. Biol. Chem.* **253,** 347-353.

Hart, G. W. (1978b). Biosynthesis of glycosaminoglycans by the separated tissues of the embryonic chick cornea. *Dev. Biol.* **62,** 78-98.

Hart, G. W., and Lennarz, W. J. (1978). Effects of tunicamycin on the biosynthesis of glycosaminoglycans by embryonic chick cornea. *J. Biol. Chem.* **253,** 5795-5801.

Hart, G. W., Brew, K., Grant, G. A., Bradshaw, R. A., and Lennarz, W. J. (1979). Primary structural requirements for the enzymatic formation of the N-glycosidic bond in glycoproteins. *J. Biol. Chem.* **254,** 9747-9753.

Hascall, V. C., and Riolo, R. L. (1972). Characteristics of the protein keratan sulfate core and of keratan sulfate prepared from bovine nasal cartilage proteoglycan. *J. Biol. Chem.* **247,** 4529-4538.

Hascall, V. C., and Sajdera, S. W. (1969). Protein polysaccharide complex from bovine nasal cartilage: The function of glycoprotein in the formation of aggregates. *J. Biol. Chem.* **244,** 2384-2396.

Hascall, V. C., and Sajdera, S. W. (1970). Physical properties and polydispersity of proteoglycan from bovine nasal cartilage. *J. Biol. Chem.* **245,** 4920-4930.

Hassell, J. R., Newsome, D. A., Krachmer, J. H., and Rodriques, M. M. (1980). Macular corneal dystrophy: Failure to synthesize a mature keratan sulfate proteoglycan. *Proc. Natl. Acad. Sci. U.S.A.* **77,** 3705-3709.

Hay, E., and Revel, J. P. (1969). "Fine Structure of the Developing Avian Cornea. I. Monographics in Developmental Biology" (A. Wolski and P. S. Chen, eds.), Karger, Basel.

Hedbys, B. O. (1961). The role of polysaccharides in corneal swelling. *Exp. Eye Res.* **1,** 81-91.

Heinegård, D. (1977). Polydispersity of cartilage proteoglycans. *J. Biol. Chem.* **252,** 1980-1989.

Heinegård, D., and Axelsson, I. (1977). The distribution of keratan sulphate in cartilage proteoglycans. *J. Biol. Chem.* **252,** 1971-1979.

Heinegård, D., and Hascall, V. C. (1974a). Aggregation of proteoglycans. III. Characteristics of the proteins isolated from trypsin digests of aggregates. *J. Biol. Chem.* **249,** 4250-4256.

Heinegård, D., and Hascall, V. C. (1974b). Characterization of chondroitin sulfate isolated from trypsin chymotrypsin digests of cartilage proteoglycans. *Arch. Biochem. Biophys.* **165,** 427-441.

Heringa, G. C., Leyns, W., and Weidinger, A. (1940). Water absorption of cornea. *Acta Neerl. Morph. Norm. Pathol.* **3,** 196-201.

Hirano, S., Hoffman, P., and Meyer, K. (1961). The structure of keratosulfate of bovine cornea. *J. Org. Chem.* **26,** 5064-5069.

Hodson, S. (1971). Why the cornea swells. *J. Theor. Biol.* **33,** 419-427.

Höök, M., Lindahl, U., Hellen, A., and Backstrom, G. (1975). Biosynthesis of heparin. *J. Biol. Chem.* **250,** 6065-6071.

Immers, J., and Runnstrom, I. (1965). Further studies of the effects of deprivation of sulfate on the early development of the sea urchin *Paracentrotus lividus. J. Embryol. Exp. Morphol.* **14,** 289-305.

Jansson, L., Höök, M., Wasteson, Å., and Lindahl, U. (1975). Biosynthesis of heparin. V. Solubilization and partial characterization of *N*- and *O*-sulphotransferases. *Biochem. J.* **149,** 49-55.

Jaques, L. B. (1980). Heparins—Anionic polyelectrolyte drugs. *Pharmacol. Rev.* **31,** 161-166.

Järnfelt, J., Rush, J., Li, Y.-T., and Laine, R. (1978). Erythroglycan, a high molecular weight glycopeptide with the repeating structure [galactosyl-(1-4)-2-deoxy-2-acetamidoglucosyl(1-3)] comprising more than one-third of the protein-bound carbohydrate of human erythrocyte stroma. *J. Biol. Chem.* **253,** 8006-8009.

Jones, S. T., and Zimmerman, L. E. (1959). Macular dystrophy of the cornea (Groenouw type II): Clinicopathologic report of two cases with comments concerning its differential diagnosis from lattice dystrophy (Bibber-Haab-Dimmer). *Am. J. Ophthalmol.* **47,** 1-16.

Jones, S. T., and Zimmerman, L. E. (1961). Histopathologic differentiation of granular, macular and lattice dystrophies of the cornea. *Am. J. Ophthalmol.* **51,** 394-410.

Karp, R., and Solurish, M. (1974). Acid mucopolysaccharide metabolism, the cell surface, and primary mesenchyme cell activity in the sea urchin embryo. *Dev. Biol.* **41,** 110-123.

Kimura, J., Osdoby, P., Caplan, A., and Hascall, V. (1978). V. Electron microscopic and biochemical studies of proteoglycan polydispersity in chick limb bud chondrocyte cultures. *J. Biol. Chem.* **253,** 4721-4729.

Kimura, J., Hardingham, T., and Hascall, V. (1980). Assembly of newly synthesized proteoglycan and link protein into aggregates in cultures of chondrosarcoma chondrocytes. *J. Biol. Chem.* **255,** 7134-7143.

Kimura, J., Hardingham, T., Hascall, V., and Solursh, M. (1979). Biosynthesis of proteoglycans and their assembly into aggregates in cultures of chondrocytes from the swarm rat chondrosarcoma. *J. Biol. Chem.* **254,** 2600-2609.

Kinoshita, S. (1969). Periodical release of heparin-like polysaccharide within cytoplasm during cleavage of sea urchin egg. *Exp. Cell Res.* **56,** 59-43.

Kinoshita, S. (1974). Some observations on a protein-mucopolysaccharide complex found in sea urchin embryos. *Exp. Cell Res.* **85,** 31-40.

Kinoshita, S., and Yoshii, K. (1979). The role of proteoglycan synthesis in the development of sea urchins. II. The effect of administration of exogenous proteoglycan. *Exp. Cell Res.* **124,** 361-369.

Kleinman, H., Silbert, J., and Silbert, C. (1975). Heparan sulfate of skin fibroblasts grown in culture. *Connect. Tissue Res.* **4,** 17-23.

Klintworth, G. K. (1977a). The contribution of morphology to our understanding of the pathogenesis of experimentally induced corneal vascularization. *Invest. Ophthalmol.* **16,** 281-285.

Klintworth, G. K. (1977b). The cornea—Structure and macromolecules in health and disease. *Am. J. Pathol.* **89,** 718-808.

Klintworth, G. K., and Smith, C. F. (1976). A comparative study of extracellular sulfated glycosaminoglycans synthesized by rabbit corneal fibroblasts in organ and confluent cultures. *Lab Invest.* **35,** 258-263.

Klintworth, G. K., and Smith, C. F. (1977). Macular corneal dystrophy: Studies of sulfated glycosaminoglycans in corneal explant and confluent stromal cell cultures. *Am. J. Pathol.* **89,** 167-182.

Klintworth, G. K., and Vogel, F. S. (1964). Macular corneal dystrophy: An inherited acid mucopolysaccharide storage disease of the corneal fibroblast. *Am. J. Pathol.* **45,** 565-576.

Kosher, R., and Church, R. (1975). Stimulation of *in vitro* somite chondrogenesis by procollagen and collagen. *Nature (London)* **258,** 327-329.

Kosher, R. A., and Searls, R. L. (1973). Sulfated mucopolysaccharide synthesis during the development of *Rana pipiens. Dev. Biol.* **32,** 50-68.

Kosher, R., Lash, J., and Minor, R. (1973). Environmental enhancement of *in vitro* chondrogenesis. *Dev. Biol.* **35,** 210-220.

Kraemer, P. M. (1971). Heparan sulfates of cultured cells. I. Membrane-associated and cell-sap species in Chinese hamster cells. *Biochemistry* **10,** 1437-1445.

Kraemer, R., and Coffey, D. (1970). Interaction of natural and synthetic polyanions with mammalian nuclei. *Biochem. Biophys. Acta* **224,** 553-567.

Kraemer, R., and Smith, D. A. (1973). High molecular-weight heparan sulfate from the cell surface. *J. Cell Biol.* **59,** 117a (Abstr. No. 353).

Lash, J., and Burger, M., eds. (1977). "Cell and Tissue Interactions." Raven, New York.

Lash, J., and Vasan, N. (1977). *In* "Cell and Tissue Interactions" (J. Lash and M. Burger, eds.). Raven, New York.

Laurent, T. C. (1970). Structure of high uronic acid. *In* "Chemistry and Molecular Biology of the Intercellular Matrix" (E. A. Balazs, ed.), Vol. 2, pp. 703-732. Academic Press, New York.

Laurent, T. C., and Anseth, A. (1961). Studies on corneal polysaccharides. II. Characterization. *Exp. Eye Res.* **1,** 99-105.

Lennarz, W. J. (1980). "The Biochemistry of Glycoproteins and Proteoglycans." Plenum, New York.

Levene, P., and López-Suárez, J. (1918). Mucins and mucoids. *J. Biol. Chem.* **36,** 105-126.

Lindahl, U. (1970). Structure of heparin, heparan sulfate and their proteoglycans. *In*

"Chemistry and Molecular Biology of the Intercellular Matrix" (E. A. Balazs, ed.), Vol. 2, pp. 943-960. Academic Press, New York.

Lindahl, R., Bäckström, G., Jansson, L., and Hallén, A. (1973). Biosynthesis of heparin. II. Formation of sulfamino groups . *J. Biol. Chem.* **248,** 7234-7241.

Lindahl, R., Jacobsson, I., Höök, M., Bäckström, G., and Feingold, D. S. (1976). Biosynthesis of heparin. VI. Loss of C-50 hydrogen during conversion of D-glucuronic to L-iduronic acid residues. *Biochem. Biophys. Res. Commun.* **70,** 492-499.

Lindahl, U. (1976). Structure and biosynthesis of L-iduronic acid-containing glycosaminoglycans. *In* "MTP International Review of Science: Organic Chemistry Series Two—Carbohydrate Chemistry" (G. O. Aspinall, ed.), Vol. 7, pp. 283. Butterworth, London.

Lindahl, U., and Axelsson, O. (1971). Identification of iduronic acid as the major sulfated uronic acid of heparin. *J. Biol. Chem.* **246,** 74-82.

Lindahl, U., and Höök, M. (1978). Glycosaminoglycans and their binding to biological macromolecules. *Annu. Rev. Biochem.* **47,** 385-417.

Lindahl, U., Höök, M., Bäckström, G., Jacobsson, I., Riesenfeld, J., Malmström, A., Rodén, L., and Feingold, D. S. (1977). Structure and biosynthesis of heparin-like polysaccharides. *Fed. Proc. Fed. Am. Soc. Exp. Biol.* **36,** 19-23.

Linker, A., and Hovingh, P. (1973). The heparitin sulfates (heparan sulfates). *Carbohyd. Res.* **29,** 41-62.

Lippman, M. (1968). *In* "Epithelial-Mesenchymal Interactions" (R. Gleischmajer and R. Billingham, eds.), pp. 208-229. Williams and Wilkins, Baltimore, Maryland.

Lloyd, P., Roberts, G., and Lloyd, K. (1960). *Biochem. J.* **75,** 14.

Lohmander, S., DeLuca, S., Nilsson, B., Hascall, V., Caputo, C., Kimura, J., and Heinegård, D. (1980). Oligosaccharide on proteoglycans from the swarm rat chrondrosarcoma. *J. Biol. Chem.* **255,** 6084-6091.

Lowther, D. A., and O'Toole, B. P. (1967). The interaction between acid mucopolysaccharide-protein complexes and tropocollagen. *In* "Symposium on Fibrous Proteins," p. 229. Butterworth, Australia.

McKusick, V. A. (1972). "Hereditable Disorders of Connective Tissue." Mosby, St. Louis, Missouri.

McKusick, V., Neufeld, E., and Kelly, R. (1978). *In* "The Metabolic Basis of Inherited Disease" (J. Stanbury, J. Wyngaarden, and D. Fredreckson, eds.), pp. 1282-1307. McGraw-Hill, New York.

Malmström, N., Fransson, L., Höök, M., and Lindahl, V. (1975). Biosynthesis of dermatan sulfate. I. Formation of L-iduronic acid residue. *J. Biol. Chem.* **250,** 3419-3425.

Manasek, F. (1973). *In* "Developmental Regulation Aspects of Cell Differentiation" (S. Conrad, ed.), pp. 193-218. Academic Press, New York.

Marshall, R. D. (1972). Glycoproteins. *Annu. Rev. Biochem.* **41,** 673-702.

Mathews, M. B. (1967). Biophysical aspects of acid mucopolysaccharides relevant to connective tissue structure and function. *Int. Acad. Pathol. Monogr., No.* 7 304-329.

Mathews, M. B. (1975). "Connective Tissue, Macromolecular Structure Evolution." Springer-Verlag, Berlin and New York.

Mathews, M. B., and Cifonelli, J. A. (1965). Comparative biochemistry of keratosulfates. *J. Biol. Chem.* **240,** 4140-4145.

Mathews, M. B., and Lozaityte, I. (1958). Sodium chondroitin sulfate-protein complexes of cartilage. I. Molecular weight and shape. *Arch. Biochem. Biophys.* **74,** 158-174.

Maurice, D. M. (1957). The structure and transparency of the cornea. *J. Physiol. (London)* **136,** 263-286.

Meier, S., and Hay, E. (1973). Synthesis of sulfated glycosaminoglycans by embryonic corneal epithelium. *Dev. Biol.* **36**, 318–331.

Meier, S., and Hay, E. (1974a). Stimulation of extracellular matrix synthesis in the developing cornea by glycosaminoglycans. *Proc. Natl. Acad. Sci. U.S.A.* **71**, 2310–2313.

Meier, S., and Hay, E. (1974b). Control of corneal differentiation by extracellular materials: Collagen as a promoter and stabilizer of epithelial stromal production. *Dev. Biol.* **38**, 249–270.

Meier, S., and Hay, E. (1975). Control of corneal differentiation *in vitro* by extracellular matrix. *In* "Extracellular Matrix Influences on Gene Expression" (Y. Slavkin and R. Greulich, eds.), pp. 185–196. Academic Press, New York.

Meyer, K., and Anderson, B. (1965). The chemical specificity of the mucopolysaccharides of the cornea. *Exp. Eye Res.* **4**, 346–348.

Meyer, K., and Chaffee, E. (1940). The mucopolysaccharide acid of the cornea and its enzymatic hydrolysis. *Am. J. Ophthalmol.* **23**, 1320–1325.

Meyer, K., Davidson, E., Linder, A., and Hoffman, P. (1956). The acid mucopolysaccharides of connective tissue. *Biochim. Biophys. Acta* **21**, 506–518.

Meyer, K., Linder, A., Davidson, E., and Weissmann, B. (1953). The mucopolysaccharide acid of the cornea and its enzymatic hydrolysis. *Am. J. Ophthalmol.* **205**, 611–616.

Moczar, M., and Moczar, E. (1972). Structural macromolecules of the corneal stroma. *Is. J. Med. Sci.* **8**, 1545–1548.

Moczar, E., Moczar, M., and Robert, L. (1969). Isolation and characterization of the glycopeptides of the structural glycoprotein of corneal stroma. *Life Sci.* (*Pt. II*) **8**, 757–762.

Morgan, G. (1966). Macular dystrophy of the cornea. *Br. J. Ophthalmol.* **50**, 57–67.

Mörner, C. (1894). Untersuchunger der protein Substanzen in den lichtbrechenden Medien des Auges. II. Hornhaut und Sclera. *Z. Physiol. Chem.* **18**, 213–218.

Muir, H., and Hardingham, T. E. (1975). Structure of proteoglycans. *MTP Int. Rev. Sci. Biochem.* **5**, 153–222.

Muthiah, P., Stuhlsatz, H., and Greiling, H. (1974). Composition of corneal proteoglycans. *Hoppe-Seyler's Z. Physiol. Chem.* **355**, 924–934.

Nakashima, Y., Ferrante, N., Jackson, R., and Pownall, H. (1975). The interaction of human plasma glycosaminoglycans with plasma lipoproteins. *J. Biol. Chem.* **250**, 5386–5392.

Nakazawa, K., and Suzuki, S. (1975). Purification of keratan sulfate-endogalactosidase and its action on keratan sulfates of different origin. *J. Biol. Chem.* **250**, 912–917.

Neufeld, E. F., Lim, R. W., and Shapiro, L. J. (1975). Inherited disorders of lysosomal metabolism. *Annu. Rev. Biochem.* **44**, 357–376.

Nevo, Z., and Dorfman, A. (1972). Stimulation of chondromucoprotein synthesis in chondrocytes by extracellular chondromucoprotein. *Proc. Natl. Acad. Sci. U.S.A.* **69**, 2069–2072.

Oegema, T. R., Hascall, V. C., and Dziewiakowski, D. D. (1975). Isolation and characterization of proteoglycans from the swarm rat chondrosarcoma. *J. Biol. Chem.* **250**, 6151–6159.

Ohnishi, T., Oshima, E., and Ohtsuka, M. (1975). Effect of liver cell coat acid mucopolysaccharide on the appearance of density-dependent inhibition in hepatoma cell growth. *Exp. Cell Res.* **93**, 136–142.

Olivecrona, T., Egelrud, T., Iverius, P. H., and Lindahl, U. (1977). Evidence for an ionic binding of lipoprotein lipase to heparin. *Biochem. Biophys. Res. Commun.* **43**, 524–529.

Pasternak, C. A., and Pirie, A. (1964). Sulphate activation in the cornea. *Exp. Eye Res.* **3,** 365-366.

Pasternak, C. A., Humphries, S. K., and Pierie, A. (1963). The activation of sulphate by extracts of cornea and colonic mucosa from normal and vitamin A-deficient animals. *Biochem. J.* **86,** 382-384.

Pessac, B., and Defendi, V. (1972). Cell aggregation role of acid mucopolysaccharides. *Science* **175,** 898-902.

Phelps, C. F. (1975). The intercellular matrix. *In* "Structure of Fibrous Biopolymers" (E. D. T. Atkins and A. Keller, eds.), pp. 53-67. Butterworth, London.

Pratt, R., Larsen, M., and Johnston, M. (1975). Migration of cranial neural crest cells in a cell-free hyaluronate-rich matrix. *Dev. Biol.* **44,** 298-305.

Praus, R., and Brettschneider, I. (1971). Glycosaminoglycans in the developing chicken cornea. *Ophthalmic Res.* **2,** 367-373.

Praus, R., and Dohlman, C. H. (1969). Changes in the biosynthesis of corneal glycosaminoglycans during wound healing. *Exp. Eye Res.* **8,** 69-76.

Praus, R., and Goldman, J. N. (1970). Glycosaminoglycans in the nonswelling corneal stroma of dogfish shark. *Invest. Ophthalmol.* **9,** 131-136.

Praus, R., and Goldman, J. N. (1971). Glycosaminoglycans in human corneal buttons removed at keratoplasty. *Ophthalmic Res.* **2,** 223-230.

Rees, D. A. (1975). *MTP Int. Rev. Sci. Biochem.* **5,** 1-41.

Roblin, R., Albert, S. O., Gelb, N. A., and Black, P. H. (1975). Cell surface changes correlated with density-dependent growth inhibition, glycosaminoglycan metabolism in 3T3, SV 3T3, and Con A selected revertent cells. *Biochemistry* **14,** 347-356.

Rodén, L. (1970). Biosynthesis of acidic glycosaminoglycans (mucopolysaccharides). *In* "Metabolic Conjugation and Metabolic Hydrolysis" (W. H. Fishman, ed.), Vol. 2, pp. 345-442. Academic Press, New York.

Rodén, L. (1980). Structure and metabolism of connective tissue proteoglycans. *In* "The Biochemistry of Glycoproteins and Proteoglycans" (W. J. Lennarz, ed.), pp. 267-371. Plenum, New York.

Rodén, L., and Horowitz, M. I. (1978). Proteoglycans and structural glycoproteins. *In* "The Glycoconjugates" (M. I. Horowitz and W. Pigman, eds.), Vol. II, p. 3. Academic Press, New York.

Rodén, L., and Schwartz, N. B. (1975). Biosynthesis of connective tissue proteoglycans. *MTP Int. Rev. Sci. Biochem.* **5,** 95-152.

Rodén, L., Baker, J. R., Cifonelli, J. A., and Mathews, M. B. (1972). Isolation and characterization of connective tissue polysaccharides. *Methods Enzymol.* **28,** 73-140.

Rosen, O., Hoffman, P., and Meyer, K. (1960). Enzymatic hydrolysis of keratosulfate. *Fed. Proc. Fed. Am. Soc. Exp. Biol.* **19,** 147. (Abstr.)

Rosenberg, R. D. (1977). Biologic actions of heparin. *Semin. Hematol.* **14,** 427-440.

Rosenberg, L., Hellmann, W., and Kleinschmidt, A. K. (1975). Electron microscopic studies of proteoglycan aggregates from bovine articular cartilage. *J. Biol. Chem.* **250,** 1877-1883.

Saiga, H., and Kinoshita, S. (1976). Changes of chromatin structure induced by acid mucopolysaccharides. *Exp. Cell Res.* **102,** 143-152.

Sajdera, S. W., and Hascall, V. C. (1969). Protein polysaccharide complex from bovine nasal cartilage: A comparison of low and high shear extraction procedures. *J. Biol. Chem.* **244,** 77-87.

Sato, R. (1952). Biochemical studies on carbohydrates. CLVIII. Karl Meyer's mucoitinsulfuric and hyaluronosulfuric acids. *J. Exp. Med. Japan* **56,** 357-363.

Satoh, C., Duff, R., Rapp, F., and Davidson, E. A. (1973). Production of mucopolysaccharides by normal and transformed cells. *Proc. Natl. Acad Sci. U.S.A.* **70,** 54-56.

Schachter, H. (1978). Glycoprotein biosynthesis. *In* "The Glycoconjugates" (M. I. Horowitz and W. Pigman, eds.), Vol. II, p. 87. Academic Press, New York.

Schachter, H., and Rodén, L. (1973). The biosynthesis of animal glycoproteins. *In* "Metabolic Conjugation and Metabolic Hydrolysis" (W. H. Rishman, ed.), Vol. 3, p. 1. Academic Press, New York.

Scher, I., and Hamerman, D. (1972). Isolation of human synovial-fluid hyaluronate by density-gradient ultracentrifugation and evaluation of its protein content. *Biochem. J.* **126,** 1073–1080.

Schiller, S., Slover, G., and Dorfman, A. (1961). A method for the separation of acid mucopolysaccharides: Its application to the isolation of heparin from the skin of rats. *J. Biol. Chem.* **236,** 983–987.

Seno, N. (1974). Isolation and characterization of a new disaccharide disulfate: 2-acetoamide-2-deoxy-3-*O*(2- or 3-O-sulfo-β-D-glucopyranosyluronic acid)-4-*O*-sulfo-D-galactose. *Biochim. Biophys. Acta* **343,** 423–426.

Seno, N., Meyer, K., Anderson, B., and Hoffman, P. (1965). Variations in keratosulfates. *J. Biol. Chem.* **240,** 1005–1010.

Seno, N., Anno, K., Yaegashi, Y., and Okuyama, T. (1975). Microheterogeneity of chondroitin sulfates from various cartilage. *Connect. Tissue Res.* **3,** 87–96.

Silbert, J. E., Kleinman, H., and Silbert, C. (1975). Heparin and heparin-like substances of cells. *In* "Heparin: Structure, Function and Clinical Implications" (R. A. Bradshaw and S. Wessler, eds.), pp. 51–60. Plenum, New York.

Slavkin, H. C. (1972) "The Comparative Molecular Biology of Extracellular Matrices." Academic Press, New York.

Slavkin, H. C., and Greulich, R. C. (1975). "Extracellular Matrix Influences on Gene Expression." Academic Press, New York.

Smelser, G. K. (1962). Corneal hydration, comparative physiology of fish and mammals. *Invest. Ophthalmol.* **1,** 11–32.

Smelser, G. D., and Ozanics, V. (1956). Studies on the differentiation of the cornea and sclera of the rabbit. *Anat. Rec.* **124,** 362–367.

Smelser, G. K., and Ozanics, V. (1957). Distribution of radioactive sulfate in the developing eye. *Am. J. Ophthalmol.* (*Pt. 2*) **44,** 102–110.

Smelser, G. K., and Ozanics, V. (1960). "Transparency of the Cornea." Thomas, Springfield, Illinois.

Solurish, M., Vaerewyck, S., and Reiter, R. (1974). Depression by hyaluronic acid of glycosaminoglycan synthesis by cultured chick embryo chondrocytes. *Dev. Biol.* **41,** 233–244.

Speziale, P., Speziale, M. S., Galligani, L., and Balduini, C. (1978). Interactions between different corneal proteoglycans. *Biochem. J.* **173,** 935–939.

Spiro, M. J. (1977). Presence of a glucuronic acid-containing carbohydrate unit in human thyroglobulin. *J. Biol. Chem.* **252,** 5424–5430.

Stuhlsatz, H. *et al.* (1971). The structure of proteoglycans from the cornea, *Hoppe-Seyler's Z. Physiol. Chem.* **352,** 289–303.

Sugiyama, K. (1972). *Dev. Growth Differ.* **14,** 63–69.

Suzuki, S. (1960). Isolation of novel disaccharides from chondroitin sulfates. *J. Biol. Chem.* **236,** 3580–3587.

Svejcar, J., and Robertson, W. (1967). Microseparation and determination of mammalian acidic glycosaminoglycans (mucopolysaccharides). *Anal. Biochem.* **18,** 333–350.

Tang, L., Rosenberg, L., Reiner, A., and Poole, R. (1979). Proteoglycans from bovine nasal cartilage. *J. Biol. Chem.* **254,** 10523–10531.

Taylor, R. L., Shively, J. E., Conrad, H. E., and Cifonelli, J. A. (1973). The uronic acid composition of heparins and heparan sulfates. *Biochemistry* **12,** 3633–3637.

Teng, C. (1966). Macular dystrophy of the cornea: A histochemical and electron microscopic study. *Am. J. Ophthalmol.* **62,** 436-454.

Toole, B. (1972). Hyaluronate turnover during chondrogenesis in the developing chick limb and axial skeleton. *Dev. Biol.* **29,** 321-329.

Toole, B. (1976). Binding and precipitation of soluble collagens by chick embryo cartilage proteoglycan. *J. Biol. Chem.* **251,** 895-897.

Toole, B. P., and Gross, J. (1971). The extracellular matrix of the regenerating newt limb: Synthesis and removal of hyaluronate prior to differentiation. *Dev. Biol.* **25,** 57-77.

Toole, B. P., and Lowther, D. A. (1968). Dermatan sulphate-protein: Isolation from and interaction with collagen. *Arch. Biochem. Biophys.* **128,** 567-578.

Toole, B. P., and Trelstad, R. L. (1971). Hyaluronate production and removal during corneal development in the chick. *Dev. Biol.* **26,** 28-35.

Trelstad, R. L., and Coulombre, A. J. (1971). Morphogenesis of the collagenous stroma in the chick cornea. *J. Cell Biol.* **50,** 840-858.

Trelstad, R., Hayashi, K., and Toole, B. (1974). Epithelial collagens and glycosaminoglycans in the embryonic cornea. *J. Cell Biol.* **62,** 815-830.

Tremblay, M., and Dubé, I. (1973). Macular dystrophy of the cornea: Ultrastructure of two cases. *Can. J. Ophthalmol.* **8,** 47-53.

Varandani, P., Wolfe, G., and Johnson, B. (1960). Function of vitamin A in the synthesis of 3′-phosphoadenosine-5′-phosphosulfate. *Biochem. Biophys. Res. Commun.* **3,** 97-100.

Waechter, C., and Lennarz, W. J. (1976). The role of polyprenol-linked sugars in glycoprotein synthesis. *Annu. Rev. Biochem.* **45,** 95-112.

Wasteson, Å., Uthne, K., and Westermark, B. (1973). A novel assay for the biosynthesis of sulphated polysaccharide and its application to studies on the effects of somatomedin on cultured cells. *Biochem. J.* **136,** 1069-1074.

Wiebkin, O. W., and Muir, H. (1973). The inhibition of sulphate incorporation in isolated adult chondrocytes by hyaluronic acid. *Fed. Eur. Biochem. Soc. Let.* **37,** 42-47.

Wiebkin, O., W. Hardingham, T., and Muir, H. (1975). *In* "Dynamics of Connective Tissue Macromolecules" (P. Burliegh and A. Poole, eds.), pp. 81-104. North-Holland, Amsterdam.

Woodin, A. (1952). The corneal mucopolysaccharide. *Biochem. J.* **51,** 319-330.

Wortman, B. (1960). Metabolism of sulfate by beef and rabbit cornea. *Am. J. Physiol.* **198,** 779-783.

Wortman, B. (1961). Enzymic sulfation of corneal mucopolysaccharides by beef cornea epithelial extract. *J. Biol. Chem.* **236,** 974-978.

Wortman, B., and Locke, R. K. (1969). Sodium-potassium dependent inorganic ^{35}S-sulfate incorporation into cornea. *Invest. Ophthalmol.* **8,** 150-155.

Yue, B. Y. J. T., and Baum, J. L. (1979). The synthesis of glycosaminoglycans by cultures of corneal stromal cells from patients with keratoconus. *J. Clin. Invest.* **63,** 545-551.

Yue, B. Y. J. T., Baum, J. L., and Silbert, M. E. (1976). The synthesis of glycosaminoglycans by cultures of rabbit corneal endothelial and stromal cells. *Biochem. J.* **158,** 567-573.

Yue, B. Y. J. T., Baum, J. L., and Silbert, J. E. (1978). Synthesis of glycosaminoglycans by cultures of normal human corneal endothelial and stroma cells. *Invest. Ophthalmol.* **17,** 523-527.

Zilversmit, D. (1973). A proposal linking atherogenesis to the interaction of endothelial lipoprotein lipase with triglyceride-rich lipoprotein. *Circ. Res.* **33,** 633-638.

2

Transport of Ions and Metabolites across the Corneal Endothelium

MICHAEL V. RILEY

I. INTRODUCTION

The cornea, because it is avascular, receives nearly all its nutritional requirements from the aqueous humor and excretes its waste products into this same fluid. Thus the endothelium, which covers the posterior

Cell Biology of the Eye

ISBN 0-12-483180-X

surface of the tissue, is the cell layer which controls the movement of substances into and out of the tissue. Moreover, this cell layer is also responsible for the control of fluid movement and thereby regulates the hydration of the stroma, a critical factor in determining the transparency of the tissue. This chapter briefly reviews the structure and function of the mammalian cornea and then examines the transport processes that maintain the functional integrity of the tissue. Most physiological studies have been performed on the rabbit cornea, and biochemical studies chiefly on the corneas from ox and rabbit.

II. CORNEAL STRUCTURE AND FUNCTION

The cornea consists of three major parts, the epithelium, the stroma, and the endothelium. Its total thickness is about 0.52 mm in the human, 0.4 mm in the rabbit, and 0.8 mm in the ox, and its diameter in these species is about 11, 13, and 25 mm, respectively. Together with the sclera, it forms the outer, roughly spherical, coat of the eye, and it provides about 75% of the total dioptric power. The anterior surface of the cornea is covered by the tear film which is about 6 μm thick and has a superficial oily layer which limits evaporation from the cornea and provides a good optical surface.

A. Epithelium

The epithelium is the outermost layer of the tissue and generally constitutes 10–15% of the total thickness. In rabbit, human, and monkey it consists of about six layers of cells, the basal ones columnar, situated on a basement membrane, and the more superficial layers progressively more flattened. The epithelium is in dynamic equilibrium, mitosis in the basal layer giving rise to cells which, over a period of several days, are pushed anteriorly and then desquamated into the tear film.

B. Stroma

The stroma accounts for about 90% of the thickness of the tissue and provides the mechanical strength required to withstand intraocular pressure and external trauma. It is a connective tissue, composed of keratocytes and ordered sheets of collagen fibrils which lie parallel to the surface of the tissue and are attached to proteoglycans, all embedded in an acid polysaccharide matrix. The cells are responsible for synthesis and turnover of the extracellular materials and are spread throughout the layer, occupying about 3% of its volume. It is the polyelectrolyte nature of the sulfated and carboxylic saccharide moieties that accounts for the imbibition pressure of the corneal stroma and the tendency of the tissue to swell when placed in water or isotonic solutions (see Hedbys, 1961; Hedbys *et al.*, 1963).

C. Endothelium

The endothelium is a single layer of cells lying on a thickened basement membrane (Descemet's membrane) at the posterior surface of the cornea and bordering the aqueous humor. The cells are about 5 μm thick and 20 μm in diameter and form a hexagonal pattern when viewed in a flat preparation (Sperling, 1977) or by means of a specular microscope (Fig. 1A) (Maurice, 1968). Although mitosis can take place

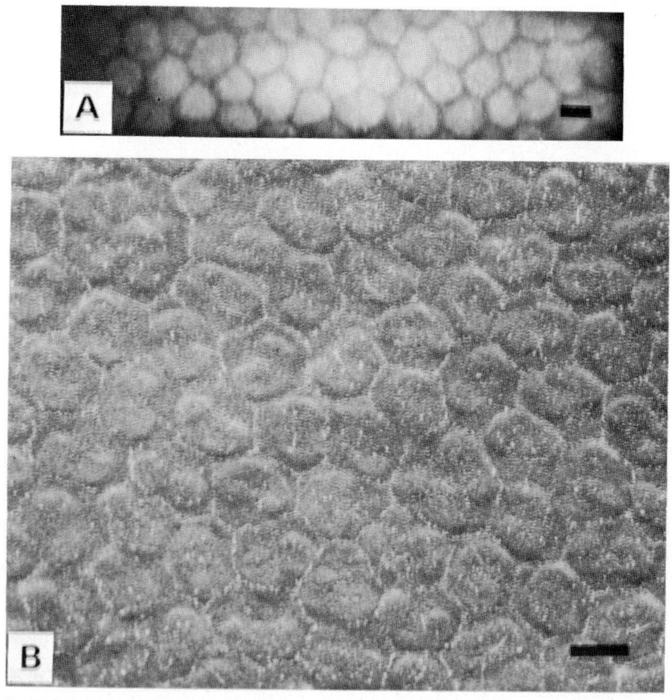

Fig. 1A–C. [pp. 55 and 56.] Structure of the rabbit corneal endothelium *in vitro*. (A) As seen in the specular microscope during perfusion with bicarbonate–Ringer's solution (bar, 10 μm). (B) As seen by scanning electron microscopy following perfusion (bar, 10 μm). Note the regular appearance of the cell borders and the smooth surface.

as a result of injury (Van Horn *et al.,* 1977), it is believed that the cells normally do not divide. Loss of cells through aging (and from surgical trauma) results in the spreading of cells to cover the defects and, in many cases, a measurable decrease in cell density (Laule *et al.*, 1978; Bourne and O'Fallon, 1978; Waltman and Cozean, 1979). In contrast, Descemet's membrane, which is secreted by the endothelial cells, con-

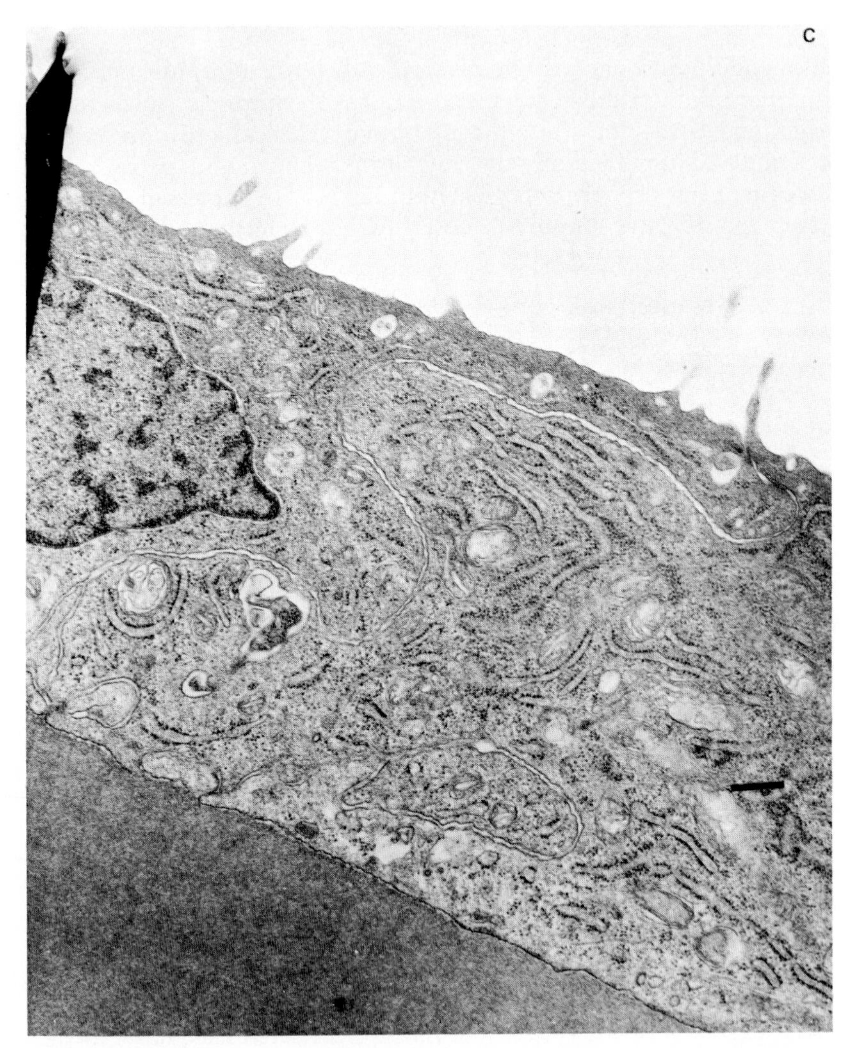

Fig. 1C. As seen by transmission electron microscopy (bar, 0.5 μm). (Micrographs kindly supplied by Dr. H. F. Edelhauser.)

tinues to increase in thickness throughout life, ranging from 5 to 10 μm in humans.

The posterior, or apical, surface of the endothelial cells, in direct contact with the aqueous humor, usually appears smooth and slightly convex by transmission electron microscopy, although a number of cells appear to be ciliated (Wolf, 1968) or to possess microvilli when viewed by scanning electron microscopy (Fig. 1B; Blümcke and Morgenroth, 1967). The intercellular spaces are narrow, generally of a fairly uniform diameter of about 30 nm (Kreutziger, 1976), and gap junctions are found toward the aqueous humor end where the endothelial cells form a lip overlying the adjacent cell (Fig. 1C). At the tip of these overlapping areas, additional junctions have been reported, apparently formed from elements of the terminal bar (Hirsch *et al.*, 1977). Neither type of junction appears to form a continuous barrier in freeze-fracture studies (Okinami *et al.*, 1976), and other studies have demonstrated rapid penetration of the intercellular spaces from the aqueous humor by horseradish peroxidase and lanthanum (Kaye *et al.*, 1973a; Ottersen and Vegge, 1977).

D. Transparency

The cornea must be transparent in order that light may be transmitted to the retina without distortion, yet the refractive inhomogeneity of the stromal constituents would be expected to cause extensive scattering of light. Maurice (1957) has shown that it is the ordered arrangement of the collagen fibrils at distances smaller than the wavelength of light that is responsible for the elimination of scattering and transmission of close to 100% of the incident light. Scattering, hence opacity, occurs when this order is disturbed, either by swelling of the acid polysaccharide matrix in which the fibrils are embedded or by the thicker and irregular-sized fibrils of scar tissue, or when other inhomogeneities of sufficient magnitude occur, such as interlamellar lakes. Since the matrix, like all polyelectrolyte gels, absorbs fluid, control of hydration of the stroma is essential. It will be seen that in mammals a major function of the endothelium is to maintain this equilibrium, allowing a limited access of fluid and also eliminating it by active transport. In other classes of animals, control of corneal swelling may be dealt with by other methods, such as the sutures found in elasmobranchs that run perpendicular to the lamellae (Payrau *et al.*, 1964; Goldman and Benedek, 1967) and the viscous aqueous humor and low corneal permeability in certain teleosts (Smelser, 1962; Edelhauser *et al.*, 1965). In amphibia, corneal swelling is controlled by active transport, but in this class the active process is found in the epithelium (Zadunaisky and Lande, 1971). Recently, Fischer and Zadunaisky (1977) have proposed that some teleost corneas may also exhibit active transport, basing their conclusion on the electrical characteristics of the tissue.

III. ACCESS OF METABOLITES

Since no blood vessels penetrate the normal cornea, the supply of metabolites is determined by diffusional or active exchanges at the surfaces of the tissue, that is, the epithelium, endothelium, and limbal circumference. The epithelium is a relatively impermeable layer, and diffusion is limited by the long intercellular path from the superficial cells to the basal cell layer and by the multiple hydrophobic barriers of the cell membranes. Apart from the diffusion of oxygen and carbon dioxide and the specific transport of sodium and chloride ions, there is little exchange of polar (hydrophilic) substances across this layer. Glucose and lactate, for instance, are present in the tear film, but their transfer across the epithelium is less than 4% of that at the endothelial surface (Riley, 1969).

Diffusion from the limbal capillaries plays a major or minor role, depending chiefly upon the size of the molecules concerned. Maurice and Watson (1965) have shown that small proteins diffuse slowly across the entire cornea, while larger proteins, particularly lipoproteins, are restricted to the periphery of the tissue by the resistance of the matrix. Small, readily diffusible molecules rapidly reach the endothelial surface where they are washed out into the aqueous humor; their contribution to the central cornea is, therefore, negligible (Maurice and Riley, 1970).

For all practical purposes, then, the aqueous humor is the source or sink for amino acids and glucose and its metabolites, and all exchanges of any magnitude take place across the endothelium. Descemet's membrane presents no significant barrier to diffusion, and the diffusibility of small molecules in the stroma is approximately 50% of that in free solution (Maurice, 1961).

A. Amino Acids

The cells of most tissues have concentrations of amino acids higher than those of the surrounding extracellular fluid, accumulating the substrates by means of active transport. The same is true of the epithelial cells of the cornea, the most pronounced concentration ratios (cell fluid/aqueous humor) being those of acidic amino acids, particularly taurine, and to a lesser extent neutral amino acids (Reddy, 1970). The basic amino acids, ornithine, arginine, histidine, and lysine, do not appear to be concentrated in the epithelium, their concentrations being no higher than those in the aqueous humor.

The concentrations of amino acids in the endothelial cells and keratocytes, unlike those of the epithelium, cannot be measured directly, and conclusions regarding the ability of these cells to concentrate amino acids depend upon experimental studies with radioactive tracers. It has been shown that, in the isolated, intact cornea, the stroma can accumulate acidic and neutral, but not basic, amino acids (Riley *et al.*, 1973). Moreover, this accumulation was inhibited by ouabain and was saturable, indicating that the uptake was due to a sodium-dependent active transport process. Nevertheless, the accumulation in the stroma was slight compared to that in the epithelium, and it could not be ascertained whether the keratocytes accounted for the total uptake in this layer or whether both the cells and the extracellular matrix held a higher concentration than the bathing medium. If the latter circumstance were to be found, then active transport *across* the endothelium would be required.

Uptake studies with isolated stroma proved inconclusive, because removal of the cellular layers altered accumulation by the keratocytes. However, permeability studies, with dual perfusion of the endothelial and stromal surfaces, showed that the flux of amino acids across the endothelium was equal in the aqueous humor-to-stroma and stroma-to-aqueous humor directions, establishing that there was no unidirectional transport system across this cell layer (Table I; Riley, 1977a). Thoft and Friend (1972) reached the same conclusion from *in vivo* studies on α-aminoisobutyric acid transfer from the blood to the aqueous humor and cornea of the rabbit. Both sets of findings confirm that the small accumulation in the stroma in the *in vitro* studies was due to high uptake by the keratocytes, equivalent to that found in the epithelial cells, while the concentration in the extracellular matrix was equal to or less than that of the bathing medium.

The isotopic fluxes of acidic and neutral amino acids across the endothelium were reduced in the two directions when excess unlabeled substrate was added to the perfusing medium, indicating that the transfer of these amino acids was by a carrier-mediated process. This facilitated diffusion of acidic and neutral amino acids enhances their

TABLE I

Permeability of the Corneal Endothelium to ^{14}C-Labeled Urea and Amino Acids[a]

	Urea		Aspartate		Glycine		Arginine	
	A → S	S → A	A → S	S → A	A → S	S → A	A → S	S → A
Control	63 ± 1.5[b]	67 ± 2.0	41 ± 1.1	44 ± 1.6	73 ± 2.7	73 ± 1.9	26 ± 0.4	31 ± 0.9
20 m*M*[c]	65 ± 1.8	68 ± 2.1	28 ± 0.9	34 ± 0.6	41 ± 1.4	—	29 ± 0.9	—
Ouabain	—	—	39 ± 1.1	38 ± 2.5	44 ± 0.8	44 ± 1.6	29 ± 2.2	39 ± 2.5

[a] From Riley (1977a). A → S indicates that flux from perfusing medium (aqueous humor) to stroma was measured, and S → A, the reverse.
[b] Values are means ± SEM of six experiments and are in centimeters per hour × 10^3.
[c] Addition of 20 m*M* unlabeled compound with tracer.

rate of exchange at the cell membranes, hence their rate of transfer from the aqueous humor to the stroma. For the basic amino acids there is no evidence of transfer other than by simple diffusion.

The measured values (or estimated ones, based on molecular size) for the permeability of the endothelium to amino acids, together with their known concentrations in the aqueous humor, allow estimates to be made of the rate at which these substrates may be supplied to the epithelium. A comparison can then be made with the estimated requirements of protein synthesis essential for cell renewal (determined by quantitative analysis of hydrolysates of epithelial cell protein), and it is found that diffusion across the endothelium, ranging from 0.04 to 0.07 cm/hr, provides a supply that exceeds the calculated demand in almost every case (Maurice and Riley, 1970). Aspartate is the only exception, requiring a permeability of 0.06 cm/hr to meet the demand, a value about 40% greater than the measured value of 0.043 cm/hr. The supply of this amino acid might, therefore, be considered deficient or marginal, though it must be recognized that the requirement is derived from a somewhat crude estimate. Thoft and Friend (1972), on the other hand, have argued that the permeability of the endothelium (hence supply) is about five times greater than the values used above, but failure to establish steady-state conditions in the aqueous humor and stroma when making the isotopic flux measurements means that the data are unsatisfactory for estimating permeability.

Other demands for amino acid supply seem to be small by comparison with the needs of the epithelium (see Maurice and Riley, 1970). Collagen, which constitutes more than 70% of the stromal protein, has a half-life in mature tissues estimated at more than 100 days, and the growth rate of Descemet's membrane is negligible. The turnover rate of the proteoglycans has not been determined, but it is probably about 10-fold faster than for the much larger collagen molecule; the acid polysaccharides have a half-life of about 10 days, but their amino acid content is very small. Soluble proteins also have a rapid turnover of about 8 days but account for less than 5% of total stromal protein. However, the most important consideration is that, whatever the turnover rate of these stromal macromolecules may be, their breakdown would not necessarily lead to a net loss of amino acids as occurs in desquamated epithelial cells. A negligible loss of radioactivity was found in corneas 16 months after they were pulse-labeled with [^{14}C]-glycine (Smelser *et al.*, 1965), indicating effective reutilization of the incorporated substrate or, alternatively, a half-life of stromal collagen much exceeding 100 days.

It may be concluded that the amino acids required by the cornea are derived almost entirely by simple or facilitated diffusion from the aqueous humor across the endothelium. In cases of marginal or deficient supply, it is possible that the transaminase activities of the cornea (Mimura *et al.*, 1963) can serve to transfer amino groups from surplus amino acids to keto acids such as oxaloacetate.

Finally, it should be noted that active accumulation of amino acids within the endothelial cells is not ruled out by the foregoing, but it must necessarily be a symmetrical process, occurring at both the apical and basolateral membranes. Evidence for such a process can be found in the inhibition of glycine fluxes by ouabain, suggesting accumulation by a sodium-dependent process (see Table I; Riley, 1977a), and in the demonstration of γ-glutamyltransferase in the rabbit endothelium (Reddy and Unakar, 1973), which suggests that accumulation could take place via the Meister cycle.

B. Glucose

Glucose is the primary source of metabolic energy in the cornea, and it also provides the carbon skeleton of a large number of other components, particularly the sulfated and amino sugars of the acid polysaccharides. Its rate of consumption in the rabbit cornea is approximately 0.5 μmol/cm^2/hr, about 85% of this being accounted for by conversion to lactate through aerobic glycolysis (Riley, 1969). Nearly half of this glucose metabolism takes place in the epithelium, although on a weight basis the endothelium is the most active cell layer in the oxidation of glucose (Freeman, 1972).

The measured concentrations of glucose in the aqueous humor and corneal layers in the rabbit, ox, and human are shown in Table II. These values, together with the observation of Hale and Maurice (1969) that the steady-state level of glucose in the stroma of isolated corneas is about 80% of that in the bathing medium over the range 0.6–7.5 m*M*, suggest that glucose is distributed passively in the cornea. In agreement with this, it was found that, when isolated corneas were perfused with solutions containing [^{14}C]glucose or nonmetabolized 3-*O*-methylglucose, there was no part of the cornea where the concentrations in tissue water were higher than those in the bathing medium (M. V. Riley, unpublished).

There are several studies in which the permeability of the endothelium to glucose has been measured or may be calculated, but all are subject to certain limitations, which are discussed below. It will be seen that, as for some amino acids, the supply is only slightly in excess of the tissue's requirements. Although glycogen is present in both the

TABLE II

Glucose Concentrations in Aqueous Humor and Cornea[a]

Species	Aqueous humor	Endothelium	Stroma	Epithelium	Reference
Ox	3.5	1.7[b]	1.6	1.9[b]	Reim *et al.* (1970)
Human	3.8	—	1.7	1.9[b]	
Rabbit	6.9	—	4.5	3.6	Schütte and Reim (1976)
Rabbit	6.1	—	4.7	—	Reim *et al.* (1972)
Rabbit	—	—	—	5.9[b]	Reim *et al.* (1966)
Rabbit	6.5	—	4.7	—	Thoft *et al.* (1971a)
Rabbit	6.0	—	4.4	—	Thoft *et al.* (1971b)
Rabbit	7.1	—	5.8	—	Hale and Maurice (1969)

[a] Values are in millimoles per liter of fluid.
[b] Derived by assuming 83% water in tissue wet weight.

epithelium (Smelser and Ozanics, 1953) and endothelium (Malinin and Bernstein, 1979) of the cornea, at steady state there can obviously be no net contribution to the glucose supply from this reserve.

Hale and Maurice (1969) measured the permeability of the rabbit endothelium to sugars by following the efflux of tracer from preloaded, intact, isolated corneas into a medium perfusing the posterior surface. They found that arabinose, a molecule slightly smaller than glucose but of similar shape, had a permeability of about 0.06 cm/hr, which is close to that expected for its size. Glucose and 3-*O*-methylglucose were washed out twice as fast as arabinose, which suggested that these sugars might be transported by a carrier-mediated mechanism. When the experiments were done at 5°C, these sugars showed a fourfold decrease in efflux rate, whereas a simple diffusion process would have undergone only a twofold change, as was found for arabinose. It was concluded that glucose and 3-*O*-methylglucose crossed the endothelium by facilitated diffusion, with a permeability of 0.12 cm/hr. However, the authors were puzzled by the exact equivalence of the data for these two sugars, for the metabolism of glucose should have contributed to its rate of loss from the stroma and, therefore, its measured turnover rate should have been faster than that of nonmetabolized 3-*O*-methylglucose. It seems that this metabolic factor, which they calculated should have had an incremental effect on the efflux rate equal to that due to facilitated diffusion, may have had less influence than expected. Since the load-up period with the tracer was only 15 min, equal to only the half-time for saturation of the stroma by small molecules (Mishima and Trenberth, 1968), the anterior portion of the cornea, particularly the metabolically active epithelium, would not have acquired an equilibrium concentration of the tracer. Therefore, the cornea would not have metabolized labeled glucose at the expected rate, and the difference in efflux rate for the two sugars was probably within the experimental limits of the method.

A value for endothelial permeability can also be calculated from the data of Thoft *et al.* (1971b) who measured the flux of glucose across the deepithelialized anterior surface of the stroma in the living eye. The net flux was 0.57 μmol/cm²/hr, and the concentration of glucose in the aqueous humor and stroma averaged 6.7 and 2.1 m*M*, respectively. From the equation

$$P\ (C_a - C_s) = J_{a\text{-}s} \tag{1}$$

where P is the permeability (cm/hr), C_a and C_s are the solute concentrations in the aqueous and stromal fluid (μmol/cm³), and J is the flux (μmol/cm²/hr), a value for P of 0.12 cm/hr is derived (Table III). A higher value is obtained using data from another series of experiments from the same laboratory (Thoft *et al.*, 1971a), where the flux was the same but the difference in glucose concentration was found to be only 3.2 m*M* instead of 4.6 m*M*, yielding $P = 0.18$ cm/hr. These values can be compared with another estimate of endothelial permeability made in a similar manner by putting J equal to glucose utilization of the cornea at steady state. Glucose utilization in the rabbit may be set at 0.5 μmol/cm²/hr and the aqueous and stromal glucose concentrations, derived from a number of studies by Reim and Thoft, at 6.4 and 4.6 m*M*, respectively (Table II). The permeability calculated in this manner is 0.28 cm/hr.

The actual consumption of glucose by the tissue influences the values for P calculated by both the above methods and, when allowed for, it results in a decrease in the higher value and an increase in the lower

TABLE III

Permeability of the Rabbit Corneal Endothelium to Glucose[a]

Calculated value	Corrected for metabolism	Reference
12	—	Hale and Maurice (1969)
12	17	Thoft *et al.* (1971a)
18	26	Thoft *et al.* (1971b)
28	24	Riley (1969)

[a] Values are in centimeters per hour × 10^3.

value. Thus, the flux at the anterior stromal surface may underestimate that which crosses the endothelium by an amount equal to that utilized by the stroma. This amount can be estimated as follows: The endothelium utilizes 13% of the total glucose (Section V), and the stroma and epithelium consume the remainder equally (i.e., 43% each), since Freeman (1972) showed their respiration to be equal and Langham (1954) showed that the ratios of respiratory and glycolytic activity in the two layers were the same. Therefore, the correction would increase the value for J by an amount equal to 43% of the total utilization. This would change the permeability calculated from flux measurements from 0.18 to 0.26 cm/hr. Conversely, the metabolic demand is probably overstated by 13%, since the glucose used by the endothelial cells does not have to cross the entire cell layer. The resulting estimate of P by this method is 0.24 cm/hr.

This last value is clearly a minimum, though still about fourfold higher than expected for simple diffusion, confirming that glucose transport is carrier-mediated. It allows a supply of glucose to the stroma and epithelium adequate only for normal metabolism, with no capacity for meeting an excess demand such as would obtain under anaerobic conditions. Even though epithelial glycogen stores are known to be rapidly depleted when the cornea is deprived of oxygen or subject to other trauma by contact lenses (Smelser and Ozanics, 1953; Burns *et al.,* 1971; Uniacke and Hill, 1972; Thoft and Friend, 1975), there must be some excess glucose available for its replenishment after the crisis. It seems clear, however, that such a reserve capacity is not great, since the maximum endothelial permeability is most probably not greater than the high estimates of 0.26–0.28 cm/hr, which could provide a surplus of not more than 30% above the steady-state demand. Indeed, whereas depletion of glycogen may be complete within 2–4 hr, restoration of the original levels requires 8–12 hr (Lowther and Hill, 1974).

C. Oxygen and Carbon Dioxide

Unlike the water-soluble metabolites, oxygen and carbon dioxide diffuse readily across the corneal layers, including the epithelium. Oxygen, therefore, is supplied largely from the atmosphere, where its partial pressure, 155 mm Hg, is about three times higher than in the aqueous humor.

Barr *et al.* (1976) claim that all layers of the cornea receive oxygen solely from the atmosphere. Their conclusion is based upon measured oxygen fluxes across the tissue, and they predict that in the living eye there is a net flow of oxygen from the atmosphere into the aqueous humor. On the other hand, Fatt *et al.* (1974) calculated from permeability and consumption data that a flux through the entire cornea would occur only when the atmospheric oxygen pressure was 200 mm Hg or above. Moreover, Kwan *et al.* (1972) observed a lower oxygen tension in the stroma than in the aqueous humor, and Grote and Zander (1976) provided a basis for this finding from theoretical considerations, both studies concluding that the supply to the endothelium must be derived from the aqueous humor. This conclusion is consistent with experimental observations on the dependence of corneal function on oxygen supply. Mishima *et al.* (1969) showed that isolated corneas, exposed to moist air at the epithelial surface, swelled when deprived of oxygen in the medium perfusing the posterior surface of the tissue. With a similar experimental design, Riley (1969) found that the perfusing medium provided 20% of the total oxygen consumed by the tissue, the remainder being derived from the air. Since Freeman (1972) determined that the endothelium utilized about 22% of the total oxygen required, it is clear that, under normal conditions of oxygen tension in the air and in the aqueous humor, the endothelium obtains its entire oxygen supply from the aqueous humor.

Fatt and Bieber (1968) have calculated that all the carbon dioxide produced by the cornea is removed by diffusion across the anterior surface, since the atmospheric concentration is very low compared to that of the aqueous humor. However, it is possible that the carbon dioxide produced by the endothelium, approximately 2 $\mu l/cm^2/hr$, plays a role in the active transport process of these cells and could be removed as bicarbonate by transfer to the aqueous humor (Section IV,B).

D. Lactate

The cornea produces lactate from glucose at a rapid rate, despite the ready availability of oxygen, especially to the epithelium. There is no significant loss of lactate to the tear film, and all that is produced, including that from the epithelium, can be accounted for by the flux across the endothelial surface (Riley, 1972). Reim *et al.* (1972) have found that the concentration of lactate in the rabbit stroma is between 12 and 19 m*M*, and in the aqueous humor of the same rabbits 8.5 m*M*, this value being in close agreement with other determinations (Langham, 1954; Riley, 1972). By using the same formula as for glucose in the preceding section, but with a factor (r = 0.65) allowing for the different activity of anions in the stroma relative to that in the aqueous humor (see Maurice, 1951),

$$P(C_s/r - C_a) = J_{s-a} \tag{2}$$

a value of 0.07 cm/hr is obtained for the permeability coefficient of the endothelium to lactate. This is the same value as for sodium and bicarbonate ions and fits with theoretical expectations, indicating that lactate leaves the cornea by simple diffusion.

E. Effect of Aqueous Flow Rate

It should be noted, in relation to all the above metabolites, that the flow rate of the aqueous humor is a critical factor in their supply or removal by this fluid. A decrease in the flow rate would cause a buildup of the lactate concentration in the aqueous humor and thereby diminish the net diffusion of lactate across the endothelium and result in increased corneal concentrations. For glucose the supply to the cornea would be diminished by a slower flow rate, since its concentration in the aqueous would remain equal to that of the plasma; for the amino acids the outcome might be different, since if active transport by the ciliary epithelium were to continue, it would result in higher concentrations in the reduced volume of aqueous humor secreted and the total available to the cornea would remain constant. Oxygen availability would probably be decreased, resulting in an increase in glycolysis in the endothelium.

IV. WATER AND IONS

It was established by the experiments of Mishima and Kudo (1967) that the endothelium of the mammalian cornea possessed a fluid pump capable of removing water from the stroma, thereby controlling the thickness and transparency of the tissue. The subsequent work of Maurice (1972; Barfort and Maurice, 1974) has demonstrated directly that the endothelial cells transport fluid equivalent to 10 times their own volume per hour against a pressure gradient of at least 20 mm Hg. This transport from stroma to aqueous humor counterbalances the movement of fluid (called the leak) that takes place in the opposite direction as a result of the imbibition pressure of the stromal matrix. Failure of the pump leads to corneal swelling. Similarly, an increase in the leak into the stroma also leads to swelling, since the pump has a maximum rate which limits its output. In either case, swelling continues until a new steady state is reached when the imbibition pressure falls to a level at which the leak is reduced enough to be again balanced by the pump. In the open eye there is an additional flux of water caused by evaporative loss from the tears. This makes the tear film approximately 10% hypertonic to the stromal matrix and aqueous humor and, in the rabbit, results in a flow of water of about 3 $\mu l/cm^2/hr$ across the entire cornea from the aqueous humor to the tears (Mishima and Maurice, 1961a). In the human, this evaporative loss of water results in a steady-state thickness about 4% less in the open eye than in the closed eye (Mishima and Maurice, 1961b; Mandell and Fatt, 1965).

A. Permeability

These observations indicate that there is a substantial movement of fluid across the endothelial cell layer. Donn *et al*. (1963), using tritiated water, found the permeability of the rabbit endothelium to be 8 μm/sec, compared to 3 μm/sec for the epithelium. More recently, Liebovitch *et al.* (1981), measuring water flow induced by osmotic gradients, have reported an even higher value of 30 μm/sec. Measurements of hydraulic conductivity, also derived in osmotic experiments, range from 2×10^{-12} cm^3/dyne/sec (Liebovitch *et al*. 1981) to 1.5×10^{-11} cm^3/dyne/sec (Mishima and Hedbys, 1967).

Despite its high permeability, the endothelium presents a significant barrier to entry of water, as shown by the high rate of stromal swelling when the cells are damaged or removed. The obstruction of the membrane to the flux of tritiated water, relative to that of an equivalent thickness of saline solution, probably lies between the values of 50 (Donn *et al.,* 1963) and 360 (Mishima and Trenberth, 1968). A swelling rate of 40–50 μm/hr is observed when the fluid pump is inhibited by ouabain, and this is considered the maximum water flux from the aqueous humor to the stroma permitted by the normal endothelium, since under these conditions there is a normal endothelial architecture with well-preserved gap junctions (Kaye and Donn, 1965), and the hydraulic conductivity and permeability to urea and sucrose are unchanged (Table IV; Trenberth and Mishima, 1968). Swelling rates two- to threefold greater than this are found when the cornea is perfused with calcium-free solutions (Kaye *et al.,* 1968, 1973b). Electron microscopy shows that in this case the gap junctions open and the cells become hemispherical, with intercellular spaces widened to the

TABLE IV

Hydraulic Conductivity and Permeability of Rabbit Corneal Endothelium[a]

Condition	Hydraulic conductivity, L_p (mm/mOsm/min × 10^4)	Permeability (cm/hr × 10^3)	
		Urea	Sucrose
Control	2.3	75	23
Ouabain (10^{-3}–10^{-5} *M*)	2.2	75	32
Ca^{2+}-free	—	155	—

[a] From Trenberth and Mishima (1968) and Kaye *et al.* (1968).

extent that Descemet's membrane is almost exposed (Fig. 2; Kaye *et al.*, 1968). Rapid swelling also follows perfusion of the endothelium with diamide, which again disrupts cell-to-cell contact (Edelhauser *et al.*, 1976). After complete removal of the endothelium *in vivo,* Ytteborg and Dohlman (1965) showed that the stroma swelled at about 300 μm/hr.

The permeability of the endothelium to ions is also high, Maurice (1951) finding a value of 0.07 cm/hr for sodium and other small ions, more than 100 times greater than for the epithelium. Urea penetrates the endothelium at a similar rate (Mishima and Trenberth, 1968; Riley, 1977a), as do several amino acids, but ions and molecules the size of sucrose or larger show decreasing permeability coefficients (Tables I and IV). Nevertheless, Maurice and Watson (1965) showed that serum albumin could cross the endothelial barrier at a slow but measurable rate, and Mishima and Trenberth (1968) measured the rate of loss of inulin across the endothelium at 0.004 cm/hr, 1/20 of the rate for urea or sodium. Molecules equal to or greater than raffinose in size all have the same osmotic effect at the endothelium and are therefore assigned a reflection coefficient of 1. Sodium and other monovalent ions have a reflection coefficient of 0.6, while glucose and sucrose have values between 0.9 and 1.

The resistance to diffusion of ions and small molecules presented by Descemet's membrane is equivalent to that of the bulk stroma, only about twice that in free solution. The resistance to water movement, however, is probably greater, although conflicting results were obtained when stromal swelling rates were compared in the presence and absence of this membrane. Mishima and Hedbys (1967) reported that little difference was observed in the two conditions, but Maurice (Maurice and Giardini, 1951) and Dohlman (Ytteborg and Dohlman,

Fig. 2. Structure of rabbit corneal endothelium following perfusion with calcium-free medium or with diamide. (A) Calcium-free medium for 2 hr. Note the hemispherical shape of the cells and the stretched appearance of the pockets between the cells (bar, 10 μm). (B) As in (A). The intercellular space has opened almost all the way to Descemet's membrane, yet the intracellular appearance is normal (bar, 1 μm). (C) After perfusion with 10^{-4} *M* diamide for 15, 30, and 45 min (bar, 10 μm). (Micrographs kindly supplied by Dr. H. F. Edelhauser.)

1965) each found slower rates when Descemet's membrane was intact. Fatt (1968) made similar observations and calculated that the obstruction to water movement was 1:17:250 for stroma, Descemet's membrane, and endothelium, respectively. However, these values must be considered very approximate, since, with the accepted value for stroma relative to free solution of 2.3, the obstruction of the endothelium would be more than 500, higher than other estimates which range from 50 to 360.

There appear to be no measures of the permeability of the endothelium to divalent ions such as Mg^{2+} and Ca^{2+}, which are found in the aqueous humor of the rabbit at 1.5 and 3.5 mEq/liter, respectively (Bito, 1970), and in the corneas of various species at about 2 and 2–5 mEq/liter, respectively (see Maurice and Riley, 1970). However, their ability to cross the endothelium is evident from the studies of Obenberger and Babický (1970) on corneal calcification; after intraperitoneal injection of ^{45}Ca into rabbits and rats the tracer was found in the anterior layers of the stroma within 5 hr. The possibility should be considered that the endothelium actively removes calcium from the stroma, for Green and Friedman (1971) reported that in isolated stroma there was a rapid and extensive replacement by calcium of sodium bound to the mucopolysaccharides, and Hara (1965) found a significant increase in the stromal calcium concentration when intact corneas were maintained at 4°C over several days.

B. Active Transport

Until quite recently, the evidence for active transport of ions across the corneal endothelium has been circumstantial; because there was a movement of fluid that was metabolically dependent and inhibited by ouabain, there was probably an ion pump. Stronger evidence was provided by the observation of a potential difference across the endothelium of about 600 μV, aqueous side negative (Fischbarg, 1972; Fischbarg and Lim, 1974; Hodson, 1974; Barfort and Maurice, 1974). The potential was shown to be directly associated with the fluid transport mechanism of the endothelium; when the swelling of a fresh cornea or a decrease in the deturgescence rate of a swollen cornea was brought about by inhibition of the pump, there was a concomitant decrease in the magnitude of the potential (Fig. 3). Such inhibition was produced by the use of inhibitors of carbonic anhydrase, by ouabain, or by decreasing the concentration of sodium, potassium, or bicarbonate in the perfusing fluid. Surprisingly, chloride could be entirely replaced in the medium by nitrate, sulfate, bicarbonate, or pyroglutamate without any change in either the potential or fluid dynamics (Hodson, 1971a; Fischbarg and Lim, 1974). Active chloride transport, therefore, does not account for the potential and, unlike the case in the epithelium of the amphibian cornea (Zadunaisky and Lande, 1971), is not responsible for endothelial fluid transport.

Alternative substrates for an electrogenic, anionic pump were bicarbonate or hydroxyl ions, or organic anions synthesized in the endothelium. The last of these possibilities was also eliminated when it

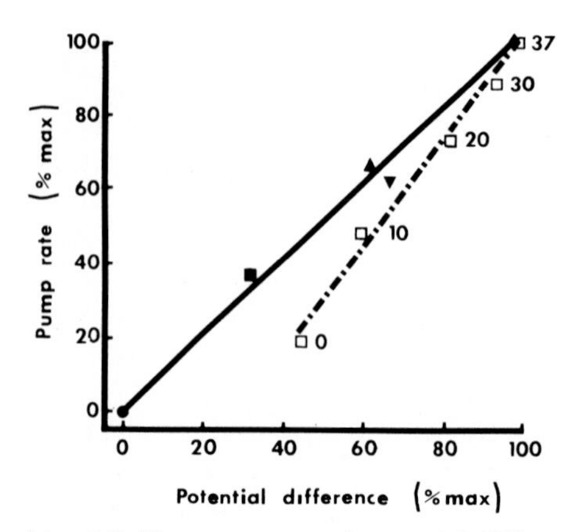

Fig. 3. Relationship of fluid pump rate and potential difference across the rabbit corneal endothelium when perfused with inhibitors of carbonic anhydrase or varying bicarbonate concentrations. Solid line represents data of Hodson and Miller (1976): ◆, 43 m*M* HCO_3^- plus 5% CO_2; ▲, 43 m*M* HCO_3^-; ▼, 43 m*M* HCO_3^- plus 5% CO_2 plus 10^{-5} *M* ethoxyzolamide; ■, 0 m*M* HCO_3^- plus 5% CO_2; ●, HCO_3^-- and CO_2-free. Broken line represents data of Fischbarg and Lim (1974) at 37, 30, 20, 10, and 0 m*M* HCO_3^-, respectively, each with 5% CO_2. Barfort and Maurice (1974) measured the relationship between fluid transport and potential difference directly, but without manipulation of the experimental conditions, and also found it to be approximately linear. The data are not included here, since the maximum (100%) transport rate could not be determined from their graph.

was shown that the most prevalent organic acid, lactate, and labellein, a polyanion synthesized from glucose (Hodson, 1971b; Hodson and Riley, 1973), were produced in the endothelium at rates of only 0.3 and 0.04×10^{-10} Eq/cm²/sec (Riley *et al.*, 1977). These rates were inadequate to carry the measured short-circuit current which has an ionic requirement of 3×10^{-10} Eq/cm²/sec.

The first direct evidence of active ion transport across the endothelium came from the work of Hodson and Miller (1976). They measured the unidirectional fluxes of bicarbonate across the endothelium and showed that the net flux from the stroma to the aqueous humor was directly proportional to the short-circuit current. The potential across the deepithelialized cornea, with a reduced hydrostatic pressure sufficient only to maintain the normal shape of the tissue, was 540 μV when the perfusing solution was a bicarbonate-Ringer's solution bubbled with CO_2. In the absence of HCO_3^-, the potential fell to 160 μV, and in the absence of CO_2 or with carbonic anhydrase inhibitors present, it fell to 340 μV. In the complete absence of CO_2 and HCO_3^-, the potential was abolished. There was a close correlation between these effects on potential difference and the changes in the swelling rates of the cornea (Fig. 3); a maximum swelling rate of 40 μm/hr (equivalent to the rate in the presence of ouabain) was obtained in the absence of HCO_3^- and CO_2 and a rate of 25 or 13 μm/hr when HCO_3^- or CO_2, respectively, was removed from the medium. The carbonic anhydrase inhibitors also caused swelling at a rate of 15 μm/hr, a value which indicates a decrease in the fluid transport rate of about 35%, virtually identical to the observations of Fischbarg and Lim (1974). Although these authors found no effect of the inhibitors on the potential, they did find that the potential and fluid pump rate decreased in almost parallel fashion in response to lowered bicarbonate concentrations.

Measurement of the unidirectional fluxes of Na^+, Cl^-, PO_4^{3-}, and HCO_3^- across the short-circuited endothelium confirmed that bicarbonate was actively transported. The flux of the other ions, measured concurrently with HCO_3^-, was reported to be symmetrical, but bicarbonate had a flux rate of approximately 9×10^{-10} Eq/cm²/sec in the stroma-to-aqueous humor direction and only 7×10^{-10} in the opposite direction (Hodson and Miller, 1976). Similar studies by Green and his colleagues (Hull *et al.*, 1977) under open-circuit conditions showed the same active bicarbonate flux of 9×10^{-10} Eq/cm²/sec, but a somewhat lower passive flux of 4×10^{-10} Eq/cm²/sec, resulting in a net flux about twice as great. (These values have been corrected for the error in corneal area in this study which was noted by Hull *et al.* 1979.) Again, additive effects on swelling were observed when low bicarbonate concentrations were combined with the presence of carbonic anhydrase inhibitors.

This apparently complementary effect of low bicarbonate concentrations and enzyme inhibitors suggested that the substrate for the pump might comprise both HCO_3^- ions in the stroma and HCO_3^- derived from CO_2 in the endothelium under the influence of carbonic anhydrase (Hodson and Miller, 1976). However, the evidence for the involvement of CO_2 cannot be considered conclusive. Green's data showed that the decrease in the net flux of bicarbonate caused by acetazolamide resulted chiefly from an increase in the passive flux of the ion and was accompanied by a concomitant increase in hydraulic conductivity. More recently, Kelly and Green (1980) showed that the net endothelial bicarbonate flux was proportional to the concentration of HCO_3^- in the perfusing medium, but at fixed HCO_3^- concentrations was independent of pCO_2 over a 10-fold range. Furthermore, Mayes and Hodson (1978) found that the actual loss of HCO_3^- ion present in a small reservoir at the anterior stromal surface was 2.1×10^{-10} Eq/

cm²/sec, identical to the net transendothelial flux observed by Hodson and Miller (1976), also suggesting that CO_2 made a negligible contribution. Mayes and Hodson (1979) calculated that less than 4% of the total CO_2 in the stromal fluid of the intact rabbit eye was in the form of CO_2, the remainder being in the form of HCO_3^-. In denuded stroma the HCO_3^- ion was at 0.93 of its concentration in the bathing medium, a value close to that expected from the Donnan distribution and comparable to the value of 0.89 for Cl^-. However, in the intact eye, while the stromal-versus-aqueous humor Cl^- ratio was slightly higher at 1.0, the HCO_3^- ratio was markedly lower, at less than 0.5. The difference between the expected and measured stromal concentrations in Donnan equilibrium with the aqueous humor, approximately 18 m*M*, is consistent with the concept of its active transport across the endothelium.

A net flux of sodium ions from stroma to aqueous humor has recently been reported (Lim, 1981; Huff and Green, 1981), a finding at variance with earlier studies (Green, 1967; Hodson and Miller, 1976) which showed unidirectional fluxes of sodium across the endothelium to be equal. The net flux was approximately 10×10^{-10} Eq/cm²/sec, somewhat higher than that found for bicarbonate, and it was decreased in the presence of ouabain but not altered when the bicarbonate concentration was reduced to as low as 5 m*M*. This is in contrast to the inhibitory effect of low sodium concentrations on the net bicarbonate flux.

It can be postulated that the fluid transport process is initiated by a sodium pump which is then coupled to a bicarbonate pump responsible for water movement. If the link between the sodium and bicarbonate systems were a one-way couple, then the observed dependence of HCO_3^- flux on Na^+ and the independence of Na^+ flux on HCO_3^- would be explained. The scheme would also be consistent with the observations that fluid transport is rapidly inhibited by ouabain, is dependent on both Na^+ and HCO_3^-, and is coupled to HCO_3^- flux.

The possibility of a "neutral" salt pump was considered by Hodson (1974) and by Fischbarg and Lim (1974) who proposed different mechanisms to account for the observed aqueous-side-negative endothelial potential. Hodson's scheme depended upon the charge characteristics of the stroma, but when it was later shown that after virtually complete removal of the stroma the potential was unchanged in magnitude and polarity, the scheme had to be abandoned (Hodson *et al.*, 1977). The model of Fischbarg and Lim (1974), on the other hand, remains compatible with the new data on sodium and bicarbonate fluxes: the salt is transported into the endothelial cells from the intercellular spaces, making them hypotonic (the "backward" pump of

Diamond and Bossert, 1968) and creating a diffusion gradient from the aqueous humor across the gap junctions; fixed negative charges at these sites then generate a diffusion potential of the observed polarity.

An alternative to this model can be proposed wherein the salt is pumped from the cell into the intercellular space. While the bulk of accompanying water flow would be toward the stroma because of the restriction of the gap junctions at the aqueous humor end of the space, the net flux of water would be determined by the relative permeabilities of the apical and basal membranes (Mathias, 1981). In Scheme I, the net flux would be from left to right, or from stroma to

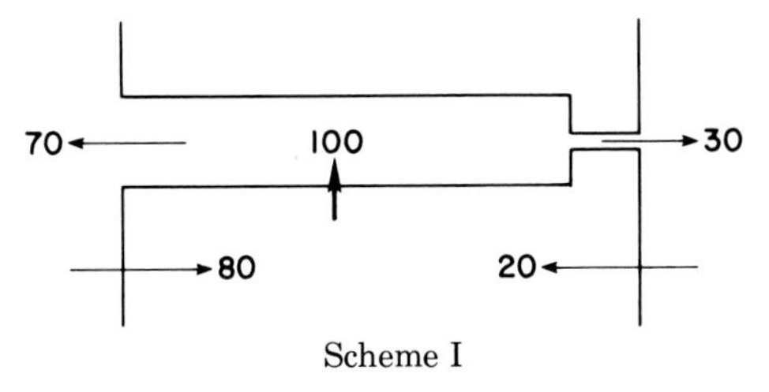

Scheme I

aqueous humor. The charge of the gap junctions would, again, influence the relative movements of the two ions and, consequently, the potential.

It has been suggested that the shape of the intercellular spaces during fluid transport could be used to distinguish between the "forward" and "backward" standing gradient models (Diamond and Bossert, 1968), but this has been questioned by Hodson and Mayes (1979). However, their conclusion rests on the assumption that, when the intact cornea is in a steady state, there is no fluid flow through the intercellular spaces. This may not be valid, since fluid could move in a cycle, still being pumped as in the net flux condition. There is, as yet, no way to resolve this question, nor any entirely coherent hypothesis of how the system is structured. The major accepted features of endothelial fluid transport, and the apparent anomalies, are summarized in Section VII.

V. METABOLISM AND ENZYMES

It is clear that the flux of ions and water, which is critical to the maintenance of normal corneal function, is an active process coupled in some manner to an energy source. This energy source, which is presumed to be ATP (though other interconvertible forms of cellular energy could also subserve this role), is provided by the metabolic activity of the cells. Langham and Taylor (1956) showed that the inhibition of glycolysis or respiration caused swelling of the cornea, and the experiments of Davson (1955) and of Harris and Nordquist (1955) with refrigerated tissue also indicated that enzymatic processes were crucial for deturgescence. Several more recent studies have shown that glucose, adenosine, glutathione, and oxygen are of particular importance in endothelial metabolism, and that ATPases, such as the Na^+, K^+- activated enzyme originally described by Skou (1957, 1965), constitute integral parts of the transport system.

A. Respiration, Glycolysis, and Shunt

Mishima and colleagues (1969) showed that an oxygen supply in the aqueous humor was a prerequisite for endothelial fluid transport; Riley (1969) found that 20% of the total oxygen uptake by the rabbit cornea was from the medium at the endothelial surface, and Freeman (1972) calculated that the respiratory activity of the endothelium was 6.5 μl oxygen/mg dry wt/hr, about five times greater than that of the epithelium. The major substrate of this oxidation appears to be glucose, since it is the major oxidizable substrate in the aqueous humor and the amount consumed by the cornea is fully accounted for in the production of lactate and consumption of oxygen (Riley, 1969). However, in studies with isolated corneas, it is clear that adenosine can substitute for glucose and even appears to have a "beneficial" effect, in that the deturgescence of swollen corneas was greater than when glucose was the substrate (Anderson *et al.*, 1973, 1974; Fischbarg *et al.*, 1977). Adenosine is most probably hydrolyzed to release ribose which is then metabolized via the pentose shunt pathway, eventually yielding hexose, whence it can enter the glycolytic and oxidative systems (Fig. 4). Adenosine, however, does not stimulate the part of the pentose shunt that generates NADPH, that is, the glucose-6-phosphate dehydrogenase and 6-phosphogluconate dehydrogenase steps. Indeed, adenosine may depress the activity of these enzymes, for Ng and Riley (1980) showed that, when adenosine was the substrate for the perfusion of corneas over a 5-hr period, the percentage of intracellular glutathione in the oxidized form was 35%, whereas with glucose as substrate it was only 21%. This ability of glucose to maintain glutathione in the relatively more reduced state indicates that the pentose shunt pathway probably operates in the endothelium under *in vivo* conditions, since in freshly isolated cells the glutathione is only 7% oxidized (Riley and Yates, 1977; Ng and Riley, 1980; Anderson and Wright, 1980a).

In isolated tissue, Geroski *et al.* (1978) demonstrated shunt activity

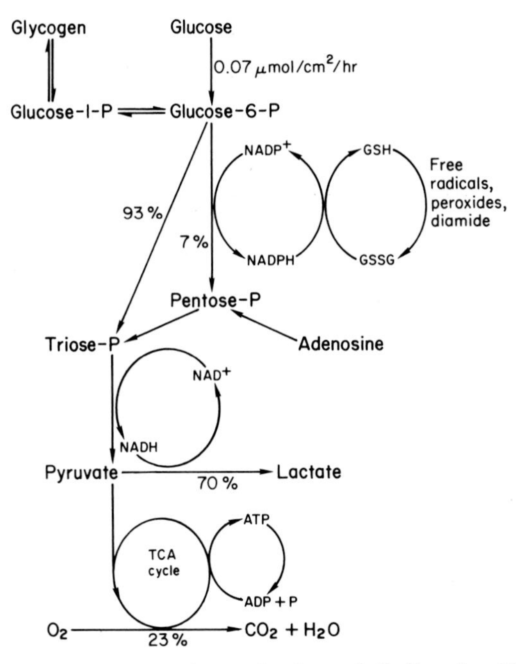

Fig. 4. Relative utilization of glucose in the endothelium by different metabolic pathways, and their relationship to glutathione metabolism. Percentages are approximate, and the fraction oxidized by the pentose pathway is presumed to be recycled.

in the endothelium, stroma, and epithelium and calculated that the percentages of glucose oxidized by the pathway were 37, 66, and 66%, respectively. In all three layers these values increased to over 80% when 0.4 m*M* diamide was added to the tissue; in the endothelium, total glucose oxidation was 60% higher than in the control, but the amount consumed by the tricarboxylic acid cycle decreased to only one-third of the control level. Although such quantitative measures of metabolism in individual layers of the cornea may be strongly influenced by the isolation procedure (e.g., scraped epithelium is essentially a homogenate, while stroma and endothelium are relatively undisturbed) and therefore may not reflect the relative activities of the cells in the intact tissue, the effect of diamide is significant. Diamide causes an acceleration of the shunt by oxidizing glutathione, which is then reduced by NADPH and glutathione reductase—the resulting increase in $NADP^+$ stimulating the two dehydrogenases (Fig. 4). This stimulation of an NADPH-generating system is considered to represent the protective response of cells to natural oxidative insults, such as free radicals and peroxides. Therefore, these results and those of Ng and Riley (1980) indicate that the shunt activity of the endothelium is important in maintaining the cellular redox balance and protecting against oxidative damage.

Hermann and Hickman (1948) showed that the cornea produced significant quantities of lactate under aerobic conditions, and Langham (1952, 1954) found that under anaerobic conditions the rate of lactate production was increased by 50%. In the epithelium it is known that the increased demand for glucose resulting from a reduced oxygen supply causes a rapid depletion of glycogen stores (Uniacke and Hill, 1972). Glycogen has also been found in the endothelium (Malinin and Bernstein, 1979), and a reduction in oxygen supply from the aqueous humor may be expected to deplete these stores in a similar manner.

Riley (1969) has determined that 85% of the total glucose utilization of the intact cornea is by aerobic glycolysis. An estimate of the fraction of glucose respired and that glycolyzed in the endothelium can be made from the oxygen data of Freeman (1972) and the lactate yield given by Riley *et al.* (1977); the values are 0.016 μmol/cm^2/hr glucose respired and 0.05 μmol/cm^2/hr glucose glycolyzed. These values and those of Geroski *et al.* (1978), summarized in Fig. 4, indicate that the metabolism of the endothelium is somewhat more aerobic than that in the remainder of the tissue, depending relatively more on the tricarboxylic acid cycle than on glycolysis or shunt. Support for this conclusion is also found in the different NAD/NADH ratios of the endothelium and epithelium, 3.6 and 2.0, respectively (M. V. Riley and E. M. Yates, unpublished). This trend is toward the pattern characteristic of tissues with a high energy need but relatively little synthetic activity.

B. Glutathione

Fluid transport in the isolated cornea was first shown to be influenced by the presence of glutathione by Dikstein and Maurice (1972), but the nature of the interaction is still elusive. Anderson *et al.* (1973) found that GSH (the reduced form), when added to a medium deficient in both glucose and adenosine, protected to a large extent against depletion of ATP in the endothelium. However, later studies by these authors (Anderson *et al.*, 1974) showed that the addition of glucose alone was equal to GSH in maintaining ATP levels and that adenosine alone was superior, which suggests that, while GSH can be metabolized to yield available energy, this role is important only under conditions of substrate deficiency. A more significant role is indicated by the improved deturgescence and survival time brought about by GSH, over and above that due to glucose and adenosine. This effect on the physiological parameters could be brought about equally well by 240 μM GSH or by 20 μM GSSG (the oxidized form) and, since such levels of GSSG were always found in solutions of GSH as a result of autoxidation, it was suggested that the oxidized compound was the active one (Anderson *et al.,* 1974). The authors speculated that membrane thiol groups might interact with the disulfide, resulting in a conformational change that would reduce membrane permeability, thus increasing net fluid transport.

The finding that 15% of the glutathione in the aqueous humor was in the oxidized form (Riley *et al.,* 1980) gave some credence to this concept of an extracellular role; on the other hand, the observations of Edelhauser *et al.* (1976) that diamide caused rapid swelling of corneas, with a clustering of endothelial cells into islands and the exposure of large areas of Descemet's membrane, clearly showed that the cornea could not withstand total oxidation of both intra- and extracellular glutathione. When the intracellular concentrations of GSH and GSSG were measured after perfusion of corneas with media of varying glutathione content or with added thiol reactants (Whikehart and Edelhauser, 1978; Ng and Riley, 1980), it was found that endothelial fluid transport was disrupted only when the total intracellular glutathione fell below one-third of the *in vivo* value. Moreover, swelling of the corneas was very rapid when the intracellular glutathione was either completely reduced (by dithiothreitol) or completely oxidized (by diamide), but no correlation could be observed between the efficiency of fluid transport and the extent of glutathione oxidation over a range from 10 to 54% (nanograms GSSG × 100/nanograms total glutathione) (Ng and Riley, 1980). Anderson and Wright (1980a) have reached the same conclusion with a different technique: with a specific inhibitor of glutathione reductase, 1,3-bis(2-chloroethyl)-1-nitrosourea (BCNU), up to 70% of the endothelial glutathione was oxidized, yet deturgescence was at the same rate as in control corneas, in which GSSG was only 6% of the total. These authors (Anderson and Wright, 1980b) also showed that *S*-methylglutathione could support fluid transport, suggesting that the thiol group may be unimportant, but this cannot be established without concurrent measures of intracellular glutathione.

The role of glutathione is, at best, unclear. There appears to be a critical level required within the endothelial cells and a need for a fraction in both the oxidized and reduced states, suggesting that it might function as a redox buffer system, combating the effects of free radicals (see Section V,A). Although the failure of BCNU to disturb endothelial function might argue against such a role, these studies were not of a duration long enough to bring about complete elimination of GSH. Glutathione may, in addition, act extracellularly, regulating the permeability or structural integrity of the cells through interaction with membrane thiol groups, and it may protect against oxidation of the thiol groups of specific enzymes such as Na^+,K^+-ATPase. It serves as an energy source only in the absence of alternative substrates, but Dikstein *et al.* (1980) have postulated that it might serve as a source of γ-aminobutyric acid (GABA) which they find supports optimal pump function in the presence of glucose and insulin. They claim that both GABA and insulin are present in the aqueous humor at effective concentrations, whereas adenosine is not, but whether these compounds more closely represent the true physiological system than glutathione and adenosine is an open question at this point.

C. Coupling of Energy and Transport

Active transport requires coupling of the transfer process of the cell membrane to a system releasing free energy. ATP may be the energy source, in which case hydrolysis ensues in an ATPase reaction, it may be electron transport, creating an electrochemical gradient of protons and hydroxyl ions, or it may be a preexisting gradient which can be collapsed to create an alternative one. In the cornea, the classical sodium-potassium-activated ATPase (Na^+,K^+-ATPase) of Skou (1957) has been found, and also a bicarbonate-stimulated ATPase. Carbonic anhydrase has also been studied in view of its association with most other tissues in which bicarbonate ions are secreted.

1. Na^+, K^+-ATPase

Following the work of Bonting *et al.* (1961) and Skou (1965), who found Na^+,K^+-ATPase activity in many fluid-transporting tissues, including the cornea, Rogers (1968) examined this tissue in more detail. He found that, despite the well-known sodium transport activity of the epithelium, the specific activity of the Na^+,K^+-ATPase in the endothelium of ox corneas was four times greater than in the epithelium. In the cat the endothelial activity was six times higher, and in the rabbit a fourfold difference has been reported (Riley, 1977b). It was also noted by Rogers that the sensitivity of the bovine enzyme to inhibition by ouabain was very similar to the dose–response curve of swelling of the rabbit cornea in the presence of this inhibitor, with half-maximal effects on both processes at about 10^{-6} M. Since Mishima (Mishima and Kudo, 1967; Trenberth and Mishima, 1968) had shown that it was the endothelium that controlled fluid transport, these findings suggested that a sodium pump was important in this process. More recently, Anderson and Fischbarg (1978) have reported a biphasic effect of ouabain, the drug producing an increase of up to 50% in the rate of stromal thinning at 10^{-10} M but reverting to the inhibitory effect at 10^{-8} M. Since stimulation of Na^+,K^+-ATPase is known to occur at low concentrations of ouabain (see Bonting, 1970), these observations further strengthen the postulated link between transendothelial fluid transport and Na^+,K^+-ATPase, activity.

The enzyme has also been studied by histochemical methods, both the 4-nitrophenylphosphatase procedure of Ernst (1972) and the ATPase procedure of Wachstein and Meisel (1957). Using the Ernst technique, Leuenberger and Novikoff (1974) demonstrated that enzyme activity was restricted solely to the lateral cell membranes of the endothelial cells, with no activity at either the basal or apical membranes (Fig. 5). Tervo and Palkama (1975), using the second method, confirmed that the ATPase activity bordered the intercellular spaces (see Kaye and Tice, 1966) and identified it as the Na^+-dependent enzyme by means of its ouabain sensitivity.

2. *Bicarbonate-Dependent ATPase*

The above findings are typical of several other fluid-transporting membranes, but it did not seem that a sodium pump could account for all the observed phenomena related to the control of corneal hydration. A bicarbonate pump, on the other hand, might be postulated to account for the net bicarbonate flux and the negative potential from the stroma to the aqueous humor. Although no such entity has been conclusively demonstrated, there are several tissues where a bicarbonate-dependent ATPase is suggested to play such a role (Durbin and Kasbekar, 1965; Simon *et al.,* 1972). The enzyme has been reported in pancreas (Simon *et al.,* 1972), salivary gland (Izutsu and Siegel, 1972), kidney (Liang and Sacktor, 1976), gastric mucosa (Kasbekar and Durbin, 1965), intestinal epithelium (van Os *et al.,* 1977), and retina (Winkler and Riley, 1977), but its function remains controversial, in part because the identity of the primary transported ion is not always clear and also because the localization of the enzyme is disputed. Some investigators find the enzyme to be located on the plasma membrane (Sachs *et al.,* 1972; Liang and Sacktor, 1976; Milutinović *et al.,* 1977; Bornancin *et al.,* 1980), but others claim that it is entirely of

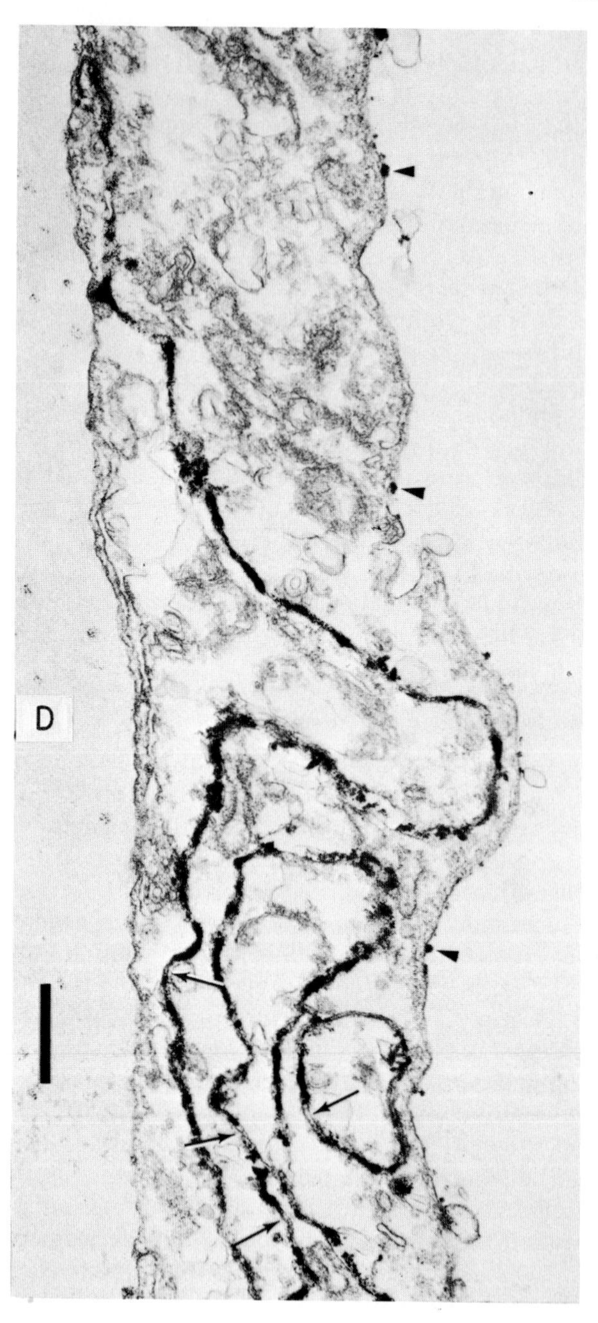
D

mitochondrial origin and can therefore play no part in transcellular ion transport (Kimelberg and Bourke, 1973; Izutsu and Siegel, 1975; van Amelsvoort *et al.*, 1977).

A bicarbonate-stimulated ATPase was found in the corneas of both rabbit and bovine eyes, with activity in the endothelium sixfold higher than in the epithelium (Riley, 1977b). In a simulated aqueous humor with 43 mM $NaHCO_3$, the Mg^{2+}-ATPase activity of rabbit endothelium was 2.0 μmol P_i liberated per hour per milligram of protein, the Na^+,K^+-ATPase activity 1.5, and the HCO_3^--ATPase activity 3.3. The HCO_3^--dependent activity was inhibited 15–25% by the carbonic anhydrase inhibitors acetazolamide, benzolamide, and ethoxyzolamide at concentrations of 10, 5, and 5 mM, respectively. The activities of both the Mg^{2+}-ATPase and the HCO_3^--ATPase were more markedly inhibited, by 80 and 70%, respectively, by either 50 mM thiocyanate or 10 mM cyanate.

The location of the enzyme in bovine endothelium has been determined by Riley and Peters (1981) using differential and density gradient centrifugation. The "plasma membrane" fraction, identified by the 10-fold purification and 35% recovery of Na^+,K^+-ATPase, possessed only 6% of the HCO_3^--ATPase activity (Table V). Moreover, this 6% was matched by an identical recovery in the plasma membrane fraction of the mitochondrial marker enzyme, cytochrome oxidase, and both enzymes showed less than a twofold increase in specific activity over that of the initial homogenate. Conversely, in the "mitochondrial" fraction, where 42% of the cytochrome oxidase was recovered, 38% of the HCO_3^--ATPase was also found, both enzymes with a fourfold purification. Only 7% of the Na^+,K^+-ATPase was found in this fraction, with a specific activity no greater than in the homogenate.

These results lead to the same conclusions reached by van Amelsvoort and others (van Amelsvoort *et al.*, 1977, 1978; Izutsu and Siegel, 1975) that the bicarbonate-dependent ATPase activity is of mitochondrial origin and is not located on the plasma membrane. Consequently, any mechanism for bicarbonate transport across the cell membrane via this enzyme cannot be analogous to sodium transport by

---

Fig. 5. Localization of the 4-nitrophenylphosphatase of rat corneal endothelium. In most areas dense deposits of reaction product obscure the plasma membranes. However, in regions indicated by arrows the plasma membranes may be seen. Arrowheads indicate some small areas with reaction product on the outer surface of the endothelium. Whether these are artifacts or not is unknown. Note, however, that such deposits are not present on the surface adjacent to Descemet's membrane (D) (bar, 0.5 μm). (From Leuenberger and Novikoff, 1974.)

TABLE V

Enzyme Activities and Recoveries in Homogenate and Fractions of Bovine Corneal Endothelium[a]

Fraction	Activity[b]		
	Na^+,K^+-ATPase	HCO_3^--ATPase	Cytochrome oxidase
Homogenate	3.6 (100)[c]	5.2 (100)	2.1 (100)
Plasma membrane	35.8 (35)	7.6 (6)	3.9 (6)
Mitochondria	3.5 (7)	18.7 (38)	8.9 (42)

[a] From Riley and Peters (1981).

[b] ATPase activities are in micromoles of P_i liberated per hour per milligram of protein. Oxidase activity is in millimoles of cytochrome *c* oxidized per hour per milligram of protein.

[c] Recovery is shown in parentheses as percentage of total activity in the homogenate.

the Na^+,K^+-ATPase, but must necessarily be an indirect one. The evidence for a connection between HCO_3^--ATPase activity and transport lies in the inhibition of both processes by thiocyanate and cyanate (Riley, 1977b) and in their parallel responses to low concentrations of bicarbonate (Fig. 6), but no mechanism has been postulated as yet.

3. *Carbonic Anhydrase*

The possible involvement of carbonic anhydrase in the flux of total CO_2 across the endothelium has already been mentioned. While inhi-

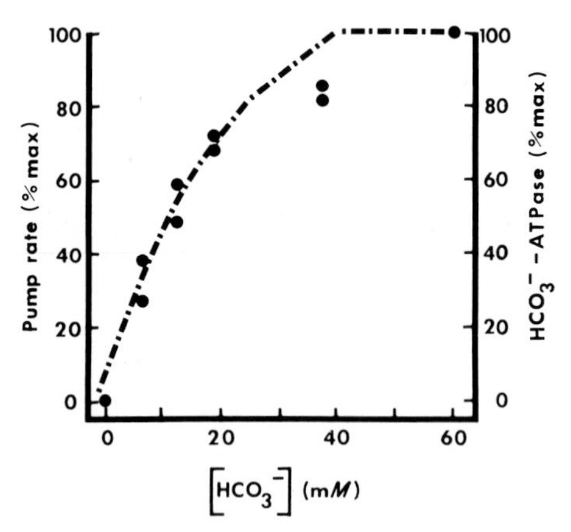

Fig. 6. Relationship of fluid pump rate and HCO_3^--dependent ATPase activity of the corneal endothelium at different bicarbonate concentrations. Broken line represents pump rate (from Fischbarg and Lim, 1974); ●, HCO_3^--ATPase activity (from Riley, 1977b).

bitors of this enzyme decrease the net movement of fluid from the stroma to the aqueous humor, the mechanism of this response is not clear. Kelly and Green (1980) argue that the HCO_3^- ion alone is transported, with little requirement for CO_2, and they further show that sulfonamide inhibitors influence the net flux of HCO_3^- chiefly via an increase in the passive flux. Moreover, this enzyme, catalyzing the reaction $OH^- + CO_2 \rightleftharpoons HCO_3^-$, has itself no capability for active transport; i.e., it is not an enzyme system that is coupled to a source of free energy, and therefore it cannot perform work directly. Energy is required for the provision of OH^- from cell water, but the nature of this step has not been elucidated. Nevertheless, carbonic anhydrase is found in all tissues secreting HCO_3^- or H^+ (Maren, 1967), ensuring an adequate supply of HCO_3^- from CO_2. In the cornea, this enzyme is found almost exclusively in the endothelium (Silverman and Gerster, 1973; Lönnerholm, 1974), but the question of its intracellular distribution is in dispute. Yokota and Waller (1975) report it to be localized in the region of the lateral cell membranes, whereas Hodson and Miller (1976) show it to be immediately adjacent to the apical cell membranes. Its association with the apical membrane is typical of other bicarbonate-secreting tissues, and the better preservation of endothelial architecture in the later study suggests that it may be the more definitive work.

VI. QUANTITATIVE ASPECTS

Any transport mechanism proposed for the movement of fluid from the stroma to the aqueous humor must be able to satisfy the observed phenomena on a quantitative as well as a qualitative basis. The presently available hypotheses propose that water flux occurs as a response to the osmotic disequilibrium created by active ion transport. The maximum rate of water flux that can take place is presumed to be given by (1) the rate of swelling of a fresh cornea upon inhibition of the pump, and (2) the rate of deturgescence of the cornea after swelling. There is good agreement between these two values, giving a rate for the fluid pump of between 45 and 65 μm/hr (Mishima and Trenberth, 1968; Maurice, 1972). If the secreted fluid were isotonic, as indicated by the data of Hodson (1974), this would require an ion flux of between 0.7 and 1.0 $\mu Eq/cm^2/hr$ (Table VI).

A second estimate of the active ion flux can be derived from the electrical parameters of the endothelium. Hodson and Miller (1976) found the resistance to be 20 $\Omega\ cm^2$ and the potential difference 540 μV, consistent with the measured short-circuit current of 28 $\mu A/cm^2$.

TABLE VI

Ion Transport and Energy Relationships in Endothelium

Ion flux required[a]	1 μEq/cm^2/hr
Measured net HCO_3^- flux	0.7–1.8 μEq/cm^2/hr
ATP yield (from oxygen use)	0.7 μmol/cm^2/hr
Maximum ATPase activity	
Na^+,K^+	0.14 μmol/cm^2/hr
HCO_3^-	0.30 μmol/cm^2/hr
Mg^{2+}	0.18 μmol/cm^2/hr

[a] Calculated from an isotonic fluid movement of 60 μm/hr or from a short-circuit current of 28 μA/cm^2.

An ion flux of approximately 1 μEq/cm^2/hr would be required to carry this current.

The case for bicarbonate as the transported ion has been made above, since it shows a net flux under short-circuit conditions, and this flux is compatible with the observed short-circuit current. The two values that have been reported for net bicarbonate flux are 0.7 and 1.8 μEq/cm^2/hr (Hodson and Miller, 1976; Hull *et al.*, 1977), which indicates that bicarbonate ion could also fulfill the quantitative requirements of the fluid pump.

Finally, the biochemical and energetic basis of fluid transport should be considered. The energy yield of the endothelium, estimated from its respiratory rate of 2 μl oxygen/cm^2/hr (Freeman, 1972) and a glycolytic component of 70% of glucose consumption, would not exceed 0.7 μmol ATP/cm^2/hr, or 6×10^{-3} cal/cm^2/hr. The energy required for bulk movement of fluid from the stroma to the aqueous humor against a pressure gradient is calculated by Maurice (1969) to be about 10^{-5} cal/cm^2/hr, while that required for an ion pump would be of a similar magnitude, or perhaps two to three times higher if the transported ion has a low reflection coefficient. Such calculations indicate that the total energy available to the endothelial cells is more than 200 times greater than that required for fluid movement. Even under anerobic conditions the energy yield is most probably 50 times more than needed.

This last observation suggests that such calculations are an oversimplification of cellular energetics. An alternative approach is to consider what supply of ATP would be required on the basis of a typical sodium pump mechanism, with the stoichiometry found in squid axon of three sodium ions transported per ATP molecule hydrolyzed. Thus, the observed ion flux of approximately 0.9 μEq/cm^2/hr would require

0.3 μmol ATP, an amount nearly one-half that estimated to be available. This is comparable to the fraction of total cellular energy production found to be utilized for active ion transport in other tissues such as brain, liver, and ciliary body (Whittam, 1962; Elshove and van Rossum, 1963; Riley, 1964). Moreover, this estimate of ATP requirement is greater than the amount that would be available under anaerobic conditions, since the Pasteur effect increases lactate production in the cornea by only 50% (Langham, 1954). Consequently, oxygen would be essential to endothelial function, as is observed.

It is worth noting that the activity of the Na^+,K^+-ATPase measured in the rabbit endothelium is 0.14 μmol ATP hydrolyzed/cm^2/hr, which suggests that this enzyme could not drive the flux of enough sodium ions to meet the requirements of the pump system in these cells, if a ratio of three ions to one ATP molecule were assumed. While the activity of the HCO_3^--ATPase is 0.3 μmol ATP hydrolyzed/cm^2/hr and could meet the demand if the same coupling ratio were assumed, it must be remembered that the mitochondrial location of this enzyme poses a difficult problem in attempting to describe a model that could function in this manner, and to date there is no definitive evidence from any tissue that the bicarbonate-stimulated ATPase directly effects active transport of this ion.

VII. CONCLUSION

The corneal endothelium is required to permit ready transfer of metabolites between the tissue and the aqueous humor, while at the same time limiting access of water to the stroma. These conflicting needs could be met, on the one hand, by a relatively impermeable barrier with multiple specific transport systems or, on the other hand, by a nonselective barrier permeable enough for nutritional needs and incorporating a mechanism to compensate for excess water influx. From the results of the studies outlined in the foregoing sections it is clear that in the mammalian eye it is the latter of these systems that has evolved. Thus, while nutrition of the tissue requires few specialized or unique processes in the endothelial cells, the maintenance of a normal level of hydration requires the operation of a complex, energy-consuming fluid pump.

The swelling that occurs when corneas are perfused with diamide, or in the absence of calcium, is shown by permeability studies and by the scanning electron microscope to be due to a failure of the barrier function of the endothelium, whereas that resulting from the presence of ouabain is due to failure of the pump. Although a balance between the two functions is clearly essential for corneal transparency, most attention has been focused on the pump, while relatively little has been established concerning the control of permeability and the nature of junctional complexes in the endothelium.

The best established features of the fluid transport system of the endothelium appear to be as follows: bicarbonate and sodium show net fluxes from the stroma to the aqueous humor and are essential to pump function; the transendothelial potential is directly related to the pump mechanism and is aqueous-side-negative; the Na^+,K^+-ATPase is located at the lateral cell membranes; the HCO_3^--ATPase is located in the mitochondria; and ouabain and carbonic anhydrase inhibitors produce complete and partial inhibition of fluid transport, respectively. The major features not yet resolved include the nature of the bicarbonate transport mechanism and its mode of coupling to an energy source, the orientation of sodium transport at the lateral membranes, the possible role of protons and CO_2, and the manner in which Na^+ and HCO_3^- transport results in a negative potential. Understanding of the overall problem of corneal hydration requires further research on these biochemical and physiological aspects of the pump and on the factors which regulate the permeability of the endothelium.

ACKNOWLEDGMENTS

I am most grateful to Drs. H. F. Edelhauser and P. M. Leuenberger for their kindness in supplying the photomicrographs, and also to Ms. Ellen Yates and Ms. Elaine Rehm for their help in preparation of the manuscript. Work in the author's laboratory has been supported by PHS grant EY 00541 from the National Eye Institute.

REFERENCES

Anderson, E. I., and Fischbarg, J. (1978). Biphasic effects of insulin and ouabain on fluid transport across rabbit corneal endothelium. *J. Physiol. (London)* **275**, 377–389.

Anderson, E. I., and Wright, D. D. (1980a). Glutathione reductase and the fluid pump of the corneal endothelium. *Proc. Int. Soc. Eye Res.* **1**, 49.

Anderson, E. I., and Wright, D. D. (1980b). Effects of *S*-methyl glutathione, *S*-methyl cysteine, and the concentration of oxidized glutathione on transendothelial fluid transport. *Invest. Ophthalmol. Vis. Sci.* **19**, 684–686.

Anderson, E. I., Fischbarg, J., and Spector, A. (1973). Fluid transport, ATP level and ATPase activities in isolated rabbit corneal endothelium. *Biochim. Biophys. Acta* **307**, 557–562.

Anderson, E. I., Fischbarg, J., and Spector, A. (1974). Disulfide stimulation of fluid transport and effect on ATP level in rabbit corneal endothelium. *Exp. Eye Res.* **19**, 1–10.

Barfort, P., and Maurice, D. (1974) Electrical potential and fluid transport across the corneal endothelium. *Exp. Eye Res.* **19**, 11-19.

Barr, R. E., Hennessey, M., and Murphy, V. G. (1976). Diffusion of oxygen at the endothelial surface of the rabbit cornea. *J. Physiol. (London)* **270**, 1-8.

Blümcke, S., and Morgenroth, M., Jr. (1967). The stereo ultrastructure of the external and internal surface of the cornea. *J. Ultrastruct. Res.* **18**, 502-518.

Bito, L. Z. (1970). Intraocular fluid dynamics. *Exp. Eye Res.* **10**, 102-116.

Bonting, S. L. (1970). Sodium-potassium activated adenosine triphosphatase and cation transport. *In* "Membranes and Ion Transport" (E. E. Bittar, ed.), Vol. 1, pp. 257-364. Wiley (Interscience), New York.

Bonting, S. L., Simon, K. A., and Hawkins, N. M. (1961). Studies on sodium-potassium-activated adenosine triphosphatase. I. Quantitative distribution in several tissues of the cat. *Arch. Biochem. Biophys.* **95**, 416-423.

Bornancin, M., de Renzis, G., and Naon, R. (1980). Cl^--HCO_3^--ATPase in gills of the rainbow trout: Evidence for its microsomal localization. *Am. J. Physiol.* **238**, R251-259.

Bourne, W. M., and O'Fallon, W. M. (1978). Endothelial cell loss during penetrating keratoplasty. *Am. J. Ophthalmol.* **85**, 760-766.

Burns, R. P., Roberts, H., and Rich, L. F. (1971). Effect of silicone contact lenses on corneal epithelial metabolism. *Am. J. Ophthalmol.* **71**, 486-489.

Davson, H. (1955). The hydration of the cornea. *Biochem. J.* **59**, 24-28.

Diamond, J. M., and Bossert, W. H. (1968). Functional consequences of ultrastructural geometry in "backwards" fluid transporting epithelia. *J. Cell Biol.* **37**, 694-702.

Dikstein, S., and Maurice, D. M. (1972). The metabolic basis to the fluid pump in the cornea. *J. Physiol. (London)* **221**, 29-41.

Dikstein, S., Neuwirth, O., and Vidal, R. (1980). Corneal endothelial pumping in the presence of insulin and GABA, *Exp. Eye Res.* **31**, 239-242.

Donn, A., Miller, S., and Mallett, N. (1963). Water permeability of the living cornea. *Arch. Ophthalmol. (Chicago)* **70**, 515-521.

Durbin, R. P., and Kasbekar, D. K. (1965). Adenosine triphosphate and active transport by the stomach. *Fed. Proc., Fed. Am. Soc. Exp. Biol.* **24**, 1377-1381.

Edelhauser, H. F., Hoffert, J. R., and Fromm, P. D. (1965). *In vitro* ion and water movement in corneas of rainbow trout. *Invest. Ophthalmol.* **4**, 290-296.

Edelhauser, H. F., van Horn, D. L., Miller, P., and Pederson, H. J. (1976). Effect of thiol-oxidation of glutathione with diamide on corneal endothelial function, junctional complexes, and microfilaments. *J. Cell Biol.* **68**, 567-578.

Elshove, A., and van Rossum, G. D. V. (1963). Net movements of sodium and potassium, and their relation to respiration, in slices of rat liver incubated *in vitro*. *J. Physiol. (London)* **168**, 531-553.

Ernst, S. A. (1972). Transport adenosine triphosphatase cytochemistry. I. Biochemical characterization of a cytochemical medium for the ultrastructural localization of a ouabain-sensitive, potassium dependent phosphatase activity in the avian salt gland. *J. Histochem. Cytochem.* **20**, 13-22.

Fatt, I. (1968). Dynamics of water transport in the corneal stroma. *Exp. Eye Res.* **7**, 402-412.

Fatt, I., and Bieber, M. T. (1968). The steady state distribution of oxygen and carbon dioxide in the *in vivo* cornea. *Exp. Eye Res.* **7**, 103-112.

Fatt, I., Freeman, D., and Lin, D. (1974). Oxygen tension distributions in the cornea: A re-examination. *Exp. Eye Res.* **18**, 357-365.

Fischbarg, J. (1972). Potential difference and fluid transport across rabbit corneal endothelium. *Biochim. Biophys. Acta* **288**, 362-366.

Fischbarg, J., and Lim, J. J. (1974). Role of cations, anions and carbonic anhydrase in fluid transport across rabbit corneal endothelium. *J. Physiol. (London)* **241,** 647-675.

Fischbarg, J., Lim, J. J., and Bourguet, J. (1977). Adenosine stimulation of fluid transport across rabbit corneal endothelium. *J. Membr. Biol.* **35,** 95-112.

Fischer, F. H., and Zadunaisky, J. A. (1977). Electrical and hydrophilic properties of fish corneas. *Exp. Eye Res.* **25,** 149-161.

Freeman, R. D. (1972). Oxygen consumption by the component layers of the cornea. *J. Physiol. (London)* **225,** 15-32.

Geroski, D. H., Edelhauser, H. F., and O'Brien, W. J. (1978). Hexose-monophosphate shunt response to diamide in the component layers of the cornea. *Exp. Eye Res.* **26,** 611-619.

Goldman, J. N., and Benedek, G. B. (1967). The relationship between morphology and transparency in the nonswelling corneal stroma of the shark. *Invest. Ophthalmol.* **6,** 574-600.

Green, K. (1967). Solute movement across the constituent membranes of the cornea. *Exp. Eye Res.* **6,** 79-92.

Green, K., and Friedman, M. H. (1971). Potassium and calcium binding in corneal stroma and the effect on sodium binding. *Am. J. Physiol.* **221,** 363-367.

Grote, J., and Zander, R. (1976). Corneal oxygen supply conditions. *Adv. Exp. Med. Biol.* **75,** 449-455.

Hale, P. N., and Maurice, D. M. (1969). Sugar transport across the corneal endothelium. *Exp. Eye Res.* **8,** 205-215.

Hara, T. (1965). Studies on the water and cation shift of the stored cornea. *Jpn. J. Ophthalmol.* **9,** 161-174.

Harris, J. E., and Nordquist, L. T. (1955). The hydration of the cornea: I. The transport of water from the cornea. *Am. J. Ophthalmol.* **40,** 100-110.

Hedbys, B. O. (1961). The role of polysaccharides in corneal swelling. *Exp. Eye Res.* **1,** 81-91.

Hedbys, B. O., Mishima, S., and Maurice, D. M. (1963). The imbibition pressure of the corneal stroma. *Exp. Eye Res.* **2,** 99-111.

Hermann, H., and Hickman, F. H. (1948). Further experiments on corneal metabolism in respect to glucose and lactic acid. *Bull. Johns Hopkins Hosp.* **82,** 260-272.

Hirsch, M., Renard, G., Faure, J.-P., and Pouliquen, Y. (1977). Study of the ultrastructure of the rabbit corneal endothelium by the freeze-fracture technique: Apical and lateral junctions. *Exp. Eye Res.* **25,** 277-288.

Hodson, S. (1971a). Evidence for a bicarbonate-dependent sodium pump in corneal endothelium. *Exp. Eye Res.* **11,** 20-29.

Hodson, S. (1971b). Macromolecular synthesis in corneal endothelium. *Exp. Eye Res.* **11,** 15-19.

Hodson, S. (1974). The regulation of corneal hydration by a salt pump requiring the presence of sodium and bicarbonate ions. *J. Physiol. (London)* **236,** 271-302.

Hodson, S., and Mayes, K. R. (1979). Intercellular spaces in chemically fixed corneal endothelia are related to solute pump activity not to solvent coupling. *J. Physiol. (London)* **294,** 627-634.

Hodson, S., and Miller, F. (1976). The bicarbonate ion pump in the endothelium which regulates the hydration of rabbit cornea. *J. Physiol. (London)* **263,** 563-577.

Hodson, S., and Riley, M. V. (1973). A sensitive method for measuring the charge of water soluble polyanions. *Biochim. Biophys. Acta* **304,** 236-239.

Hodson, S., Miller, F., and Riley, M. (1977). The electrogenic pump of rabbit corneal endothelium. *Exp. Eye Res.* **24,** 249-253.

Huff, J. W., and Green, K. (1981). Demonstration of active sodium transport across the isolated rabbit corneal endothelium. *Curr. Eye Res.* **1,** 113-114.

Hull, D. S., Green, K., Boyd, M., and Wynn, H. R. (1977). Corneal endothelium bicarbonate transport and the effect of carbonic anhydrase inhibitors on endothelial permeability and fluxes and corneal thickness. *Invest. Ophthalmol. Vis. Sci.* **16,** 883-892.

Hull, D. S., Green, K. and Buyer, J. (1979). Corneal endothelial bicarbonate fluxes following storage in moist chamber, MK medium, and MK medium with added hydrocortisone. *Invest. Ophthalmol. Vis. Sci.* **18,** 484-489.

Izutsu, K. T., and Siegel, I. A. (1972). A microsomal HCO_3^--stimulated ATPase from the dog submandibular gland. *Biochim. Biophys. Acta* **284,** 478-484.

Izutsu, K. T., and Siegel, I. A. (1975). Bicarbonate ion-ATPase in rat liver cell fractions. *Biochim. Biophys. Acta* **382,** 193-203.

Kasbekar, D. K., and Durbin, R. P. (1965). An adenosine triphosphatase from frog gastric mucosa. *Biochim. Biophys. Acta* **105,** 472-482.

Kaye, G. I., and Donn, A. (1965). Studies on the cornea. IV. Some effects of ouabain on pinocytosis and stromal thickness in the rabbit cornea. *Invest. Ophthalmol.* **4,** 844-852.

Kaye, G., and Tice, L. W. (1966) Studies on the cornea. V. Electron microscopic localization of adenosine triphosphatase activity in the rabbit cornea in relation to transport. *Invest. Ophthalmol.* **5,** 22-32.

Kaye, G. I., Mishima, S., Cole, J. D., and Kaye, N. W. (1968). Studies on the cornea. VII. Effects of perfusion with a Ca^{++}-free medium on the corneal endothelium. *Invest. Ophthalmol.* **7,** 53-66.

Kaye, G. I., Sibley, R. C., and Hoefle, F. B. (1973a). Recent studies on the nature and function of the corneal endothelial barrier. *Exp. Eye Res.* **15,** 585-613.

Kaye, G. I., Hoefle, F. B., and Donn, A. (1973b). Studies on the cornea. VIII. Reversibility of the effects of *in vitro* perfusion of the rabbit corneal endothelium with calcium-free medium. *Invest. Ophthalmol.* **12,** 98-113.

Kelly, G., and Green, K. (1980). Influence of bicarbonate and CO_2 on rabbit corneal transendothelial bicarbonate fluxes. *Exp. Eye Res.* **30,** 641-648.

Kimelberg, H. K., and Bourke, R. S. (1973). Properties and localization of bicarbonate-stimulated ATPase activity in rat brain. *J. Neurochem.* **20,** 347-359.

Kreutziger, G. O. (1976). Lateral membrane morphology and gap junction structure in rabbit corneal endothelium. *Exp. Eye Res.* **23,** 285-293.

Kwan, M., Niinikoski, J., and Hunt, T. K. (1972). *In vivo* measurements of oxygen tension in the cornea, aqueous humor and anterior lens of the open eye. *Invest. Ophthalmol.* **11,** 108-114.

Langham, M. (1952). Utilization of oxygen by the component layers of the living cornea. *J. Physiol.* (*London*) **117,** 461-470.

Langham, M. E. (1954). Glycolysis in the cornea of the rabbit. *J. Physiol.* (*London*) **126,** 396-403.

Langham, M. E., and Taylor, I. S. (1956). Factors affecting the hydration of the cornea in the excised eye and the living animal. *Br. J. Ophthalmol.* **40,** 321-340.

Laule, A., Cable, M. K., Hoffman, C. E., and Hanna, C. (1978). Endothelial cell population changes of human cornea during life. *Arch. Ophthalmol. (Chicago)* **96,** 2031-2035.

Leuenberger, P. M., and Novikoff, A. B. (1974). Localization of transport adenosine triphosphatase in rat cornea. *J. Cell Biol.* **60,** 721-731.

Liang, C. T., and Sacktor, B. (1976). Bicarbonate-stimulated ATPase in the renal proximal tubule luminal (brush border) membrane. *Arch. Biochem. Biophys.* **176,** 285-297.

Liebovitch, L. S., Fischbarg, J., and Koatz, R. (1981). Osmotic water permeability of rabbit corneal endothelium and its dependence on ambient concentration. *Biochim. Biophys. Acta* **646,** 71-76.

Lim, J. J. (1981). Sodium transport across the rabbit corneal endothelium. *Curr. Eye Res.* **1,** 255-258.

Lönnerholm, G. (1974). Carbonic anhydrase in the cornea. *Acta Physiol. Scand.* **90,** 143-152.

Lowther, G. E., and Hill, R. M. (1974). Corneal epithelium: Recovery from anoxia. *Arch. Ophthalmol. (Chicago)* **92,** 231-234.

Malinin, G. I., and Bernstein, H. (1979). Histochemical demonstration of glycogen in corneal endothelium. *Exp. Eye Res.* **28,** 381-385.

Mandell, R. B., and Fatt, I. (1965). Thinning in the human cornea on awakening. *Nature (London)* **208,** 292-293.

Maren, T. H. (1967). Carbonic anhydrase: Chemistry, physiology and inhibition. *Physiol. Rev.* **47,** 597-781.

Mathias, R. T. (1981). A balanced gradient mechanism for isotonic transport. *Invest. Ophthalmol. Vis. Sci. Suppl.* **20,** 31.

Maurice, D. M. (1951). The permeability to sodium ions of the living rabbit's cornea. *J. Physiol. (London)* **122,** 367-391.

Maurice, D. M. (1957). The structure and transparency of the cornea. *J. Physiol. (London)* **136,** 263-286.

Maurice, D. M. (1961). The use of permeability studies in the investigation of submicroscopic structure. *In* "The Structure of the Eye" (G. K. Smelser, ed.), pp. 381-392. Academic Press, New York.

Maurice, D. M. (1968). Cellular membrane activity in the corneal endothelium of the intact eye. *Experientia* **24,** 1094-1095.

Maurice, D. M. (1969). The cornea and sclera. *In* "The Eye" (H. Davson, ed.), 2nd ed., Vol. 1, pp. 489-600. Academic Press, New York.

Maurice, D. M. (1972). The localization of the fluid pump in the cornea. *J. Physiol. (London)* **221,** 43-54.

Maurice, D. M., and Giardini, A. A. (1951). Swelling of the cornea *in vivo* after the destruction of its limiting layers. *Br. J. Ophthalmol.* **35,** 791-797.

Maurice, D. M., and Riley, M. V. (1970). The cornea. *In* "Biochemistry of the Eye" (C. N. Graymore, ed.), pp. 1-104. Academic Press, New York.

Maurice, D. M., and Watson, P. G. (1965). The distribution and movement of serum albumin in the cornea. *Exp. Eye Res.* **4,** 355-363.

Mayes, K. R., and Hodson, S. (1978). Local osmotic coupling to the active trans-endothelial bicarbonate flux in the rabbit cornea. *Biochim. Biophys. Acta* **514,** 286-293.

Mayes, K. R., and Hodson, S. (1979). An *in vivo* demonstration of the bicarbonate ion pump of rabbit corneal endothelium. *Exp. Eye Res.* **28,** 699-707.

Milutinović, S., Sachs, G., Haase, W., and Schulz, I. (1977). Studies on isolated subcellular components of cat pancreas. I. Isolation and enzymatic characterization. *J. Membr. Biol.* **36,** 253-279.

Mimura, Y., Morisue, T., and Kotzuka, Y. (1963). Transaminase activity in bovine cornea. *Folia Ophthalmol. Jpn.* **14,** 314-319.

Mishima, S., and Hedbys, B. O. (1967). The permeability of the corneal epithelium and endothelium to water. *Exp. Eye Res.* **6,** 10-32.

Mishima, S., and Kudo, T. (1967). *In vitro* incubation of rabbit cornea. *Invest. Ophthalmol.* **6,** 329-339.

Mishima, S., and Maurice, D. M. (1961a). The oily layer of the tear film and evaporation from the corneal surface. *Exp. Eye Res.* **1**, 39-45.

Mishima, S., and Maurice, D. M. (1961b). The effect of normal evaporation on the eye. *Exp. Eye Res.* **1**, 46-52.

Mishima, S., and Trenberth, S. M. (1968). Permeability of the corneal endothelium to nonelectrolytes. *Invest. Ophthalmol.* **7**, 34-43.

Mishima, S., Kaye, G. I., Takahashi, G. H., Kudo, T., and Trenberth, S. M. (1969). The function of the corneal endothelium in the regulation of corneal hydration. *In* "The Cornea" (M. E. Langham, ed.), pp. 207-235. Johns Hopkins Press, Baltimore, Maryland.

Ng, M. C., and Riley, M. V. (1980). Relation of intracellular levels and redox state of glutathione to endothelial function in the rabbit cornea. *Exp. Eye Res.* **30**, 511-517.

Obenberger, J., and Babický, A. (1970). Experimental corneal calcification: A study of transport of ^{45}Ca into the aqueous and cornea. *Ophthalmol. Res.* **1**, 187-192.

Okinami, S., Ohkuma, M., and Tsukahara, I. (1976). Freeze-fracture replica of the rat cornea. *Acta Ophthalmol.* **54**, 743-754.

Ottersen, O. P., and Vegge, T. (1977). Ultrastructure and distribution of intercellular junctions in corneal endothelium. *Acta Ophthalmol.* **55**, 69-78.

Payrau, P., Offret, G., Faure, J.-P., Hamean, J. P., and Pouliquen, Y. (1964). Sur la structure lamellaire des cornées de certains poissons. *Arch. Ophthalmol. (Paris)* **24**, 685-690.

Reddy, V. N. (1970). Distribution of free amino acids and related compounds in rabbit cornea. *Ophthalmol. Res.* **1**, 48-57.

Reddy, V. N., and Unakar, N. J. (1973). Localization of gamma-glutamyl transpeptidase in rabbit lens, ciliary process and cornea. *Exp. Eye Res.* **17**, 405-408.

Reim, M., Schmidt, F., and Meyer, D. (1966). Die Metabolite des Energie liefernden Stoffwechsels in Hornhautepithel verschiedener Säugetiere. *Ber. Dtsch. Ophthalmol. Ges.* **67**, 164-171.

Reim, M., Foerster, K. H., and Cattepoel, H. (1970). Some criteria of the metabolism in the donor cornea. *Ophthalmol. Proc. XXI Int. Cong. Part I,* pp. 728-733.

Reim, M., Boeck, H., Krug, P., and Venske, G. (1972). Aqueous humor and cornea stroma metabolite levels under various conditions. *Ophthalmol. Res.* **3**, 241-250.

Riley, M. V. (1964). Inhibition by ouabain of glycolysis in ciliary body. *Nature (London)* **204**, 380.

Riley, M. V. (1969). Glucose and oxygen utilization by the rabbit cornea. *Exp. Eye Res.* **8**, 193-200.

Riley, M. V. (1972). Intraocular dynamics of lactic acid in the rabbit. *Invest. Ophthalmol.* **11**, 600-607.

Riley, M. V. (1977a). A study of the transfer of amino acids across the endothelium of the rabbit cornea. *Exp. Eye Res.* **24**, 35-44.

Riley, M. V. (1977b). Anion-sensitive ATPase in rabbit corneal endothelium and its relation to corneal hydration. *Exp. Eye Res.* **25**, 483-494.

Riley, M. V., and Peters, M. I. (1981). The localization of the anion-sensitive ATPase activity in corneal endothelium, and its relation to bicarbonate transport. *Biochim. Biophys. Acta* **644**, 251-256.

Riley, M. V., and Yates, E. M. (1977). Glutathione in the epithelium and endothelium of bovine and rabbit cornea. *Exp. Eye Res.* **25**, 385-389.

Riley, M. V., Campbell, D., and Linz, D. H. (1973). Entry of amino acids into the rabbit cornea. *Exp. Eye Res.* **15**, 677-681.

Riley, M. V., Miller, F., Hodson, S., and Linz, D. (1977). Elimination of anions derived from glucose metabolism as substrates for the fluid pump of rabbit corneal endothelium. *Exp. Eye Res.* **24,** 255–261.

Riley, M. V., Meyer, R. F., and Yates, E. M. (1980). Glutathione in the aqueous humor of human and other species. *Invest. Ophthalmol. Vis. Sci.* **19,** 94–96.

Rogers, K. T. (1968). Levels of ($Na^+ + K^+$)-activated and Mg^{2+}-activated ATPase activity in bovine and feline corneal endothelium and epithelium. *Biochim. Biophys. Acta* **163,** 50–56.

Sachs, G., Shah, G., Strych, A., Cline, G., and Hirschowitz, B. I. (1972). Properties of ATPase of gastric mucosa. III. Distribution of HCO_3^--stimulated ATPase in gastric mucosa. *Biochim. Biophys. Acta* **266,** 625–638.

Schütte, E., and Reim, M. (1976). Fructose metabolism of the cornea. *Ophthalmol. Res.* **8,** 434–437.

Silverman, D. N., and Gerster, R. (1973). The detection and localization of carbonic anhydrase in the rabbit cornea. *Exp. Eye Res.* **17,** 129–136.

Simon, B., Kinne, R., and Sachs, G. (1972). The presence of a HCO_3^--ATPase in pancreatic tissue. *Biochim. Biophys. Acta* **282,** 293–300.

Skou, J. C. (1957). The influence of some cations on an ATPase from peripheral nerves. *Biochim. Biophys. Acta* **23,** 394–401.

Skou, J. C. (1965). Enzymatic basis for active transport of Na^+ and K^+ across cell membrane. *Physiol. Rev.* **45,** 596–617.

Smelser, G. K. (1962). Corneal hydration: Comparative physiology of fish and mammals. *Invest. Ophthalmol.* **1,** 11–32.

Smelser, G. K., and Ozanics, V. (1953). Structural changes in corneas of guinea pigs after wearing contact lenses. *Arch. Ophthalmol. (Chicago)* **49,** 335–340.

Smelser, G. K., Polack, F. M., and Ozanics, V. (1965). Persistence of donor collagen in corneal transplants. *Exp. Eye Res.* **4,** 349–354.

Sperling, S. (1977). Combined staining of corneal endothelium by alizarine red and trypane blue. *Acta Ophthalmol.* **55,** 573–580.

Tervo, T., and Palkama, A. (1975). Electron microscopic localization of adenosine triphosphatase (Na,K-ATPase) activity in the rat cornea. *Exp. Eye Res.* **21,** 269–279.

Thoft, R. A., and Friend, J. (1972). Corneal amino acid supply and distribution. *Invest. Ophthalmol.* **11,** 723–727.

Thoft, R. A., and Friend, J. (1975). Biochemical aspects of contact lens wear. *Am. J. Ophthalmol.* **80,** 139–145.

Thoft, R. A., Friend, J., and Dohlman, C. H. (1971a). Corneal glucose concentration. *Arch. Ophthalmol. (Chicago)* **85,** 467–472.

Thoft, R. A., Friend, J., and Dohlman, C. H. (1971b). Corneal glucose flux. *Arch. Ophthalmol. (Chicago)* **86,** 685–691.

Trenberth, S. M., and Mishima, S. (1968). The effect of ouabain on the rabbit corneal endothelium. *Invest. Ophthalmol.* **7,** 44–52.

Uniacke, C. A., and Hill, R. M. (1972). The depletion course of epithelial glycogen with corneal anoxia. *Arch. Ophthalmol. (Chicago)* **87,** 56–59.

van Amelsvoort, J. M. M., de Pont, J. J. H. H. M., Stols, A. L. H., and Bonting, S. L. (1977). Is there a plasma-membrane-located anion-sensitive ATPase? II. Further studies on rabbit kidney. *Biochim. Biophys. Acta* **471,** 78–91.

van Amelsvoort, J. M. M., Jansen, J. W. C. M., dePont, J. J. H. H. M., and Bonting, S. L. (1978). Is there a plasma membrane-located anion-sensitive ATPase? IV. Distribution of the enzyme in rat pancreas. *Biochim. Biophys. Acta* **512,** 296–308.

Van Horn, D. L., Sendele, D. D., Seidemann, S., and Buco, P. J. (1977). Regenerative capacity of the corneal endothelium in rabbit and cat. *Invest. Ophthalmol. Vis. Sci.* **16,** 597-613.

van Os, C. H., Mircheff, A. K., and Wright, E. M. (1977). Distribution of bicarbonate-stimulated ATPase in rat intestinal epithelium. *J. Cell Biol.* **73,** 257-260.

Wachstein, M., and Meisel, E. (1957). Histochemistry of hepatic phosphatases at a physiological pH. *Am. J. Clin. Pathol.* **27,** 13-23.

Waltman, S. R., and Cozean, C. H., Jr. (1979). The effect of phacoemulsification on the corneal endothelium. *Ophthalmic Surg.* **10,** 31-33.

Whikehart, D. R., and Edelhauser, H. F. (1978). Glutathione in rabbit corneal endothelia: The effects of selected perfusion fluids. *Invest. Ophthalmol. Vis. Sci.* **17,** 455-464.

Whittam, R. (1962). The dependence of the respiration of brain cortex on active cation transport. *Biochem. J.* **82,** 205-212.

Winkler, B. S., and Riley, M. V. (1977). Na^+,K^+ and HCO_3^--ATPase activity in retina: Dependence on calcium and sodium. *Invest. Ophthalmol. Vis. Sci.* **16,** 1151-1154.

Wolf, J. (1968). Inner surface of regions in the anterior chamber taking part in the regulation of the intraocular tension, including the demonstration of the covering viscous substance. *Doc. Ophthalmol.* **25,** 113-149.

Yokota, S., and Waller, W. K. (1975). Electron microscopic localization of carbonic anhydrase (CA) activity in rabbit cornea. *Albrecht von Graefes Arch. Klin. Exp. Ophthalmol.* **147,** 145-152.

Ytteborg, J., and Dohlman, C. H. (1965). Corneal edema and intraocular pressure. I. Animal experiments. *Arch. Ophthalmol. (Chicago)* **74,** 374-381.

Zadunaisky, J. A., and Lande, M. A. (1971). Active chloride transport and control of corneal transparency. *Am. J. Physiol.* **221,** 1837-1844.

3

Growth Factors, Effect on Corneal Tissue

D. GOSPODAROWICZ AND L. GIGUÈRE

Cell Biology of the Eye

ISBN 0-12-483180-X

I. INTRODUCTION

A tantalizing objective pursued actively for the past 80 yr by cell biologists has been to reconstruct from single cells maintained *in vitro* the corresponding functional tissue. This requires not only that cells obtained from a given tissue and maintained singly *in vitro* proliferate actively, but also that upon reaching confluence they reconstruct the tissue to which they belong *in vivo* in such a way that both the morphological characteristics of the tissue and its function are preserved. If such tissues could be maintained *in vitro,* then both theoretical and practical applications could be considered. In the case of a tissue whose *in vivo* location is not readily accessible, one could find an *in vitro* model which would make it possible to explore under defined conditions and in a totally controlled environment the real capacities and contributions of this tissue to the welfare of the organism *in vivo.* One could thereby explore the physiological and metabolic properties of the tissue as well as study factors involved in the control of its proliferation and differentiation. Such tissues reconstructed from single cells *in vitro* would also constitute an endless source of material for replacing their *in vivo* counterparts incapacitated following disease.

In recent years, the discovery of mitogenic factors which maintained cells in an active stage of proliferation, thereby preventing their precocious senescence *in vitro,* has made it possible to achieve these objectives. In this chapter we will concentrate upon what is presently known about the role of growth factors in the control of proliferation of the various cell layers of the cornea when the cells are maintained under either *in vivo* or *in vitro* conditions.

II. THE CORNEA: STRUCTURE AND DEVELOPMENT

Although deceptively simple at first glance, the cornea is a complex organ composed of three different tissues separated from each other by acellular layers which have the biochemical characteristics of basement membranes. The organization and interrelationship of each tissue with the others is essential for the proper embryonic development as well as for the proper function of the cornea during the life of the individual.

The adult cornea is composed of five different layers, namely (from outermost to innermost), the corneal epithelium, the acellular Bowman's membrane, the corneal stroma, the acellular Descemet's membrane, and the corneal endothelium (Hogan *et al.*, 1971). The corneal epithelium is composed of several cell layers. The outer layer is in contact with the tear film and is composed of terminally differentiated cells no longer capable of dividing, while the basal cell layer rests on a well-defined basement membrane and proliferates actively. Between these outermost layers are several cell layers in various stages of differentiation. Beneath the corneal epithelial basement membrane is a collagenous acellular layer called Bowman's membrane, and beneath it is the corneal stroma, composed of collagenous lamellae running parallel to each other and to the corneal surface. Fibroblasts are found throughout the corneal stroma, lying between the lamellae and occasionally extending into them. Below the corneal stroma another acellular layer of collagenous material called Descemet's membrane is found. It represents the basement membrane produced by the corneal endothelium. The innermost layer of the cornea is an endothelial cell monolayer composed of highly contact-inhibited and flattened cells with an hexagonal configuration that can be seen clearly by staining with alizarin red or silver nitrate, both of which stain the intercellular border (Stocker, 1971). The apical cell surface of the corneal endothelium is exposed to the aqueous humor.

The embryonic development of the cornea is complex and results from a series of inductive interactions between germ layers (for a review, see Hay and Revel, 1969). After detachment of the lens from the overlying ectoderm, the presumptive corneal epithelium is at first two cells thick. Under the inductive influence of the lens, the basal layer hypertrophies and assumes the morphology of a secretory tissue, producing an extracellular corneal matrix: the primary stroma. Following formation of the primary stroma, the endothelium of the cornea migrates into spaces between the primary stroma and the lens. The corneal endothelium, although it was first thought to be derived from the primary mesenchyme closely associated with the developing blood vessels in the vicinity of the lip of the octic cup (Hay and Revel, 1969), is in fact derived from the neural crest (Noden, 1980). The migrating endothelial cells flatten out and migrate on top of the inner surface of the primary stroma. Dividing repeatedly until they become a monolayer, they then adopt the configuration of a cuboidal endothelium with a free surface facing the lens. The space between the lens and the endothelium enlarges and gives rise to the anterior chamber. The endothelium then produces a basement membrane on its basal surface, which thickens and later becomes the Descemet's membrane.

During this process the primary stroma begins to swell. It is likely that the edema of the stroma triggers the subsequent fibroblast invasion. This invasion proceeds rapidly, and mesenchymal cells derived from the secondary mesenchyme penetrate between the lamellae of the primary stroma but not into the area of the primary stroma immediately adjacent to the epithelium. This zone is later destined to become the definitive acellular Bowman's membrane.

Development of the cornea is henceforth dominated by the activities of the newly differentiated fibroblasts. As a result of their secretory activity, numerous collagen fibrils are added to the corneal stroma in a highly organized fashion which follows the blueprint represented by the lamellar organization of the inner part of the primary stroma. This results in a three-dimensional, orthogonal organization of the collagenous lamellae. Although opaque at the outset, the cornea starts to lose water rapidly. This could result from a loss of ability of the stroma to bind water and the newly acquired ability of the corneal endothelium to perform its pump function. It could also result from the active synthesis of glycosaminoglycans (chondroitin and heparan sulfates), which first occurs in the posterior layer of the mesodermal stroma and spreads toward the epithelial surface.

The developing cornea provides several examples of tissue interactions that might be more easily approached experimentally, especially with respect to the role of extracellular materials in cell migration and cell proliferation. After an epithelial-epithelial interaction represented by the interaction between the basal surface of the lens and the presumptive corneal epithelium, two interactions occur in the cornea in which mesenchymal cells seem to respond to the products of the epithelia. In the first of these, the vascular mesenchyme at the limbus of the eye inserts itself between the optic cup and the ectoderm and migrates along the posterior surface of the primary corneal stroma. In the second interaction, changes consisting principally of the disruption of collagen fibrils and swelling of the extracellular interstices may be the result of actions taken by the endothelium and epithelium and seem to be involved in bringing about invasion of the stroma by fibroblasts (Hay and Revel, 1969).

The different embryological origins of the various tissues composing the cornea make possible a study of the ability of various growth factors to stimulate the proliferation of cells originating from embryological territories as diverse as the ectoderm and the primary and secondary mesenchymes. It also allows one to study how the extracellular material produced during differentiation of the various cell layers can later affect the proliferative ability of the cells, as well as their response to various growth factors. (The effect of growth factors on other ocular tissues is discussed by Reddan in another chapter in this volume.) Finally, the various tissues which compose the cornea have widely varying abilities to regenerate after being wounded. While the ability of the epithelium is the greatest, that of the endothelium is the lowest. The endothelium therefore offers an excellent model for testing whether growth factors stimulate the proliferation of tissues which otherwise have a low potential *in vivo* for regeneration.

III. VARIOUS CLASSES OF GROWTH FACTORS INVOLVED IN THE CONTROL OF PROLIFERATION OF CORNEAL CELLS

Among the growth factors shown to influence the proliferation of the various cell types present in the cornea are epidermal growth factor (EGF), insulin-like growth factors (IGFs), fibroblast growth factor (FGF), eye-derived growth factor (EDGF), and mesodermal growth factor (MGF).

A. Epidermal Growth Factor

EGF was identified more than 20 yr ago by S. Cohen on the basis of an observation made while working on the purification of NGF from mouse submaxillary gland (Cohen, 1959, 1962). When a crude extract of this organ was injected into mouse neonates, it resulted in precocious eyelid opening and tooth eruption. These events, which can take place as early as 6 days after birth in experimental animals, versus 11–13 days after birth in untreated littermates, provided the basis for a bioassay later used for the purification of mouse EGF (mEGF).

A single-chain polypeptide, mEGF is composed of 53 amino acids with an isoelectric point of 4.6 and a molecular weight of 6045 (Savage *et al.*, 1972; Taylor *et al.*, 1972). It is present in the submaxillary gland as a high-molecular-weight dimer complex in which one EGF peptide chain is associated with a specific arginine esteropeptidase—the EGF-binding protein. This binding protein is involved in the production of active EGF from a larger precursor molecule with a molecular weight of 9000 (Frey *et al.*, 1979). On the basis of a radioligand assay, a factor with all the biological properties of mEGF has been isolated from human urine (hEGF). Although capable of competing to the same extent as mEGF for the receptor sites present on fibroblasts, it differs from mEGF in its immunological cross-reactivity with mEGF antibodies, which is 1000-fold lower than that of mEGF (Cohen and Carpenter, 1975). It also has a different amino acid composition, a more neutral isoelectric point, and a lower molecular weight than mEGF (Cohen and Carpenter, 1975). However, hEGF and mEGF, as indicated by their cross-reactivity in the radioreceptor assay, can be structurally related, thereby raising the possibility that hEGF and mEGF could have evolved from a common precursor. The similarity between mEGF and hEGF has been further demonstrated by the unexpected finding of Gregory (1975), who reported that, when the amino sequence of human urogastrone was compared to that of mEGF, both molecules had 37 amino acid residues in common out of a total complement of 53. Since both urogastrone and hEGF have the same ability to compete with [^{125}I]mEGF in radioreceptor assays (Hollenberg and Gregory, 1976), it is likely that hEGF and urogastrone are one and the same. Their identity is further supported by the ability of both factors to elicit nearly identical biological responses in humans and mice. Urogastrone can promote early eyelid opening in mice, and mEGF can block the release of HCl from the gastric mucosa.

EGF has been shown to be a potent mitogen for a variety of cultured cells of ectodermal and mesodermal origin (Carpenter and Cohen, 1979). Among the cells of ectodermal origin which respond to it are a variety of keratinocytes derived from corneal skin, conjunctival, lung, or pharyngeal tissue (Sundell *et al.*, 1975; Sun and Green, 1977). In all cases, EGF markedly stimulates their proliferation and leads to enhanced keratinization and squame production. It also delays the ultimate senescence of the cells, thereby increasing their culture lifetime (Rheinwald and Green, 1977). Because of these effects, the culture of epidermal keratinocytes, particularly those of the human, has been greatly improved (Green *et al.*, 1979). Single cultured cells can now generate *in vitro* stratified colonies that ultimately fuse and form an epithelium that is a reasonable approximation of the epidermis. Among the cells of mesodermal origin which respond to EGF in culture are granulosa cells, corneal endothelial cells, vascular smooth muscle cells, chondrocytes, and fibroblasts (Gospodarowicz *et al.*, 1978c,d,e). This last cell type has been used extensively, as will be detailed later, to study the extensive sequence of events that are part of the mitogenic response produced by EGF (for a recent review, see Carpenter and Cohen, 1979). *In vivo*, EGF has been shown to be a potent mitogen for corneal epithelial cells.

B. Insulin-like Growth Factors

The insulin-like growth factors or somatomedins are a family of polypeptides that have been purified from human plasma. They include IGF I (formerly known as NSILA I), IGF II, somatomedins A and C, and multiplication-stimulating activity (MSA). These factors have in common the following properties (Daughaday, 1977; Rechler and Nissley, 1977; Zapf *et al.,* 1978): They are single-chain polypeptides with a molecular weight of 7500, all exhibit an insulin-like activity when tested on adipocytes *in vitro,* and all can stimulate anabolic processes in cartilage *in vitro.* Although they interact weakly with insulin receptors, they can all strongly interact with IGF receptors. In plasma, they are associated with globulin carrier proteins. As far as their immunoreactivity is concerned, IGF I and somatomedins A and C cross-react extensively in radioimmunoassays; lesser cross-reactivity has been noted for IGF II.

The amino acid sequences of IGFs and II have recently been determined (Zapf *et al.,* 1978). Both polypeptides show strong homology of sequence with human proinsulin. It is therefore likely that IGFs I and II arose from an insulin gene duplication occuring prior to the vertebrates. The chemical similarity of the IGFs and insulin is consistent with their similar biological reactivities and receptor cross-reactivities. Somatomedin C, which is indistinguishable from IGF I in various radioreceptor assays (Van Wyk *et al.,* 1980), exhibits impressive similarities to IGF I in chemical structure (Svoboda *et al.,* 1980). Associated with the family of IGFs is MSA, which is composed of polypeptides synthesized by an established cell line (BRL 3A) derived from the liver of a normal 5-week-old female Buffalo rat and is capable of growing in serum-free medium (Rechler *et al.,* 1978; Nissley and Rechler, 1978; Moses *et al.*, 1979; Nissley *et al.*, 1979).

The difficulties in purifying somatomedin from plasma and the limited supply of pure somatomedin have limited studies on its *in vivo* effect. Rothstein and Van Wyk have recently reported that the injection of pure somatomedin into the dorsal lymph sacs of hypophysectomized frogs results in a striking proliferative response on the part of the lens epithelium (Rothstein *et al.*, 1980). The experiments were of too short a duration, however, to determine an effect on linear growth (Rothstein *et al.,* 1980). In cartilage, somatomedin stimulates a generalized *in vitro* growth response which includes chondrocyte proliferation as well as synthesis of collagens and proteoglycans (Van Wyk *et al.*, 1974).

Insulin-like growth factors (somatomedin, MSA, and IGF) when present alone are only weak mitogens for cultured cells and require previous exposure of the cells to competence factors such as FGF in order to be active. Therefore, although insulin-like growth factors support cell proliferation, they do not induce it. Nevertheless, they are as important as growth factors, since without them cells committed to proliferate will not progress through their cycle (Stiles *et al.*, 1979a,b; Pledger and Wharton, 1980).

C. Fibroblast Growth Factor

The observation that pituitary extracts and partially purified pituitary hormone preparations are mitogenic for various cultured cell types (Holley and Kiernan, 1968; Clark *et al.*, 1972; Corvol *et al.*, 1972; Armelin, 1973) has led to the purification from pituitary tissue of FGF (Gospodarowicz, 1974, 1975; Gospodarowicz *et al.*, 1978a). This peptide has a molecular weight of 13,400 and an isoelectric point of 9.6 (Gospodarowicz, 1974, 1975; Gospodarowicz *et al.*, 1978a). During the course of the purification of pituitary FGF, it was realized that brain tissue contained a similar activity (Gospodarowicz, 1974). When purified, brain FGF was found to be related to myelin basic protein (MBP). Of the two mitogenic polypeptides isolated from brain tissue, brain FGF-1, which had the lower biological activity, had an amino acid composition and primary structure similar to that corresponding to residues 44–166 of MBP, while the more potent brain FGF-2 had a primary structure corresponding to that of residues 44–153 of MBP (Westfall *et al.*, 1978). When injected with adjuvants into guinea pigs, brain-derived FGF induced clinical and histological signs of autoimmune encephalomyelitis. Whereas brain-derived FGF cross-reacted immunologically with basic myelin, it failed to cross-react with pituitary FGF (Westall *et al.*, 1978). This tends to demonstrate that, although brain and pituitary FGF have a similar spectrum of activity on diverse cell types, they are nevertheless separate entities (Gospodarowicz *et al.*, 1978d,e; Westall *et al.*, 1978).

In vivo, brain FGF has been shown to stimulate the proliferation of blastemal cells in frogs and *Triturus viridescens* and could therefore contribute to the regeneration process in lower vertebrates by eliciting the initial recruitment of cells in an area of amputation injury (Gospodarowicz *et al.*, 1978d; Mescher and Gospodarowicz, 1979; Gospodarowicz and Mescher, 1980a,b). It has also been shown to be an angiogenic factor *in vivo* (Gospodarowicz *et al.*, 1979a). Whether FGF plays a physiological role in developmental processes has yet to be investigated.

The biological effects of FGF *in vitro* can be seen with respect to cell proliferation, cell differentiation, and cell senescence. Both pituitary and brain FGF have been shown to be potent mitogens for mesoderm-derived cells (Gospodarowicz *et al.,* 1978c,d,e; Gospodarowicz, 1979). Until now, they have mostly been used *in vitro* to develop new cell lines (Gospodarowicz and Zetter, 1977; Gospodarowicz, 1979), particularly from vascular and corneal endothelia, since endothelial cells originating from these tissues are most sensitive to FGF (Gospodarowicz *et al.,* 1977a, 1978e, 1979b, 1980b). FGF can be observed to be mitogenic both for cells seeded at clonal density and for low-density cultures (Gospodarowicz and Zetter, 1977). Although the response of the cells can vary widely, depending on the culture conditions (types of sera and media used), the addition of FGF to the cultures of most mesoderm-derived cells results in a greatly reduced average doubling time, which in the case of vascular endothelial cells, for example, can drop from 72 to 18 hr (Gospodarowicz *et al.,* 1978e; Duthu and Smith, 1980; Gospodarowicz and Lui, 1981). This results primarily from a shortening of the G_1 phase of the cell. FGF has also been shown to affect the phenotypic expression of some cell types (Vlodavsky and Gospodarowicz, 1979; Vlodavsky *et al*., 1979). This is a particularly interesting characteristic of FGF, since it has made possible the long-term culturing of various cell types which otherwise would rapidly lose their normal phenotypes in culture when passaged repeatedly at a low cell density. This biological effect of FGF has been best studied using vascular endothelial cell cultures cloned and maintained in the presence of FGF and then deprived of it for various time periods (Vlodavsky and Gospodarowicz, 1979; Vlodavsky *et al*., 1979).

FGF can also significantly delay the ultimate senescence of cultured cells. In the case of granulosa cells, the addition of FGF to either clonal granulosa cell lines or mass cultures can extend their lifespan in culture from 10 to 60 generations (Gospodarowicz and Bialecki, 1978). Adrenal cortex cell lines cloned in the presence of FGF show a similar dependence during their limited lifespan, and its removal from the culture medium results not only in a greatly extended doubling time but also in rapid senescence (Simonian and Gill, 1979; Simonian *et al.,* 1979). In the case of vascular (Gospodarowicz *et al.,* 1978e; Gospodarowicz, 1979; Duthu and Smith, 1980) and corneal endothelial cells (Gospodarowicz *et al.,* 1981a), FGF has been shown to extend the lifespan of the cultures greatly, the effect being best observed with corneal endothelial cells which, when maintained in the absence of FGF have a lifespan of 20–30 generations, while in its presence they can proliferate for 200 generations (Gospodarowicz *et al.,* 1981a).

D. The Eye-Derived Growth Factor

Particularly relevant to the control of proliferation of various cell types of the eye has been the observation by Arruti and Courtois (1978) that neutral extracts of adult retina can stimulate the growth and modify the morphology of lens epithelial cells *in vitro*. Such extracts are also able to stimulate cells from different origins and species (vascular and corneal endothelial cells, myoblasts, chondrocytes, neuroblastoma cells, keratinocytes). In this respect, it is similar in many ways to FGF and EGF, while it differs from them for some target cells such as fibroblasts, for which it is toxic. Further studies have shown that similar activity can be found in other ocular tissues. The highest growth-promoting capacities were found in extracts of iris, pigmented epithelium with choroid, and vitreous body. While the nature of all these extracts has not yet been determined, the fact that they are prepared in a similar way and that they have a similar growth-promoting activity has led Barritault *et al.* (1981) to postulate that there is in the eye an ubiquitous growth factor called EDGF which may play an important role in physiology and pathology of the eye.

E. The Mesodermal Growth Factor

Weimar and Haraguchi (1975) have described the partial purification from mouse submaxillary glands of a growth factor called MGF. This agent has been shown to stimulate the hypertrophy and mitosis of corneal stroma cells (Rich *et al.*, 1979; Weimar, 1979; Weimar and Haraguchi, 1979; Weimar *et al.*, 1980; Squires and Weimar, 1980). It has also been shown to stimulate the healing process in organ culture of wounded rabbit and human corneal endothelium. The relationship between MGF and EGF, which is also present in mouse submaxillary gland, is at present unclear. In particular, the effect of EGF antibody on the growth-promoting effect of MGF has not been reported.

IV. EFFECTS OF GROWTH FACTORS ON THE PROLIFERATION AND DIFFERENTIATION OF CORNEAL ENDOTHELIAL CELLS *IN VITRO*

A. Proliferation of Low-Density Corneal Endothelial Cell Cultures Exposed or Not Exposed to Growth Factors

Significant limitations on the culture of corneal endothelial cells have in the past been imposed by the slow doubling time of these cultures, which can be passaged only at a high cell density if precocious senescence is to be avoided (Mannagh and Irving, 1965; Perlman and Baum, 1974; Perlman *et al.*, 1974). Even in this case, though, most, if not all corneal endothelial cell cultures thus far established have shown a short lifespan and early signs of dedifferentiation (Mannagh and Irving, 1965; Perlman and Baum, 1974; Perlman *et al.*, 1974). However, the use of mitogens such as FGF and EGF, which are mitogenic *in vitro* for corneal endothelial cells, has greatly facilitated the culture of these cells (Gospodarowicz *et al.*, 1977a, 1978b, 1981a), since low-density cell populations can, when maintained in their presence, proliferate actively and thereby yield an unlimited supply of this cell type for further studies on cell differentiation.

Although neither FGF nor EGF is required to establish primary cultures of bovine corneal endothelial cells provided that the cells are plated at a high enough cell density, if cultures are started at clonal cell density, the development of a monolayer will depend on their presence in the culture medium (Gospodarowicz *et al.*, 1977a, 1978b, 1981a). In media supplemented with 10% serum alone, small colonies develop from cell aggregates during the first few days, but the cells look unhealthy, proliferate slowly, and quickly become senescent. However, if as little as 50 ng/ml of FGF or EGF is added to the culture, the population doubling time is reduced to as little as 18 hr, and such cultures can be propagated when seeded at either a high (up to 1:1000) or low split ratio. Upon reaching confluence, the cells adopt a morphological configuration similar to that of the confluent culture from which they originated (Fig. 1).

The response of corneal endothelial cell cultures to FGF has been shown to be a function of the age of the culture as well as of the density to which the confluent cultures are split. While high-density, early-passage cultures did not require FGF in order to become confluent, after a few passages, and especially if cultures were split at a high split ratio (1:100–500), the need for FGF became apparent, since in its absence cells proliferated poorly and rapidly became senescent (Gospodarowicz *et al.*, 1981a). The mitogenic effect of EGF on long-term cultures was not as marked as that of FGF, since after 20–30 generations cells no longer responded to it (Gospodarowicz *et al.*, 1981a).

Cultured corneal endothelial cells have also been shown to respond to EDGF with increased cell proliferation (Barritault *et al.*, 1981). Although these studies have been less extensive than those with FGF, EDGF has been shown to have similar biological properties (shortening of the cell cycle, allowing growth at clonal density and delaying senescence in culture). Using an organ culture system rather than tissue culture, Weimar *et al.* (1980) and Rich *et al.* (1979) have observed a positive healing effect of MGF on wounded rabbit and human corneal

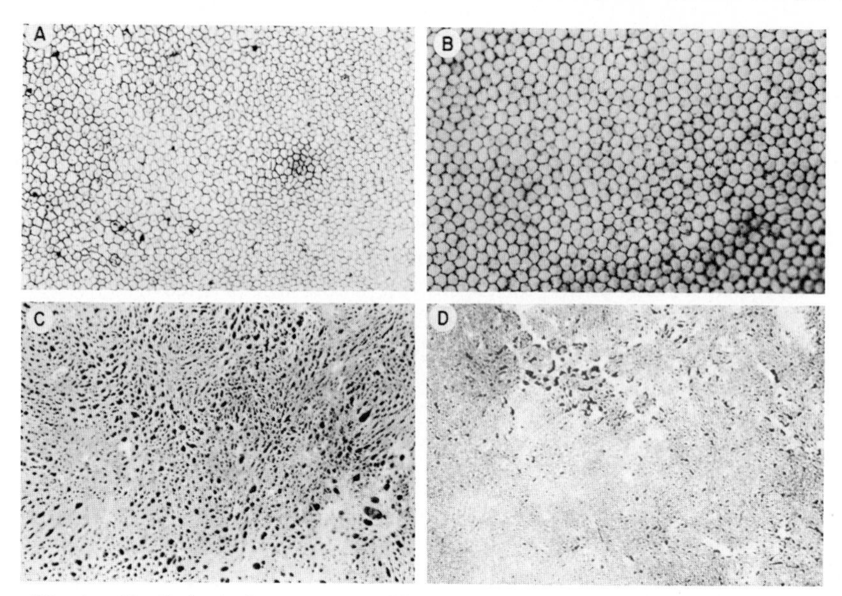

Fig. 1. Morphological appearance of long-term corneal endothelial cell culture and synthesis of the basement membrane *in vitro*. (A) Corneal endothelial cell cultures maintained for 1 month in tissue culture were stained with alizarin red. The intercellular border stained bright red, and the polygonal and contact-inhibited morphology of the cell monolayer are readily apparent. (B) The morphology can be compared to that of the endothelium of a corneal button stained with alizarin red just after excision. (C and D) Immunofluorescence of the basement membrane produced by corneal endothelial cell cultures. Confluent corneal endothelial cell cultures were treated with 0.5% Triton X-100 to remove the cell layer and to expose the underlying extracellular matrix. The immunofluorescence pattern of the extracellular matrix when exposed to fibronectin or antilaminin (D) followed by incubation with fluorescein-conjugated anti-rabbit IgG is shown.

endothelium. This effect is similar to that reported for EGF and FGF by Gospodarowicz and Greenburg (1979a), who used organ cultures of bovine and cat corneas.

B. Morphological Differentiation of Long-Term Confluent Cultures Grown in the Presence or Absence of Growth Factors

Confluent cultures of corneal endothelial cells grown in the presence of FGF exhibited all the attributes of endothelial cells *in vivo,* including synthesis of a basal lamina. As shown in Fig. 2, an extracellular matrix typical of basement membrane *in vivo* is seen underneath the cell monolayer. Pinocytotic vesicles with an electron-dense coat are present near the base of these cells; it is presumed that these vesicles

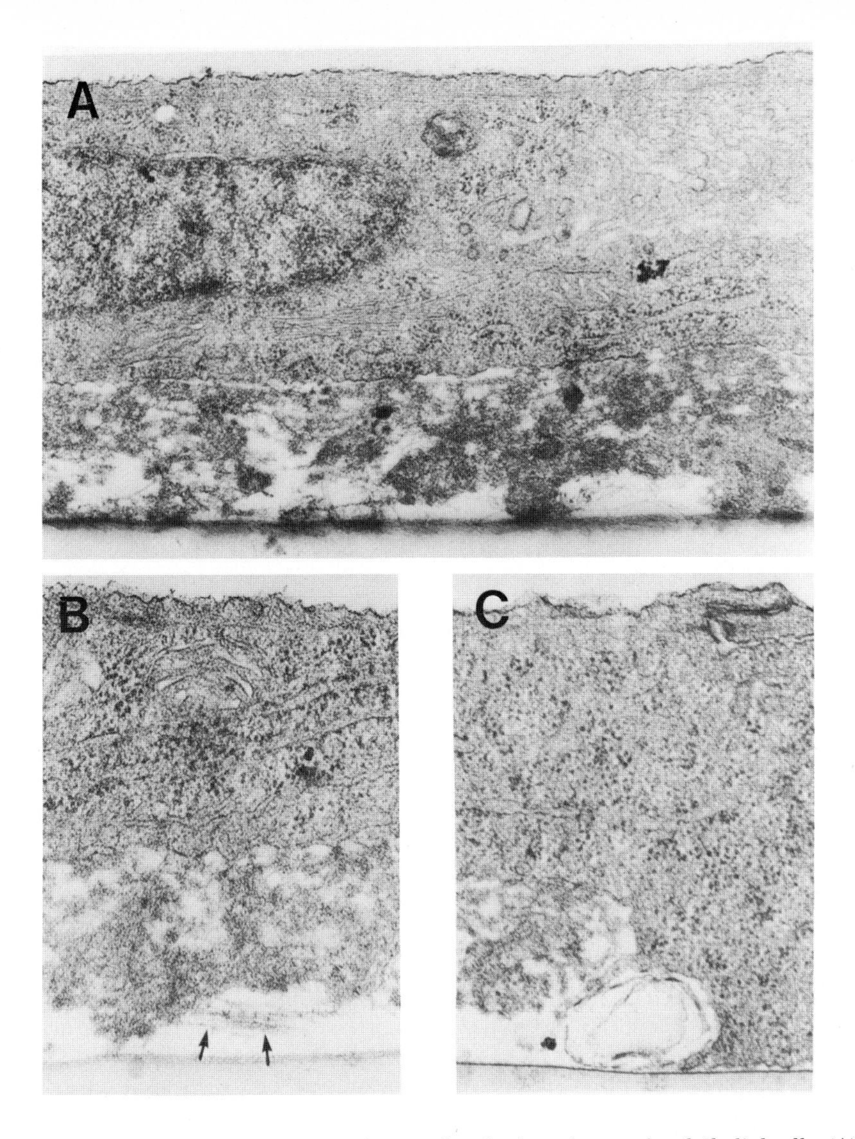

Fig. 2. Transmission electron micrographs of cultured corneal endothelial cells. (A) Bovine corneal endothelium. The nucleus is shown on the left. Deposited collagenous fibrils accumulate between the endothelial cell and the dish. The apical cell zone is organelle-free and filamentous. The basal side contains mitochondria, rough endoplasmic reticulum, free ribosomes, and vesicles probably involved in the secretion of collagen and other protein (×36,800). (B) A portion of a bovine corneal endothelial cell showing rough endoplasmic reticulum cisternae and a mitochondrion with typical orientation of its cristae mitochondriales. Collagenous material is deposited between the cell base and the dish. Larger collagen fibrils with striations are seen near the dish (arrows). Rough endoplasmic reticulum is more abundant in these cells than in the same cells *in vivo*, perhaps because these cells are in the process of forming a new Descemet's membrane (×43,200). (C) A typical junctional complex is shown near the apical cell surface on the right side of the illustration. Note the presence of a "terminal web" along the apical surface of the cell (×43,200).

are involved in secretion of the amorphous material found underneath the basal cell membrane. Furthermore, cultured bovine corneal endothelial cells *in vitro* show a distinct polarity in the localization of their organelles. Bundles of microfilaments run parallel to the apical plasma membrane and delimit a zone free of organelles. A junctional complex is present at the apex of the cells. Microplicae and villi characterize the apical cell membrane as well as a marginal fold found adjacent to the junctional complex (Gospodarowicz *et al.,* 1979d). The cell mitochondria exhibit the unique arrangement of the cristae mitochondriales which is characteristic of corneal endothelial cells (Fig. 2).

Confluent cultures of corneal endothelial cells exhibit, as *in vivo,* an assymetry of cell surfaces. While the apical cell surface is a nonthrombogenic surface to which platelets do not bind (Gospodarowicz *et al.,* 1979d,e), the basal cell surface is involved in the synthesis of a highly thrombogenic basal lamina which, when examined by immunofluorescence, is composed of collagen type III and, to a lesser extent, of collagen type IV (Gospodarowicz and Tauber, 1980; Gospodarowicz *et al.,* 1981a). Chemical analysis of the collagen types synthesized by bovine corneal endothelial cells has led to the conclusion that type III collagen is the major component both deposited in the extracellular matrix and secreted into the medium (Tseng *et al.,* 1981). The basement membrane collagens, types IV and V, are also found in each compartment, though the latter is associated preferentially with the cell matrix. The I/III/(IV + V) ratio of collagens synthesized by corneal endothelial cells is 3:16:1 (Tseng *et al.,* 1981). Associated with the extracellular matrix collagens are proteoglycans composed mostly of heparan sulfate and dermatan sulfate proteoglycans (10%) (Gospodarowicz and Fujii, 1981; Neno *et al.,* 1982). For a more detailed discussion, see the chapter by Hart (this volume). Also present in large quantities are two glycoproteins which have been held responsible for cell attachment to the basal lamina. These are fibronectin and laminin (Gospodarowicz *et al.,* 1979d, 1981b). Laminin is also probably involved in corneal endothelial cell migration (Gospodarowicz *et al.,* 1981b).

The presence of both laminin and collagen type IV, as well as of fibronectin, in the extracellular matrix produced by cultured corneal endothelial cells has led to the conclusion that this matrix has the characteristics of a basement membrane (Gospodarowicz and Tauber, 1980; Gospodarowicz *et al.,* 1981b). Indeed, previous studies have demonstrated that laminin is an antigen present within the lamina lucida portion of basement membranes produced by various epithelia *in vivo,* while collagen type IV and fibronectin are localized to the lamina densa (Foidart and Reddi, 1980; Foidart *et al.,* 1980).

In contrast to cultures grown in the presence of FGF, cultures maintained in its absence lose within three passages their ability to form at confluence a monolayer of closely apposed, flattened cells. Instead, the cultures are composed of large, overlapping cells which are no longer contact-inhibited (Gospodarowicz *et al.,* 1981a). Parallel to these changes in cell morphology, the apical cell surface becomes thrombogenic, as reflected by an increase in platelet-binding capacity (Gospodarowicz *et al.,* 1979d,e, 1981a). Likewise, marked changes in the distribution and appearance of cell surface proteins such as fibronectin and collagen can be observed. Fibronectin and collagen, which in confluent and highly organized cultures grown in the presence of FGF are detected only in the basal and not in the apical cell surface, now appear in both basal and apical cell surfaces (Gospodarowicz *et al.,* 1979e, 1981a).

These observations suggest that corneal endothelial cells maintained in the absence of FGF exhibit, in addition to a much slower growth rate, morphological as well as structural alterations that mostly involve changes in the composition and distribution of the basal lamina. This raises the possibility that the basal lamina produced by these cells could have an effect on their ability to proliferate and to express their normal phenotype once confluent (Gospodarowicz and Tauber, 1980).

C. Involvement of the Basal Lamina Produced by Corneal Endothelial Cells in the Control of Cell Proliferation

Cell migration and growth *in vivo* are the result of a complex balance between cell-cell and cell-substrate interactions. The forces which combine to modulate the cell shape may either permit or prevent cell proliferation and differentiation (Thompson, 1961; Maroudas, 1973; Folkman, 1977; Gospodarowicz *et al.,* 1978b, 1979f). Following its original proposal by Grobstein (1955), a role for cell/substrate interactions in the control of cell proliferation and morphogenesis has been demonstrated (for review, see Wessells, 1964; 1977; Grobstein, 1967, 1975; Kratochwil, 1972). In the case of epithelial tissues with a high rate of cell turnover, such as the epidermis or the corneal epithelium, active cell proliferation is restricted to their basal layer, composed of tall, columnar cells. These cells are in close contact with a basal lamina. In contrast, cells in the upper layers, which have lost their ability to proliferate and gradually adopt a flattened configuration, are no longer in contact with the basal lamina. Thus, contact between the cells and their substrate rather than contact between cells could, *in vivo,* have a permissive influence on cell proliferation. Likewise, mammalian cells maintained under tissue culture conditions require, in order to proliferate and to express their normal phenotype, not only nutrients and growth factors (Holley and Kiernan,

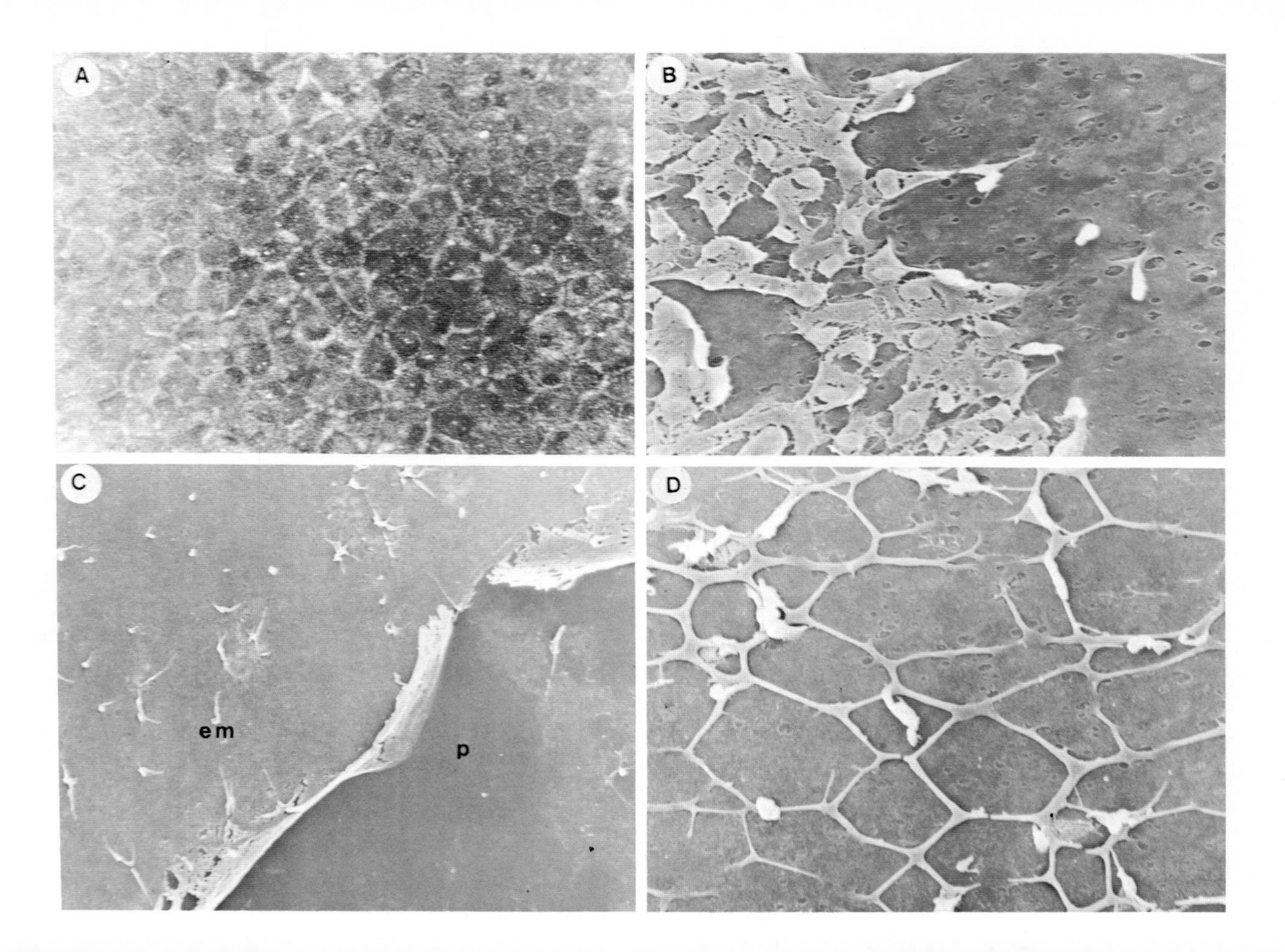

A
B
C
D
em
p

1968; Temin *et al.*, 1972) but also an appropriate physical substrate upon which they can attach and spread (for review, see Bornstein and Sage, 1980; Gospodarowicz and Tauber, 1980). Recent investigations have suggested that *in vitro* cell–substrate interactions may play other roles in addition to cell anchorage (Murray *et al.*, 1979; Rollins and Culp, 1979; Wicha *et al.*, 1979). The substrate upon which a cell rests can dictate its shape and may play a role in its response to serum factors (Folkman and Moscona, 1978) and a differential response to various growth factors (Wessels, 1964; Folkman and Moscona, 1978). Components of the basal lamina produced by cells have also been shown to be involved in cell differentiation and the control of gene expression (Hay, 1977; Sun and Green, 1977; Bernfield, 1978; Yang *et al.*, 1979), as well as in cell attachment (Murray *et al.*, 1979; Rollins and Culp, 1979; Wicha *et al.*, 1979). Yaoi and Kanaseki (1972), as well as Weiss *et al.* (1975), using microexudate carpets from chick or mouse embryo cells, have demonstrated a role for this material in cell proliferation. The substrate upon which cells rest *in vitro* could therefore be a decisive element in their proliferative response to various factors.

Identification of the components present within the basal lamina which could be involved in controlling cell proliferation either *in vivo* or *in vitro* has been made difficult mostly by its intricate nature. Because the *in vitro* reconstruction of the basal lamina from its isolated native components into the correct highly ordered structure it represents would be a formidable task, advantage was taken of the fact that cultured corneal endothelial cells have the ability to produce a basal lamina *in vitro* which closely resembles the basement membrane upon which cells rest *in vivo* (Gospodarowicz *et al.*, 1979d,e). This membrane can easily be denuded from the cell monolayer which covers it by exposing confluent cultures to either a detergent (0.5% Triton X-100 in phosphate-buffered solution) or weak alkali (20 mM NH_4OH in water) and subsequent washing (Gospodarowicz and Ill, 1980a,b; Gospodarowicz *et al.*, 1980a, 1981c) (Fig. 3).

Fig. 3. Scanning electron microscopy of a monolayer of bovine corneal endothelial cells before and after exposure to Triton X-100. A monolayer composed of polygonal, highly flattened, closely apposed cells can be seen in (A) (×600). After the monolayer has been treated with Triton X-100, it is composed of nuclei and cytoskeletons which no longer attach firmly to the extracellular matrix. In some areas the extracellular matrix has been exposed (B, ×600). Washing the dishes with phosphate-buffered saline removed the cytoskeleton and exposed the extracellular matrix present underneath the cells (C, ×200). The plate has been scratched with a needle to expose the plastic (p) to which the extracellular matrix (em) strongly adheres. In some areas, heavy deposits of extracellular matrix material form hexagonal ridges (D, ×600) on top of an amorphous material covering the whole dish.

When the ability of the basal almina to support cell proliferation was analyzed using either rabbit or bovine corneal endothelial cells seeded at low cell density in the presence of high (10%) concentrations of serum, cells maintained on basal lamina-coated dishes proliferated even faster than cells maintained on plastic and exposed to FGF (Gospodarowicz and Ill, 1980c; Gospodarowicz *et al.,* 1981a). It can therefore be concluded that, when the proliferation of corneal endothelial cells from two different species is compared, cells maintained on plastic proliferate poorly and FGF is needed in order for the cultures to become confluent. In contrast, when similar cultures are maintained on basal lamina, they proliferate actively and no longer require FGF in order to become confluent. Similar results were observed when lens epithelial cells were maintained on plastic versus a basal lamina. Although the cells maintained on plastic hardly proliferated, yielding within a few days a population composed of large, binucleated cells, the cells plated on basal lamina proliferated actively, with an average doubling time for lens epithelial cells of 15 hr during their logarithmic growth phase (Gospodarowicz and Ill, 1980c; Gospodarowicz *et al.,* 1981a).

When the ability of plasma versus serum to sustain corneal endothelial cell proliferation was analyzed as a function of substrate (plastic versus a basal lamina), cultures maintained on plastic proliferated at a lower rate when exposed to plasma than when exposed to serum. In both cases, FGF was required in order for cells to become confluent. When similar cultures were maintained on a basal lamina, they proliferated equally well regardless of whether they were exposed to plasma or serum, and FGF was no longer required. One can therefore conclude that the simple change in substrate from plastic to basal lamina will restore the sensitivity of these cells to agents present in plasma and that the basal lamina has a permissive effect on cell growth, since plasma is required in order for cells to proliferate actively (Gospodarowicz *et al.*, 1980c, 1981a).

The permissive effect of the basal lamina produced by corneal endothelial cells on cell proliferation has also been observed in the case of granulosa cells (Gospodarowicz *et al.*, 1980a; Savion *et al.*, 1981b), adrenal cortex cells (Gospodarowicz *et al.,* 1980a), vascular smooth muscle cells (Gospodarowicz and Ill, 1980a; Gospodarowicz *et al.,* 1981c), and vascular endothelial cells (Gospodarowicz and Ill, 1980b; Gospodarowicz *et al.,* 1980b). Although these various cell types when maintained on plastic require serum and/or FGF in order to proliferate, they no longer require FGF when maintained on a basal lamina, and the addition of plasma alone is sufficient to make them proliferate at an optimal rate. The permissive effect of the basal lamina produced by corneal endothelial cells is therefore not restricted to the corneal endothelium but also extends to cell types of unrelated origin.

These observations raise the possibility that, although the final effect of FGF is that of a mitogen (Gospodarowicz *et al.*, 1978c,d,e), its action could be indirect. FGF could either replace the cellular requirement for a substrate such as the basal lamina or could induce the synthesis and secretion of the basal lamina produced by the cells. This could in turn make the cells sensitive to factors present in plasma. That the latter alternative could occur finds support in previous observations that FGF could control the production by vascular endothelial cells of extracellular and cell surface components such as fibronectin and various collagen types (Vlodavsky and Gospodarowicz, 1979; Vlodavsky *et al.*, 1979; Greenburg *et al.*, 1980). Since sparse cultures of bovine corneal endothelial cells proliferate poorly when maintained on plastic but not when maintained on a basal lamina, it may be that, at a very low cell density, cultures maintained on plastic are unable to produce enough extracellular material to support further growth. The mitogenic effect of FGF on these cells could therefore be the result not only of an increase potential of the cells for synthesizing their basal lamina (Gospodarowicz and Tauber, 1980; Gospodarowicz and Greenburg, 1981).

These observations emphasize how drastically one can modify the proliferative response of a given cell type to serum factors in accordance with the substrate upon which the cells are maintained. It is possible that the lack of response of different cell types maintained under tissue culture conditions to agents responsible *in vivo* for their proliferation and differentiation could be attributed to the artificial substrate, whether plastic or glass, upon which the cells rest and which limits their ability to produce a basal lamina either of the correct type or in adequate amounts (Gospodarowicz, 1980).

The ways in which the basal lamina exerts its permissive effect on cell proliferation can only be the object of speculation. One possible effect is to modify the cell shape in order to make it responsive to factor(s) to which the cells do not respond unless they adopt an appropriate shape. Recently, Folkman and Moscona (1978), using vascular endothelial cells maintained on tissue culture dishes coated with an agent which modifies the adhesiveness of the cells to the dish, were able to control precisely the cellular shape in morphologies ranging from highly flattened to almost spheroidal. When the extent of cell spreading was correlated with DNA synthesis or cell growth, it was found to be highly coupled. Whereas greatly flattened cells responded to serum factors, spheroidal cells no longer responded and intermediate degrees of response could be observed depending on how flattened the cells were. Likewise, with corneal epithelial cells, changes in cell shape which depend on the substrate upon which the cells are maintained correspond to drastically altered sensitivities of the cells to EGF versus FGF (Gospodarowicz *et al.*, 1977b, 1978b, 1979f). An attractive hypothesis proposed by Yaoi and Kanaseki (1972) is that the basal lamina plays a key role in mitosis and facilitates cytokinesis. This hypothesis is based on their observation that, while both high- and low-density cultures maintained on an extracellular matrix exhibit a high rate of DNA synthesis and a high mitotic index, only high-density cultures maintained on plastic have both a high rate of DNA synthesis and a high mitotic index. In contrast, low-density cultures maintained on plastic, although they have a high rate of DNA synthesis, have a low mitotic index, thus suggesting that cells do not enter into mitosis. It is therefore likely that, while plastic provides a foreign substrate upon which cells can attach tenaciously and spread in a vain attempt to phagocytose it, a basal lamina provides a natural substrate that the cells recognize and upon which they can undergo the characteristic changes in morphology (rounding up) occurring at mitosis. These morphological changes probably reflect the rearrangement of the cellular cytoskeleton, so that cells can go through the cleavage steps, giving rise to two progeny cells instead of undergoing endomitosis.

D. Plasma Factors Involved in the Control of Proliferation of Bovine Corneal Endothelial Cells *in Vitro*

In order to investigate the factors present in plasma which could affect the proliferation of bovine corneal endothelial cells, the innovative approach of Gordon Sato and his colleagues, who have shown that the serum or plasma requirement for the growth of a number of cell lines can be satisfied by the addition of specific hormones and growth factors to synthetic media (Sato and Reid, 1978; Bottenstein *et al.*, 1979; Barnes and Sato, 1980), was followed. Further incentive to follow this approach was provided by previous observations which indicated that bovine corneal endothelial cells maintained on basal lamina-coated dishes had a much lower requirement for either serum or plasma in order to proliferate actively than when they were maintained on plastic (Gospodarowicz and Ill, 1980c). It is therefore possible that these cells, when maintained on basal lamina-coated dishes, could survive and still be responsive to plasma factors even when maintained in a well-defined synthetic medium supplemented with neither plasma nor serum. Among the factors likely to promote the growth of bovine corneal endothelial cells in serum-free conditions are FGF and EGF, which have been shown to be mitogenic for cultures exposed to either plasma or serum (Gospodarowicz *et al.*, 1977a). Insulin and transferrin are also likely candidates, since these two agents are required for the growth of most cell types cultured in serum-free medium (Sato and Reid, 1978; Bottenstein *et al.*, 1979). Other less well-studied factors are represented by high-density lipoproteins (HDLs) and low-density lipoproteins (LDLs), both of which have been shown to be mitogenic for several cell types when cultures are exposed to lipoprotein-deficient serum (Ross and Glomset, 1973; Brown *et al.*, 1976; Fischer-Dzoga *et al.*, 1976) or to serum-free medium (Tauber *et al.*, 1980, 1981).

When the effects of these various factors were compared, it was observed that bovine corneal endothelial cells seeded on basal lamina-coated dishes in the total absence of serum could actively proliferate when exposed to a synthetic medium supplemented with transferrin, HDLs, insulin, and FGF or EGF (Fig. 4). There was an absolute requirement for transferrin, since in its absence cells seeded in the total absence of serum did not respond to HDLs, insulin, and FGF or EGF added either alone or in combination. This absolute requirement for transferrin may reflect either its role as the main iron-carrying protein present in plasma or its ability to detoxify the medium by chelating trace amounts of toxic metals which could have been present in the medium (Barnes and Sato, 1980). Since transferrin, HDLs, insulin, and EGF are present in plasma (FGF alone being absent), they may be part of the plasma constituents involved in control of the proliferation of bovine corneal endothelial cells when such cells are maintained on basal lamina-coated dishes and exposed to plasma.

The exposure of bovine corneal endothelial cell cultures to a synthetic medium supplemented with HDLs, transferrin, insulin, and FGF or EGF not only ensures their active proliferation but also allows them to be subcultured repeatedly and to undergo 40 generations in the total absence of serum before senescing (Giguère *et al.*, 1982) (Fig. 5). Under these conditions, the longevity of the cultures is clearly dependent on the use of a basal lamina as substrate, since cultures maintained on plastic and exposed to similar conditions cannot be passaged even once. The effect of a basal lamina in extending the life span of bovine corneal endothelial cell cultures is even more impressive in the case of cells exposed to a synthetic medium supplemented with an optimal concentration of serum (Fig. 5). While cultures maintained on plastic and exposed to serum can be maintained at best for 20 generations, cultures maintained on a basal lamina can be passaged for 160 generations. It may therefore be concluded that the basal lamina upon which the cells were maintained not only made them sensitive to fac-

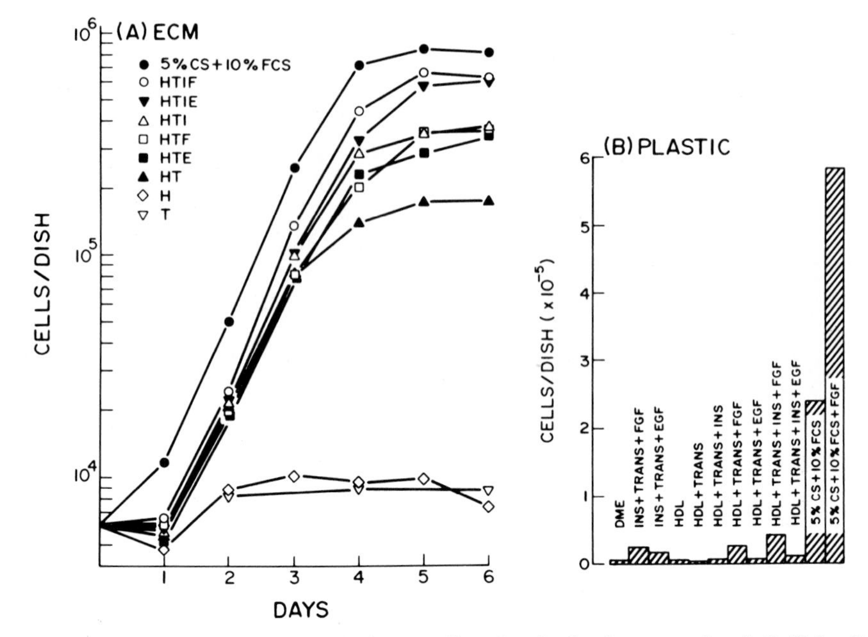

Fig. 4. Comparison of the growth rate of low-density bovine corneal endothelial cell cultures maintained on extracellular matrix-coated dishes and exposed to DMEM supplemented with either 5% calf serum or 10% fetal calf serum, or to medium supplemented with HDLs, transferrin, insulin, and FGF or EGF, added either singly or in combination. (B) Same as (A), except that the cultures were maintained on plastic and their final density was compared after 6 days. (A) Bovine corneal endothelial cells were seeded at 1×10^4 cells per 35-mm dish on extracellular matrix-coated dishes in the presence of DMEM supplemented with 5% calf serum (CS) or 10% fetal calf serum (FCS); HDLs, transferrin, insulin, and FGF (HTIF); HDLs, transferrin, insulin, and EGF (HTIE); HDLs, transferrin, and insulin (HTI); HDLs, transferrin, and FGF (HTF); HDLs, transferrin, and EGF (HTE); HDLs and transferrin (HT); or HDLs (H) and transferrin (T) alone. The concentration of HDLs added was 500 μg protein/ml; transferrin was 10 μg/ml; insulin, FGF, and EGF were 5 μg/ml, 100 ng/ml, and 25 ng/ml, respectively. HDLs and transferrin were added only once at day 0, while FGF was added every other day and insulin every 4 days. At daily intervals, triplicate plates representing each condition were trypsinized and counted. The standard deviation for the different determinations did not exceed 10% of the mean.

(B) Bovine corneal endothial cells were seeded as described above, but on plastic instead of ECM-coated dishes. Cultures were exposed to DMEM supplemented with 5% calf serum, 10% fetal calf serum with or without FGF, or DMEM supplemented with HDLs, transferrin, insulin, and FGF or EGF (1),HDLs and transferrin with EGF,FGF, or insulin present (2), HDLs and transferrin (3), HDLs alone (4), insulin, transferrin, and FGF or EGF (5), and DMEM alone (6). The final densities of the cultures kept under various conditions were then compared after 6 days in culture. The concentrations and schedule of addition of transferrin, HDLs, insulin, and FGF or EGF were as in (A).

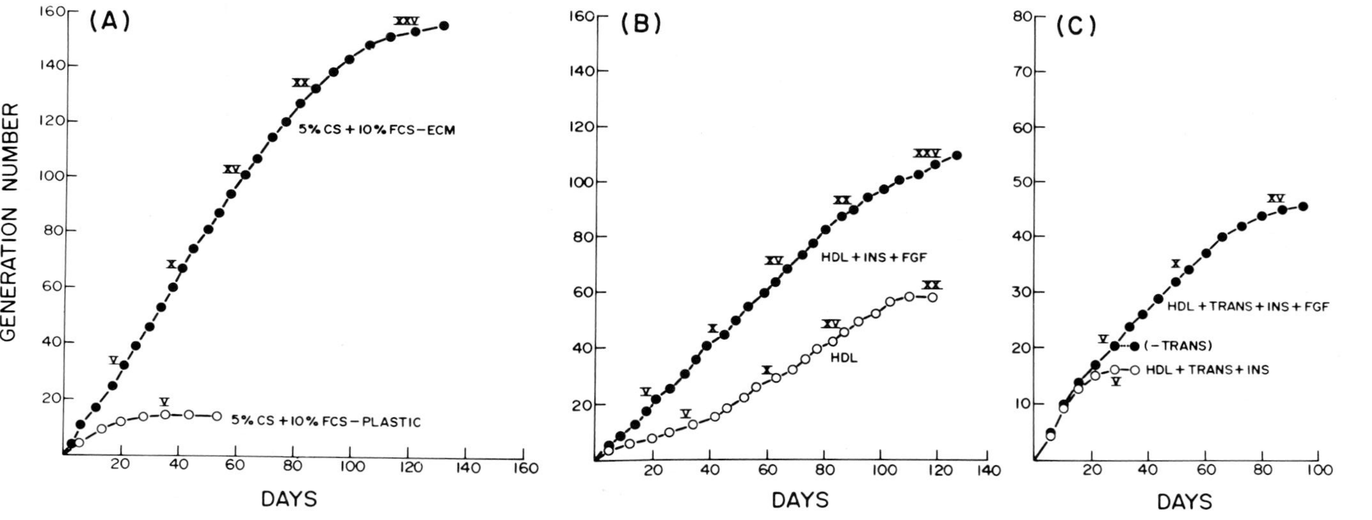

Fig. 5. (A) Comparison of the effect of the substrate (extracellular matrix (versus plastic) upon which bovine corneal endothelial cells are maintained on the life span of the cell cultures. The cells were maintained and passaged on plastic dishes (○) or on dishes coated with extracellular matrix (●). Cultures were exposed to DMEM supplemented with 5% calf serum (CS) or 10% fetal calf serum (FCS). (B) Culture life span of bovine corneal endothelial cell cultures seeded in the presence of serum on extracellular matrix-coated dishes and exposed to DMEM supplemented with HDLs alone (○) or HDLs, insulin, and FGF (●). HDL (250 μg protein/ml) was added only once at each transfer, while FGF (100 ng/ml) was added every other day. Insulin (2 μg/ml) was added every fourth day. The number of generations was determined from the initial cell density 8 hr after seeding and the number of cells harvested at each transfer. Each point represents a single transfer. Roman numerals indicate the passage number. (C) Life span of bovine corneal endothelial cell cultures seeded in the absence of serum on matrix-coated dishes. Cultures were exposed to DMEM supplemented with transferrin (10 μg/ml), HDLs (500 μg/ml), and insulin (5 μg/ml) (○), or HDLs, transferrin, insulin, and FGF (100 ng/ml) (●). Similar life spans were observed when cultures were passaged in the presence of EGF (50 ng/ml) instead of FGF (unpublished data). The schedule of addition of the various factors was as described. As soon as transferrin was omitted from the medium (−TRANS), cells stopped dividing and could no longer be passaged.

tors present in the serum but also delayed to a considerable extent their ultimate senescence when they were exposed to a synthetic medium supplemented with either well-defined factors or with serum. Furthermore, since the life span of cultures exposed to serum is longer than that of cultures exposed to a combination of transferrin, HDLs, insulin, and FGF or EGF, it is likely that there are other factor(s) present in serum which can further prevent their senescence (Giguère *et al.,* 1982).

One important conclusion from these studies is that cells in cultures seem to respond to growth-promoting agents which govern their proliferation early in their ontogeny. This conclusion can be reached on the basis of a comparison of the growth requirements of bovine *corneal endothelial* cells and *vascular endothelial* cells maintained and passaged in total absence of serum (Tauber *et al.,* 1981). While one might expect that both cell types would have the same growth requirements (HDL and transferrin), corneal endothelial cells additionally required the presence of insulin and EGF in order to grow at an optimal rate (Giguere *et al.,* 1982). Their growth requirements are therefore similar to those of vascular smooth muscle cells (Gospodarowicz *et al.,* 1981c) and demonstrate that endothelial cells from different organs do not necessarily have similar growth requirements. This may reflect their different embryological origins. Indeed, while the vascular endothelium is the first tissue to develop in the embryo and is derived from the primary mesenchyme (Hay and Revel, 1969), *both* vascular smooth muscle cells from the aortic arch (Le Douarin, 1980) and corneal endothelial cells (Noden, 1980) develop later and originate from the neural crest.

The ability to maintain, in a synthetic medium supplemented with defined factors, actively proliferating bovine corneal endothelial cells which at confluence can still exhibit the morphological characteristics of the corneal endothelium *in vivo* or those of confluent cultures maintained in the presence of serum and FGF-supplemented media (Fig. 6) could help to shed some light on the *in vivo* growth requirements of this cell type. It could also result in a cell model which more closely mimics the *in vivo* situation, since *in vivo* the corneal endothelium is exposed to the aqueous humor which, depending on the species considered, has a protein concentration which is only 0.1–1% that of plasma.

V. THE USE OF CORNEAL ENDOTHELIAL CELL CULTURES IN STUDIES ON THEIR DIFFERENTIATED PROPERTIES

The development in culture of corneal endothelial cells has led to the study of some of its properties which would otherwise have been dif-

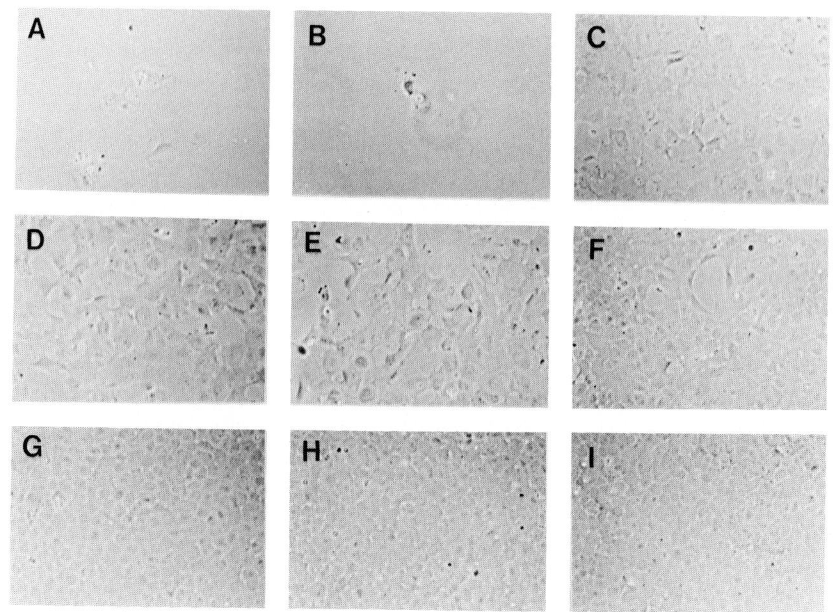

Fig. 6. Morphological appearance of bovine corneal endothelial cells maintained on extracellular matrix-coated dishes and exposed to HDLs (A), transferrin (B), HDLs plus transferrin (C), HDLs, transferrin, and insulin (D), HDLs, transferrin, and FGF (E), HDLs, transferrin, and EGF (F), HDLs, transferrin, insulin, and FGF (G), HDLs, transferrin, insulin, and EGF (H), or 5% calf serum plus 10% fetal calf serum (I). Cultures were seeded and maintained as described in Fig. 4A. Concentrations and the schedule of addition of the various factors to the culture media were the same as those described in Fig. 4A. Pictures were taken on day 5(phase-contrast, ×100).

ficult to study *in vivo*. Among these are its roles in hemostasis and as a selective permeability barrier. The functionality of cultured cells has also been tested in transplantation studies.

A. The Corneal Endothelium and Hemostasis

Hemostasis, or the cessation of blood flow, is the result of complex interactions between multiple factors leading to blood coagulation. Among the key factor(s) involved in blood clotting is the protease thrombin, which serves as an initiator of the biochemical and functional changes seen in platelets early in hemostasis and in the conversion of fibrinogen into fibrin. One of the principal factors involved in the disappearance of blood clots or fibrinolysis is the protease plasmin, which appears as a result of the activation of plasminogen. This factor has a great affinity for fibrin and adsorbs to it during blood clot formation. The activation of plasminogen into plasmin can be controlled by various tissue kinases, one of which is the plasminogen activator present in the vascular endothelium. Clotting in blood vessels following various types of injury often has a beneficial effect, since it interrupts blood flow, thereby preventing hemorrhage. In contrast, when it occurs in the anterior chamber of the eye, it has only undesirable effects. In addition to preventing proper vision, adhesion of a fibrin clot to the endothelium could cause cell death and could damage the endothelium, which, because it has a limited ability to regenerate, would result in cell death and its permanent denudation.

The hemostasis properties of the corneal endothelium and the mechanism by which the cells could prevent clot formation or accelerate the fibrinolysis process are difficult, if not impossible, to study *in vivo*. Cultured corneal endothelial cells have therefore been used to study how their endothelium could prevent blood clot formation. Studies have focused on the interaction of thrombin with the endothelium (Fehrenbacher *et al.,* 1979; Isaacs *et al.,* 1981; Savion *et al.,* 1981a; Shuman *et al.*, 1981), as well as on the generation of plasmin from plasminogen catalyzed by corneal endothelial cells.

Studies on the interaction of thrombin with the corneal endothelial cell surfaces have demonstrated the presence of specific receptor sites to which thrombin can be linked covalently (Isaacs *et al.,* 1981). Following binding to the cell surface, rapid internalization occurs which proceeds at a rate of 4 ng thrombin per 10^6 cells/hr. Following internalization, the thrombin receptor complexes are degraded within the lysosomal system and degradation products are released into the medium. In addition to being able to link thrombin covalently to specific cell surface binding sites, corneal endothelial cells also release thrombin-binding sites into the medium (Savion *et al.*, 1981a). Similar thrombin-binding sites can also be detected *in vivo* in the aqueous humor (unpublished results). These soluble sites are capable of binding thrombin. This leads to the formation of a complex of molecular weight 77,000 capable of binding to the cells and of being internalized and degraded. The interaction of thrombin with corneal endothelial cells is therefore unique in three respects: (1) the ability of thrombin to bind covalently to its receptor sites on the surface of corneal endothelial cells, (2) the ability of the cells to internalize and degrade such complexes rapidly, and (3) the ability of the cells to release free thrombin receptor binding sites into the medium, where *in vivo* they can at all times bind thrombin and control its concentration in either the medium or aqueous humor. These three characteristics of the interaction of thrombin with the corneal endothelium can result in the rapid inactivation of thrombin generated within the aqueous humor, thereby limiting the risk of blood clot formation (Isaacs *et al.*, 1981; Savion *et al.*, 1981a; Shuman *et al.*, 1981).

Equally remarkable is the ability of corneal endothelial cells to lyze fibrin clots rapidly. This is due to the direct activation of plasminogen by these cells through the mediation of plasminogen activator (Fehrenbacher *et al.*, 1979). This kinase is far more active in corneal endothelial cells than in vascular endothelial cells and is capable of rapidly generating high concentrations of plasmin which can then dissolve the fibrin clot. Also, unlike the membrane-bound plasminogen activator present in vascular cells, that of corneal endothelial cells is present in the cytosol and upon cell death could be released in a soluble form into either the medium or the aqueous humor, where it could diffuse into various parts of the eye such as the trabecular meshwork complex in order further to dissolve fibrin clots located far away from the endothelium.

Finally, corneal endothelial cells, like vascular cells, can synthesize prostacyclins, which are potent inhibitors of platelet aggregation (Gospodarowicz *et al.*, 1980b,c; Gospodarowicz and Tauber, 1980).

One could therefore conclude that corneal endothelial cells are uniquely prepared to defend themselves against fibrin clot formation and can prevent it in at least three different ways: by neutralizing thrombin generated during the clot formation process, by preventing platelet aggregation, and by rapidly dissolving the fibrin clot.

B. The Corneal Endothelium and Its Selective Permeability Barrier Function

The formation of a highly confluent and fully organized corneal endothelial cell monolayer is associated with changes in the cell surface membrane which impede the process of adsorptive endocytosis and allow cells to form an efficient block against the internalization of various macromolecules such as LDLs (Goldminz *et al.*, 1979). The barrier formed by the confluent endothelium is correlated with a massive accumulation, mainly in the pericellular matrix, of disulfide cross-linked fibronectin, which might stiffen the cell membrane and, as observed at confluence, restrict the lateral mobility of various cell surface receptors (Gospodarowicz *et al.*, 1979d,e). Confluent endothelial cell monolayers will therefore no longer internalize LDL receptor complexes. From these studies the notion emerged that the organization of a tissue, such as the strict nonoverlapping, highly flattened morphology of the endothelium which leads to the adoption of a two-dimensional cell monolayer configuration, could in turn dictate its physiological corneal properties, one of these properties being to act as a selective barrier between the aqueous humor and the corneal stroma (Goldminz *et al.,* 1979). Other aspects of the corneal endothelium and transport are treated by Riley (this volume).

C. The Transplantation *in Vivo* of Cultured Corneal Endothelial Cells

Of all cell types currently maintained in tissue culture, one of the most attractive, for purposes of transplantation, is the corneal endothelial cell. Studies on the *in vivo* transplantation of cultured cells had been hindered in the past by two factors. One was that, unless one uses animals such as "nude" mice, which lack an immunological system, the transplanted cells are ordinarily rejected by the host within a short period of time. The second factor was that it is extremely difficult to evaluate the performance of transplanted cells *in vivo* as far as their differentiation and integration into the host is concerned. In most cases, to evaluate such functions one has to kill the host, thereby ending the experiment. In the case of *in vivo* transplantation of cultured corneal endothelial cells, however, both these factors could be obviated.

The corneal endothelium *in vivo* is found at the interface between the corneal stroma and the aqueous humor, where it forms the inner lining of the anterior chamber and of the cornea. This location is among the most privileged immunological sites which exist *in vivo* and thereby severely limits, when compared to other sites of transplantation, the possibility of graft rejection. The second factor could also be obviated if one takes into account the physiological function of the corneal endothelium. This tissue, which functions as a transport pump, maintains the corneal stroma in a permanent state of deturgescence. It thereby prevents corneal edema and ensures the transparency of the corneal stroma. If the endothelium is no longer functional as a result of increased retention of water in the stroma, corneal thickness increases and corneal edema will develop. Therefore, there is a simple test for evaluating the physiological performance of the transplanted endothelium. If it performs its normal function, the cornea of the host will stay clear, corneal thickness will be normal, and no edema will develop. If the transplanted endothelium fails to perform its normal function, the cornea of the host will become edematous and increase in thickness within a matter of days.

Studies on corneal endothelial cell transplantation, in addition to their theoretical implications, may also have considerable potential for further clinical applications, since if corneal endothelial cells grown in tissue culture for a considerable number of generations remained functional, they could be used to replace *in vivo* either a malfunctioning endothelium or areas with extensive endothelial denudation resulting from an injury. This would be of particular importance in species such as primates whose corneal endothelium has a limited regenerative capacity.

Studies done by Maurice *et al.* (1979), Jumblatt *et al.* (1978), and Gospodarowicz *et al.* (1979b,c) have demonstrated that such transplantations are feasible. Using cultured rabbit corneal endothelial cells, Jumblatt *et al.* (1978) have reported successful transplantations into homologous recipients (rabbits eye). Gospodarowicz *et al.*, 1979b,c) have reported the successful transplantation of bovine corneal endothelial cells into heterologous recipients (rabbits and cats). In both cases, the transplanted cells resumed their normal functions and were capable of maintaining the corneal stroma of the recipient in a permanent stage of deturgescence. Identification of the transplanted cells after various periods of time in the eye of the recipient was made either by putting the cells back into culture and karyotyping them (Gospodarowicz *et al.*, 1979b,c,e; Gospodarowicz and Greenburg, 1979b) or by autoradiography of *en face* preparations of transplanted endothelium. In the latter case, cells were labeled with [^{3}H]thymidine before transplantation (Jumblatt *et al.*, 1978).

One of the main implications resulting from such studies has been the demonstration that cultured cells maintained *in vitro* and passaged repeatedly for long periods of time can stay fully differentiated and resume their normal function when put back in an *in vivo* environment. Our ability to transplant heterologous endothelium in various species could also increase our understanding of the factors which control their proliferation *in vivo*. Such factors may be intrinsic to the cells, which may lose their ability to proliferate early in their life, or they may be found in the limiting environment (aqueous humor) to which the cells are exposed and/or in the substrate upon which the cells rest and which may or may not favor their proliferation. This could be explored by transplanting cultured cells into various species and analyzing how they react to a wound situation.

VI. THE CONTROL BY GROWTH FACTORS OF CORNEAL STROMAL FIBROBLASTS

The stroma, which constitutes approximately 90% of the corneal thickness, is a dense body of collagenous connective tissue with a highly regular organization. Approximately 5% of the bulk is stroma cells (keratocytes) (Hogan *et al.,* 1971). There is no evidence that keratocytes of a mature, healthy cornea divide, and turnover of corneal collagen is negligible also. However, many keratocytes can be labeled with [^{3}H]thymidine in the late fetal stage in mice and during the first postnatal week or 10 days. Apparently, cell multiplication ceases early in life. Metabolic activity is minimal in mature keratocytes as well, but during wound healing they become extremely active and altered in morphology (La Tessa and Ross, 1964). Should a subpopulation of keratocytes be lost as a result of injury in the adult, the remaining keratocytes will divide to replace those lost (Harris, 1965). The effect of growth factors on keratocyte proliferation *in vivo* has been studied only in the context of the wound-healing process. Weimar and Haraguchi (1979) have reported that, in organ culture, MGF stimulates fibroblast formation, mitosis, and migration into the dead cell zone adjacent to the wound edge in rat (Weimar, 1979) and rabbit corneal buttons (Rich *et al.,* 1979). With the exception of this study, no serious attempt has yet been made to define the role of other growth factors in the wound-healing process or their eventual role in development of the corneal stroma in the early stages of embryogenesis.

Only a few reports on the effect of growth factors on the proliferation of keratocytes *in vitro* have been made. Among these are the observations of Zetter *et al.* (1977), who studied the effect of EGF and FGF on the proliferation of cultured rabbit keratocytes. While EGF had little effect on subconfluent cultures, FGF had a mitogenic effect which was potentiated by insulin and dexamethasone. The protease thrombin had a pronounced synergistic effect when added together with FGF, while it had little or no effect when added with EGF. This led to the conclusion that cultured keratocytes were under the control of FGF and thrombin and responded poorly to EGF whether or not thrombin was present.

The paucity of either *in vivo* or *in vitro* data on the effect of other growth factors on the ability of stromal keratocytes to proliferate under either *in vivo* or *in vitro* conditions reflects their inherent ability, when activated, to proliferate actively under both conditions. *In vivo,* keratocytes located at the edges of a wound proliferate readily and actively. It is therefore difficult to evaluate whether growth factors would have a significant effect on a process which already proceeds at nearly maximal speed. Likewise, even in sparse cultures keratocytes proliferate actively and are in fact one of the main cellular contaminants present in either corneal endothelial or epithelial cell cultures. This ability to proliferate, which is already optimal, is therefore not improved by the addition of growth factors. In order to observe a significant effect of growth factors, one has to maintain the cells under suboptimal conditions, and extrapolation of such studies to *in vivo* situations, where environmental conditions are supposed to be optimal, is therefore hazardous at best.

VII. THE CONTROL OF PROLIFERATION OF CORNEAL EPITHELIAL CELLS BY GROWTH FACTORS

The corneal epithelium, which is of ectodermal origin, occupies 7% of the total corneal thickness. It rests on a basement membrane, itself in close contact with Bowman's membrane, which separates it from the mesenchymal stroma. The epithelium is composed of three groups of cells: (1) a single, highly organized row of basal cells localized next to the basement membrane; these cells are tall and columnar and are linked to each other by the zona occludens and adherens, which act as a sort of belt, encircling the entire cell and binding it firmly to neighboring cells; the highest mitotic rate is found in this cell layer, which has a turnover rate of 7 days; (2) a group of winged or polygonal cells forming an intermediate zone two to three cells thick; these cells are no longer in contact with the basement membrane and no longer divide; (3) finally, two or three layers of flattened, platelike surface cells covering the intermediate layer (Hogan *et al.,* 1971).

Epithelial cell layers of the cornea are known to respond to growth factors *in vivo.* The effect of EGF on the regeneration of epithelial cells of the cornea maintained in organ culture and *in vivo* was first reported independently by Frati *et al.* (1972) and by Savage and Cohen (1973). They demonstrated that EGF stimulated the proliferation of rabbit corneal epithelium both in organ culture and *in vivo* and that the topical application of this factor to the eyes of rabbits with corneal wounds resulted in marked hyperplasia of the epithelial layers as well as acceleration of the wound-healing process. It is likely that the substrate (basement membrane) upon which the basal cell layer rests is a crucial factor in allowing the cells to proliferate actively and to respond to various growth factors. That this could be the case can be inferred from studies dealing with the mitogenic effect of EGF on other epithelia, such as skin epithelia (Cohen, 1965). In the case of chick skin, for example, when the isolated epidermis is maintained under organ culture conditions, an EGF effect can be observed provided that the basal cell layer rests on a Millipore filter, collagen, or killed dermis, but not when it rests on plastic. The importance of cell attachment to a proper substrate for the response of basal cells to EGF can further be demonstrated by inversion of the basal and peridermal layers. When the chick epidermis is isolated from the dermis and cultured on a filter with the basal cell layer attached to it, EGF is highly mitogenic (Cohen, 1965). If the epidermis is inverted so that its upper surface is in close contact with the filter while the basal cells become the apical cell layer, EGF is no longer active. This suggests that either the upper epidermal cells do not attach effectively to the filter or, if they do attach, they are no longer sensitive to EGF. It also suggests that the loss of sensitivity of the basal cell layer to EGF could be directly linked to its inability to attach to the proper substrate. Similar conclusions were made using corneal rather than skin epithelium as a target tissue (Gospodarowicz *et al.,* 1977a, 1978b, 1979f).

Since epithelial cells of the cornea can also be maintained easily in tissue culture, i.e., under conditions in which they are no longer in contact with their mesenchymal element or their basement membrane, the corneal epithelium represents an ideal cell type for a comparison of the sensitivity to growth factors of cells associated or dissociated from their stroma.

Individual or aggregated basal bovine corneal epithelial cells attached to the bottoms of plastic tissue culture dishes and formed colonies which grew slowly as a monolayer of flattened cells. When the growth-promoting effects of FGF and EGF were tested using the incorporation of [^{3}H]thymidine into DNA or an increase in cell number as a criterion, it was observed that only FGF was capable of stimulating both initiation of DNA synthesis and cell proliferation. In contrast, the addition of EGF had no effect on [^{3}H]thymidine incorporation in these cultures, nor did it have any effect on cell proliferation (Gospodarowicz *et al.,* 1977a).

One likely explanation for the lack of effect of EGF on cultured corneal epithelial cells is that cells maintained under such conditions lose their EGF receptor sites. That this is not the case has been demonstrated by the observation that corneal epithelial cells maintained in tissue culture specifically bind EGF (half-maximal binding at 7×10^{-10} M, total binding capacity 250 pg/10^6 cells) and are capable of internalizing and degrading such EGF receptor complexes (Gospodarowicz *et al.,* 1977a). Therefore, although corneal epithelial cells exhibit specific EGF receptor sites and can process EGF receptor complexes in ways similar to those described for other cell types which do respond to EGF, cultured corneal epithelial cells will not respond to it.

In contrast to the situation encountered under tissue culture conditions, when corneas from bovine fetuses were maintained in organ culture, the addition of EGF to the culture medium resulted, as already reported by Savage and Cohen (1973), in marked enhancement of the proliferation of the epithelium, while FGF had little or no effect (Gospodarowicz *et al.,* 1977a).

A comparative study of the mitogenic responses of bovine corneal epithelium to FGF and EGF under different culture conditions (tissue versus organ culture) therefore leads to the conclusion that, although corneal epithelial cells in tissue culture are responsive to FGF but not to EGF when maintained under organ culture conditions, EGF becomes a strong mitogen, whereas FGF is far less potent.

The most obvious difference between corneal epithelial cells maintained in tissue culture versus organ culture is in the substrate upon which the cells rested. In tissue culture, the epithelial cells proliferated on plastic as a collection of individual cells which had lost their orientation to one another and had a flattened appearance, while in organ culture, as *in vivo,* the basal cell layer of the corneal epithelium rested on Bowman's membrane, which is composed of collagen associated with proteoglycans. Since it had already been shown by others that the epidermal cells of the skin or the cornea proliferated and differentiated in response to EGF when maintained in tissue culture on collagen (Liu and Karasek, 1978)) or on a feeder layer of fibroblasts (Rheinwald and Green, 1977; Sun and Green, 1977), a comparison was thus made between the effect of collagen and that of fibroblasts on the mitogenic response of corneal epithelial cells to EGF.

When corneal epithelial cells were plated on collagen-coated dishes, the addition of FGF to the culture did not result in any marked increase in the rate of cell proliferation, but the cultures responded to EGF with both a marked increase in cell number and an increased rate of keratinization which reflected the increase in cell number. It was also observed that a noticeable change in the shape of the cells occurred when they were maintained on collagen as opposed to plastic. On plastic, the cells remained flattened, while on collagen they were rounded. After their first week in culture, cells maintained on collagen formed a monolayer of packed, tall, and columnar cells. Multiple layers began to form during the second week, the second and third cell layers having the appearance of winged cells. During the third week, keratinization occurred in the upper cell layers, as indicated by the appearance of keratinized cellular envelopes and positive histochemical staining with rhodamine B (Gospodarowicz *et al.,* 1978b, 1979f). Basal cells maintained on collagen could therefore mimic the complete sequence of events occurring in the corneal epithelium (differentiation into winged and plated cells), while they did not progress beyond the stage of a monolayer when maintained on plastic (Gospodarowicz *et al.*, 1977a, 1978b).

It is probable that the primary factor determining the proliferation of basal cells is whether or not the cells are in contact with their basement membrane, which can be produced *in vitro* by the corneal epithelium itself when the latter is maintained under the proper conditions. Dodson and Hay (1974) have demonstrated that isolated corneal epithelium grown on killed lens capsules produces a collagenous stroma *in vitro* reminiscent of that formed *in vivo,* while Meier and Hay (1974a,b) have reported that any type of collagenous substrate tested supports the efforts of the corneal epithelium to produce a stroma. On glass, plastic, and substrates of noncollagenous proteins, however, the corneal epithelium fails to produce a stroma.

This result further emphasizes the role of the substrate (either the basement membrane produced by corneal epithelial cells or Descemet's membrane produced by corneal endothelial cells) in determining cellular response to growth-promoting agents.

The corneal stroma could also play a significant role in supporting the growth of cultured epithelial cells. As shown by Sun and Green (1977), with proper fibroblast support, such as lethally irradiated 3T3 cell monolayers, human epidermal cells (keratinocytes) can be grown serially in cell culture, where they show many of the differentiation markers characteristic of this cell type *in vitro*. Single cells give rise to colonies, each forming a stratified squamous epithelium containing multiplying and differentiating cells. The cells synthesize keratins and grow in size, eventually forming cornified (cross-linked) envelopes. The final stages of epidermal differentiation are best observed by placing the cells in suspension—a condition which does not permit them to grow. Instead, they become permeable to trypan blue and their keratin filaments become detergent-insoluble, forming cross-linked envelopes.

Human and rabbit corneal epithelial cells seeded with lethally irradiated 3T3 cells were shown to respond in a similar manner and formed expanding colonies that displaced the 3T3 cells from the vessel surface. In the presence of EGF, the cells of large colonies of corneal keratinocytes sustain their growth rate much better than in its absence (Fig. 7). Such an effect of EGF is to be expected from its action on intact corneal epithelium (Frati *et al.*, 1972; Savage and Cohen, 1973). Conjunctival epithelial cells were similar to corneal epithelial cells in their colony-forming properties and EGF responsiveness. When examined under a phase-contrast microscope, the basal cells of colonies of corneal epithelial cells seemed similar to those of epidermal cells in their shape and mosaic-like arrangement (Fig. 8). The stratified struc-

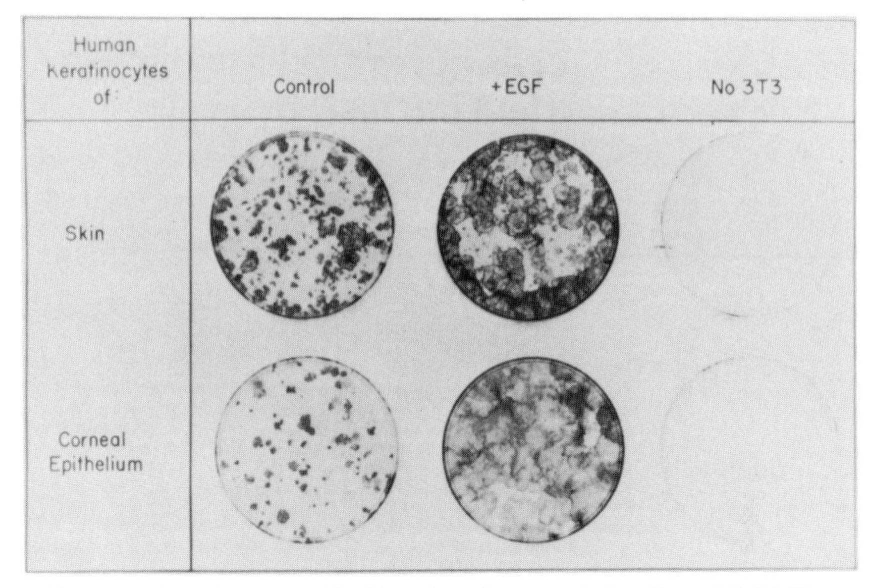

Fig. 7. Colony formation, fibroblast dependence, and the effect of EGF. Human corneal epithelial cells (5×10^4, strain B, donor aged 25 hr, secondary culture) and epidermal cells (5×10^4, strain M, newborn donor culture) were plated in 60-mm dishes with or without 4×10^5 lethally irradiated 3T3 cells. To some dishes, DGF was added to a final concentration of 15 ng/ml, starting on the third day after plating. After a total of 11 days the cultures were fixed with 10% Formalin and stained with rhodanile blue. (From Sun and Green, 1977, reprinted with the permission of the authors.)

ture of the colonies, and the silver-staining properties of the superficial cells were also very similar (Sun and Green, 1977).

This similar phenotype exhibited by the cultured keratinocytes of skin, conjunctival, or corneal origin, is contrary to what is observed *in vivo* and might best be explained by either the absence of specific fibroblast instruction or the presence of common instruction provided by the feeder layer. The possibility exists that any one of these keratinocyte types, especially in the human, when placed at the *in vivo* site of another one, will conform in phenotype, even after birth, to the other keratinocyte type and generate the site-specific epithelium. This was analyzed by Doran and Sun (1979) and by Sun *et al*. (1980), who transplanted cultured corneal epithelial cells into athymic mice and compared their differentiation to that of transplanted epithelial cells derived from either the skin or the esophagus. Under these conditions, all three transplanted epithelia faithfully regained their *in vivo* differentiated states, as judged by both morphological and histochemical

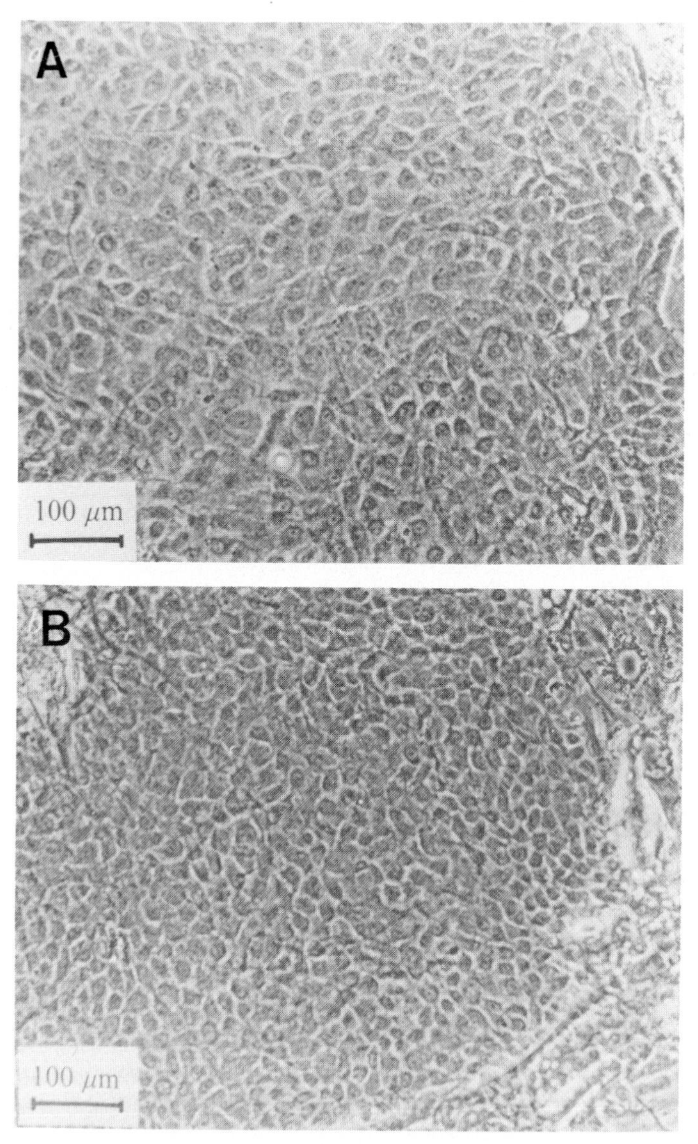

Fig. 8. Single colonies of corneal epithelial and epidermal cells. Primary colony of corneal epithelial cells of strain B(A) and secondary colony of epidermal cells of strain M (B) 11 days after inoculation. The cultures were fixed with buffered 10% Formalin and photographed directly without staining. One margin of each colony appears on the right-hand side of the picture. Phase-contrast microscopy. (From Sun and Green, 1977, reprinted with the permission of the authors.)

criteria; thus the injected (cultured) epidermal cells formed a keratinized cyst, corneal epithelial cells formed a nonkeratinized cyst, and esophageal epithelial cells formed a parakeratinized cyst. Hence the epithelia of skin, cornea, and esophagus do not assume the same morphological differentiation even when provided with identical *in vivo* environments, although when cultured *in vitro* they lose most of their *in vivo* differentiative properties and become less distinguishable (Sun and Green, 1977; Sun *et al.*, 1980). Such results suggest that these three epithelia have diverged significantly from each other during development and that intrinsic divergence plays a major role in regulating their *in vivo* differentiated states. On the other hand, the fact that epithelium cultivated from a given tissue could faithfully regain its *in vivo* differentiated states under appropriate environmental conditions demonstrates that, under certain conditions, external modulation may also play an important role in regulating epithelial differentiation.

VIII. CONCLUSION

Growth factors have been shown to affect, at least *in vitro,* the proliferation of all cell types derived from the various tissues which compose the cornea. They have proved to be useful tools in developing corneal cells in culture, particularly corneal endothelial cells. Although studies on the mode of action of growth factors are still in their infancy, they have provided insight into the role and importance of the various basement membranes upon which corneal cells rest in supporting their proliferation and eventual differentiation.

The number of biological problems which can be studied using only cultured corneal endothelial cells is truly astonishing. The following is a list of only some of the questions which can now be explored using cultured corneal endothelial cells: the cause(s) and mechanisms leading to the acquisition of cell surface polarity; the biochemical characteristics of nonthrombogenic surfaces; the mechanisms underlying the inhibition of cell proliferation, so well exemplified by the corneal endothelium *in vivo;* the involvement of various cytoskeletal elements in controlling corneal endothelial cell shape and the reorganization of these elements when cells pass from an active proliferative stage to a resting stage, as reflected in the formation of a confluent monolayer composed of tightly packed and nonoverlapping cells; the cell surface proteins involved in active cell migration versus cell attachment; the relationship between cell shape and a permeability barrier function; the kind of basal lamina produced by cells and how this production is related to cell attachment, migration, proliferation, and differentiation; and the cause(s) of cell senescence *in vitro*, the role of the basal lamina in cell senescence, and how growth factors can delay this senescence.

In the field of pharmacology, cultured corneal endothelial cells can be used to test various drugs used to combat ocular disease and to test their side effects. In the field of surgery, one can foresee the day when a diseased and incapacitated endothelium is replaced by cells grown in tissue culture. Because of the progress which will, in all likelihood, be made in the next few years in the culture of corneal endothelial cells, it is likely that the long-term culture of human corneal endothelial cells will be established. Provided that adequate controls of the karyotypes of these cells are made, and provided that their preneoplastic transformation in tissue culture is carefully assessed, one could use cultured endothelial cells as an endless supply to replace this tissue *in vivo* when it is no longer functional. This is even more likely in view of the foreseen shortage of corneas available for transplantation. This predicted shortage will be due primarily to the increased longevity of the population, which will result in both a diminished supply of young cornea donors and in an increased demand for cornea transplants on the part of older individuals. The transplantation of cultured corneal endothelial cells for experimental purposes can be considered not only on a homologous basis but on a heterologous basis as well. This is mostly due to the unique location of the corneal endothelium, which protects it from immediate rejection.

Cultured corneal epithelial cells can also be used for the study of factors involved in the control of their proliferation and differentiation. Such studies could help us to understand how corneal epithelial cells differ from other epithelia, such as skin epithelium, and the role environmental or stromal factors can play in supporting their differentiated characteristics.

However, the *in vivo* significance of growth factors other than EGF is currently not as clear as their role *in vitro* in controlling cell proliferation. Although EGF has been shown to promote regeneration of the corneal epithelium, there have as yet been no systematic studies *in vivo* on the role of other growth factors in promoting the wound-healing processes of either the corneal epithelium or endothelium, nor have there been any studies on their possible role in the developmental processes of the ocular system. It is likely that these areas of investigation will be broached in the coming years, so that we will know whether growth factors should be considered only as tools to be used *in vitro* in the development of new cell types of the ocular system or whether they have similar effects *in vivo,* thereby qualifying as physiological agents.

REFERENCES

Armelin, H. A. (1973). Pituitary extracts and steroid hormones in the control of BALB/c 3T3 cell growth. *Proc. Natl. Acad. Sci. U.S.A.* **70,** 2702-2706.

Arruti, C., and Courtois, Y. (1978). Morphological changes and growth stimulation of bovine epithelial lens cells by a retinal extract *in vitro*. *Exp. Cell Res.* **117,** 283-295.

Barnes, D., and Sato, G. (1980). Serum-free cell culture: A unifying approach. *Cell* **22,** 649-655.

Barritault, D., Arruti, C., and Courtois, Y. (1981). Is there an ubiquitous growth factor in the eye? Proliferation induced in different cell types by eye-derived growth factor(s). *Differentiation (Berlin)* (in press).

Bernfield, M. R. (1978). The cell periphery in morphogenesis. *Int. Congr. Ser. Excerpta Med., No. 423 1978*, pp. 111-125.

Bornstein, P., and Sage, H. (1980). Structurally distinct collagen types. *Annu. Rev. Biochem.* **49,** 957-1003.

Bottenstein, J., Hayashi, I., Hutchings, S., Masui, H., Mather, J., McClure, D., Ohasa, S., Rizzino, A., Sato, G., Serrero, G., Wolfe, R., and Wu, R. (1979). The growth of cells in serum free hormone-supplemented media. *Methods Enzymol.* **58,** 94-109.

Brown, G., Mahley, R., and Assmann, G. (1976). Swine aortic smooth muscle cells in tissue culture: Some effects of purified swine lipoproteins on cell growth and morphology. *Circ. Res.* **39,** 415-424.

Carpenter, G., and Cohen, S. (1979). Epidermal growth factor. *Annu. Rev. Biochem.* **48,** 193-218.

Clark, J. F., Jones, K. L., Gospodarowicz, D., and Sato, G. H. (1972). Hormone dependent growth response of a newly established rat ovarian cell line. *Nature (London), New Biol.* **236,** 180-182.

Cohen, S. (1959). Purification and metabolic effects of a nerve growth-promoting proteins from snake venom. *J. Biol. Chem.* **234,** 1129-1137.

Cohen, S. (1962). Isolation of a mouse submaxillary gland protein accelerating incisor eruption and eyelid opening in the new-born animal. *J. Biol. Chem.* **237,** 1555-1562.

Cohen, S. (1965). The stimulation of epidermal cell proliferation by a specific protein (EGF). *Dev. Biol.* **12,** 394-407.

Cohen, S., and Carpenter, G. (1975). Human epidermal growth factor: Isolation and chemical and biological properties. *Proc. Natl. Acad. Sci. U.S.A.* **72,** 1317-1321.

Corvol, M. T., Malemud, C. J., and Sokoloff, L. (1972). A pituitary growth-promoting factor for articular chondrocytes in monolayer culture. *Endocrinology (Baltimore)* **90,** 262-271.

Daughaday, W. H. (1977). Hormonal regulation of growth by somatomedin and other tissue growth factors. *Clin. Endocrinol. Metab.* **6,** 117-135.

Dodson, J. W., and Hay, E. D. (1974). Secretion of collagen by corneal epithelium. II. Effect of the underlying substratum on secretion and polymerization of epithelial cell products. *J. Exp. Zool.* **189,** 51-65.

Doran, T. I., and Sun, T. T. (1979). The specificity of rabbit corneal and skin epithelial cells. *Invest. Ophthalmol. Vis. Sci.,* **18,** Suppl. p. 8 (Abstr.).

Duthu, G. S., and Smith, J. R. (1980). *In vitro* proliferation and lifespan of bovine aorta endothelial cells: Effect of culture conditions and fibroblast growth factor. *J. Cell. Physiol.* **103,** 385-392.

Fehrenbacher, L., Gospodarowicz, D., and Shuman, M. A. (1979). Synthesis of plasminogen activator by bovine corneal endothelial cells. *Exp. Eye Res.* **29,** 219-228.

Fischer-Dzoga, K., Fraser, R., and Wissler, R. W. (1976). Stimulation of proliferation in stationary primary cultures of monkey and rabbit aortic smooth muscle cells. 1. Effects of lipoprotein fractions of hyperlipemic serum and lymph. *Exp. Mol. Pathol.* **24,** 346-359.

Foidart, J. M., and Reddi, A. H. (1980). Immunofluorescent localization of type IV collagen and laminin during endochondral bone differentiation and regulation by pituitary growth hormone. *Dev. Biol.* **75,** 130-136.

Foidart, J. M., Bere, E. W., Yaar, M., Rennard, W. I., Gullino, M., Martin, G. R., and Katz, S. I. (1980). Distribution and immunoelectron microscopic localization of laminin, a noncollagenous basement membrane glycoprotein. *Lab. Invest.* **42,** 336-342.

Folkman, J. (1977). Conformational control of cell and tumor growth. *Recent Adv. Cancer Res. Cell Biol., Mol. Biol., Tumor Virol.* **1,** 119-130.

Folkman, J., and Moscona, A. (1978). Role of cell shape in growth control. *Nature (London)* **273,** 345-349.

Frati, L., Daniele, S., Delogu, A., and Covelli, I. (1972). Selective binding of the epidermal growth factor and its specific effect on the epithelial cells of the cornea. *Exp. Eye Res.* **14,** 135-149.

Frey, P., Forand, R., Maciag, T., and Shooter, E. M. (1979). The biosynthetic precursor of epidermal growth factor and the mechanism of its processing. *Proc. Natl. Acad. Sci. U.S.A.* **76,** 6294-6298.

Giguere, L., Cheng, J., and Gospodarowicz, D. (1982). Factors involved in the control of proliferation of bovine corneal endothelial cells maintained in serum-free medium. *J. Cell. Physiol.* **110,** 72-80.

Goldminz, D., Vlodavsky, I., Johnson, L. K., and Gospodarowicz, D. (1979). Contact inhibition in the regulation of the permeability barrier function of the corneal endothelium: Correlation with the appearance of a fibronectin meshwork. *Exp. Eye Res.* **29,** 331-351.

Gospodarowicz, D. (1974). Localization of a fibroblast growth factor and its effect alone and with hydrocortisone on 3T3 cell growth. *Nature (London)* **249,** 123-127.

Gospodarowicz, D. (1975). Purification of a fibroblast growth factor from bovine pituitary. *J. Biol. Chem.* **250,** 2515-2519.

Gospodarowicz, D. (1979). Fibroblast and epidermal growth factors: Their uses *in vivo* and *in vitro* in studies on cell functions and cell transplantation. *Mol. Cell. Biochem.* **25,** 79-110.

Gospodarowicz, D. (1980). The extracellular matrix and the control of cell proliferation. *In* "Neurosecretion and Brain Peptides: Implications for Brain Function and Neurological Disease" (J. B. Marin *et al.,* eds.), Vol. 28, pp. 243-262. Raven, New York.

Gospodarowicz, D., and Bialecki, H. (1978). The effects of the epidermal and fibroblast growth factor on the replicative lifespan of bovine granulosa cells in culture. *Endocrinology (Baltimore)* **103,** 854-858.

Gospodarowicz, D., and Fujii, D. K. (1981). The extracellular matrix and the control of cell proliferation and differentiation. *In* "Cellular Responses to Molecular Modulators," Miami Winter Symp. Vol. 18 (L. W. Mozes *et al.*, eds.), pp. 113-135. Academic Press, New York.

Gospodarowicz, D., and Greenburg, G. (1979a). The effects of epidermal and fibroblast growth factors on the repair of corneal endothelial wounds in bovine corneas maintained in organ culture. *Exp. Eye Res.* **28**, 147-157.

Gospodarowicz, D., and Greenburg, G. (1979b). The coating of bovine and rabbit corneas denuded of their endothelium with bovine corneal endothelial cells. *Exp. Eye Res.* **28**, 249-265.

Gospodarowicz, D., and Greenburg, G. (1981). Growth control of mammalian cells. Growth factors and extracellular matrix. *In* "The Biology of Normal Human Growth," Proceedings of the First Karolinska Institute Nobel Conference (M. Ritzen *et al.*, eds.), pp. 1-21. Raven, New York.

Gospodarowicz, D., and Ill, C. R. (1980a). Do plasma and serum have different abilities to promote cell growth? *Proc. Natl. Acad. Sci. U.S.A.* **77**, 2726-2730.

Gospodarowicz, D., and Ill, C. R. (1980b). The extracellular matrix and the control of proliferation of vascular endothelial cells. *J. Clin. Invest.* **65**, 1351-1364.

Gospodarowicz, D., and Ill, C. R. (1980c). The extracellular matrix and the control of proliferation of corneal endothelial and lens epithelial cells. *Exp. Eye Res.* **31**, 181-199.

Gospodarowicz, D., and Lui, G.-M. (1981). Effect of substrata and fibroblast growth factor on the proliferation *in vitro* of bovine aortic endothelial cells. *J. Cell. Physiol.* **109**, 69-81.

Gospodarowicz, D., and Mescher, A. (1980a). Fibroblast growth factor and the control of vertebrate regeneration and repair. *Ann. N. Y. Acad. Sci.* **339**, 151-174.

Gospodarowicz, D., and Mescher, A. (1980b). Fibroblast growth factor and vertebrate regeneration. *Adv. Neurol.* **29**, 149-171.

Gospodarowicz, D., and Tauber, J.-P. (1980). Growth factors and extracellular matrix. *Endocr. Rev.* **1**, 201-227.

Gospodarowicz, D., and Zetter, B. (1977). The use of fibroblast and epidermal growth factors to lower the serum requirement for growth of normal diploid cells in early passage: A new method for cloning. *Jt. WHO/IABS Symp. Stand. Cell Substrates Prod. Virus Vaccine, Proc. Symp, 62nd, 1976* pp. 109-130.

Gospodarowicz, D., Mescher, A. L., and Birdwell, C. R. (1977a). Stimulation of corneal endothelial cell proliferation *in vitro* by fibroblast and epidermal growth factors. *Exp. Eye Res.* **25**, 75-89.

Gospodarowicz, D., Mescher, A. L., Brown, K., and Birdwell, C. R. (1977b). The role of fibroblastic growth factor and epidermal growth factor in the proliferative response of the corneal and lens epithelium. *Exp. Eye Res.* **25**, 631-649.

Gospodarowicz, D., Bialecki, H., and Greenburg, G. (1978a). Purification of the fibroblast growth factor activity from bovine brain. *J. Biol. Chem.* **253**, 3736-3743.

Gospodarowicz, D., Greenburg, G., and Birdwell, C. R. (1978b). Determination of cellular shape by the extracellular matrix and its correlation with the control of cellular growth. *Cancer Res.* **38**, 4155-4171.

Gospodarowicz, D., Mescher, A. L., and Birdwell, C. R. (1978c). Control of cellular proliferation by the fibroblast and epidermal growth factors. *Natl. Cancer Inst. Monogr. No. 48,* pp. 109-130.

Gospodarowicz, D., Mescher, A. L., and Moran, J. (1978d). Cellular specificities of fibroblast growth factor and epidermal growth factor. *35th Symp. Soc. Dev. Biol.* pp. 33-61.

Gospodarowicz, D., Greenburg, G., Bialecki, H., and Zetter, B. (1978e). Factors involved in the modulation of cell proliferation *in vivo* and *in vitro:* The role of fibroblast and epidermal growth factors in the proliferative response of mammalian cells. *In Vitro* **14**, 85-118.

Gospodarowicz, D., Bialecki, H., and Thakral, T. K. (1979a). The angiogenic activity of the fibroblast and epidermal growth factor. *Exp. Eye Res.* **28**, 501-514.

Gospodarowicz, D., Greenburg, G., and Alvarado, J. (1979b). Transplantation of cultured bovine corneal endothelial cells to rabbit cornea: Clinical implications for human studies. *Proc. Natl. Acad. Sci. U.S.A.* **76**, 464-468.

Gospodarowicz, D., Greenburg, G., and Alvarado, J. (1979c). Transplantation of cultured bovine corneal endothelial cells to species with nonregenerative endothelium: The cat as an experimental model. *Arch. Ophthalmol. (Chicago)* **97**, 2163-2169.

Gospodarowicz, D., Greenburg, G., Vlodavsky, I., Alvarado, J., and Johnson, L. K. (1979d). The identification and localization of fibronectin in cultured corneal endothelial cells: Cell surface polarity and physiological implications. *Exp. Eye Res.* **29**, 485-509.

Gospodarowicz, D., Vlodavsky, I., Greenburg, G., Alvarado, J., and Johnson, L. K. (1979e). Studies on atherogenesis and corneal transplantation using cultured vascular and corneal endothelia. *Rec. Progr. Horm. Res.* **35**, 375-448.

Gospodarowicz, D., Vlodavsky, I., Greenburg, G., and Johnson, L. K. (1979f). Cellular shape is determined by the extracellular matrix and is responsible for the control of cellular growth and function. *Cold Spring Harbor Conf. Cell Proliferation* **6**, 561-592.

Gospodarowicz, D., Delgado, D., and Vlodavsky, I. (1980a). Permissive effect of the extracellular matrix on cell proliferation *in vitro*. *Proc. Natl. Acad. Sci. U.S.A.* **77**, 4094-4098.

Gospodarowicz, D., Vlodavsky, I., and Savion, N. (1980b). The extracellular matrix and the control of proliferation of vascular endothelial and vascular smooth muscle cells. *J. Supramol. Struct.* **13**, 339-372.

Gospodarowicz, D., Vlodavsky, I., Savion, N., and Tauber, J.-P. (1980c). The control of proliferation and differentiation of vascular endothelial cells by fibroblast growth factor. *Soc. Gen. Physiol. Ser.* **35**, 1-38.

Gospodarowicz, D., Vlodavsky, I., and Savion, N. (1981a). The role of fibroblast growth factor and the extracellular matrix in the control of proliferation and differentiation of corneal endothelial cells. *Vision Res.* **21**, 87-103.

Gospodarowicz, D., Greenburg, G., Foidart, J.-M., and Savion, N. (1981b). The production and localization of laminin in cultured vascular and corneal endothelial cells. *J. Cell. Physiol.* **107**, 173-183.

Gospodarowicz, D., Hirabayashi, K., Giguere, L., and Tauber, J.-P. (1981c). Control of proliferation of vascular smooth muscle cells by high and low density lipoproteins, insulin, and by fibroblast and epidermal growth factors. *J. Cell Biol.* **89**, 568-578.

Green, H. (1977). Terminal differentiation of cultured human epidermal cells. *Cell* **11**, 405-416.

Green, H., Kehinde, O., and Thomas, J. (1979). Growth of cultured human epidermal cells into multiple epithelia suitable for grafting. *Proc. Natl. Acad. Sci. U.S.A.* **76**, 5665-5668.

Greenburg, G., Vlodavsky, I., Foidart, J.-M., and Gospodarowicz, D. (1980). conditioned medium from medium from endothelial cell cultures can restore the normal phenotypic expression of vascular endothelium maintained *in vitro* in the absence of fibroblast growth factor. *J. Cell. Physiol.* **103**, 333-347.

Gregory, H. (1975). Isolation and structure of urogastrone and its relationship to epidermal growth factor. *Nature (London)* **257**, 325-327.

Grobstein, C. J. (1955). Inductive interaction in the development of the mouse metanephros. *J. Exp. Zool.* **130**, 319-340.

Grobstein, C. (1967). Mechanisms of organogenetic tissue interaction. *Natl. Cancer Inst. Monogr.* **26**, 279-299.

Grobstein, C. (1975). Developmental role of intercellular matrix: Retrospective and prospective. *In* "Extracellular Matrix Influences on Gene Expression" (H. C. Slavkin and R. C. Grenlich, eds.), pp. 9-16 and 804-814. Academic Press, New York.

Harris, J. E. (1965). *In* "The Cornea First World Congress" (J. H. King and J. W. McTigue, eds.), p. 73. Butterworth, London.

Hay, E. D. (1977). Interaction between the cell surface and extracellular matrix in corneal development. *In* "Cell and Tissue Interaction" (J. W. Lash and M. M. Burger, eds.), pp. 115-137. Raven, New York.

Hay, E. D., and Revel, J. P. (1969). General outline of the development of the cornea. *In* "Fine Structure of the Developing Avian Cornea," Monographs in Developmental Biology, pp. 4-15. Karger, Basel.

Hogan, M. J., Alvarado, J. A., and Weddell, J. E. (1971). "Histology of the Human Eye. Atlas and Textbook," pp. 55-109. Saunders, Philadelphia, Pennsylvania.

Hollenberg, M. D., and Gregory, H. (1976). Human urogastrone and mouse epidermal growth factor share a common receptor site in cultured human fibroblasts. *Life Sci.* **20,** 267-274.

Holley, R. W., and Kiernan, J. A. (1968). Contact inhibition of cell division in 3T3 cells. *Proc. Natl. Acad. Sci. U.S.A. 60,* 300-304.

Isaacs, J. D., Savion, N., Gospodarowicz, D., Fenton, J. W., and Shuman, M. A. (1981). Covalent binding of thrombin to corneal endothelial cells. *Biochemistry* **20,** 398.

Jumblatt, M., Maurice, M., and McCulley, J. (1978). Transplantation of tissue-cultured corneal endothelium. *Invest. Ophthalmol. Visual Sci.* **17,** 1135-1141.

Kratochwil, K. (1972). Tissue interaction during embryonic development. *In* "General Properties in Tissue Interaction in Carcinogenesis" (D. Tarin, ed.), pp. 1-48. Academic Press, New York.

La Tessa, A. J., and Ross, M. H. (1964). Electron microscope studies of nonpenetrating corneal wounds in the early stages of healing. *Exp. Eye Res.* **3,** 298-303.

Le Douarin, N. (1980). Migration and differentiation of neural crest cells. *Curr. Top. Dev. Biol.* **16,** 31.

Liu, S.-C., and Karasek, M. (1978). Isolation and growth of adult human epidermal keratinocytes in cell culture. *J. Invest. Dermatol.* **71,** 157-172.

Mannagh, J., and Irving, A. R., Jr. (1965). Human corneal endothelium: Growth in tissue cultures. *Arch. Ophtahlmol. (Chicago)* **74,** 847.

Maroudas, N. G. (1973). Chemical and mechanical requirements for fibroblast adhesion. *Nature (London)* **244,** 353-354.

Maurice, D., McCulley, J., and Perlman, M. (1979). Development in use of cultured endothelium in corneal transplantation. *Doc. Ophthalmol. Proc. Ser.* **20,** 151-153.

Meier, S., and Hay, E. D. (1974a). Control of corneal differentiation by extracellular materials: Collagen as a promoter and stabilizer of epithelial stroma production. *Dev. Biol.* **38,** 249-264.

Meier, S., and Hay, E. D. (1974b). Stimulation of extracellular matrix synthesis in the developing cornea by glycosaminoglycans. *Proc. Natl. Acad. Sci. U.S.A.* **71,** 2310-2314.

Mescher, A., and Gospodarowicz, D. (1979). Mitogenic effect of a growth factor derived from myelin on denervated regenerates of newt forelimbs. *J. Exp. Zool.* **207,** 497-499.

Moses, A. C., Nissley, S. P., Rechler, M. M., Short, P. A., and Podskalny, J. M. (1979) The purification and characterization of multiplication stimulating activity (MSA) from media conditioned by a rat liver cell line. *In* "Somatomedins and Growth" (G. Giordano, J. J. Van Wyk, and F. Minuto, eds.), pp. 45-59. Academic Press, New York.

Murray, J. C., Stingl, G., Kleinman, H. K., Martin, G. R., and Kidwell, W. R. (1979).

Epidermal cells adhere preferentially to type IV (basement membrane) collagen. *J. Cell Biol.* **80**, 197–202.

Neno, Z., Gonzalez, R., and Gospodarowicz, D. (1982). Characterization of proteoglycans synthesized by cultured bovine corneal endothelial cells. *J. Biol. Chem.* (in press).

Nissley, S. P., and Rechler, M. M. (1978). Multiplication stimulating activity (MSA): A somatomedin-like polypeptide from cultured rat liver cells. *Natl. Cancer Inst. Monog.* **48**, 168–180.

Nissley, S. P., Rechler, M. M., Moses, A. C., Eisen, H. J., Higa, O. Z., Fennoy, I., Bruni, C. B., White, R. M., Schilling, E. E., Podskalny, J. M., and Short, P. A. (1979). *Cold Spring Harbor Conf. Cell Proliferation* **6**, 79–94.

Noden, D. M. (1980). The migration and cytodifferentiation of cranial neural cells. *In* "Current Research Trends in Prenatal Craniofacial Developments" (R. M. Pratt and R. O. Christiansen, eds.), p. 3. Elsevier, North Holland.

Perlman, M., and Baum, J. L. (1974). The mass culture of rabbit corneal endothelial cells. *Arch. Ophthalmol. (Chicago)* **92**, 235–239.

Perlman, M., Baum, J. L., and Kaye, G. I. (1974). Fine structure and collagen synthesis activity of monolayer cultures or rabbit corneal endothelium. *J. Cell Biol.* **63**, 306–308.

Pledger, W. J., and Wharton, W. (1980). Regulation of early cycle events by serum components. *In* "Control Mechanisms in Animal Cells" (L. Jimenez de Asua, ed.), p. 165. Raven, New York.

Rechler, M. M., and Nissley, S. P. (1977). Somatomedins and related growth factors. *Nature (London)* **270**, 665–666.

Rechler, M. M., Fryklund, L., Nissley, S. P., Hall, H., Podskalny, J., Skottner, A., and Moses, A. (1978). Purified human somatomedin A and rat multiplication stimulating activity. Mitogens for cultured fibroblasts that cross-react with the same growth peptide receptors. *Eurp. J. Biochem.* **82**, 5–12.

Rheinwald, J. G., and Green, H. (1977). Epidermal growth factor and the multiplication of cultured human epidermal keratinocytes. *Nature (London)* **265**, 421–424.

Rich, L. F., Weimar, V. L., Squires, E. L. *et al.* (1979). Stimulation of corneal wound healing responses *in vivo* with mesodermal growth factor. *Arch. Ophthalmol. (Chicago)* **97**, 1326–1330.

Rollins, B. J., and Culp, L. A. (1979). Preliminary characterization of the proteoglycans in the substrate adhesion sites of normal and virus-transformed murine cells. *Biochemistry* **18**, 5621–5629.

Ross, R., and Glomset, J. A. (1973). Atherosclerosis and the arterial smooth muscle cells. *Science* **180**, 1332–1339.

Rothstein, H., Van Wyk, J. J., Hayden, J. H., Gordon, S. R., and Weinsieder, A. (1980). Somatomedin C: Cell restoration of hypophysectomized animals. *Science* **208**, 410–412.

Sato, G., and Reid, L. (1978). Replacement of serum in cell cultures by hormones. *Int. Rev. Biochem.* **20**, 219–251.

Savage, C. R., Jr., and Cohen, S. (1973). Proliferation of corneal epithelium induced by epidermal growth factor. *Exp. Eye Res.* **15**, 361–372.

Savage, C. R., Jr., Inagami, T., and Cohen, S. (1972). The primary structure of epidermal growth factor. *J. Biol. Chem.* **247**, 7612–7621.

Savion, N., Isaacs, J. D., Shuman, M. A., and Gospodarowicz, D. (1981a). Internalization and degradation of thrombin and up-regulation of thrombin binding sites in cultured corneal endothelial cells. *J. Biol. Chem.* **256**, 4514–4519.

Savion, N., Lui, G.-M., Laherty, R., and Gospodarowicz, D. (1981b). Factors controlling proliferation and progesterone production by bovine granulosa cells in serum-free medium. *Endocrinology* **109,** 409–421.

Shuman, M. A., Isaacs, J. D., Maerowitz, T., Savion, N., and Gospodarowicz, D. (1981). Endothelial cell and platelet binding sites for thrombin. *Ann. N.Y. Acad. Sci.* **370,** 57–66.

Simonian, M. H., and Gill, G. N. (1979). Regulation of deoxyribonucleic acid synthesis in bovine adrenocortical cells in culture. *Endocrinology* **104,** 588–595.

Simonian, M. H., Hornsby, P. J., Ill, C. R., O'Hare, M. J., and Gill, G. (1979). Characterization of cultured bovine adrenocortical cells and derived clonal lines: Regulation of steroidogenesis and culture life span. *Endocrinology* **105,** 99–108.

Squires, E. L., and Weimar, V. L. (1980). Stimulation of repair of human corneal endothelium in organ culture by mesodermal growth factor. *Arch. Ophthalmol. (Chicago)* **98,** 1462–1466.

Stiles, C. D., Capone, G. T., Scher, C. D., Antoniades, H. M., Van Wyk, J. J., and Pledger, W. J. (1979a). Dual control of growth by somatomedin and platelet-derived growth factor. *Proc. Natl. Acad. Sci. U.S.A.* **76,** 1279–1283.

Stiles, C. D., Pledger, W. J., Van Wyk, J. J., Antoniades, H. N., and Scher, C. D. (1979b). *Cold Spring Harbor Conf. Cell Proliferation* **6,** 425–440.

Stocker, F. W. (1971). "The Endothelium of the Cornea and Its Clinical Implications," pp. 58–74. Thomas, Springfield, Illinois.

Sun, T. T., and Green, H. (1977). Cultured epithelial cells of cornea, conjunctiva, and skin: Absence of marked intinsic difference in their differentiated state. *Nature (London)* **269,** 489–492.

Sun, T. T., Doran, T. I., and Vidrich, A. (1980). An analysis of the intrinsic and external regulation of epithelial regulation. *J. Supramol. Struct. Supp.* **4,** 200. (Abstr.)

Sundell, H., Serenius, R. S., Barthe, P., Friedman, Z., Kanarek, K. S., Escobedo, M. B., Orth, D. N., and Stahlman, M. T. (1975). The effect of EGF on fetal lamb lung maturation. *Pediatr. Res.* **9,** 371–376.

Svoboda, M. E., Van Wyk, J. J., Klapper, D. G., Fellow, R. E., Grissom, F. E., and Schlueter, R. J. (1980). Purification of somatomedin C from human plasma: Chemical and biological properties, partial sequence analysis, and relationship to other somatomedins. *Biochemistry* **19,** 790–797.

Tauber, J.-P., Cheng, J., and Gospodarowicz, D. (1980). The effect of high and low density lipoproteins on the proliferation of vascular endothelial cells. *J. Clin. Invest.* **66,** 696–708.

Tauber, J.-P., Cheng, J., and Gospodarowicz, D. (1981). High density lipoproteins and the growth of vascular endothelial cells in serum-free medium. *In Vitro* **17,** 519–530.

Taylor, J. M., Mitchell, W. M., and Cohen, S. (1972). Epidermal growth factor: Physical and chemical properties. *J. Biol. Chem.* **247,** 5928–5934.

Temin, H. M., Pierson, R. W., and Dulak, N. C. (1972). The role of serum in the control of multiplication of avian and mammalian cells in culture. *In* "Growth, Nutrition and Metabolism of Cells in Culture" (G. H. Rothblat and J. Cristofalo, eds.), pp. 49–81. Academic Press, New York.

Thompson, D. (1961). *In* "On Growth and Form" (J. T. Bonner, ed.), pp. 1–64. Cambridge Univ. Press, London and New York.

Tseng, S. C., Savion, N., Stern, R., and Gospodarowicz, D. (1981). Characterization of collagens synthesized by bovine corneal endothelial cell cultures. *J. Biol. Chem.* **256,** 3361–3365.

Van Wyk, J. J., Underwood, L. E., Hintz, R. L., Clemmons, D. R., Voina, S. J., and Weaver, R. P. (1974). The somatomedins: A family of insulin-like hormones under growth hormone control. *Rec. Prog. Horm. Res.* **36,** 259–318.

Van Wyk, J. J., Svobada, M. E., and Underwood, L. E. (1980). Evidence from radioligand assays that Sm-C and IGF-I are similar to each other and different from other somatomedins. *J. Clin. Endocrinol. Metab.* **50**, 206–208.

Vlodavsky, I., and Gospodarowicz, D. (1979). Structural and functional alterations in the surface of vascular endothelial cells associated with the formation of a confluent cell monolayer and with the withdrawal of fibroblast growth factor. *J. Supramol. Struct.* **12**, 73–114.

Vlodavsky, I., Johnson, L. K., Greenburg, G., and Gospodarowicz, D. (1979). Vascular endothelial cells maintained in the absence of fibroblast growth factor undergo structural and functional alterations that are incompatible with their *in vivo* differentiated properties. *J. Cell Biol.* **83**, 468–486.

Weimar, V. L. (1979). Activation of initial wound healing responses in rat corneas in organ culture by mesodermal growth factor. *Invest. Ophthalmol.* **18**, 532–535.

Weimar, V. L., and Haraguchi, K. H. (1975). A potent new mesodermal growth factor from mouse submaxillary gland: A quantitative, comparative study with previously described submaxillary gland growth factors. *Physiol. Chem. Phys.* **7**, 7–21.

Weimar, V. L., and Haraguchi, K. H. (1979). Stimulation of mitosis and hypertrophy of corneal stromal cells by mesodermal growth factor: Influence of endothelium and epithelium. *Doc. Ophthalmol.* **18**, 325–331.

Weimar, V. L., Squires, E. L., and Knox, R. J. (1980). Acceleration of rabbit corneal endothelial wound healing by mesodermal growth factor. *Invest. Ophthalmol.* **19**, 350–361.

Weiss, L., Poste, G., Mackearnin, A., and Willett, K. (1975). Growth of mammalian cells on substrates coated with cellular microexudates: Effect on cell growth at low population densities. *J. Cell Biol.* **64**, 135–145.

Wessels, N. K. (1964). Substrate and nutrient effects upon epidermal basal cell orientation and proliferation. *Proc. Natl. Acad. Sci. U.S.A.* **52**, 252–259.

Wessles, N. K. (1977). *In* "Tissue Interaction and Development" (J. W. Lash, and M. M. Burger, eds.), pp. 213–229. W. A. Benjamim, Inc., Reading, Massachusetts.

Westall, F. C., Lennon, V. A., and Gospodarowicz, D. (1978). Brain-derived fibroblast growth factor: Identity with a fragment of the basic protein of myelin. *Proc. Natl. Acad. Sci. U.S.A.* **75**, 4675–4678.

Wicha, M. S., Liotta, L. A., Garbisa, S., and Kidwell, W. R. (1979). Basement membrane collagen requirements for attachment and growth of mammary epithelium. *Exp. Cell Res.* **124**, 181–190.

Yang, J., Richards, J., Bowman, P., Guzman, R., Eham, J., McCormick, K., Hamamoto, S., Pitelka, D., and Nandi, S. (1979). Substained growth and three-dimensional organization of primary mammary tumor. *Proc. Natl. Acad. Sci. U.S.A.* **76**, 3401–3405.

Yaoi, Y., and Kanaseki, T. (1972). Role of microexudate carpet in cell division. *Nature (London)* **237**, 283–285.

Zapf, J., Rinderknecht, E., Humbel, R. E., and Froesch, E. R. (1978). Non-suppressible insulin-like activity (NSILA) from human serum: Recent accomplishments and their physiologic implications. *Metabolism* **27**, 1803–1815.

Zetter, B. R., Sun, T. T., Chen, L. G., and and Buchanan, J. M. (1977). Thrombin potentiates the mitogenic response of cultured fibroblasts to serum and other growth promoting agents. *J. Cell. Physiol.* **92**, 233–240.

4

Ontogeny and Localization of the Crystallins in Eye Lens Development and Regeneration

DAVID S. McDEVITT AND SAMIR K. BRAHMA

I. INTRODUCTION: THE CRYSTALLINS

A. History

Berzelius (1830) was the first to investigate the proteins of the eye lens. He determined that the bovine lens contained 36% (wet weight) protein, which he termed crystallins. He considered it to be a single, tissue-specific protein. In 1894, Mörner, in his pioneering study on the

Cell Biology of the Eye

ISBN 0-12-483180-X

proteins of the bovine lens, reported the isolation of four different protein fractions from the lens: an insoluble fraction which he called albuminoid, and three (water) soluble fractions which he termed α-crystallin, β-crystallin, and albumin. (Membrane and other insoluble lens proteins, which are beyond the scope of this chapter, are treated in the chapter by Bettelheim and Siew in this volume.) Most of the workers after Mörner up until the present also used the bovine lens as experimental material because of its size and availability. Woods and Burky (1927) and Burky and Woods (1928a,b) used isoelectric precipitation to fractionate the soluble lens proteins into α-, β-, and γ-crystallin, the latter being the classical albumin fraction of Mörner (Krause, 1933). This tidy classification was confirmed by early electrophoresis analysis: Hesselvik (1939) and Viollier *et al.* (1947) used free electrophoresis, and Francois *et al.* (1954) and Bon and Nobel (1958) employed paper electrophoresis. But the introduction of more sensitive techniques since that time has permitted realization that the soluble protein composition of the eye lens is indeed much more complex. The operational use of α-, β-, and γ-crystallins has proved convenient, however, providing classes in which to place soluble lens proteins of similar physicochemical and/or immunological characteristics. In addition, sauropsidans have been found to possess another, immunologically unique, class of crystallins, δ-crystallins (Zwaan and Ikeda, 1965); and γ-crystallins have not been identified in these organisms (McDevitt and Croft, 1977). Thus, the crystallins, comprising 90% of the soluble lens proteins, have come to be classified generally on the basis of their net surface charge and molecular weight: The α-crystallins have the highest molecular weight and greatest electrophoretic mobility, and the γ-crystallins the lowest molecular weight and electrophoretic mobility, with the β-crystallins occupying a rather indeterminate middle ground. The molecular aspects of the (mammalian) crystallins have been recently reviewed by Harding and Dilley (1976) and Bloemendal (1977), and by Clayton (1974) in a comparative treatment.

B. Current Biochemical Knowledge

Most of our knowledge of the molecular aspects of the crystallins is derived from studies on mammalian lenses, especially bovine and human. When appropriate, reference will be made to other vertebrate crystallins. Primary fractionation of the crystallins is usually accomplished by gel-filtration chromatography of a lens homogenate supernatant on agarose or cross-linked dextrans, resulting in a classical profile (in order of elution from a G-200 column) of α- (including high-molecular-weight crystallins), β- (high-molecular-weight), β- (low-molecular-weight), and γ-crystallins. The proportion and degree of separation of the crystallin classes varies widely, however, among the different mammals investigated (e.g., Ramaekers *et al.*, 1979); individual variation within species cannot (fortuitously) be detected, even with the extremely sensitive technique of isoelectric focusing (Bours, 1980).

α-Crystallin (bovine) is composed of polypeptides with molecular weights of approximately 2×10^4. Two main types of polypeptide chains may be distinguished, the acidic A chains and the more basic B chains. The molecular weight of the α-crystallin aggregate is variable; newly synthesized α-crystallin contains only A_2 and B_2 species as aggregates with a approximate molecular weight of 7×10^6 (Spector *et al.*, 1968; Stauffer *et al.*, 1974), while altered A and B chains and higher-molecular-weight α-crystallin species can also be present (Spector *et al.*, 1971). These can also be found in the human lens (Jedziniak *et al.*, 1975) and are thought to be due to posttranslational changes as a result of aging. They are usually detected as subunits with sodium dodecyl sulfate (SDS) electrophoresis techniques. Thus a 24K α subunit, as well, is detectable in the rat (Cohen *et al.*, 1978); and old normal human and cataractous human α-crystallin has as many as five B and six A chains (Roy and Spector, 1976a), indicative of degradation and/or deamidation. Apparently these changes are *not* caused by an enzymatic mechanism (van Kleef *et al.*, 1975). High-molecular-weight crystallin aggregates [associated with aging and senile cataractogenesis (Spector *et al.*, 1971; Jedziniak *et al.*, 1975; and Chapter 6 of this volume)] are composed exclusively of α-crystallin subunits in bovine, but in rabbit and human (Roy and Spector, 1976b; Cohen *et al.*, 1978) β-crystallin polypeptides are present in this fraction as well. The first crude characterization of amphibian α-crystallin, as two fractions on DEAE-cellulose chromatography, was reported by McDevitt (1967). de Jong *et al.* (1976) have reported briefly on the subunit composition of purified α-crystallin from, among other phyla, three amphibians, all anurans. All showed essentially similar subunit patterns on alkaline urea-acrylamide gel electrophoresis, which were almost identical to those seen in the cow. In addition, molecular-weight determinations and peptide mapping also established homology between amphibian and bovine α-crystallins. Amphibian α-crystallins are *not* purified by a one-step passage through G-200, as is possible for many mammalian lens protein preparations. Some β-crystallins coelute with anuran (*Rana pipiens*, unpublished) and uordelan (*Notophthalmus viridescens*, McDevitt, 1981) α-crystallins.

β-Crystallin is also physically heterogeneous, in fact very heterogeneous, and this class has been notoriously difficult to purify and characterize with any degree of confidence, with reports ranging to 15 components in the chick lens (Clayton and Truman, 1974). Operationally, the situation has been simplified in mammals by the demonstration of two distinct β-crystallin populations of quite different molecular weight (Zigler and Sidbury, 1973; Herbrink and Bloemendal, 1974). These two classes have been termed β_H and β_L for the higher- and lower-molecular-weight forms, respectively, in bovine lens. *Three* β-crystallin size classes have recently been distinguished in the human lens (Kramps *et al.*, 1976; Zigler *et al.*, 1980). Again the only, albeit crude, characterization of β-crystallins from other than mammalian or chick lens until recently (McDevitt 1981) was that of McDeivtt (1967) as two fractions isolated on DEAE-cellulose chromatography of *R. pipiens* lens proteins. Zigler and Sidbury (1976) reported on the β-crystallins of fish, frog, and turtle, using Sephadex G-200 chromatography. Each was found to have the β_H and β_L classes, comprised in large part of identical polypeptides. Only partial immunochemical identity with mammalian β_H and β_L was established, however. The β-crystallins of *N. viridescens,* the eastern spotted newt, can be separated on G-200 into β_H and β_L fractions, with β_H containing six polypeptides and β_L 7 polypeptides, and values on SDS electrophoresis of 17.5K–36K (McDevitt, 1981). Seven β-crystallin polypeptides can be derived from adult chick lens and exhibit SDS values of 19K–35K (Ostrer and Piatigorsky, 1980). Zigler (1978) has recently reported the existence of age-related changes in β-crystallins similar to those seen in α-crystallins.

The γ-crystallins have been found to be the simplest class of crystallins physicochemically. Bjork (1961) found that calf γ-crystallin isolated by Sephadex G-75 chromatography was homogeneous when examined by ultracentrifugation but heterogeneous on electrophoresis. This indicated that γ-crystallin was in fact a mixture of proteins of similar molecular weight (approximately 20,000) displaying slight differences in net charge, a condition since confirmed in broad outline in rabbit (Hines and Olive, 1970), rat (Zigman, 1969), human (Spector *et al.*, 1973; Dilley and Harding, 1975), and the amphibian *R. pipiens* (McDevitt, 1967; McDevitt *et al.*, 1969). The number of γ-crystallin components resolvable by ion-exchange chromatography ranges as high as seven, but four seem to recur consistently and give virtually identical peptide maps, N-terminals (Gly-Glu), and characteristic amino acid compositions (low alanine and lysine values) (Harding and Dilley, 1976). One γ-crystallin, termed fraction II (calf), has been completely sequenced by Croft (1972).

The γ-crystallins of the adult eastern spotted newt, *N. viridescens*, have recently been separated and characterized (McDevitt, 1981). Only a single, distinct band can usually be seen on SDS electrophoresis, comigrating with the B-chain polypeptide (as has been reported for bovine γ-crystallin). The other low-molecular-weight entities seen by some other workers in calf (Kabasawa *et al.*, 1977a) could not be identified in newt γ-crystallins. Peptide mapping and amino acid composition analysis of newt γ-crystallins revealed similarities to calf γII-crystallin (Croft, 1972) (Fig. 1). Aging effects have recently been noted in γ-crystallins of the human lens (Kabasawa *et al.*, 1977b); all these crystallin classes in mammals thus exhibit such changes, earlier attributed to artefact production (Bloemendal, 1972).

δ-Crystallin, originally termed first important soluble crystallin

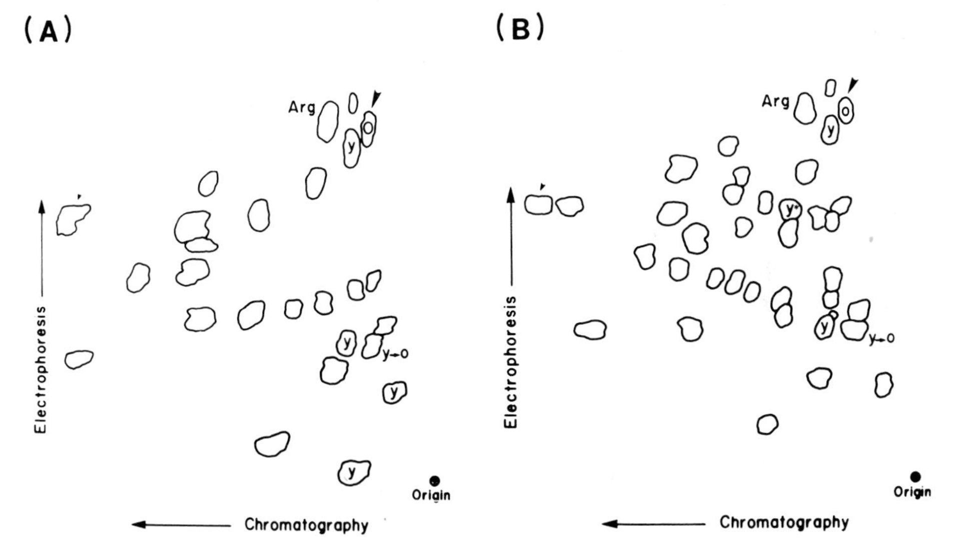

Fig. 1. Peptide maps of tryptic digests of calf γII-crystallin (A) (Croft, 1972) and *N. viridescens* total γ-crystallins (B) (McDevitt, 1981), oxidized with performic acid. Low-voltage electrophoresis was performed with pH 3.6 pyridine–acetate buffer; chromatography was performed at 90° opposite the electrophoresis direction in the solvent BAWP (butan-1-ol–acetic acid–water–pyridine, 15:3:12:10, v/v). Large and small arrowheads indicate N-terminal (Gly-Lys) and C-terminal peptides (Val-Met-Asp-Phe-Tyr), respectively, identified in calf γII-crystallin.* (in B): this peptide found only in calf γIII-crystallin.

(FISC) in birds because of its primacy in their lens development, amounts to 60–80% of the soluble proteins of the lens synthesized throughout chick development (Piatigorsky *et al.*, 1972a). It has a native molecular weight of 680,000 and exhibits both a 48K and a 50K subunit on SDS–urea electrophoresis (Reszelbach *et al.*, 1977). Recombinant DNA technology has now permitted the investigation of two closely related δ-crystallin genes (Bhat *et al.*, 1980), each presumably coding for one of the two polypeptides (above). Interestingly enough, there is some evidence that may be interpreted to indicate the presence of δ-crystallin gene sequences in other than sauropsidans, although they are not expressed (Williams and Piatigorsky, 1979).

In addition to the major classes of α-, β-, γ-, and δ-crystallins, other less well-definable crystallins have been described.

In the bovine lens van Dam and Ten Cate (1966) observed a fast-moving crystallin fraction that exhibited the highest mobility during electrophoresis in alkaline buffer. They termed it pre-α-crystallin (analogous to prealbumin). Later, van de Broek *et al.* (1973) named this fraction fast-migrating (FM) crystallin because it had a low molecular weight and its amino acid composition did not show any similarity to that of bovine α-crystallin.

In the lens of some anuran and urodele amphibians a fast-moving protein, pre-α-crystallin, has also been reported to be present (Campbell *et al.*, 1968; Brahma and van Doorenmaalen, 1969; McDevitt and Collier, 1975). This protein fraction was first detected immunologically in frog lens by Zwaan (1963). Brahma and van Doorenmaalen (1976) prepared specific antiserum against pre-α-crystallin from *Rana esculenta* and studied its ontogeny during lens development in *R. temporaria* (Brahma, 1977; described in Section II, B,2).

S. K. Brahma and D. S. McDevitt (unpublished) also tested *R. esculenta* pre-α-crystallin antiserum immunologically against lens extracts from several anuran and urodele amphibians. Surprisingly, pre-α-crystallin could not be detected in the lens extracts of *Rana tigrina, R. palustris, R. catesbeiana,* and *R. adspersa*. It was also absent in three species of the family Bufonidae tested, namely, *Bufo bufo, B. marinus,* and *B. americanus,* in several urodele amphibians, and also in *Discoglossus pictus* and *Xenopus laevis* [both wild type (+/+) and periodic albino mutant (a^p/a^p)]. It thus appears that pre-α-crystallin is not present in all members of the family Ranidae; this raises some interesting questions regarding the classification of this lens protein.

In the avian (pigeon) lens Rabaey *et al.* (1972) discovered another class of crystallin with intermediate electrophoretic mobility; they termed it LM (low-molecular-weight) protein or "γ-crystallin." McDevitt and Croft (1977) isolated and characterized this crystallin fraction from pigeon lens and found that pigeon "γ-crystallin" did not satisfy any of the commonly accepted criteria for bovine or other γ-crystallins; physical analysis suggested it was a β-crystallin, possibly related to the low-molecular-weight β_s-crystallin described by van Dam (1966).

In subvertebrates, structural lens proteins are also found to be present in cephalopods. But vertebrate crystallins and invertebrate lens proteins do not cross-react immunologically; this will be discussed in Section IV.

C. Presence in the Pineal

Circumstantial evidence for the rare expression of crystallins in other than classical ocular tissues has been reported by McDevitt (1972a) in the lens of the median (parapineal) eye of adult *Anolis carolinensis*, the American chameleon. The pineal complex arises embryonically as an evagination from the *dorsal* aspect of the diencepha-

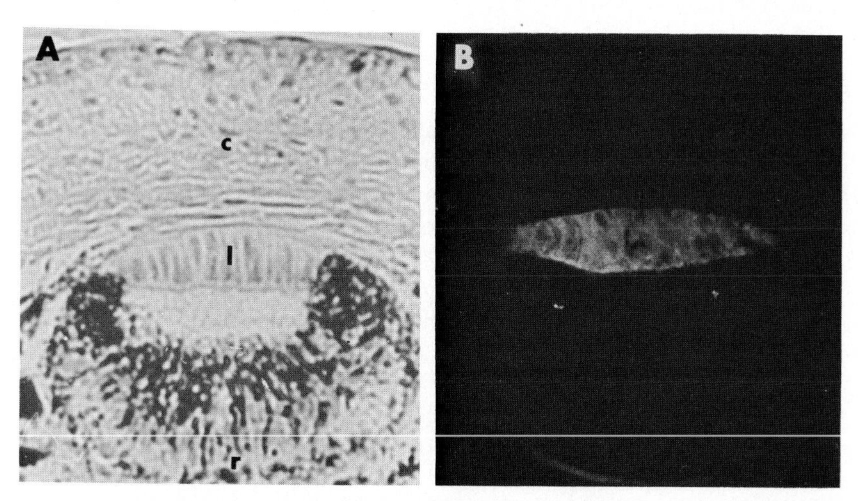

Fig. 2. (A) Tungsten light photomicrograph of a cross section of the median eye of *A. carolinensis*. l, Lens; r, retina; c, cornea. (B) Darkfield fluorescence photomicrograph of a cross section of the median eye of *A. carolinensis*. The section, comparable to that in (A), was treated with antibody to crystallins of the lateral eye with the indirect immunofluorescence technique. Detached lens tissue in the lumen of the median eye appears as two small, strongly fluorescing areas. [From McDevitt, *Science* 175, 763–764 (1972a). Copyright, American Association for the Advancement of Science.]

lon roof, and the lens is derived from neurectoderm rather than epidermal ectoderm, as is the lateral eye lens (for review, see Eakin, 1973). The adult pineal complex (median eye) reaches its greatest histological similarity to the lateral eye in the iguanid lizards. Antibody specific for α-, β-, and δ-crystallins of the adult lateral eye lens of *Anolis* is localized by the immunofluoresence technique only in the lens of the median eye of *A. carolinensis* (Fig. 2). This sharing of tissue-specific antigenic determinants suggests a possible evolutionary relation between the two structures based on biochemical as well as previously reported anatomical criteria.

The role of the crystallins in the eye lens may be described as a dual one. They are important for their contribution as structural proteins to the shape, transparency, and refractive index attained by the lens, alterations in which result in a cataractous condition. This topic will be dealt with in detail by Bettelheim and Siew in another chapter in this volume. The crystallins also serve as markers of a specific tissue differentiation and cell (epithelium→fiber) differentiation; this will be discussed further in Section II.

II. CRYSTALLINS AND DEVELOPMENT OF THE LENS

A. Embryonic Development of the Eye Lens in Vertebrates

Development of the vertebrate eye lens continues to be one of the most extensively investigated problems in developmental biology, i.e., analysis of the causal relationship in the early embryo between the optic vesicle and the overlying ectoderm. As early as 1901, Spemann demonstrated the role of the optic vesicle in lens development in *Rana fusca,* as did Lewis (1904) by transplantation experiments in *R. palustris.* Since then many experiments have been carried out on a variety of organisms, especially amphibians, and a number of review articles on this topic have been written (for recent reviews, see Reyer, 1977; McAvoy, 1980b). Lewis (1907) and LeCron (1907), in fact, have proposed that the degree of lens differentiation is proportional to the period of contact between the optic vesicle and the lens ectoderm.

There are, however, records showing "free" lens development in certain bony fishes and also in some anuran and urodele amphibians in the absence of the optic vesicle (Mencel, 1903; Stockard, 1910; Spemann, 1912; Balinsky, 1951; Jacobson, 1955; Ten Cate, 1956; Mangold, 1961). Most of these free lenses were small and poorly differentiated compared to a normally developed lens. Woerdemann

(1941, 1953) has reported that, for normal differentiation of the lens, continued influence of the eye cup or a longer contact between the optic vesicle and the lens ectoderm seems necessary. Liedke (1955) has suggested that lens induction is a continuous process beginning at neurulation, with continued influence of the retina necessary for final differentiation of the lens.

Lens induction is no longer believed to be a single-step process, but rather a chain of consecutive interactions extending over a long period. Coulombre (1965) discussed the sequence of individual steps in differentiation and morphogenesis of the lens. According to Jacobson (1966) the pharyngeal endoderm, heart mesoderm, and retina act synergistically and sequentially on a population of two small groups of ectodermal cells that will ultimately develop into a lens.

The eye arises as a lateral protrusion of the diencephalon which becomes a distinct optic vesicle (this and the subsequent stages of eye development are illustrated in Fig. 3 and described in Table I). Soon after this contact, the distal wall of the optic vesicle invaginates to form a double-walled optic cup having a thick inner and a thin outer wall. The inner wall differentiates into the neural retina and the outer wall into pigmented epithelium. The inner layer of ectoderm in contact with the optic vesicle thickens to form a palisade-like structure known as the lens placode; this is the earliest stage of lens formation. Along with the invagination of the optic vesicle into the optic cup, the lens placode also begins to invaginate, forming a vesicle; in birds and mammals, a lens pit, open to the exterior, is present before lens vesicle formation. This invagination into a lens vesicle is independent of the invagination of the optic vesicle (Coulombre, 1965). A lens vesicle is known to be present in some amphibians and in all birds and mammals so far investigated, while in the bony fishes and in the majority of amphibians it is represented by a round mass of cells lacking an internal cavity (Lopaschov and Stroeva, 1964).

The cells of the anterior wall of the lens vesicle facing the surface ectoderm initially comprise the external layer, later called the lens epithelium; when fully formed this epithelium is made up of a single layer of columnar cells lying below the lens capsule. The cells of the posterior wall of the vesicle facing the neural retina thicken while elongating, ultimately obliterating the cavity of the lens vesicle. These cells are known as the primary lens fibers. It has been shown that the neural retina is the source of a factor which promotes fiber cell elongation and is also responsible for the polarity of the lens (Coulombre and Coulombre, 1963; Philpott and Coulombre, 1968). Recently, Beebe *et al.* (1980) have isolated a protein fraction from the vitreous body of the

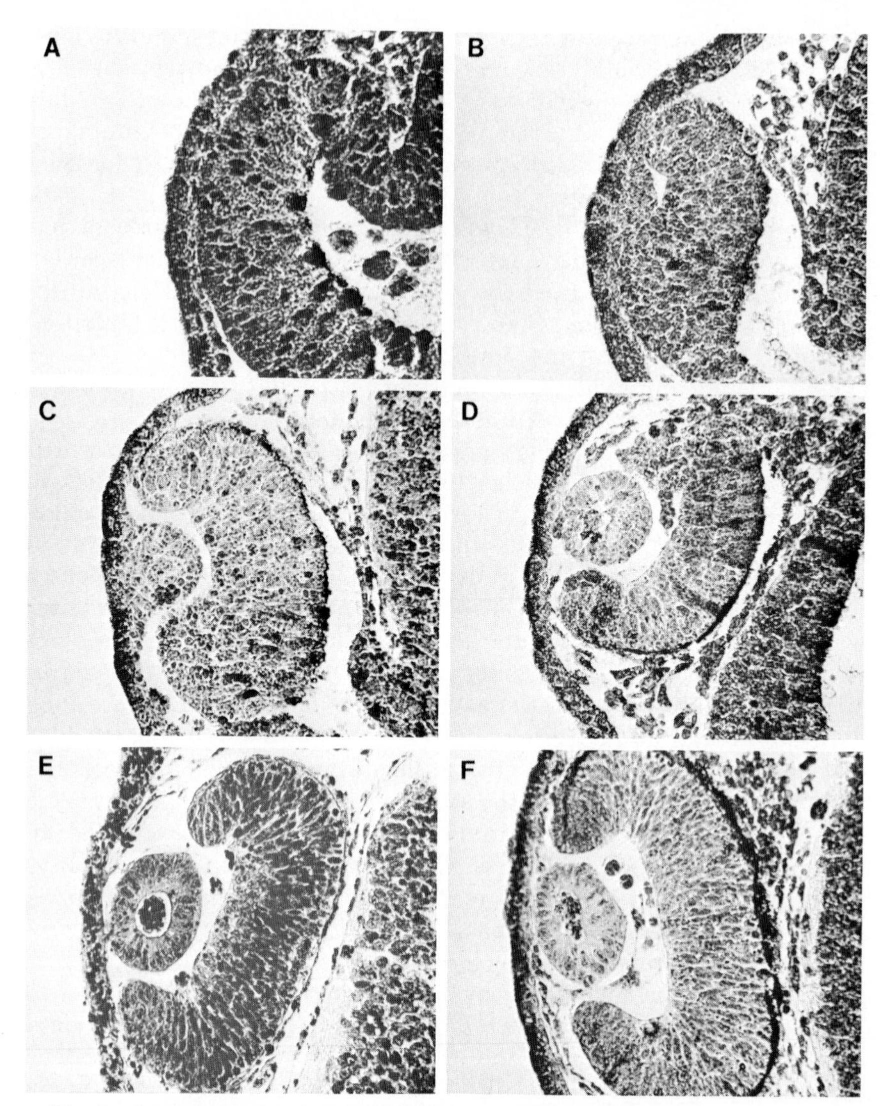

Fig. 3. Lens developmental stages I–X (embryonic *R. pipiens*; for a description of each stage, see Table I). (A) Stage I; (B) stage II; (C) stage III; (D) stage IV; (E) stage V; (F) stage VI; (G) stage VII; (H) stage VIII; (I) stage IX; (J) stage X. (Reproduced from McDevitt *et al.*, 1969.)

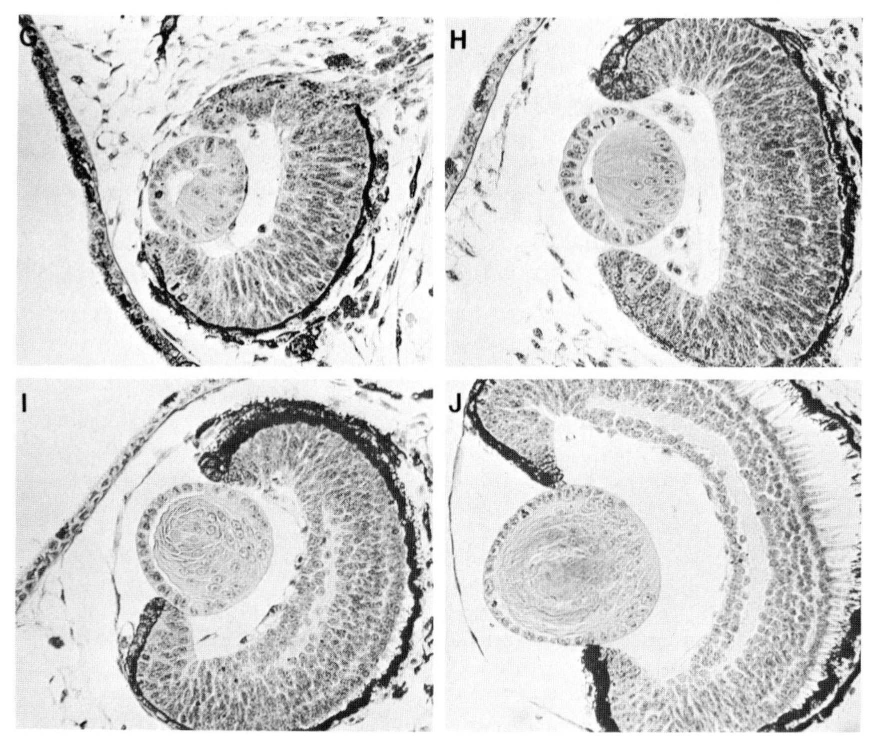

Fig. 3. (*Continued*)

chick embryo which promotes lens fiber differentiation, termed by them "lentropin."

The primary lens fibers, which are laid down during early development, are gradually displaced toward the center of the body of the lens, resulting in a compact area known as the nucleus of the lens. It is formed of primary fibers; the newly developed (secondary) fibers are found in the cortex of the lens and are formed by elongating epithelial cells added on at the lens equator.

Very little is known about the actual mechanism of interaction between the optic vesicle and the lens ectoderm or of the nature of the putative substance(s) involved in lens induction, although it is generally believed that lens induction may involve a transfer of substance(s) from the optic vesicle to the lens ectoderm. Weiss (1950) has proposed that, during the critical period of contact between the optic vesicle and the lens ectodern, an oriented reorganization of the presumptive lens plate in relation to the retinal plate occurs, resulting in a selective

TABLE I

Lens Developmental Stages of *R. pipiens*, the Leopard Frog[a]

Stage I	The optic vesicle is in apparent contact with the prospective lens ectoderm, which is not significantly thicker than the surrounding ectoderm (Fig. 3A).
Stage II	Thickening of the prospective lens ectoderm leads to formation of the lens placode. This thickening is mainly observed in the internal layer of the ectoderm. The invagination of the external (lateral) wall of the optic vesicle transforms the optic vesicle into a double-layered cup (Fig. 3B).
Stage III	A solid hemisphere of the internal layer cells of the lens placode, which fits into the expanding depression of the optic cup, has been formed. The thickening of the external layer (prospective retina) of the optic cup is accompanied by thinning of its internal layer (prospective pigment layer) (Fig. 3C).
Stage IV	The lens placode has become a sphere, within which a small central cavity can be recognized (a vesicle). The sphere is connected with the ectodermal layer through a small number of cells. The cells forming the lens vesicle show signs of extrusion of pigment granules which are abundant in all ectodermal cells as well as in the lens placode cells (Fig. 3D).
Stage V	The lumen of the lens vesicle has been enlarged, and a number of pycnotic cells and pigment granules are enclosed in it (Fig. 3E).
Stage VI	The lens vesicle is completely separated from the surface ectoderm and simultaneously shows thickening of its internal wall. The cells in the thickened part of the wall have elongated along the mediolateral axis. The pigmentation of the prospective pigment layer of the optic cup has visibly progressed (Fig. 3F).
Stage VII	The thickening of the internal wall of the lens vesicle has led to a heavily stratified condition of the originally single-layered epithelium. The lumen of the lens vesicle now assumes a concave curvature toward the median plane of the embryo (Fig. 3G).
Stage VIII	An increase in the diameter of the lens vesicle has been accompanied by further thickening of the internal wall, which considerably obliterates the lumen. The external wall is now applied to the enlarging sphere derived from the internal wall. Thus, the histological structure of a lens is rapidly being approached (Fig. 3H).
Stage IX	The external wall of the lens is flattened to form a regular single-layered epithelium (lens epithelium) that is closely applied to the sphere of fiber-differentiating cells. A transitional zone between the lens epithelium and the complex of fiber cells can be located internally to the equator (subequatorially) (Fig. 3I).
Stage X	Further flattening of the lens epithelium and closer application of the latter to the growing fiber mass accompany an increase in the diameter of the lens. In the internal core of the fiber mass, cells have lost nuclei. A number of young secondary fiber cells can be recognized in the transitional zone (Fig. 3J).

[a] From McDevitt *et al.* (1969).

adhesion and formation of the linkage between "matching molecules" across the contact surface. The early experiments of McKeehan (1951) indicated that contact might be a prerequisite factor in the mechanism of lens induction. Electron microscopic studies on lens development in chick (Hunt, 1961; Weiss and Jackson, 1961; Silver and Wakey, 1974; Hendrix and Zwaan, 1975) and in mouse (Cohen, 1961) have revealed, however, the existence of an interepithelial space or interlayer between the optic vesicle and the lens ectoderm without any evidence of cellular contact across the space. There are other reports indicating that physical contact is not a prerequisite for lens induction (McKeehan, 1958; Muthukkaruppan, 1965; Karkinen-Jääskeläinen, 1978; van de Starre, 1978). McAvoy (1980a) nevertheless observed cytoplasmic processes in the interspace between the optic vesicle and the presumptive lens ectoderm during lens development in rat, but the actual function of these processes could not be determined.

McKeehan (1956) suggested a transfer of RNA from the optic vesicle to the lens ectoderm during lens induction based upon his observation of an increased RNA concentration per cell in the induced lens with a simultaneous decrease in RNA per cell in the retina. He, however, could not decide whether the RNA was transferred undegraded or its degradation products concentrated in the induced lens. Sirlin and Brahma (1959) showed for the first time direct transfer of radioactive substances from the optic vesicle to the induced lens using *DL*-3-phenyl[2-^{14}C] alanine by autoradiography, but they could not determine the nature of the substance transferred that might be involved in lens induction, or rule out simple diffusion.

Hunt (1961) confirmed the results of McKeehan (1956) as he observed interepithelial clouds which according to him could represent a mechanism transferring specific RNA particles from the optic vesicle to the lens ectoderm. He observed a statistically significant change in the relative abundance of RNA particles between the two reacting tissues. Weiss and Jackson (1961) also confirmed the results of McKeehan (1956) in that cytoplasmic basophilia increased in the prospective lens cells, but they failed to observe the presence of any such small particles within the interlayer during the entire period of lens induction. According to Silver and Wakey (1974), the electron-dense clouds described by Hunt (1961) are extralaminar basement membrane materials probably involved in the production of collagenous cellular fibrils and are not characteristic of the developing eye.

Babcock (1961) has reported that, during lens induction in *X. laevis*, there is a rise and a subsequent fall in the RNA content of the optic vesicle with a simultaneous increase in RNA in the lens primordium, as determined histochemically. The increase in basophilia or RNA noted in all the above could not be determined as a cause or result of induction, however.

A fully developed lens is made up of an outer single layer [in all vertebrates (Srinivasan, 1960; Srinivasan and Harding, 1961)] of epithelium and inner fiber cells. The epithelium, on the anterior face, can be divided into a central region, where mitosis is very rare, and a germinative region, where the cells show relatively high mitotic rates. Here the cells divide and gradually migrate toward the peripheral area, known as the equatorial or transitional zone, where the cells begin to elongate to form the fibers (Papaconstantinou, 1967). Fiber cell formation represents an irreversible stationary phase, while the cells of the central epithelial region which normally do not divide are known to be in reversible G_0 phase (Harding *et al.,* 1971).

Fiber cell differentiation has been shown to be characterized by the initiation of synthesis of a specific lens protein class, γ-crystallins, in vertebrates where γ-crystallin is known to be present (Papaconstantinou, 1965; Takata *et al.,* 1965; McDevitt *et al.,* 1969; Shubert *et al.,* 1970; McDevitt and Brahma, 1973, 1977, 1979; Brahma and McDevitt, 1974a,b; McAvoy, 1978a; Konyukhov *et al.,* 1978). The fiber cells are added to the body of the lens throughout the life span of the animal. In their final phase of differentiation, the fiber cell nuclei undergo pycnosis, degenerate, and finally disappear in a strict spatial and temporal pattern (chick, Modak and Perdue, 1970). Maturation of the lens fibers and their denucleation process have been investigated in various mammalian lenses by Srinivasan and Iwamoto (1973), Kuwabara and Imaizumi (1974), and Kuwabara (1975).

Brini *et al.* (1962) have reported that in differentiating embryonic chick lens fiber cells there is a gradual decrease in cell organelles. Microtubules have been shown to play a role in the elongation of fiber cells and become prominent in lens fiber differentiation in the developing vertebrate lens (Byers and Porter, 1964; Piatigorsky *et al*., 1972a,b). Kuwabara (1968, 1975) has reported that in the deep bow zone of the lens fibers more numerous microtubules are found than in the actively elongating cells at the lens equator.

B. Ontogeny and Localization of the Crystallins

1. *Fish*

References to crystallins in fish eye lens development are few. Rabaey (1965) reported that, in the lenses of several genera of Osteichthyes, the low-molecular-weight, slow-moving (electrophoretically) proteins were predominant and were relatively much more abundant and sometimes unique in young and embryonic fish. This fraction was probably what we know now as the γ-crystallin fraction. Zeitlin and McDevitt (1972), in an immunofluorescence study, noted in *Hemigrammus caudovittatus* and *Brachydanio rerio* embryos a first positive reaction to an anti-total crystallins antibody at the stage of elongation of the prospective primary fiber cells. In further support of this finding, the lens rudiment of the blind cave fish, *Anoptichthys jordani,* which develops normally until the vesicle stage and then becomes atretic along with the rest of the eye, never exhibited any positive reaction for crystallins. Thus γ-crystallins are circumstantially suggested as one of the first crystallins to appear in fish lens morphogenesis.

2. Amphibia

In amphibia α-, β-, and γ-crystallins are known to be present. Using classical immunological methods Ogawa (1965) investigated the appearance of lens crystallins in developing *Triturus pyrrhogaster* embryos. He reported that α-crystallin was the first of the crystallins to appear, followed by β- and γ-crystallins, but the techniques applied by Ogawa (1965) have been criticized by McDevitt *et al.* (1969). With antibodies against total soluble lens proteins, Takata *et al.* (1964) obtained the first positive immunofluorescence reaction in *T. pyrrhogaster* at the lens vesicle stage, with the lens epithelium demonstrating a positive reaction at a later stage. Since the antibodies were directed against total soluble lens proteins, it was impossible to determine which of the three crystallin classes appeared first. McDevitt (1972b) reported that, during normal lens development in *R. pipiens,* β- and γ-crystallins were first detectable immunologically, while α-crystallin was the last to appear.

The indirect immunofluorescence staining method has been used frequently since 1969 to study the ontogeny of lens crystallins in several anuran and urodele amphibians during normal development. In most of these investigations, antibodies directed against total soluble lens proteins as well as γ-crystallins have been used in homologous and heterologous combinations (McDevitt *et al.,* 1969; Nöthiger *et al.,* 1971; McDevitt and Brahma, 1973, 1977, 1979; Brahma and McDevitt, 1974a). Results from all these experiments clearly show that γ-crystallin is one of the first of the crystallins to appear during normal lens development and is confined to fiber cells only (Fig. 4) (McDevitt *et al.,* 1969). Recently, McDevitt and Brahma (1981a) used α-, β-, and

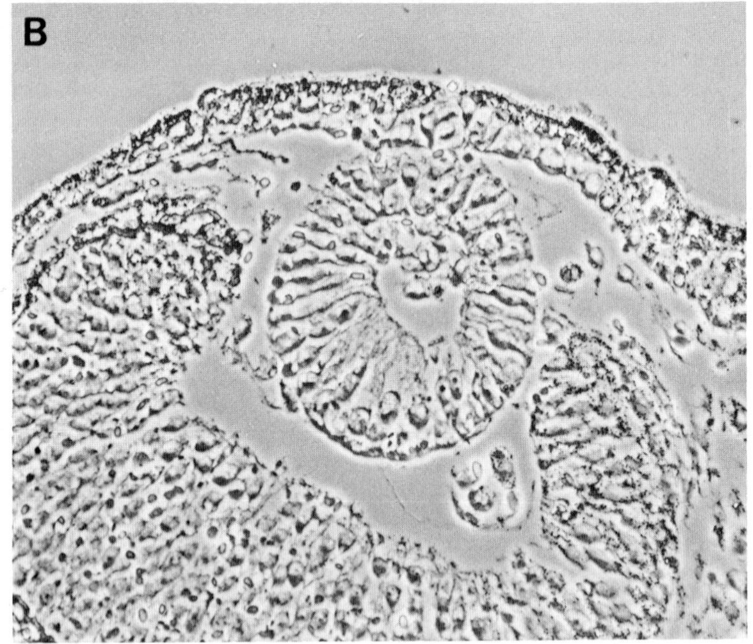

Fig. 4. Immunofluorescence photomicrographs of sections through the eye region of embryonic *R. pipiens* treated with antibody to γ-crystallins. (A and C) Lens developmental stages VI and X, respectively. (A) The first stage at which a positive reaction, only in the prospective primary fiber cells, could be obtained. (C) Immunofluorescence

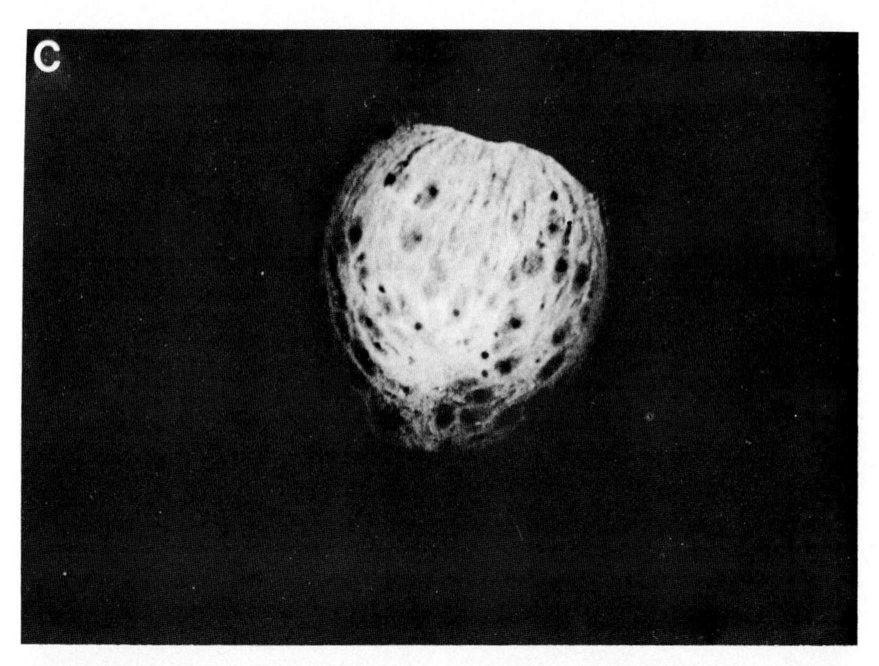

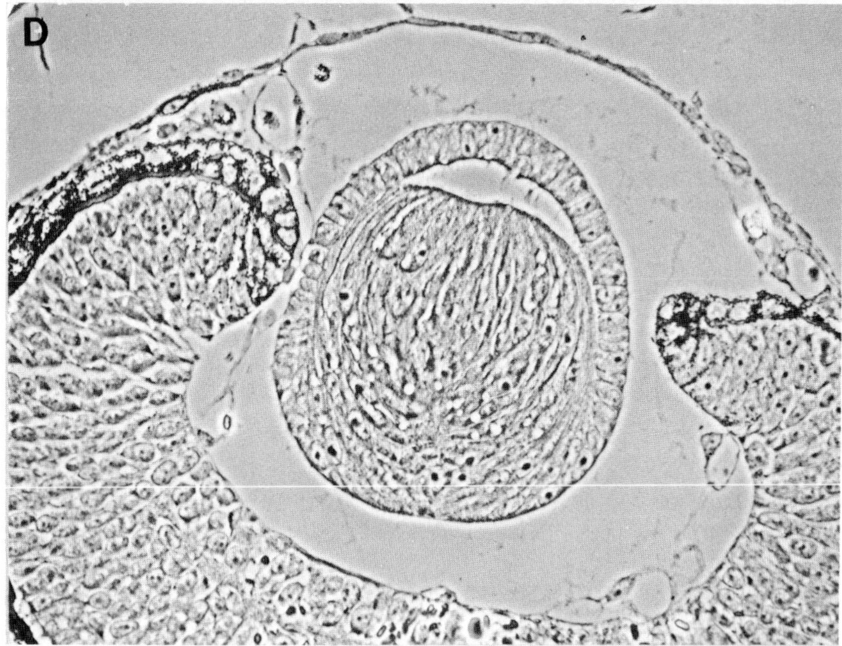

was present in primary and secondary fiber cells but still absent in the lens epithelium. The space seen between the fiber mass and epithelium is artefactual. (B and D) Phase-contrast photomicrographs of sections identical to (A) and (C), as a histological reference. (From McDevitt *et al.*, 1969, with the permission of Academic Press.)

γ-crystallin antibodies derived from *N. viridescens* adult lenses to study the ontogeny of these crystallins during normal development in the same species. This was the first use in amphibia of homologous antibodies specific for each of the crystallin classes. The results revealed that β-crystallins were the first to appear during normal lens development (Fig. 5), in a few prospective primary lens fiber cells even before their histodifferentiation. γ-Crystallins appeared at a slightly

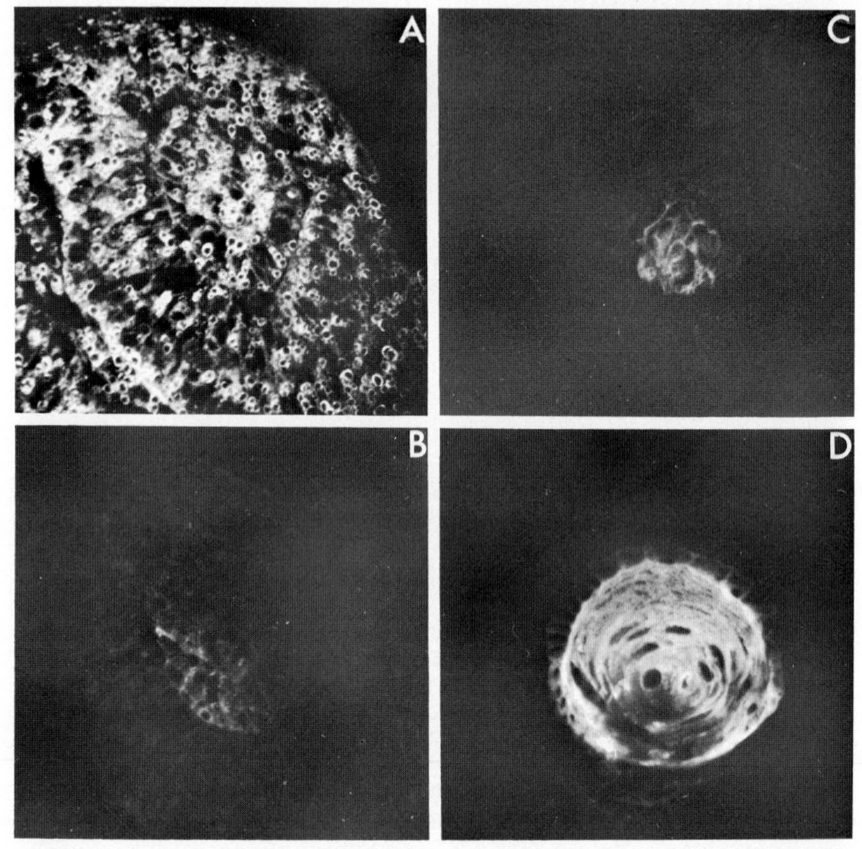

Fig. 5. Dark-field photomicrographs of sections through the eye region of embryonic *N. viridescens* treated with antibody to *N. viridescens* total β-crystallins. The prospective cornea is at the top and the retina at the bottom; refer also to the illustrations in Fig. 3. (A) Tungsten illumination, as an identical histological reference for (B) in which the immunofluorescence for β-crystallins is first detectable in a few cells of the inner wall of the early lens vesicle (stage V). (C) Stage VIII (immunofluorescence); (D) Stage X (immunofluorescence); the single-layered epithelium is now positive for β-crystallins (the large nuclei of the lens epithelial cells are negative).

later stage also in the fiber cells, while α-crystallins appeared last, in a few primary fiber cells. These authors failed to detect α-crystallins in the epithelium as late as the premetamorphic stage. Using the same three antibody types Brahma and McDevitt (in preparation) found that, in developing *Ambystoma mexicanum* lens, both β- and γ-crystallins appeared simultaneously, followed by α-crystallins. α-Crystallin could not be detected in the lens epithelium even from the 5.5-cm axolotl (Fig. 6). It thus appears that, in *A. mexicanum,* the lens epithelium from animals as advanced as 5.5 cm contains only β-crystallin. This same α-crystallin antibody is capable, however, of recognizing α-crystallin in the epithelium of the *regenerating* lens of *N. viridescens* (McDevitt and Brahma, 1981b, discussed in Section III,C).

In normally developing (+/+) and periodic albino mutant (a^p/a^p) (Hoperskaya, 1975) *X. laevis,* McDevitt and Brahma (1973, 1979) reported that the first positive immunofluorescence reaction for γ-crystallins could be detected earlier in the wild type than in the mutant. These authors did not detect γ-crystallins in the epithelium of these two types of larvae. Mikhalov and Korneev (1980), however, reported no difference in the ontogeny of lens crystallins of +/+ and a^p/a^p mutants of *X. laevis*. They also reported a positive immunofluorescence reaction with a γ-crystallin antiserum in fibers and in the epithelium of the developing lens. Examination of the immunofluorescence photographs demonstrating the presence of γ-crystallin in the epithelium reveals equal background fluorescence in the cornea and regions of the optic cup, so that these results are difficult to interpret.

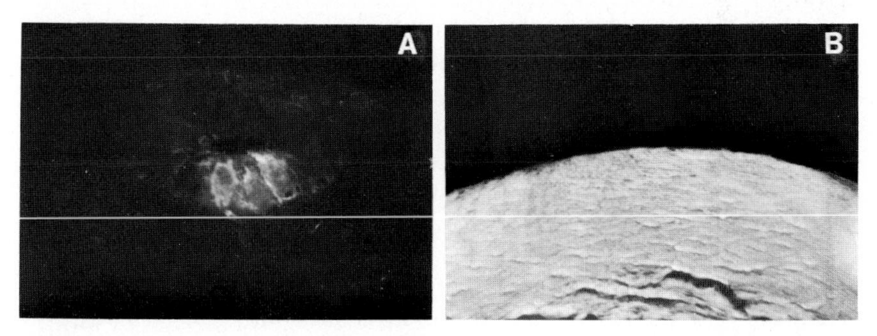

Fig. 6. (A) Immunofluorescence photomicrograph of a section through the stage VI lens of embryonic *A. mexicanum* exposed to *N. viridescens* α-crystallin antibody. (B) Lens from a 5.5-cm *A. mexicanum* larva exposed to α-crystallin antibody. Even at this age, the epithelium did not show any positive reaction. (S. K. Brahma and D. S. McDevitt, unpublished.)

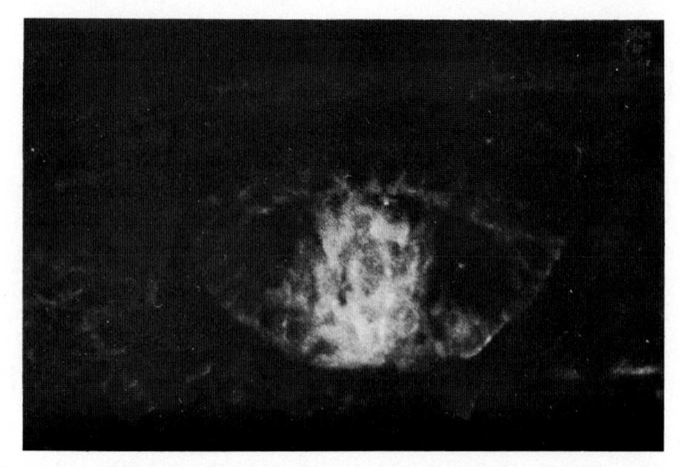

Fig. 7. Immunofluorescence photomicrograph of a section through a stage VII lens (McDevitt *et al.*, 1969; Fig. 1) of *R. temporaria* exposed to *R. esculenta* pre-α antibody. The first positive reaction appeared at this stage, after γ-crystallin, and, like γ-crystallin, was localized in the fiber cell area. (From Brahma, 1977.)

In addition, they found that γ-crystallins appeared at a rather advanced stage of morphogenesis. These authors have suggested that some γ-crystallins and perhaps β-crystallins appear first and that α-crystallins are the last to appear.

With monospecific *R. esculenta* pre-α-crystallin antiserum, Brahma and van Doorenmaalen (1976) obtained the first positive immunofluorescence reaction in normally developing *R. temporaria* lens immediately after the appearance of the γ-crystallins (Brahma and McDevitt, 1974a). It appeared at stage VII (McDevitt *et al.*, 1969); unlike γ-crystallin, it was found to be present in the entire fiber cell area (Fig. 7). In later stages of lens development it was found to be confined to the primary and secondary fibers, and the epithelium did not show any positive reaction with this antibody.

Immunofluorescence studies with the lens-*regenerating* systems from urodele and anuran amphibians will be dealt with in Section III,C and D.

3. Birds

Almost all our knowledge regarding the ontogeny of lens crystallins in the avian lens comes from the chick, where three major classes of lens crystallins, α-, β-, and δ-crystallins, are known to be present. β-Crystallins have the lowest molecular weight; they were also earlier termed the "long line" (Zwaan, 1963) and (confusingly) γ-crystallin

(Maisel and Langman, 1961). β-Crystallins in chick can be operationally grouped into anodally and cathodally migrating groups (Clayton, 1970, 1974). In addition to the three above-mentioned crystallins, another, low-molecular-weight, lens crystallin has been isolated and characterized from various avian species. It has been called LM or γ-crystallin (Rabaey *et al.*, 1972), and tentatively β_s by McDevitt and Croft (1977), and is usually present in very small quantities.

Indirect immunofluorescence studies with monospecific antibodies to δ-crystallin have confirmed the earlier reports of Rabaey (1962) that it is the first of the crystallins to appear during lens development in the chick; it is detectable in the cells of the lens vesicle nearest the prospective retina (Zwaan and Ikeda, 1968; Brahma and van Doorenmaalen, 1971; McDevitt and Clayton, 1979). In the embryonic pigeon and duck lens δ-crystallin was also found, using immunofluorescence techniques, to be the first crystallin to appear (S. K. Brahma and H. van de Starre, unpublished). Using antibodies to total β-crystallin, Zwaan and Ikeda (1968) have shown that it appears at about 56 hr, after δ-crystallin. There is, however, a discrepancy regarding the ontogeny of the anodal and cathodal β-crystallins in the chick lens. Waggoner *et al.* (1976) reported that cathodal β-crystallins appeared first in the cytoplasm of the lens cells adjacent to the retina at 72 hr, while anodal β-crystallins appeared in the entire fiber cell area of a 5-day lens. McDevitt and Clayton (1979), on the other hand, detected both anodal and cathodal β-crystallins at much earlier stages than Waggoner *et al.* (1976); the difference could have been due to the specificities and/or avidities of the antibodies used. α-Crystallin has been shown to be the last of the crystallins to appear in chick lens fibers, being first detectable in the 3½-day lens (Zwaan and Ikeda, 1968; Brahma and Van Doorenmaalen, 1971; McDevitt and Clayton, 1979).

McDevitt and Clayton (1979) also examined the ontogeny of the lens crystallins in a strain of chicks known as Hy-1 (Clayton, 1975), which demonstrate a high growth rate and anomalous development of the eye lens leading to hyperplasia of the epithelium, annular pad, and often fibers. They observed a striking difference in the ontogeny of α-, β-, and δ-crystallins in normal and Hy-1 embryos, seemingly correlated with the abnormal histogenesis of the lens and resulting in a change in the regulation of ontogeny of the crystallins.

With antibodies to LM crystallin, Brahma *et al.* (1972) found that the first immunofluorescence reaction for LM crystallin in pigeon, an especially rich source, appeared in the fiber cells of the 4-day embryo lens (Fig. 8). In late embryonic and posthatch lenses it could be detected in the epithelium as well. Using the same antibody, Brahma and

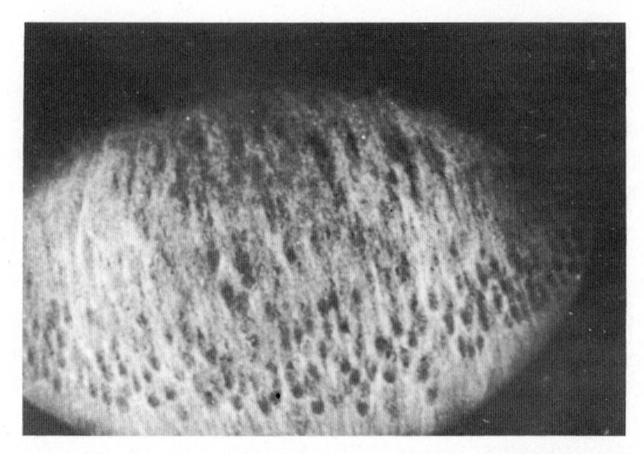

Fig. 8. Immunofluorescence photomicrograph of a section through the eye lens of a 4-day embryonic pigeon (*Columba livia*) exposed to pigeon lens low-molecular-weight (LM) protein antibody. This first positive reaction was detected only in the fiber cells at this stage; later it appeared in the epithelium as well as in the annular pad region. (From Brahma *et al.*, 1972.)

van Doorenmaalen (1973) found that in the *chick* lens this crystallin could be immunologically detected from 7-day posthatch lens and later.

To the authors' knowledge, there are no reports concerning crystallins in development of the reptilian lens.

4. *Mammals*

In the mammalian lens three main classes of proteins, α-, β-, and γ-crystallins, are known to be present (Section I).

α-Crystallin is known to be the first crystallin detectable immunologically (Rabaey, 1965; Barabanov, 1966). Vermorken and Bloemendal (1978) have suggested that it could be used as a marker for terminal differentiation in bovine lens, since a change occurs in the subunit composition of α-crystallin during the transition from epithelial to fiber cells. According to Vermorken *et al.* (1977), the onset of fiber cell differentiation is also accompanied by the formation of β-crystallin chains B_{1a}, B_5, and B_p (principal), although it has been shown by Papaconstantinou (1967) that fiber cell differentiation in bovine lens is correlated with the onset of γ-crystallin synthesis, present only in the fiber cells.

Using the indirect immunofluorescence staining method, Shubert *et al.* (1970) reported that, in the embryonic rat lens, γ-crystallin appeared first in the primary fiber cells; later it could also be detected, but to a degree not above background, in the epithelium of late embryonic stages. McAvoy (1978a) found that in 1-day-old rat lens α-crystallin was present in the epithelium and γ-crystallin in the fiber cells only; he questioned the purity of the γ-crystallin antiserum used by Shubert *et al.* (1970) and its presence in the epithelium. Recently McAvoy (1980c) reported that rat lens epithelium in culture synthesized β- and γ-crystallins upon elongation in the presence of neural retina. In one of his earlier papers, McAvoy (1978b) reported that in the embryonic rat lens α-crystallins appeared first in a few cells of the lens pit and, when the lens vesicle was formed, both β- and γ-crystallins appeared simultaneously in some cells at the posterior part of the vesicle. He did not observe any β- or γ-crystallins in the epithelium. Recently Maschino *et al.* (1980) obtained a positive reaction with anti-rat total lens protein antiserum at the beginning of invagination of the lens placode. This is the earliest evidence of crystallin ontogeny in rat lens, and according to the authors is probably due to the presence of α-crystallin.

van de Kamp and Zwaan (1973) obtained the first positive immunofluorescence reaction with antibodies to total mouse lens crystallins in the basal cytoplasm of a few centrally located cells of the deeply invaginated lens placode of the mouse embryo, but they could not ascribe the result to any specific crystallin class. Konyukhov *et al.* (1978) reported, using immunofluorescence, that in developing mouse lens α-crystallin appeared first, followed by γ-crystallin, and that β-crystallin appeared last. These authors did not observe any immunofluorescence reaction in the epithelium and in the equatorial zone with anti-γ-crystallin antibodies, in agreement with others. They also examined crystallins in embryos of the ZRDGT strain of mice with hereditary anopthalmia. They observed α-crystallin in the cells of the reabsorbed placode, while Zwaan (1975) failed to obtain any positive immunofluorescence in the lens rudiment of the aphakic mouse.

III. CRYSTALLINS AND REGENERATION OF THE LENS

A. Historical Perspective and Cellular Events of Lens Regeneration from the Dorsal Iris

As early as 1781, Bonnet reported regeneration of a small but normal eye from a portion of the eye after its partial removal in the adult newt *Triturus cristatus*. Blumenbach (1787) found that, when the eye was totally removed, no regeneration of the eye occurred, whereas incomplete removal resulted in a small, intact eye. About a century later, Philippaux (1880) confirmed these earlier results on eye regeneration. Colucci (1891) was the first to examine lens regeneration in adult *T. cristatus* from histological preparations of lens regenerates and observed that they developed from the inner lamina of the dorsal iris and passed through several stages during regeneration. Unaware of the work of Colucci, Gustav Wolff (1895, 1901, 1904) published a series of articles on lens regeneration in both larval and adult *Triturus taeniatus* and *T. cristatus,* describing in great detail changes in the iris during the process of lens regeneration. Although lens regeneration from dorsal iris is also known as Wolffian regeneration, the credit for its discovery should rightfully be given to Colucci. It is known that this capacity for lens regeneration from dorsal iris is confined to a group of urodele amphibians [Stone, 1967; Berardi and McDevitt (*E. bislineata*), 1981]. The topic of lens regeneration per se has been reviewed recently by Reyer (1977) and Yamada (1977).

Sato (1940) classified lens regenerates from *T. taeniatus* and *T. pyrrhogaster* into 13 morphological stages according to the histological changes observed during the entire course of regeneration. Zalokar (1944) and Reyer (1948, 1950) used this classification of Sato for *T. cristatus* and *T. viridescens,* respectively, with slight modifications. Reyer (1954) divides these 13 stages of lens regeneration into four periods: (1) latent period, (2) initial period, (3) period of lens fiber differentiation, and (4) period of growth. The first period includes stages I and II; the second, stages III–VI; the third, stages VII–XI, and the last, stages XII and XIII. Yamada (1967a) supplemented light microscopic observation of lens regeneration with the results obtained from electron microscopic studies (Eguchi, 1963 1964; Karasaki, 1964) and defined 11 stages, omitting the last 2 stages since histogenesis was essentially complete.

For reference purposes, the stages of lens regeneration after Yamada (1967a) are described briefly below and illustrated in Fig. 9 (with permission of the author and Academic Press.)

Stage I The dorsal iris at the pupillary margin swells, and the inner and outer laminae separate.

Stage II The nuclei of the iris epithelial cells become spherical, and a reduction in pigment granules occurs at the middorsal margin.

Stage III This results in a group of depigmented cells at the pupillary margin (inner lamina).

Stage IV These nonpigmented cells form a hollow vesicle with its cavity continuous with the intralaminar space.

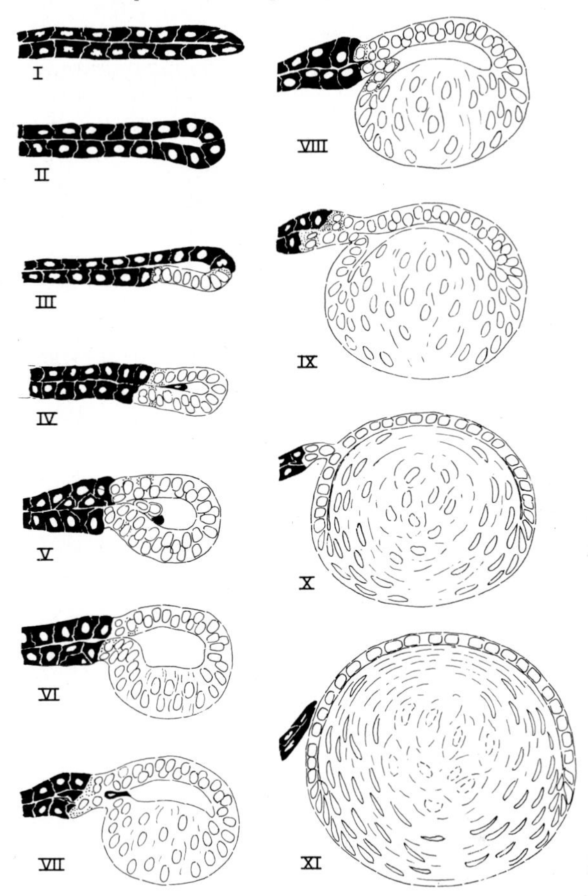

Fig. 9. Diagrams showing morphological stages of Wolffian lens regeneration (from dorsal iris) in adult *N. viridescens*, the eastern spotted newt. Sections are through the middorsal pupillary margin of the iris, oriented perpendicular to the main body axis. The cornea is above, the retina below, and dorsal is to the left. The pigmented iris cells are indicated by black cytoplasm, and depigmented regenerate cells by white cytoplasm. Incomplete depigmentation is shown as dotted areas. Iris stroma cells and ameboid cells are not pictured. (From Yamada, 1967a.)

Stage V Cells of the internal layer of the newly formed vesicle thicken and elongate. They no longer divide and start synthesizing lens-specific γ- and β-crystallins.

Stage VI More cells are added to the lens vesicle by cell multiplication, as well as by the addition of depigmented cells. The central part of the internal layer thickens by cell elongation.

Stage VII Cells of the central part of the internal wall of the lens vesicle continue elongating to form primary fibers; some begin synthesis of α-crystallins. More cells are added to the lens vesicle from the pigmented epithelium via connection with the pigmented area by a stalk. As a result of thickening and elongation of the internal layer, the inner surface projects into the lumen of the lens vesicle (fiber hillock).

Stage VIII The lumen of the vesicle is nearly obliterated, and secondary fibers begin to differentiate from the equatorial zone.

Stage IX The entire external layer of the lens and proximal part of the internal layer become flattened. New fiber cells are continually added at the equatorial zone.

Stage X The single-layered epithelium is thus formed from the external layer as well as from the proximal part of the internal layer. Elongating secondary fibers completely surround the core of primary fibers, with the lens still connected to the stalk.

Stage XI The lens grows by continual addition of secondary fibers. The connection between the iris (stalk) and the lens is lost, and lens histogenesis is basically complete. The primary fibers become acidophilic, and their nuclei later disappear.

Histologically, the normal newt iris consists of a two-layered iris epithelium, with a stroma on its anterior face. These laminae each contain a single layer of epithelial cells with their cytoplasm packed with pigment granules—melanosomes. According to Eguchi (1963) these pigment granules are fewer and larger in the inner lamina. At the pupillary margin the two laminae are joined. In the adult, one can observe a change from pigmented to unpigmented epithelium at the border between the iris and ciliary body (Fig. 10).

In normal adult iris, the epithelial cells do not divide, and the earliest sign of cell transformation after lens removal is depigmentation of the cells of the iris epithelium, a good marker for the onset of transformation of the iris into lens (Yamada, 1977). According to Wolff (1895), the melanosomes are transferred directly into leukocytes which crowd the eye chamber after lens removal. Fischel (1900), on the other hand, believed that the melanosomes were discharged and taken up by leukocytes. Eguchi (1963) has reported that "special ameboid cells" resembling leukocytes are connected with the depigmentation and suggested two mechanisms of depigmentation:

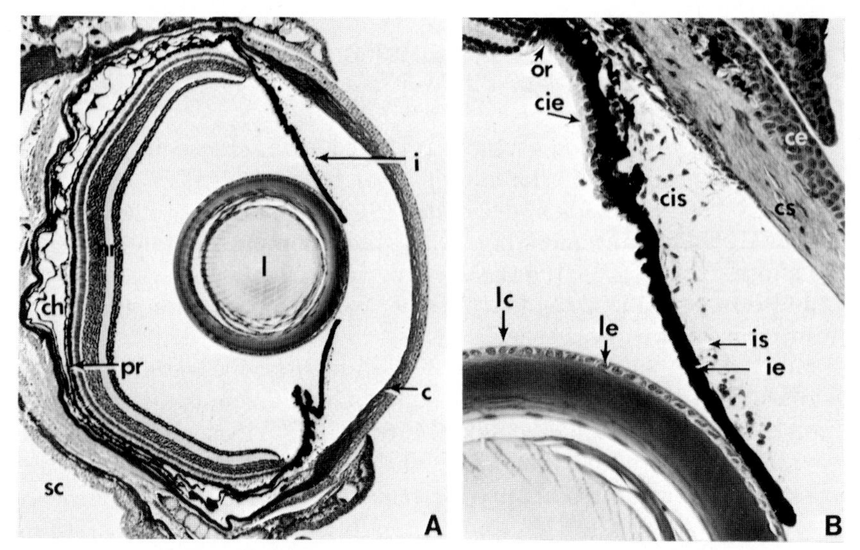

Fig. 10. (A) Vertical, meridional section of the eye of the newt *N. viridescens*. c, Cornea; l, lens; i, iris; nr, neural retina; pr, pigmented retina; ch, choroid; sc, sclera. (B) Detail of cornea, lens, dorsal iris, and ciliary body. le, Lens epithelium; ie, iris epithelium; is, iris stroma; cie, ciliary epithelium; cis, stroma of ciliary body; or, ora serrata; ce, corneal epithelium; cs, corneal stroma (substantial propria). (From Reyer, 1977, with the permission of the author and Springer-Verlag.)

(1) The melanosomes are taken up by the ameboid cells by pinocytosis, and (2) the melanosomes are discharged as a mass from the iris cells into the lumen without being taken up by the ameboid cells. Later, Yamada and Dumont (1972) identified the ameboid cells of Eguchi (1963) as macrophages of monocytic origin which could not be found in the normal eye. After autophagy and exocytosis the melanosomes from the regenerating dorsal iris are taken up by the macrophages by phagocytosis (Yamada *et al.*, 1978).

The earliest cellular event following depigmentation in the dorsal iris is the ultrastructural change in the cell nucleolus (Eguchi, 1964; Karasaki, 1964; Dumont *et al.*, 1970). The granular component of the nucleolus is either scanty or absent, and the fibrous component is also smaller in a cell not engaged in rRNA synthesis. The same condition is also found in normal iris. Within 2 days after lens removal these two nucleolar components increase, with a preferential increase in the granular components, associated with the enhancement of rRNA synthesis (Reese *et al.*, 1969). The retinal cells taking part in the formation of the lens enter into a DNA synthesis phase (Eisenberg and Yamada, 1966; Yamada and Roesel, 1969; Eguchi and Shingai, 1971; Reyer, 1971). As these cells differentiate into fibers, they are withdrawn from the cell cycle.

Autoradiographic studies have revealed an increase in RNA and protein synthesis in the dorsal iris after lens removal (Yamada and Karasaki, 1963; Yamada and Takata, 1963). According to Yamada and Takata (1963) enhancement of [^{3}H]leucine incorporation takes place in two phases, the first during the early period of depigmentation and the second from stage IV to X. During fiber cell differentiation a low level of incorporation was observed.

Concomitant with the process of depigmentation, there is also an increase in the number of mitochondria per cytoplasmic unit area (Eguchi, 1963, 1964; Karasaki, 1964). The ribosome population reaches its maximum value when the cells enter into the fiber differentiation phase. Ribosomes which take part in protein synthesis are then associated with mRNA and form polysomes. Jurand (reported by Yamada, 1967a) observed that during and soon after depigmentation most ribosomes appeared singly or in small clusters of various sizes.

Cellular transformation of epithelial cells into fibers is accompanied by a gradual disappearance of mitochondria (Eguchi, 1963, 1964; Karasaki, 1964), as in the normally developing lens (Brini *et al.*, 1962). Jurand and Yamada (1967) have observed myelin-like or vesicular structures in the intracellular space, which are apparently related to the elimination of mitochondria (for review, see Reyer, 1977; Yamada, 1977).

Zalik and her group have been investigating the cell surface changes accompanying Wolffian lens regeneration, as well as hormonal factors involved in this phenomenon (Hornsby and Zalik, 1977; Cuny and Zalik, 1981). Cell surface components disappear from the pigmented iris during dedifferentiation after lentectomy and reappear when redifferentiation begins, presumably microtubule-mediated. The association of microtubules with lens fiber differentiation and concomitant γ-crystallin synthesis is indicated. In addition, results with newt lens regeneration *in vitro* suggest that pituitary hormones, particulary bovine thyrotropin, enable the inherent lens-forming capacity of dorsal iris to express itself. The morphology of the regenerates was not normal (as reported previously by Yamada and McDevitt, 1974), yet fiber differentiation, as indicated by γ-crystallin presence, did occur. That thyrotropin was responsible *in vivo* for this initiation of lens regeneration could not, however, be established.

It has been well-documented for several decades that lens regeneration from the dorsal iris *in vivo* is dependent upon a stimulus coming from the neural retina (for reviews, see Reyer, 1977; Yamada, 1977). Despite much research, such a factor(s) has not been identified or characterized. Iris epithelial cells in cell culture, however, can now be found to transform into lens cells without the presence of retina (Eguchi *et al.*, 1974; D. S. McDevitt and T. Yamada, unpublished). The effect of neural retina thus appears to be permissive rather than instructive, and its presence is not an absolute condition for transformation of iris epithelial cells into lens cells. Yamada and Beauchamp (1978) suggest that the factor involved in lens regeneration acts indirectly, to shorten preferentially the cell cycle time in the dorsal iris epithelial cells *in vivo* as compared to the ventral iris, thus permitting the dorsal iris epithelial cells to go through the requisite number of cell cycles for conversion into lens cells before mitosis ceases once again in the iris epithelium.

A chapter by Yamada in this volume treats the most recent cell cycle data on the regeneration of the lens as a putative mechanism to explain dorsal iris competency versus ventral iris incompetency in *in vivo* lens regeneration. Issues of transdifferention which this and other systems elicit are also discussed in the general context of the regeneration problem per se.

B. Lens Regeneration from Ventral Iris and Neural Retina

Wolffian lens regeneration has been shown to occur *only* from the dorsal iris at the pupillary margin after lens removal; there is no record of normally induced Wolffian regeneration *in vivo* from the ventral iris epithelium. The experimental results of Sato (1930, 1951) and Mikami (1941) clearly showed that no lens developed from the ventral iris when implanted in a lentectomized host. Eguchi and Watanabe (1973), however, subjected the ventral iris of *T. pyrrhogaster* to a potent carcinogen, *N*-methyl-*N*′-nitro-*N*-nitrosoguanidine; either the compound was inserted into the lentectomized eye or the ventral iris treated with the compound was reimplanted in a lentectomized eye. No carcinogenic effect could be seen, and its action appeared specific in transforming ventral iris into a lens, without any other cell type. The actual process underlying this transformation could not be ascertained, although loss *in situ* of the integrity and contiguity of the surrounding tissues, e.g., the stroma and basal lamina, cannot be ruled out. Yamada and McDevitt (1974) and Eguchi *et al.* (1974) have reported that ventral iris epithelium in culture has the ability to transform into lens cells based on immunological criteria. This indicates that tissue organization *in situ* is important in controlling the capacity for lens regeneration in urodeles.

Clayton and her group (de Pomerai *et al.*, 1977; Pritchard *et al.*, 1978; de Pomerai and Clayton, 1978) examined the interesting capability of chick neural retina to transdifferentiate into lens cells (lentoids) *in vitro*. This was restricted to cells derived from 8 to 15-day *embryonic* chicks, the most success being obtained with neural retina cells from earlier embryos. Whether this phenomenon is related to withdrawal of the *in vivo* source cells from the cell cycle (e.g., 8-day embryonic chick neural retina cells are still in the cell cycle) is not clear. The relative amounts of the crystallins synthesized differ according to such factors as age at explantation and culture medium, demonstrating nevertheless the ability of this eye tissue *in vitro* to form cell-specific products not normally produced and the influence of extrinsic factors on patterns of gene expression. Similar results were reported by Araki *et al.* (1979) and are discussed in depth by Yamada (this volume).

C. Ontogeny and Localization of Crystallins in the Classic Urodelan Lens Regeneration System

The earliest attempt to determine the antigenic or crystallin composition of the regenerating lens was made by Titova (1957) in *T. taeniatus*. Using the relatively insensitive anaphylactic technique, she found that lens antigens were lacking in the 5-day regenerate but present in the 15-day regenerate, apparently at stage IX from the illustration. No other time periods after lentectomy were tested. As a further refinement, Vyazov and Sazhina (1961), using single immunodiffusion, first detected a weak reaction for lens antigens in *T. taeniatus* regenerates 7 days after lentectomy (stage IV/V), and a definite precipitin band 11 days after lentectomy (stage VII/VIII).

Based upon immunoelectrophoretic results Ogawa (1967) has concluded that, in *T. pyrrhogaster* lens regenerates, α-crystallin appears first on day 10, β-crystallin on day 15, and γ-crystallin on day 30 after lens removal. His antiserum absorption procedure, however, probably did not remove antibodies against ubiquitous (common tissue) antigens. These results are not in agreement with those of Takata *et al.* (1965), who detected a first positive immunofluorescence reaction with *T.* (now *Notophthalmus*) *viridescens* γ-crystallin antibodies in the stage V regenerate (12–15 days after lentectomy). It was localized in approximately the same region where these authors had previously obtained (1964) a first positive reaction with an antibody against *total* soluble lens proteins. Yamada, in his monograph (1977), reported unpublished results of the late Chinami Takata indicating the simultaneous appearance of β- and γ-crystallins in the prospective primary lens fibers of the early stage V regenerate. β-Crystallin appeared in the external layer at stage VIII, and α-crystallin in both the epithelium and in the fibers at stage IX. Recently, McDevitt and Brahma (1981b) found that, in *N. viridescens* lens regenerates, β-crystallins appeared in the thickening internal layer at stage V and in the external layer at stage VIII (Fig. 11); the γ-crystallins appeared, albeit erratically, in a few cells of the internal layer at stage V and could not be detected in the epithelium. The α-crystallins appeared at stage VII in a few cells of the developing fiber region, and in the external layer or presumptive lens epithelium at stage VIII, at which time the secondary fibers have begun to form. The results for β- and γ-crystallins reported are thus in agreement with those of Takata (cited by Yamada, 1977), while observations on α-crystallins are not. Whether this is due to a difference in antibody specificity or other undetermined factors cannot now be determined (difficulties in obtaining sufficiently pure α- and β-crystallins suitable for specific antibody production were referred to parenthetically by Takata *et al.*, 1966).

D. Lens Regeneration from the Cornea

In the class Amphibia, in addition to lens regeneration from the pupillary margin of the dorsal iris (Wolffian regeneration), lens regeneration from the inner wall of the outer cornea has been recorded in one species of urodele, *Hynobius unnangso* (Ikeda, 1936, 1939), and in one species of the order Anura, *X. laevis* (in larvae only, Freeman, 1963; Waggoner, 1973). Campbell (1963), however, reportedly obtained lens regeneration in adult *X. laevis,* also from the dorsal iris, which Brahma and van Doorenmaalen (1968) could not confirm.

The actual mechanism which leads to lens regeneration from the cornea in *X. laevis* is not clearly understood. The importance of the eye cup in such regeneration is also controversial (Freeman, 1963; Campbell and Jones, 1968; Waggoner, 1973; Sologub, 1977; Reeve and Wild, 1978). According to Reeve and Wild (1978) a stimulatory factor is released from the eye cup that initiates lens regeneration. Lens regeneration is prevented when the stimulatory factor fails to reach the outer cornea because of a quick closure of the inner cornea. Bosco *et al.* (1979) has also reported that corneal transformation into lens is possible only when the inner cornea has been incised.

This type of lens regeneration from the cornea differs from that of Wolffian regeneration in that the latter can take place throughout the

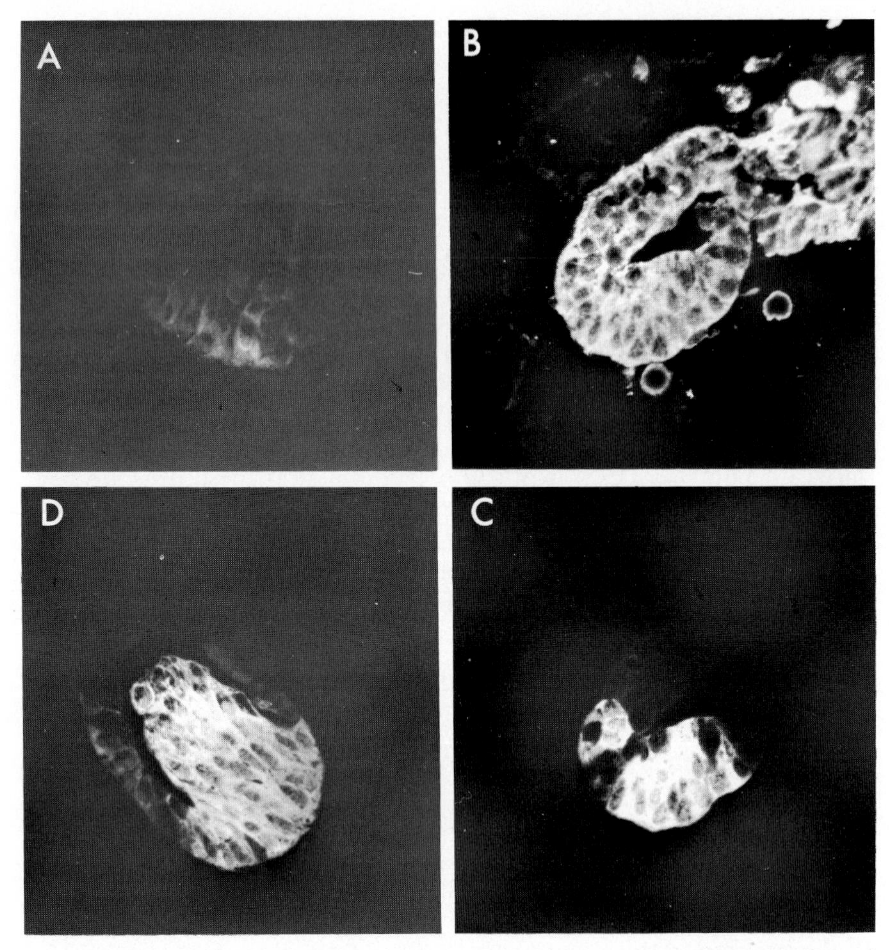

Fig. 11. Dark-field photomicrographs of sections through the regenerating lens of *N. viridescens* treated with antibody to *N. viridescens* total β-crystallins. The cornea of the lentectomized eye is at the top, retina is at the bottom, and the dorsal iris is lateral. Refer also to Fig. 9 (A) Stage V (immunofluorescence). β-Crystallins appear for the first time in a few cells of the vesicle internal layer. (B) Tungsten illumination, as an identical histological reference for (C) which shows stage VII. β-Crystallins at this stage of regenerate morphogenesis exhibit a characteristic "patchy" distribution, with negative cells sharply delimited by positive cell areas in the internal layer. (D) Stage VIII (immunofluorescence); some cells of the external layer are now weakly positive for β-crystallins.

life span of the animal. Since the source of the lens is a population of cells terminally differentiated and withdrawn from the cell cycle, this type of change has been termed "cellular metaplasia" by Yamada (1972). In the case of *X. laevis,* lens regeneration from the cornea is confined to *larval* stages only and takes place *before* differentiation of the adult cornea is complete. It is therefore not a strict case of metaplastic transformation as occurs in urodele dorsal iris epithelial cells.

Freeman (1963) divided the entire process of lens regeneration in *X. laevis* into five arbitrary stages. He also subdivided some of these into early, middle, and late stages for more precise categorization of the process of lens regeneration. A brief description of these stages is given below.

- Stage I Immediately after lens removal, cells of the inner layer of the outer cornea change from squamous to cuboidal in shape.
- Stage II These cells form a mass in the pupillary space two to three cell layers thick.
- Stage III The loose cells of the aggregate tend to orient with regard to each other, distinct from other corneal cells, and form a rudiment which occupies the space of the original lens.
- Stage IV At the beginning of this stage some of the cells furthest from the cornea show enlargement of the nuclei and nucleoli. Mitosis is seen in the cells at the periphery of the vesicle, and the cells with large nucleoli produce irregular fibers, the earliest sign of lens fiber formation. From this stage on, the regenerate is generally found to be separated from the cornea.
- Stage V Secondary fibers begin to develop from the equatorial zone of the lens epithelium, and the nucleoli of the primary fibers begin to disappear. No major histological change is observed in the lens from this stage onward except growth in size and addition of fiber cells.

Freeman (1963) observed a change in the nucleolar ratio in the cell aggregates in *X. laevis* regenerates that develop into lens and suggested that this reflected a shift in the metabolism of the cell population to initiate the process of lens morphogenesis. Overton (1965) observed an increase in the granular endoplasmic reticulum during lens regeneration, followed by its breakdown into vesicles and formation of an increased number of free ribosomes. Waggoner and Reyer (1975) have reported that DNA synthesis in *X. laevis* corneal regenerates is enhanced as early as stage II and is continued until stage IV, the onset of lens fiber differentiation. It is at this stage that Brahma and McDevitt (1974b) found the first positive immunofluorescence for crystallins. As in Wolffian lens regeneration, there appears to be a correlation between the cessation of DNA synthesis and the onset of γ-crystallin synthesis.

Using an antiserum against total soluble lens proteins, Campbell (1965) observed positive immunofluorescence in the inner layer of the cornea 1 day after lens removal. He observed positive immunofluorescence in the regenerates from stage III. With antibodies to total *X. laevis* lens proteins, and *R. pipiens* γ-crystallins, Brahma and McDevitt (1974b) observed the first positive immunofluorescence reaction in early stage IV regenerates, and localization was identical for both antibody preparations. By mid-stage IV, the external layer of the regenerate began to exhibit a weak reaction with the anti-total lens

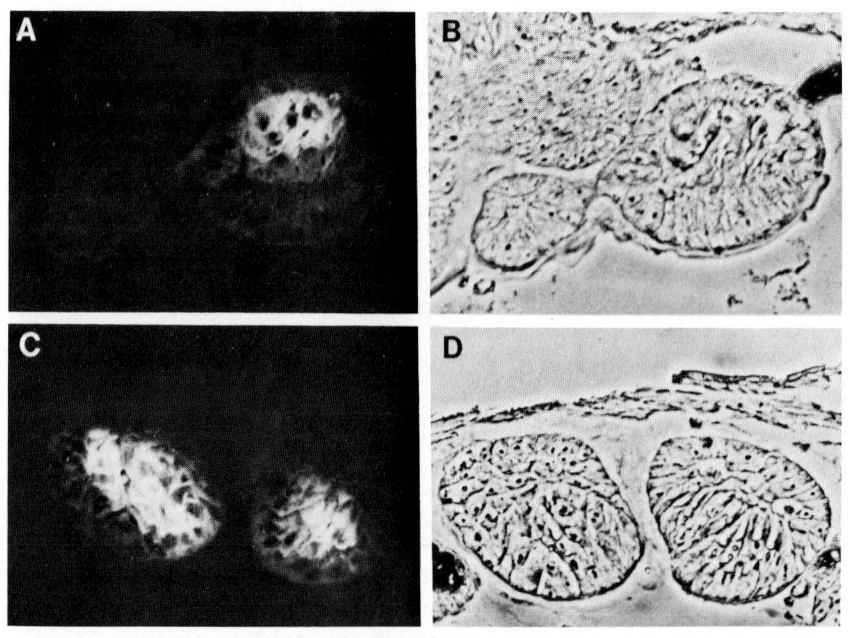

Fig. 12. (A and C) Immunofluorescence photomicrographs of occasionally occurring multiple lens regenerates in *X. laevis*, treated with antibody to *R. pipiens* γ-crystallins. In (A) only one lens regenerates, the more advanced morphologically, demonstrates positive immunofluorescence, confirming the relationship between fiber cell differentiation and the presence of γ-crystallins. In (C) *both* lens regenerates, at comparable stages of differentiation, are positive for γ-crystallins. (B and D) are identical phase-contrast photomicrographs of (A) and (C), respectively, for histological reference purposes. (From Brahma and McDevitt, 1974b.)

protein antibodies, and at mid-stage V the secondary fibers showed a positive reaction.

With the anti-γ-crystallin antibody, only the fiber cells exhibited a positive reaction. At mid-stage V, the secondary fibers showed less fluorescence intensity than the primary fibers. At no time could a positive immunofluorescence reaction be seen in the epithelium with this antibody (Fig. 12).

With his anti-total lens protein antiserum Campbell (1965) did not observe any immunofluorescence in the epithelium; he reported that most antigens were localized in the fibers. This is in contrast with the results of Brahma and McDevitt (1974b) and could be due to the specificites of the antiserum and/or the methods used.

IV. INVERTEBRATE LENS PROTEINS ("CRYSTALLINS")

The eye lens of modern cephalopods is very vertebrate-like in structure, but its development and the analogous proteins it contains are immunologically different from those of vertebrates (Wollman *et al.* 1938; Halbert and Fitzgerald, 1958). Crystallins per se have long been considered vertebrate-specific (e.g., Manski and Malinowski, 1978; Menezes, 1980). The homology between the cephalopod and vertebrate eye is a functional one; this case of convergence in evolution has been discussed by Packard (1972) in great detail. He presents evidence supporting the proposal that every feature fundamental to the vision of cephalopods can also be found in the eyes of fish.

Unlike the situation in vertebrates, the cephalopod eye appears as two thickened placodes on either side of the embryo, which invaginate to form the optic vesicle. The lens is produced from a specialized group of cells known as the ciliary body, corpus epitheliale, or lentigenic body (Williams, 1909; Arnold, 1966, 1967). Fine cytoplasmic processes (lentigenic processes) extend from the cells of the lentigenic body and fuse in the posterior chamber of the eye cup to form the lens primordium. Growth of the lens is accompanied by the addition of lentigenic processes which do not contain nuclei. A fully developed lens consists of an anterior planoconvex and a posterior subspherical section developing from the inner and outer cells of the lentigenic body.

There are relatively few studies on cephalopod lens proteins (Clayton, 1974). Brahma (1978) studied the ontogeny of lens crystallins in three genera of cephalopods, *Loligo vulgaris, Sepia officinalis,* and *Octopus vulgaris,* using homologous antibody to total soluble lens proteins. The embryos were staged according to Naef (1923). In all

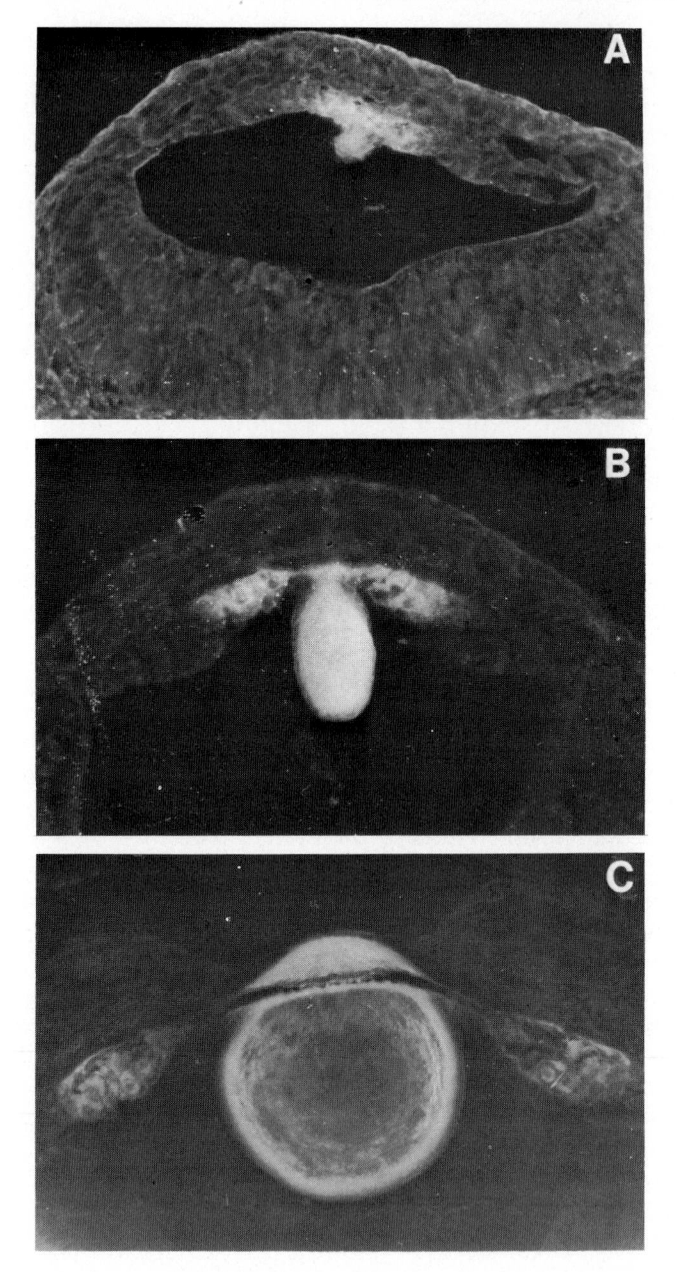
A
B
C

cases it was found that the first positive immunofluorescence appeared simultaneously in the lens and in the lentigenic cells (Fig. 13). In a fully developed lens where the two areas could be seen, a fluorescent cytoplasmic connection was observed between the lens and the lentigenic cells, suggesting a transport of lens material from the lentigenic area, possibly via the microtubules present in the lentigenic processes (Arnold, 1966, 1967).

V. CRYSTALLINS OF THE REGENERATED AND RE-REGENERATED LENS

Campbell and Truman (1977) and Brahma (1980) have investigated the crystallin composition of *X. laevis* corneally regenerated wild-type and periodic albino mutant lenses, respectively. Campbell and Truman noted a reduction in the number of β-crystallin arcs in regenerated wild-type lenses on immunoelectrophoresis, while Brahma, using isoelectric focusing analysis as well as immunoelectrophoresis, found regenerated and normally developed a^p/a^p lenses from 6-month froglets to have identical α-, β-, and γ-crystallin profiles. Whether this difference is due to the use of normal (wild-type) versus mutant *X. laevis* in the two reports, to a paucity of material (Campbell and Truman, 1977, as cited in Brahma, 1980), or to different antibody avidities to the β-crystallins (only antibodies to total crystallins were used) could not be determined.

The first analysis of the crystallin composition of fully regenerated (6 months after lentectomy) lenses from *N. viridescens* dorsal iris was recently reported by McDevitt, (1981). Sephadex G-200 superfine chromatography demonstrated the presence of an α–β-crystallin complex (i.e., coeluting), as observed in the normal adult *N. viridescens* lens. The β-crystallins, however, elute with more heterogeneity and polydispersity than those of the normal lens, with the β_H fraction much reduced. γ-Crystallins also comprise a larger percentage (52 versus 37%) of the elution profile of the regenerated lens crystallins. Thus the regenerated lens exhibits features consid-

---

Fig. 13. Immunofluorescence photomicrographs of sections through the eye lens region of developing *Loligo vulgaris* exposed to homologous total soluble lens protein antibody. (A) Stage 10 (Naef, 1923); (B) stage 15; (C) stage 18. It is interesting to note that the earliest positive reaction was observed in the lentigenic area as well as in the lens. This was observed in all stages of lens development investigated, in support of the histological-causal relationship between the lens and the lentigenic area suggested by earlier workers. (From Brahma, 1978.)

ered embryonic or immature, i.e., increased amounts of γ-crystallins and decreased amounts of β_H (Asselbergs *et al.*, 1979). Peptide maps of total γ-crystallins derived from normal and regenerated lenses (McDevitt and Croft, 1981; Fig. 14) reveal significant differences, probably reflecting the presence of different proportions of γ-crystallin subfractions. Twenty peptides were common to both regenerated and normal lenses; 3 peptides were found only in the regenerated γ-crystallins; and 4 peptides only in the γ-crystallins derived from the adult normal lens.

Isoelectric focusing and immunoelectrophoretic analysis have recently revealed (Borst *et al.*, 1981) that the re-regenerated lens (produced *after* removal of the 6-month regenerated lens) *also* contains crystallins. There appear to be qualitative and quantitative differences in the pH region ascribed by Bours (1980) to β-crystallins, compared to the originally regenerated lens. The cells that produce the second regenerated lens are by necessity those cells which took the pathway of reversion to pigmented iris epithelial cells when they first regenerated. These cells upon relentectomy *now* follow the pathway of conversion to lens cells and produce, albeit not identically, crystallins in the re-regenerate. The mode by which such consecutive diametric deci-

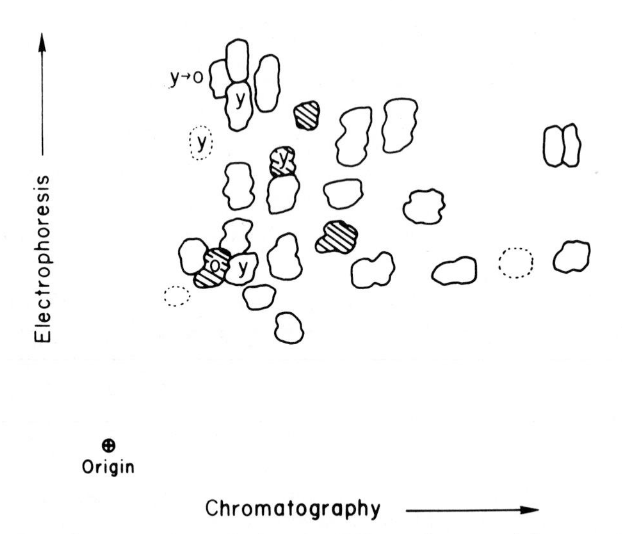

Fig. 14. Peptide maps of tryptic digests of *N. viridescens* adult normal and regenerated γ-crystallins oxidized with performic acid. Cross-hatched areas indicate peptides present only in normal γ-crystallins; broken outlines indicate peptides present only in regenerated γ-crystallins; solid outlines indicate peptides common to both. Conditions as in Fig. 1.

sions are made is not known but should provide an intriguing model for the regulation of cell differentiation.

Thus, the results from studies on lens regeneration in the classical urodelan system indicate that conversion of terminally differentiated iris tissue into lens tissue is accomplished biochemically as well as morphologically. The degree of fidelity, although not absolute, imposed by the differentiation process on morphogenetic events in normal and regenerative lens development is remarkably similar. It is also obvious that the alteration in genome expression in Wolffian lens regeneration (reversion to an amitotic state) is not critical for reutilization of this genome. It remains for new approaches in molecular biology to elucidate the mechanism(s) by which new information is made available to the iris epithelial cell, allowing its transformation into a lens cell after lentectomy.

ACKNOWLEDGMENTS

We are grateful to Aleid Weimar, Arnold Wachter, Th. Hulskes, Kathleen McWilliams, and Sharon DiRienzo for their contributions to this work. In addition, we thank Mary Jane Loop (Biology Division, Oak Ridge National Laboratory) and Randall Reyer for providing us with original photographs for use in preparing Figs. 9 and 10. Much of the research reported herein was supported by NIH grant EY-02534 from the National Eye Institute (D.S.McD.).

REFERENCES

Araki, M., Yanagida, M., and Okada, T. S. (1979). Crystallin synthesis in lens differentiation in cultures of neural retinal cells of chick embryos. *Dev. Biol.* **69,** 170–181.

Arnold, J. M. (1966). On the occurrence of microtubules in the developing lens of the squid *Loligo peali. J. Ultrastruct. Res.* **14,** 534–539.

Arnold, J. M. (1967). Fine structure of the development of the cephalopod lens. *J. Ultrastruct. Res.* **17,** 525–543.

Asselbergs, F. A. M., Koopmans, M., van Venrooij, W. J., and Bloemendal, H. (1979). β-Crystallin synthesis in *Xenopus* oocytes. *Exp. Eye Res.* **28,** 475–482.

Babcock, R. G. (1961). Ribonucleic acid and lens induction in *Xenopus laevis* (Daudin). Thesis, University of Virginia, Charlottesville, Virginia.

Balinsky, B. I. (1951). On the eye cup-lens correlation in some South African amphibians. *Experientia* **7,** 180–181.

Barabanov, V. H. (1966). Formation of organ-specific antigens of the lens in ontogenesis in mice. *Byull. Eksp. Biol. Med.* **62,** 82–85. (Transl.)

Beebe, D. C., Feagans, D. E., and Jebens, H. A. H. (1980). Lentropin: A factor in vitreous humor which promotes lens fiber cell differentiation. *Proc. Natl. Acad. Sci. U.S.A.* **77,** 490–493.

Berardi, C., and McDevitt, D. S. (1982). Lens regeneration from the dorsal iris in *Eurycea bislineata*, the two-lined salamander. *Experientia* (in press).

Berzelius, J. J. (1830). "Larobok i Kemien," Vol. 6, p. 512. P. A. Norstedt och Söner, Stockholm.

Bhat, S. P., Jones, R. E., Sullivan, M. A., and Piatigorsky, J. (1980). Chicken lens crystallin DNA sequences show at least two δ-crystallin genes. *Nature (London)* **284**, 234–238.

Bjork, I. (1961). Studies on γ crystallin from calf lens. I. Isolation by gel filtration. *Exp. Eye Res.* **1**, 145–154.

Bloemendal, H. (1972). Alpha crystallin. Structural aspects and biosynthesis. *Acta Morphol. Neerl. Scand.* **10**, 197–213.

Bloemendal, H. (1977). The vertebrate eye lens. *Science* **197**, 127–138.

Blumenbach, W. (1787). Specimen physiologiae comparatae inter animatea caldi et frigidi sanguinis. *Comment. Soc. Reg. Sci., Göttingen* **8**, 95.

Bon, W. F., and Nobel, P. C. (1958). Electrophoresis of the proteins in the eye lens of calf and cow. *Rec. Trav. Chim. Pays-Bas* **77**, 813.

Bonnet, C. (1781). Sur les reproductions des salamandres. *Oeuvres Hist. Natur. Philos.* **11**, 175–179.

Borst, D. E., Di Rienzo, S. M., and McDevitt, D. S. (1981). Sequential differentiative decisions in newt iris epithelial cells. *J. Cell Biol.* **91**, 28a.

Bosco, L., Filoni, S., and Cannata, S. (1979). Relationship between eye factors and lens formation in the cornea and pericorneal epidermis of larval *Xenopus laevis*. *J. Exp. Zool.* **209**, 261–282.

Bours, J. (1980). Species specificity of the crystallins and the albuminoid of the aging lens. *Comp. Biochem. Physiol.* **65B**, 215–222.

Brahma, S. K. (1977). Ontogeny and localization of pre-α crystallin antigen in *Rana temporaria* lens development. *Exp. Eye Res.* **25**, 311–315.

Brahma, S. K. (1978). Ontogeny of lens crystallins in marine cephalopods. *J. Embryol. Exp. Morphol.* **46**, 111–118.

Brahma, S. K. (1980). Isofocusing and immunoelectrophoretic studies of soluble eye lens proteins from regenerated and normally developed *Xenopus laevis*. *Exp. Eye Res.* **30**, 269–275.

Brahma, S. K., and Lancieri, M. (1980). Isofocusing and immunological investigations on Cephalopod lens proteins. *Exp. Eye Res.* **29**, 671–678.

Brahma, S. K., and McDevitt, D. S. (1974a). Ontogeny and localization of gamma crystallins in *Rana temporaria, Ambystoma mexicanum* and *Pleurodeles waltlii* normal lens development. *Exp. Eye Res.* **19**, 379–387.

Brahma, S. K., and McDevitt, D. S. (1974b). Ontogeny and localization of the lens crystallins in *Xenopus laevis* lens regeneration. *J. Embryol. Exp. Morphol.* **45**, 271–281.

Brahma, S. K. , and van Doorenmaalen, W. J. (1968). Studies on lens regeneration in *Xenopus laevis*. *Experientia* **24**, 519–521.

Brahma, S. K., and van Doorenmaalen, W. J. (1971). Immunofluorescence studies of chick lens FISC and α-crystallin antigens during lens morphogenesis and development. *Ophthalmol. Res.* **2**, 344–357.

Brahma, S. K., and van Doorenmaalen, W. J. (1973). Ontogeny of γ-crystallin in the chick lens. *Exp. Eye Res.* **17**, 341–346.

Brahma, S. K., and van Doorenmaalen, W. J. (1976). Preparation of specific antiserum against *Rana esculenta* pre-α crystallin. *Experientia* **32**, 930–931.

Brahma, S. K., Rabaey, M., and van Doorenmaalen, W. J. (1972). Ontogeny and localization of γ-crystallin antigen in the developing pigeon (*Columba livia*) lens. *Exp. Eye Res.* **14,** 130–134.

Brini, A., Porte, A., and Stoeckel, M. E. (1962). Étude au microscope électronique de quelques problems d'embryologie ocularies chez l'embryon de poulet á des stades précoces. *Bull. Soc. Fr. Ophthalmol.* **75,** 192–209.

Burky, E. L., and Woods, A. C. (1928a). Lens protein—New methods for the preparation of beta crystallin. *Arch. Ophthal.* (*N. Y.*) **57,** 41–45.

Burky, E. L., and Woods, A. C. (1928b). Lens protein—The isolation of a third (gamma) crystallin. *Arch. Ophthal.* (*N. Y.*) **57,** 464–466.

Byers, B., and Porter, K. R. (1964). Oriented microtubules in elongating cells of the developing lens rudiment after induction. *Proc. Natl. Acad. Sci. U.S.A.* **52,** 1091–1099.

Campbell, J. C. (1963). Lens regeneration from iris, retina and cornea in lentectomized eye of *Xenopus laevis. Anat. Rec.* **145,** 214–215.

Campbell, J. C. (1965). An immunofluorescent study of lens regeneration in larval *Xenopus laevis. J. Embryol. Exp. Morphol.* **13,** 171–179.

Campbell, J. C., and Jones, K. W. (1968). The *in vitro* development of the lens from cornea of larval *Xenopus laevis. Dev. Biol.* **17,** 1–15.

Campbell, J. C., and Truman, D. E. S. (1977). Variation in differentiation in the regenerating lens of *Xenopus laevis. Exp. Eye Res.* **25,** 99–100.

Campbell, J. C., Clayton, R. M., and Truman, D. E. S. (1968). Antigens of the lens of *Xenopus laevis. Exp. Eye Res.* **7,** 4–10.

Clayton, R. M. (1970). Problems of differentiation in the vertebrate lens. *Curr. Top. Dev. Biol.* **5,** 115–180.

Clayton, R. M. (1974). Comparative aspects of lens proteins. *In* "The Eye" (H. Davson, ed.), Vol. 5, pp. 399–494. Academic Press, New York.

Clayton, R. M. (1975). Failure of growth regulation of the lens epithelium in strains of fast growing chicks. *Genet. Res.* **25,** 70–83.

Clayton, R. M., and Truman, D. E. S. (1974). The antigenic structure of chick β-crystallin subunits. *Exp. Eye Res.* **18,** 495–506.

Cohen, A. I. (1961). Electron microscopic observations of the developing mouse eye. I. Basement membranes during early development and lens formation. *Dev. Biol.* **3,** 297–316.

Cohen, L. H., Westerhuis, L. W., de Jong, W. W., and Bloemendal, H. (1978). Rat α-crystallin A chains with an insertion of 22 residues. *Eur. J. Biochem.* **89,** 259–266.

Colucci, V. L. (1891). Sulla regenerazione parziale dell 'occhio nei tritoni: Istogenesi e suilluppo—Studio sperimentale. *Mem. R. Accad. Bologna Sez. Sci. Nat.* (*Ser. 5*) **1,** 167–203.

Coulombre, J., and Coulombre, A. J. (1963). Lens development: Fibre elongation and lens orientation. *Science* **142,** 1489–1490.

Coulombre, A. J. (1965). The Eye. *In* "Organogenesis" (R. L. De Haan and H. Ursprung, eds.), pp. 219–251. Holt, New York.

Croft, L. R. (1972). The amino acid sequence of γ crystallin (fraction II) from calf lens. *Biochem. J.* **128,** 961–970.

Cuny, R., and Zalik, S. E. (1981). Effects of bovine pituitary hormone preparations on newt lens regeneration *in vitro*: Stimulation by thyrotropins. *Dev. Biol.* **81,** 23–35.

de Jong, W. W., Terwindt, E. C., and Groenewoud, G. (1976). Subunit compositions of vertebrate α crystallins. *Comp. Biochem. Physiol.* **55B,** 49–56.

de Pomerai, D. I., and Clayton, R. M. (1978). Influences of embryonic stages on the transdifferentiation of chick neural retina cells in culture. *J. Embryol. Exp. Morphol.* **47,** 179-193.

de Pomerai, D. I., Pritchard, D. J., and Clayton, R. M. (1977). Biochemical and immunological studies of lentoid formation in cultures of embryonic chick neural retina and day-old chick lens epithelium. *Dev. Biol.* **60,** 416-427.

Dilley, K. F., and Harding, J. J. (1975). Changes in proteins of the human lens in development and aging. *Biochim. Biophys. Acta* **386,** 391-408.

Dumont, J. N., Yamada, T., and Cone, M. V. (1970). Alteration of nucleolar ultrastructure in iris epithelial cells during initiation of Wolffian lens regeneration. *J. Exp. Zool.* **174,** 187-204.

Eakin, R. M. (1973). "The Third Eye." Univ. of California Press, Berkeley, California.

Eguchi, G. (1963). Electron microscopic studies on lens regeneration. I. Mechanism of depigmentation of the iris. *Embryologia* **8,** 45-62.

Eguchi, G. (1964). Electron microscopic studies on lens regeneration. II. Formation and growth of lens vesicle and differentiation of lens fibres. *Embryologia* **8,** 247-287.

Eguchi, G., and Shingai, R. (1971). Cellular analysis on localization of lens forming potency in the newt iris epithelium. *Dev. Growth Differ.* **13,** 337-349.

Eguchi, G., and Watanabe, K. (1973). Elicitation of lens formation from the "ventral iris" epithelium of the newt by a carcinogen *N*-methyl-*N'*-nitro-*N*-nitrosoguanidine. *J. Embryol. Exp. Morphol.* **30,** 63-71.

Eguchi, G., Abe, S., and Watanabe, K. (1974). Differentiation of lens-like structure from newt iris epithelial cells *in vitro*. *Proc. Natl. Acad. Sci. U.S.A.* **71,** 5052-5056.

Eisenberg, S., and Yamada, T. (1966). A study of DNA synthesis during the transformation of the iris into lens in the lentectomized newt. *J. Exp. Zool.* **162,** 353-368.

Fischel, A. (1900). Über die Regeneration der Linse. *Anat. Hefte* **14,** 1-256.

Francois, J., Wieme, R., Rabaey, M., and Neetens, A. (1954). L'électrophorèse sur papier des protéines hydrosolubles du cristallin. *Experientia* **10,** 79-80.

Freeman, G. (1963). Lens regeneration from the cornea in *Xenopus laevis*. *J. Exp. Zool.* **154,** 39-66.

Halbert, S. P., and Fitzgerald, P. L. (1958). Studies on immunologic organ specificity of ocular lens. *Am. J. Ophthalmol.* **46,** 187-195.

Halbert, S. P., and Manski, P. (1963). Organ specificity with special reference to the lens. *Prog. Allergy* **7,** 107-186.

Harding, C. V., Reddan, J. R., Unakar, N. J., and Bagchi, M. (1971). Control of cell division in the ocular lens. *Int. Rev. Cytol.* **31,** 215-300.

Harding, J. J., and Dilley, K. J. (1976). Structural proteins of the mammalian lens: A review with emphasis on changes on development, aging and cataract. *Exp. Eye Res.* **22,** 1-73.

Hendrix, R. W., and Zwaan, J. (1975). The matrix of the optic vesicle-presumptive lens interface during induction of the lens in the chick embryo. *J. Embryol. Exp. Morphol.* **33,** 1023-1049.

Herbrink, P., and Bloemendal, H. (1974). Studies on β crystallin. I. Isolation and partial characterization of the principal polypeptide chain. *Biochim. Biophys. Acta* **336,** 370-382.

Hesselvik, L. (1939). An electrophoretical investigation on the proteins of the eye lens and the vitreous body. *Skand. Arch. Physiol.* **82,** 151-156.

Hines, M. C., and Olive, J. A. (1970). On the components of rabbit lens γ crystallin. *Life Sci.* (*Pt. 2*) **9,** 1101-10.

Hoperskaya, O. A. (1975). The development of animals homozygous for a mutation causing periodic albinism (a^p) in *Xenopus laevis*. *J. Embryol. Exp. Morphol.* **34,** 253-264.

Hornsby, S., and Zalik, S. E. (1977). Redifferentiation, cellular elongation and the cell surface during lens regeneration. *J. Embryol. Exp. Morphol.* **39,** 23-43.

Hunt, H. H. (1961). A study of the fine structure of the optic vesicle and lens placode of the chick embryo during induction. *Dev. Biol.* **3,** 175-209.

Ikeda, Y. (1936). Beiträge zur Fräge der Fähigkeit zur Linsenregeneration bei einer Art von *Hynobius* (*Hynobius unnangso* Tago). *Arb. Anat. Inst. Kais Jpn. Univ. Sendai* **18,** 17-50.

Ikeda, Y. (1939). Zur Frage der Linsenpotenz der Hornaut in spätenbryonalen und larvalen Stadien bei einer Art von *Hynobius* (*Hynobius unnangso* Tago). *Arb. Anat. Inst. Kais Jpn. Univ. Sendai* **22,** 27-52.

Jacobson, A. G. (1955). The roles of the optic vesicle and other head tissues in lens induction. *Proc. Natl. Acad. Sci. U.S.A.* **41,** 522-525.

Jacobson, A. G. (1966). Inductive process in embryonic development. *Science* **152,** 25-34.

Jedziniak, J. A., Kinoshita, J. H., Yates, E. M., and Benedek, G. B. (1975). The concentration and localization of heavy molecular weight aggregates in aging normal and cataractous human lenses. *Exp. Eye Res.* **20,** 367-369.

Jurand, A., and Yamada, T. (1967). Elimination of mitochondria during Wolffian lens regeneration. *Exp. Cell Res.* **46,** 636-638.

Kabasawa, I., Tsunematsu, Y., Barber, G. W., and Kinoshita, J. H. (1977a). Low molecular weight proteins of the bovine lenses. *Exp. Eye Res.* **24,** 437-448.

Kabasawa, I., Barber, G. W., and Kinoshita, J. H. (1977b). Aging effects and some properties on the human lens low molecular weight proteins. *Jpn. J. Ophthalmol.* **21,** 87-97.

Karasaki, S. (1964). An electron microscopic study of Wolffian lens regeneration in adult newt. *J. Ultrastruct. Res.* **11,** 246-273.

Karkinen-Jääskeläinen, M. (1978). Transfilter lens induction in avian embryo. *Differentiation (Berlin)* **12,** 31-37.

Konyukhov, B. V., Malinina, N. A., Platonov, E. S., and Yakovlev, M. I. (1978). Immunohistochemical study of crystallin synthesis in mouse lens morphogenesis. *Biol. Bull. Acad. Sci. USSR (Engl. Transl.)* **5,** 397-405.

Kramps, H. A., Hoenders, H. J., and Wollensak, J. (1976). Protein changes in the human lens during development of senile nuclear cataracts. *Biochim. Biophys. Acta* **434,** 32-43.

Krause, A. C. (1933). Chemistry of the lens. II. Composition of beta crystallin, albumin (gamma crystallin) and capsule. *AMA Arch. Ophthalmol.* **9,** 617-624.

Kuwabara, T. (1968). Microtubules in the lens. *Arch. Ophthal. (Chicago)* **79,** 189-195.

Kuwabara, T. (1975). The maturation of the lens cells: A morphologic study. *Exp. Eye Res.* **20,** 427-443.

Kuwabara, T., and Imaizumi, M. (1974). Denucleation process of the lens. *Invest. Ophthalmol.* **13,** 973-981.

LeCron, W. L. (1907). Experiments on the origin and differentiation of the lens in *Ambystoma. Am. J. Anat.* **6,** 245-257.

Lewis, W. H. (1904). Experimental studies on the development of the eye in amphibia. I. On the origin of the lens. *Rana palustris. Am. J. Anat.* **3,** 505-536.

Lewis, W. H. (1907). III. On the origin and differentiation of the lens. *Am. J. Anat.* **6,** 473-509.

Liedke, K. B. (1955). Studies on lens induction in *Ambystoma punctatum. J. Exp. Zool.* **130,** 353-380.

Lopaschov, G. V., and Stroeva, O. G. (1964). "Development of the Eye" (Transl. from Russian). Academy of Sciences of the U.S.S.R. (Official publication).

McAvoy, J. W. (1978a). Cell division, cell elongation and the coordination of crystallin gene expression during lens morphogenesis in the rat. *J. Embryol. Exp. Morphol.* **45**, 271-281.

McAvoy, J. W. (1978b). Cell division, cell elongation and distribution of α-, β-, and γ-crystallins in the rat lens. *J. Embryol. Exp. Morphol.* **44**, 149-165.

McAvoy, J. W. (1980a). Cytoplasmic processes interconnect lens placode and optic vesicle during eye morphogenesis. *Exp. Eye Res.* **31**, 527-543.

McAvoy, J. W. (1980b). Induction of the eye lens. *Differentiation (Berlin)* **17**, 137-149.

McAvoy, J. W. (1980c). β and γ crystallin synthesis in rat lens epithelium explanted with neural retina. *Differentiation (Berlin)* **17**, 85-91.

McDevitt, D. S. (1967). Separation and characterization of the lens proteins of the amphibian, *Rana pipiens*. *J. Exp. Zool.* **164**, 21-30.

McDevitt, D. S. (1972a). Presence of lateral eye lens crystallins in the median eye of the American chameleon. *Science* **175**, 763-764.

McDevitt, D. S. (1972b). Sequential appearance of α, β and γ crystallins in embryonic and metamorphic *Rana pipiens*. *Experientia* **28**, 1215-1217.

McDevitt, D. S. (1981). The crystallins of normal and regenerated newt eye lenses. *In* "The Regulability of the Differential State." Plenum, New York. In press.

McDevitt, D. S., and Brahma, S. K. (1973). Ontogeny and localization of the crystallins during *Xenopus laevis* embryonic lens development. *J. Exp. Zool.* **186**, 127-140.

McDevitt, D. S., and Brahma, S. K. (1977). Eye lens development and γ crystallins in *Discoglossus pictus*. *Experientia* **33**, 1087-1089.

McDevitt, D. S., and Brahma, S. K. (1979). Embryonic appearance of α, β and γ crystallins in the periodic albinism (a^p) mutant of *Xenopus laevis*. *Differentiation (Berlin)* **14**, 107-112.

McDevitt, D. S., and Brahma, S. K. (1981a). Ontogeny and localization of the α, β and γ crystallins in newt eye lens development. *Dev. Biol.* **84**, 449-454.

McDevitt, D. S., and Brahma, S. K. (1981b). α, β and γ crystallins in the regenerating lens of *Notophthalmus viridescens*. *Exp. Eye Res.* (*in press*).

McDevitt, D. S., and Clayton, R. M. (1979). Ontogeny and localization of the crystallins during lens development in normal and Hy-1 (hyperplastic lens epithelium) chick embryos. *J. Embryol. Exp. Morphol.* **50**, 31-45.

McDevitt, D. S., and Collier, C. R. (1975). The lens proteins of eastern North American salamanders, and their application to urodelan systematics. *Exp. Eye Res.* **21**, 1-8.

McDevitt, D. S., and Croft, L. R. (1977). On the existence of γ crystallin in the bird lens. *Exp. Eye Res.* **25**, 473-481.

McDevitt, D. S. and Croft, L. R. (1981). Gene expression in Wolffian lens regeneration. *Submitted for publication.*

McDevitt, D. S., Meza, I., and Yamada, T. (1969). Immunofluorescence localization of the crystallins in amphibian lens development, with special reference to the gamma crystallins. *Dev. Biol.* **19**, 581-607.

McKeehan, M. S. (1951). Cytological aspects of embryonic lens induction in the chick. *J. Exp. Zool.* **117**, 31-64.

McKeehan, M. S. (1956). The relative ribonucleic acid content of lens and retina during lens induction in the chick. *Am. J. Anat.* **99**, 131-155.

McKeehan, M. S. (1958). Induction of portion of the chicken lens without contact with the optic cup. *Anat. Rec.* **132**, 297-305.

Maisel, H., and Langman, J. (1961). An immuno-embryological study on the chick lens. *J. Embryol. Exp. Morphol.* **9**, 191-201.

Mangold, O. (1961). Linsen in augenlosen Köpfen und Isolaten von *Triturus alpestris*. *Acta Morphol. Acad. Sci. Hung.* **10**, 153-176.

Manski, W., and Malinowski, K. (1978). The evolutionary sequence and quantities of different antigenic determinants of calf lens alpha crystallin. *Immunochemistry* **15**, 781-786.

Maschino, F., Tamboise, E., and Croiseille, Y. (1980). Mise en évidence d'antigéne spécifiques aux stades précoces de la différenciation du cristallin chez l'embryon de rat. *Arch. Anat. Microsc. Morphol. Exp.* **69**, 233-240.

Mencel, E. (1903). Ein Fall von beiderseitiger Augenlinsenausbildung während der Abwesenheit von Augenblasen. *Arch. Entwicklungsmech. Org.* **16**, 328-339.

Menezes, M. R. (1980). Immunochemical analyses of soluble lens proteins in some marine fishes. *Indian J. Mar. Sci.* **9**, 63-65.

Mikami, Y. (1941). Experimental analysis of the Wolffian lens-regeneration in adult newt *Triturus pyrrhogaster*. *Jpn. J. Zool.* **9**, 269-302.

Mikhalov, A. T., and Korneev, A. Ya. (1980). The lens proteins in adult and embryos of the periodic albino mutant of *Xenopus laevis*. *Wilhelm Roux's Arch. Dev. Biol.* **189**, 155-163.

Modak, S. P., and Perdue, S. W. (1970). Terminal lens cell differentiation. I. Histological and microspectrophotometric analysis of nuclear degeneration. *Exp. Cell Res.* **59**, 43-56.

Morner, C. T. (1894). Untersuchungen der Protein-substanzen in den lichtbrechenden Medien des Auges. *Hoppe-Seyler's Z. Physiol. Chem.* **18**, 61-106.

Muthukkaruppan, V. (1965). Inductive tissue interaction in the development of the mouse lens *in vitro*. *J. Exp. Zool.* **159**, 269-288.

Naef, A. (1923). *Flora ae fauna di Golfo di Napoli* **35** (monogr.).

Nöthiger, R., McDevitt, D. S., and Yamada, T. (1971). Comparison of γ crystallins from frog and newt. *Exp. Eye Res.* **12**, 94-98.

Ogawa, T. (1965). Appearance of lens antigens during the embryonic development of the newt. *Embryologia* **8**, 345-361.

Ogawa, T. (1967). The similarity between antigens in the embryonic lens and in the lens regenerate of the newt. *Embryologia* **9**, 295-305.

Ostrer, H., and Piatigorsky, J. (1980). β-Crystallins of the adult chicken lens: Relatedness of the polypeptides and their aggregates. *Exp. Eye Res.* **30**, 679-689.

Overton, J. (1965). Changes in cell fine structure during lens regeneration in *Xenopus laevis*. *J. Cell Biol.* **24**, 211-222.

Packard, A. (1972). Cephalopod and fish: The limits of convergence. *Biol. Rev.* **47**, 241-304.

Papaconstantinou, J. (1965). Biochemistry of the bovine lens proteins. II. The γ crystallins of the adult bovine, calf and embryonic lenses. *Biochim. Biophys. Acta* **107**, 81-90.

Papaconstantinou, J. (1967). Molecular aspects of lens cell differentiation. *Science* **156**, 338-346.

Philippaux, J. M. (1880). Note sur la production de l'oeil chez la salamandre aquatique. *Gaz. Med. Paris* **51**, 453-457.

Philpott, G. W., and Coulombre, A. J. (1968). Cytodifferentiation of precultured embryonic chick lens epithelial cells *in vitro* and *in vivo*. *Exp. Cell Res.* **52**, 140-146.

Piatigorsky, J., Webster, H. deF., and Craig, S. P. (1972a). Protein synthesis and ultrastructure during the formation of embryonic chick lens fibres *in vivo* and *in vitro*. *Dev. Biol.* **27**, 176-189.

Piatigorsky, J., Webster, H. deF., and Wollberg, M. (1972b). Cell elongation in the cultured embryonic chick lens epithelium with and without protein synthesis: Involvement of microtubules. *J. Cell Biol.* **55**, 82-92.

Pritchard, D. J., Clayton, R. M., and de Pomerai, D. I. (1978). "Transdifferentiation" of chicken neural retinas into lens and pigment epithelium in culture: Controlling influences. *J. Embryol. Exp. Morphol.* **48,** 1-28.

Rabaey, M. (1962). Electrophoretic and immunoelectrophoretic studies on the soluble proteins in the developing lens of birds. *Exp. Eye Res.* **1,** 310-316.

Rabaey, M. (1965). Lens proteins during embryonic development of different vertebrates. *Invest. Ophthalmol.* **4,** 560-578.

Rabaey, M., Rikkers, I., and De Mets, M. (1972). Low molecular weight protein (γ-crystallin) in the lens of birds. *Exp. Eye Res.* **14,** 208-213.

Ramaekers, F. C. S., van Kan, P. L. E., and Bloemendal, H. (1979). A comparative study of β-crystallins from ungulates, whale and dog. *Ophthalmol. Res.* **11,** 143-153.

Reese, D. H., Puccia, E., and Yamada, T. (1969). Activation of ribosomal RNA synthesis in initiation of Wolffian lens regeneration. *J. Exp. Zool.* **170,** 259-268.

Reeve, J., and Wild, A. E. (1978). Lens regeneration from cornea of larval *Xenopus laevis* in the presence of the lens. *J. Embryol. Exp. Morphol.* **48,** 205-214.

Reszelbach, R., Shinohara, T., and Piatigorsky, J. (1977). Resolution of two distinct chick δ-crystallin bands by polyacrylamide gel electrophoresis in the presence of sodium dodecyl sulfate and urea. *Exp. Eye Res.* **25,** 583-593.

Reyer, R. W. (1948). An experimental study of lens regeneration in *Triturus viridescens viridescens.* I. Regeneration of a lens after lens extripation in embryos and larvae of different ages. *J. Exp. Zool.* **107,** 217-267.

Reyer, R. W. (1950). An experimental study of lens regeneration in *Triturus viridescens viridescens.* II. Lens development from the dorsal iris in the absence of embryonic lens. *J. Exp. Zool.* **113,** 317-353.

Reyer, R. W. (1954). Regeneration of the lens in amphibian eye. *Q. Rev. Biol.* **29,** 1-46.

Reyer, R. W. (1971). DNA synthesis and incorporation of labeled iris cells into the lens during lens regeneration in adult newts. *Dev. Biol.* **24,** 214-245.

Reyer, R. W. (1977). The amphibian eye: Development and regeneration . *In* "Handbook of Sensory Physiology. VII/5. The Visual System in Evolution," (F. Crescitelli, ed.), pp. 309-390. Springer-Verlag, Berlin and New York.

Roy, D., and Spector, A. (1976a). Human alpha crystallin. III. Isolation and characterization of protein from normal infant lenses and old lens peripheries. *Invest. Ophthalmol.* **15,** 394-399.

Roy, D., and Spector, A. (1976b). High molecular weight protein from human lenses. *Exp. Eye Res.* **22,** 273-279.

Sato, T. (1930). Beitrage zur Analyse der Wolffschen Linsenregeneration. I. *Arch. Entwicklungsmech. Org.* **122,** 451-493.

Sato, T. (1940). Vergleichende Studies über die Geschwindigkeit der Wolffsches Linsenregeneration bei *Triton taeniatus* und bei *Diemyctylus pyrrhogaster. Wilhelm Roux' Arch. Entwicklungsmech. Org.* **140,** 470-613.

Sato, T. (1951). Über die linsenbildende Fahigkeit des Pigmentepithels bei *Diemyctylus pyrrhogaster.* 1. Pigmentepithel aus dorsalen Augenbereich. *Embryologia* **1,** 21-57.

Shubert, E. D., Trevithick, J. R., and Hollenberg, M. J. (1970). Localization of γ-crystallins in the developing lens of the rat. *Can. J. Ophthalmol.* **5,** 353-365.

Silver, P. H. S., and Wakey, J. (1974). Fine structure, origin and fate of extracellular materials in the interspace between the presumptive lens and presumptive retina of the chick embryo. *J. Anat.* **118,** 19-31.

Sirlin, J. L., and Brahma, S. K. (1959). Studies on embryonic induction using radioactive tracers. II. The mobilization of protein components during induction of the lens. *Dev. Biol.* **1,** 234-246.

Sologub, A. A. (1977). Mechanism of repression and derepression of artificial transformation of pigmented epithelium into retina in *Xenopus laevis*. *Wilhelm Roux's Arch. Dev. Biol.* **182**, 277-291.

Spector, A., Wandel, T., and Li, L. K. (1968). The purification and characterization of the highly-labelled protein fraction from calf lens. *Invest. Ophthalmol.* **7**, 179-190.

Spector, A., Freund, T., Li, L. K., and Augusteyn, R. D. (1971). Age dependent changes in the structure of alpha crystallin. *Invest. Ophthalmol.* **10**, 677-686.

Spector, A., Stauffer, J., and Sigelman, J. (1973). Preliminary observations upon the proteins of the human lens. *Ciba Found. Symp.* **19**, 185-202.

Spemann, H. (1901). Über korrelation in der Entwicklung des Auges. *Verh. Anat. Ges.* **15**, 61-79.

Spemann, H. (1912). Zur Entwicklung des Wirbeltierauges. *Zool. Jahrb.* (*Abt. 3*) **32**, 393-424.

Srinivasan, B. D. (1960). Whole mount preparation of the fish lens epithelium from various species. *Biol. Bull.* (*Woods Hole, Mass.*) **119**, 339.

Srinivasan, B. D., and Harding, C. V. (1961). *In vitro* studies on thymidine incorporation and cell division in skate (*Raja erinace*). *Biol. Bull.* (*Woods Hole, Mass.*) **121**, 409.

Srinivasan, B. D., and Iwamoto, T. (1973). Electron microscopy of rabbit lens nucleoli. *Exp. Eye Res.* **16**, 9-18.

Stauffer, J., Rothschild, C., Wandel, T., and Spector, A. (1974). Transformation of alpha-crystallin polypeptide chains with aging. *Invest. Ophthalmol.* **13**, 135-146.

Stockard, C. R. (1910). The independent origin and development of the crystallin lens. *Am. J. Anat.* **10**, 393-424.

Stone, L. S. (1967). An investigation recording all salamanders which can and cannot regenerate a lens from the dorsal iris. *J. Exp. Zool.* **164**, 87-104.

Swanborn, L. (1966). The method of simultaneous electrophoresis of antiserum and antigen (immuno-osmophoresis) applied to lens antigens. *Exp. Eye Res.* **5**, 302-308.

Takata, C., Albright, J. F., and Yamada, T. (1964). Study of lens antigens in the developing newt lens with immunofluorescence. *Exp. Cell Res.* **34**, 207-210.

Takata, C., Albright, J. F., and Yamada, T. (1965). Lens fiber differentiation and gamma crystallins: Immunofluorescent study of Wolffian regeneration. *Science* **147**, 1299-1301.

Takata, C., Albright, J. F., and Yamada, T. (1966). Gamma-crystallins in Wolffian lens regeneration demonstrated by immunofluorescence. *Dev. Biol.* **14**, 382-400.

Ten Cate, G. (1956). The intrinsic embryonic development. *Verh. K. Ned. Akad. Wet.* **51**, 1-257.

Titova, I. I. (1957). A study of the antigenic properties of the crystalline lens in Wolffian regeneration. *Bull. Exp. Biol. Med.* (*Engl. Transl.*) **43**, 715-719.

Uhlenhuth, P. T. (1903). Zur Lehre von den Unterscheidungen verschiedener Eiweiszarten mitHilfe spezifischer Sera. *Festschr. 60. Geburtstag R. Koch, (Gustav Fischer, Jena)*, pp. 49-74.

van Dam, A. F. (1966). Purification and composition study of β_s-crystallin. *Exp. Eye Res.* **5**, 255-266.

van Dam, A. F. (1967). Thesis, Nijmegen University, Nijmegen, Netherlands.

van Dam, A. F., and Ten Cate, G. (1966). Isolation and some properties of α-crystallin. *Biochim. Biophys. Acta* **121**, 183-186.

van de Broek, W. G. M., Leget, J. N., and Bloemendal, H. (1973). FM-crystallins: A neglected component among lens proteins. *Biochim. Biophys. Acta* **310**, 278-282.

van de Kamp, M., and Zwaan, J. (1973). Intracellular localization of lens antigens in the developing eye of the mouse embryo. *J. Exp. Zool.* **186**, 23-32.

van der Starre, H. (1978). Biochemical investigation of lens induction *in vitro*. II. Demonstration of the induction substance. *Acta Morphol. Neerl. Scand.* **16,** 109-120.

van Kleef, F. S. M., de Jong, W. W., and Hoenders, H. J. (1975). Stepwise degradation and deamidation of the eye lens protein α crystallin in aging. *Nature (London)* **258,** 264-266.

Vermorken, A. J. M., and Bloemendal, H. (1978). α-Crystallin polypeptides as markers of lens cell differentiation. *Nature (London)* **271,** 779-781.

Vermorken, A. J. M., Herbrink, P., and Bloemendal, H. (1977). Synthesis of lens protein *in vitro:* Formation of β-crystallin. *Eur. J. Biochem.* **78,** 617-622.

Viollier, G., Labhart, H., and Süllmann, H. (1947). Elecktrophoretische Untersuchung der löslichen Linsenproteine. *Helv. Physiol. Pharmacol. Acta* **5,** C10-C12.

Vyazov, O. E., and Sazhina, M. V. (1961). Immuno-biological study of the process of lens regeneration in *Triton taeniatus. Zh. Obshch. Biol.* **4,** 305-310.

Waggoner, P. R. (1973). Lens differentiation from cornea following lens extripation or cornea transplantation in *Xenopus laevis. J. Exp. Zool.* **186,** 97-110.

Waggoner, P. R., and Reyer, R. W. (1975). DNA synthesis during lens regeneration in larval *Xenopus laevis. J. Exp. Zool.* **192,** 65-72.

Waggoner, P. R., Lieska, N., Alcala, J., and Maisel, H. (1976). Ontogeny of the chick lens β-crystallin polypeptides by immunofluorescence. *Ophthalmol. Res.* **8,** 292-301.

Weiss, P. (1950). Perspectives in the field of morphogensis. *Q. Rev. Biol.* **26,** 177-198.

Weiss, P., and Jackson, S. F. (1961). Fine structural changes associated with lens determination in the avian embryo. *Dev. Biol.* **3,** 532-554.

Williams, L. A., and Piatigorsky, J. (1979). Comparative and evolutionary aspects of δ crystallin in the vertebrate lens. *Eur. J. Biochem.* **100,** 349-357.

Williams, L. W. (1909). "The Anatomy of the Common Squid *Loligo peali* Lesueur." E. J. Brill, Leiden, Holland.

Woerdeman, M. W. (1941). Self-differentiation of the lens anlage of *Rana esculenta* after transplantation. *Acta Neerl. Morphol. Norm. Pathol.* **4,** 91-94.

Woerdemann, M. W. (1953). The differentiation of the crystallin. *J. Embryol. Exp. Morphol.* **1,** 301-305.

Wolff, G. (1895). Entwicklungsphysiologische Studien. I. Die Regeneration der Urodelelinse. *Arch. Entwicklungsmech. Org.* **1,** 380-390.

Wolff, G. (1901). Entwicklungsphysiologische Studien. II. Weitere Mittheilungen zur Regeneration der Urodelenlinse. *Arch. Entwicklungsmech. Org.* **12,** 307-351.

Wolff, G. (1904). Entwicklungsphysiologische Studien. III. Zur Analyse der Entwicklungspotenzen des Irisepithels bei *Triton. Arch. Mikrosk. Anat.* **63,** 1-9.

Wollman, E., Gonzales, P. and Ducrest, P. (1938). Recherches sérologiques sur les milieux transparents de l'oeil: Propriétés spécifiques du cristallin. *C. R. Soc. Biol.* **127,** 668-670.

Woods, A. C., and Burky, E. L. (1927). Lens protein and its fractions. *JAMA, J. Am. Med. Assoc.* **89,** 102-110.

Yamada, T. (1967a). Cellular and subcellular events in Wolffian lens regeneration. *Curr. Top. Dev. Biol.* **2,** 247-283.

Yamada, T. (1967b). Cellular synthetic activities in induction in tissue transformation. *Cell Differ., Ciba Found. Symp., 1967,* pp. 116-126.

Yamada, T. (1972). Control mechanism in cellular metaplasia. *In* "Cell Differentiation" (R. Harris, P. Allin, and D. Viza, eds.), pp. 56-60. Copenhagen, Munksgaard.

Yamada, T. (1977). Control mechanisms in cell type conversion in newt lens regeneration. *Monogr. Dev. Diol.* **13.**

Yamada, T., and Beauchamp, J. J. (1978). The cell cycle of cultured iris epithelial cells: Its possible role in cell-type conversion. *Dev. Biol.* **66,** 275-278.

Yamada, T., and Dumont, J. N. (1972). Macrophage activity in Wolffian lens regeneration. *J. Morphol.* **136,** 367-384.

Yamada, T., and Karasaki, S. (1963). Nuclear RNA synthesis in newt iris engaged in regenerative transformation into lens cells. *Dev. Biol.* **7,** 595-640.

Yamada, T., and McDevitt, D. S. (1974). Direct evidence for transformation of differentiated iris epithelial cells into lens cells. *Dev. Biol.* **38,** 104-118.

Yamada, T., and Roesel, M. E. (1969). Activation of DNA replication in the iris epithelium by lens removal. *J. Exp. Zool.* **171,** 425-432.

Yamada, T., and Takata, C. (1963). An autoradiographic study of protein synthesis in regenerative tissue transformation of iris into lens in the newt. *Dev. Biol.* **8,** 358-369.

Yamada, T., Dumont, J. N., Moret, R., and Brun, J. P. (1978). Autophagy in dedifferentiating newt iris epithelial cells *in vitro*. *Differentiation (Berlin)* **11,** 133-147.

Zalokar, M. (1944). Contribution á l'étude de la régénération du cristallin chez le *Triton*. *Rev. Suisse Zool.* **51,** 443-521.

Zeitlin, S., and McDevitt, D. S. (1972). Fluorescent antibody study of the developing lens of the blind cave fish. *J. Cell. Biol.* **59,** 375a.

Zigler, J. S., Jr. (1978). Age-related changes in the polypeptide composition of β-crystallin from bovine lens. *Exp. Eye Res.* **26,** 537-546.

Zigler, J. S., and Sidbury, J. B. (1973). Structure of calf lens β crystallin. *Exp. Eye Res.* **16,** 207-214.

Zigler, J. S., and Sidbury, J. B. (1976). A comparative study of the β crystallins of four sub-mammalian species. *Comp. Biochem. Physiol.* **55B,** 19-24.

Zigler, J. S., Jr., Horowitz, J., and Kinoshita, J. H. (1980). Human β-crystallin. I. Comparative studies on the β_1-, β_2- and β_3-crystallins. *Exp. Eye Res.* **31,** 41-55.

Zigman, S. (1969). Sulphonation of rat lens proteins. *Biochim. Biophys. Acta* **181,** 310-322.

Zwaan, J. (1963). Immunochemical analysis of the eye lens during development. Thesis, University of Amsterdam, Amsterdam, Netherlands.

Zwaan, J. (1975). Immunofluorescent studies on aphakia, a mutation of a gene involved in the control of lens differentiation in the mouse embryo. *Dev. Biol.* **44,** 306-312.

Zwaan, J., and Ikeda, A. (1965). An immunochemical analysis on the β-crystallins of the chick lens: Ontogenetic and phylogenetic aspects. *In* "Structure of the Eye," 2nd Symp. (J. Rohen, ed.), pp. 419-429. Schattauer, Stuttgart.

Zwaan, J., and Ikeda, A. (1968). Macromolecular events during differentiation of the chick lens. *Exp. Eye Res.* **7,** 301-311.

5

Transdifferentiation of Lens Cells and Its Regulation

TUNEO YAMADA

I. INTRODUCTION

At the present time, the mainstream of research on cell differentiation is the molecular biological approach, in which sequence analysis of DNA, the diversity and dynamics of mRNA, the changing pattern of cell surface macromolecules, and the role of chromatin proteins in transcriptional control are the dominating themes. Although the

Cell Biology of the Eye

ISBN 0-12-483180-X

necessity of these studies is unquestionable, the unique value of investigations into the mechanisms which control cell differentiation in intact living systems must be recognized. One example which suggests the indispensability of such investigations is the demonstration of the role of the localized cytoplasmic diversity of egg cells in determining the developmental pathway of blastomeres. No one could have predicted this situation from strictly molecular biological information. The combining of such developmental information with molecular biological principles leads us to a higher level of understanding of the ontogenetic control of cell differentiation. In this chapter we deal with the reprogramming of cell differentiation at later developmental stages at which tissue differentiation has been initiated. Information on the mechanism of the reprogramming of differentiation also provides an understanding of the type of regulation of cell differentiation actually occurring in the living system and should supplement the molecular description of cell differentiation in obtaining a biologically meaningful reconstruction of elementary events into an organized system.

Experimental embryology emerged on the biological scene in the latter part of the last century and flourished in the early part of this century. One of its major contributions has been to demonstrate that, by altering the developmental environment of a specific region of the embryo by surgical methods, without manipulating the genetic apparatus, the differentiation events occurring in normogenesis can be greatly modified, and that these modifications follow definite patterns from which one can draw conclusions concerning the principles of epigenetic control of development. Most of these modifications can be called reprogramming, because they consist of shifting the pathway of one cell group to other pathways which are in normogenesis reserved for other cell groups. The information on reprogramming of the differentiation pathway that has accumulated for more than a century is vast and divergent, but various cases of reprogramming can be classified according to the developmental stages of the cell groups in which they occur. The phenomena discussed in this chapter involve special cases of reprogramming that occurs late in development after the differentiation of tissues or cell types has been more or less expressed, and can be called transdifferentiation. During the era of surgical experimental embryology, the eye tissue provided a number of prominent examples of transdifferentiation. In recent years, there has been a surge of interest in this area because of the discovery of new systems of transdifferentiation through the application of cell culture methods. It is a striking fact that the majority of the known transdifferentiating systems have lens differentiation as the final pathway, although no explanation has been found for this observation. The main purpose of this chapter is to review these cases of transdifferentiation into lenses and to draw conclusions which may contribute to our general understanding of the regulation of differentiation operating in intact cells and tissues.

Review articles covering the area in question include those by Mangold (1931), Reyer (1954, 1977), Scheib (1965), Yamada (1967, 1977), Clayton (1970, 1978), Okada (1976, 1980), Lopashov (1977), McAvoy (1980), and Piatigorsky (1981).

Research techniques and semantics in this area are apt to be controversial. In this chapter an effort will be made to be critical in both respects. The following section serves to clarify a number of terminological problems related to the expressions used in discussing reprogramming, as well as the general concept of differentiation.

II. DEFINITION OF KEY EXPRESSIONS

First it is necessary to clarify the meaning of the term "differentiation" as used in this article. In developmental biology and cell biology differentiation is a relative concept and makes sense only when the qualities affected by it are specified. It is unrealistic to assume that cells are either differentiated or undifferentiated in a general or absolute sense. All cells of multicellular organisms exist on the basis of differential expression of their genomic information. Therefore they must be differentiated in one respect or another. In this chapter, differentiation denotes, if not otherwise indicated, the appearance and establishment of the structural and functional properties of the cells in question associated with specificities of somatic tissues and their cell types. Thus, although the zygote and its direct descendants are highly specialized because of the cell-type differentiation undergone during gametogenesis of the last generation, they are regarded here as not differentiated.

The concept of transdifferentiation (Selman and Kafatos, 1974) has been used with different degrees of precision and with various connotations. To make the dicussion in this chapter meaningful, expressions related to the reprogramming of differentiation, including transdifferentiation, must be defined and their relationships with each other clarified. It is pertinent to do this from an ontogenetic perspective, since we are dealing with the developmental control of differentiation.

In the initial stage of vertebrate ontogenesis, the polarity of the egg cell establishes localized diversities of cell properties within the clonal cell population derived from the zygote, which are later translated into the embryonic organization associated with the differentiation of tissues and cell types. However, during the early stages of development, cells retain the ability to traverse a wide range of differentiation pathways. For instance, the gastrula ectodermal cells of the amphibian embryo are able to participate in almost all pathways of somatic cell-type differentiation involving the three germ layers (Yamada, 1961), and even in the pathway of primordial germ cells (Satasurja and Nieuwkoop, 1974). The type of differentiation realized is decided by intercellular interactions and other factors of the cell environment, although it is erroneous to assume that the cell plays only a passive role. The range of realizable differentiation pathways (prospective potency) is a function of the developmental stage; it becomes restricted stepwise as cells advance in stages and converges to the normal pathway (prospective significance). This restriction of differentiation potentiality is called determination. Some time before the differentiated state of cells is expressed, the determination of differentiation is completed so that the cells have no other options for differentiation. As a rule, this determination is stable, so that even if the differentiated state is lost, for instance by induced proliferation, the cells restore the original state of differentiation whenever the conditions allow its expression. It should also be pointed out that expression of the differentiated state usually occurs stepwise, and that the stability of the differentiated state increases in parallel.

The classical data of experimental embryology demonstrate that the normal developmental program as outlined in the preceding paragraph can be shifted. But because of progressive determination, the aptitude for reprogramming steadily decreases with the progress of development. The empirical data concerning the reprogramming of embryonic cells before the final determination, and its possible mechanisms, make up an important chapter of experimental embryology. The reprogramming which occurs after determination is completed but before the differentiated state is attained is called transdetermination. The prototype of transdetermination is the alteration of the organ type of *Drosophila* imaginal disks by grafting into the adult body cavity (for review, see Hadorn, 1965, 1966; Gehring and Nöthiger, 1973). It should be pointed out that theoretically the concept of transdetermination is self-contradictory, because if differentiation has been truly determined, it should be unalterable, and if it can be altered, it cannot have been determined. However, there is a situation where a set of commonly used test methods for determination show that a cell group has been fully determined for a particular pathway corresponding to its prospective significance and has been generally accepted as determined, but a new method is found which alters the pathway of the cell group. Transdetermination of imaginal disks is such a case.

In this chapter reprogramming of the differentiation pathway after a cell group has expressed a specific differentiated state will be called transdifferentiation. This term will be used without any implication of the specific mechanism involved. Since expression of the differentiated state occurs stepwise, often in correlation with the control of cell replication, there may be different modes of transdifferentiation for the different steps of differentiation at which the alteration occurs. Furthermore, since transdifferentiation can be applied to different levels of organization (cell, cell group, tissue, etc.), as a whole it comprises a number of phenomena of divergent cell biological values. One important parameter of the transdifferentiating systems is the degree of homogeneity of the cell population in question. If the cell population is heterogeneous, it is possible that the original differentiated state is manifested by a group of cells, and that after an interval of time this group loses its differentiated state, while another cell group traverses another pathway of differentiation and expresses the second differentiated state. Even if this is the case, the system as a whole will be called a transdifferentiating system in this chapter.

Cell-type conversion is a phenomenon in which a cell that has already expressed a specific cell-type differentiation loses this differentiation and becomes committed to and expresses another cell-type differentiation (Yamada, 1977), either with or without intervening cell replication. It is clear from this definition that cell-type conversion is a specific type of transdifferentiation. Certainly there are further questions concerning the cell type and cell-type differentiation. These questions will not be discussed here because of the nature of this chapter.

"Metaplasia" is an expression originally used to denote the pathological alteration of tissue specificity *in vivo* associated with changes in physiological conditions. A well-known example of metaplasia is the option between the mucus-secreting and keratinizing conditions of a number of epithelia under the influence of vitamin A levels. This alteration in differentiation is associated with a new commitment of precursor cells which were not committed in the original differentiation (Fell and Mellanby, 1953). Hence it does not involve cell-type conversion. However, the term "metaplasia" has also been applied to the phenomenon now called cell-type conversion. The expression "cellular metaplasia" (Grobstein, 1959) was introduced to distinguish this phenomenon from the other type of metaplasia mentioned above, for which the expression "tissue metaplasia" was proposed. These are valid terms, but are falling into disuse because they are not self-explanatory and have often been used in ambiguous ways. In this chapter the term "metaplasia" will be avoided when possible. Although the expressions "metaplasia" and "transdifferentiation" are almost identical in meaning, they seem to differ somewhat in connotation. In some cases "metaplasia" is interpreted to mean that dedifferentiation or activation of a new differentiation process is involved, while "transdifferentiation" is taken to imply that a sudden switch in the differentiated state occurs. In this chapter we will ignore the connotations associated with the concepts of reprogramming and restrict the meaning of these terms to the definitions given above.

The expressions "autotypic" and "allotypic," as used in transdetermination (Gehring and Nöthiger, 1973), will be sometimes utilized. "Autotypic" denotes the type of differentiation according to the normal fate of the cells or the original differentiation state, and "allotypic" the type of differentiation not in conformity with the normal fate or the original state of differentiation.

III. IRIS EPITHELIUM AS A SOURCE OF LENS CELLS

System A: Conversion of Iris Epithelial Cells (IECs) to Lens Cells in Adult Newts

Research on this system dates back to the latter part of the last century when the regeneration of amphibian eyes was actively investigated. While studying the regenerative process with histological methods, after partial removal of the eye, Colucci (1891) became aware that the epithelium of the newt dorsal iris had the capacity to form a new lens. Independent of this discovery, Wolff (1895) observed, after complete lentectomy of newt eyes, regeneration of the lens from the dorsal iris epithelium and made detailed histological studies on this phenomenon. At the beginning of this century many workers participated in the analysis of this phenomenon, called Wolffian lens regeneration, with surgical methods, and it became one of the classical subjects of experimental embryology. A number of reviews are available on this phase of research (Mangold, 1931; Reyer, 1954, 1977). The phenomenon is known to be restricted to a small group of urodeles (Stone, 1967). The species most frequently used for research on this system belong to the genera *Triturus* (*Triton*), *Cynopus,* and *Notophthalmus,* all of which can be called newts.

When cell biological studies on this subject were initiated early in the 1960s, one of the first findings which attracted workers was a regular association of the proliferation of iris epithelial cells with lens regeneration. In adult newts, IECs are completely withdrawn from the cell cycle. When the lens is removed from the same eye, these cells start to replicate DNA from day 3, and to divide from day 4 (Reyer, 1966, 1971; Eisenberg and Yamada, 1966; Yamada and Roesel, 1969, 1971; Eguchi and Shingai, 1971). This reentry into the cell cycle is only partially synchronized, and the first peak of mitotic cells is observed on day 7. Further details of the labeling pattern of mitotic cells suggest that IECs are arrested at G_0 and go through the transition to G_1 before replicating DNA (Yamada, 1977). A rapid increase in RNA synthesis is followed by an increase in protein synthesis during the early phase of lens regeneration (Reese *et al.*, 1969; Reese, 1973; Thorpe *et al.*, 1974). Active cell proliferation in IECs starts to diminish before day 20, and no mitotic cells can be found after day 30 in *Notophthalmus* tissue. The rudiment of a new lens is laid down about day 15, and the cell proliferation in the lens epithelium continues during the whole period of growth of the regenerated lens (Yamada and Roesel, 1969).

Direct immunofluorescence involving antibodies against total crystallins or γ-crystallins, and absorbed antibodies against α- or β-crystallins, were used to detect various groups of crystallins in paraffin sections through tissues involved in lens regeneration (Takata *et al.*, 1964, 1966; Yamada, 1966). No crystallin has been detected from day 0 to day 13, during which time the middorsal margin of iris epithelium starts to show fully depigmented cells (Sato stage III). Later, when depigmented cells form the lens vesicle (Sato stage IV–V), the cells of the internal wall of the vesicle start to show β-crystallins, sometimes together with α-crystallins. Soon afterward, at Sato stage V, γ-crystallins can be detected in the same region where the primary lens fibers are now being formed. Subsequently the reactions of the three crystallins, especially that of gamma-crystallins, intensify and expand into a larger number of cells in parallel with the growth of the lens.

According to the immunofluorescence study of Ogawa (1963), who applied antilens serum absorbed with iris to unsectioned test tissue, lens antigens appear in the dorsal part of the iris soon after depigmentation, before the lens vesicle is formed. The same author (1967) has reported that α-crystallins become detectable on immunoelectrophoresis involving a similar antilens serum in regenerating iris before lens vesicle formation. Furthermore, Ogawa (1964) demonstrated that depigmented cells and lens rudiment in a lens-regenerating larval newt could be destroyed by treatment with antilens serum. These studies by Ogawa tend to indicate an earlier appearance of lens antigens than the results obtained by Takata *et al.* Since a comparison of the immunoelectrophoretic patterns obtained by Ogawa with those obtained by Takata (reproduced in Yamada, 1977) indicated the presence of prominent arcs in the α-crystallin region in Ogawa's pattern which are lacking in Takata's pattern, the difference in the outcome of the two studies is probably due to these molecules. It should also be added that Ogawa used *Cynopus* (*Triturus*) *pyrrhogaster,* and Takata *Notophthalmus* (*Triturus*) *viridescens,* as experimental organisms.

An attempt to follow proliferating iris epithelial cells *in situ* after lentectomy by labeling with [^{3}H]thymidine by intraperitoneal injection and studying the labeling pattern after various time intervals indicated that part of the progeny of IECs participates in formation of the new lens, while the rest of the progeny remained in the iris epithelium after formation of the new lens and contributed to the recovery and establishment of the epithelium (Eisenberg and Yamada, 1966; Reyer, 1971; Eguchi and Shingai, 1971). Thus two pathways can be distinguished for the progeny of IECs which are called the pathways of conversion and retrieval. Iris epithelial cells are typical melanocytes. In the normal adult condition, their cytoplasm is loaded with melanosomes which have been synthesized and accumulated in the cells. In the conversion pathway the number of melanosomes rapidly decreases, and they completely disappear from the cytoplasm before crystallins become detectable. In the retrieval pathway the number of melanosomes decreases, but no cells lose them completely. After withdrawal from the cell cycle they regain the original levels of pigmentation. That these two pathways are under epigenetic control is suggested by the fact that lens regeneration can be repeated in the same eye probably indefinitely (Nikitenko, 1939), so that the cells traversing the pathway of retrieval should be able to produce cells for the pathway of conversion in the next round of regeneration. That they do so is indicated by McDevitt and Brahma in another chapter in this volume. Further observations suggesting epigenetic control of these pathways will be discussed later.

The autoradiographic data cited above as a whole gave the impression that cell proliferation was more intensive in the conversion pathway than in the retrieval pathway. To obtain more information, the cell cycle parameters of iris epithelial cells in both pathways *in situ* were estimated by a mathematical procedure, using a computerized program, from the percentage of labeled mitotic cells as a function of time after labeling (Yamada *et al.,* 1975). The data indicated a significant difference in the total cell cycle time between two pathways, the

TABLE I

Estimates of the Duration of Cell Cycle Phases of Iris Epithelial Cells *in Situ* after Lentectomy in Adult Newts[a]

Pathway	Duration of cell cycle phases			Duration of total cell cycle
	G_1	S	G_2	
Conversion	11.14 ± 6.58	27.09 ± 3.19	7.60 ± 4.39	45.85
Retrieval	29.94 ± 11.41	40.50 ± 7.41	8.11 ± 0.54	78.55

[a] Values are in hours plus or minus the standard error. (Yamada *et al.*, 1975.)

conversion pathway having a smaller value (Table I). The minimum number of cell cycles traversed by the progeny of IECs before expressing the lens phenotype in the pathway of conversion *in situ* was estimated on the basis of the time of first mitosis, the time of appearance of lens-specific antigens as revealed by immunofluorescence studies (Takata *et al.*, 1964, 1966), and the total cell cycle time. The first mitosis was presumed to be the end of the first cell cycle which should start with the transition from G_0 and G_1. The estimation showed that the minimum cell cycle number was six (Yamada, 1977). This estimation was based on the assumption that, in the conversion pathway, cells were constantly in the cell cycle from their reentry after lentectomy until the time of crystallin appearance. This assumption is in conformity with autoradiographic observations. Further it appeared probable, but was not definitely shown, that in the retrieval pathway with a long cell cycle time cells failed to traverse six cycles before withdrawing from the cell cycle, because no reliable estimation was possible for the period of proliferation. Recent observations on segregated cultures of IECs (p. 209) have indicated a close correlation between the extent of pigmentation and the number of cell cycles traversed by the progeny, which allows an approximate determination of the number of cell cycles traversed by the degree of pigmentation (T. Yamada, unpublished). If we assume that such information obtained from cultured cells can be transferred to cells *in situ*, the number of cell cycles traversed in the retrieval pathway *in situ* should be less than four (inclusive).

Most investigators involved in experimental morphological analysis of this system shared the idea that the appearance of lens specificity depends on the inductive influence of neural retina (Mangold, 1931). This idea originated from observations that lens formation from iris epithelium depended on the presence of neural retina under various experimental conditions. If the neural retina is removed from an eye together with the lens, lens formation from iris epithelium does not occur until the regeneration of neural retina from pigmented retina has proceeded to a certain stage (Wachs, 1920; Zalokar, 1944; Stone and Steinitz, 1953; Hasegawa, 1965; Reyer, 1971). If the dorsal iris is grafted onto another body site, usually no lens formation follows unless neural retina accompanies the graft (Wachs, 1914; Ikeda, 1936; Stone, 1958; Reyer, 1966). One prominent exception to this rule is the observation of Reyer *et al.* (1973) that lens formation from dorsal iris regularly occurred when the latter was implanted onto the blastema of a limb regenerate. The concept that neural retina plays the role of an inductor in this system was found attractive in view of the fact that the embryological precursor of neural retina, the optic cup, is involved in induction of the lens in ontogenesis (for review, see McAvoy, 1980 and McDevitt and Brahma, this volume). According to a classical theory, the retina secretes a factor for lens formation which is paralyzed by the metabolism of the lens, and removal of the lens allows the retinal factor to act on the iris to form a lens (Spemann, 1905; Wachs, 1914, 1919).

Early in the century it was established that the lens could be formed by a piece of dorsal iris grafted into an optic chamber from which the lens had been removed (Wachs, 1914; Sato, 1930, 1935; Mikami, 1941) and that this occurred even if the optic chamber belonged to a species without a lens-regenerating capacity (Ikeda, 1934; Amano and Sato, 1940; Reyer, 1956). However, for technical reasons, this type of experiment has not been done with dorsal iris epithelium as the graft. Thus from these grafting experiments one can conclude that the progenitor cells for the regenerated lens are present in the dorsal iris. But whether they are in the dorsal iris epithelium or in the dorsal iris stroma cannot be determined by these experiments. The idea that dorsal iris epithelium is the source of cells forming the regenerated lens was based on descriptive data indicating that the regenerated lens rudiment grows from the dorsal margin of the iris epithelium as its extension, before it is separated as a young lens. Although these morphological data are convincing by themselves, they are not completely indisputable: In spite of the fact that iris epithelium has structural integrity as a result of the basal lamina covering all epithelial cells, it is not a completely closed system. It was found that, under the inflammatory conditions created by lentectomy, the epithelium was penetrated by macrophages and neutrophils (Yamada and Dumont, 1972) and that openings in the basal lamina could even be detected microscopically. Under such conditions, it was hard to ignore the possibility that a cell type of stroma could move into the iris epithelium, proliferate, and give rise to the lens. Thus it appeared necessary to study the system in culture to identify the progenitor cells.

First, the organ culture studies will be reviewed. Zalokar (1944) and Reese (Yamada *et al.*, 1973) succeeded in obtaining lens regeneration by culturing a whole lentectomized eye *in vitro*. Stone and Gallagher (1958) and Eguchi (1967) did not obtain any sign of lens regeneration in irises cultured *in vitro* for more than 20 days. However, some of the cultured irises produced lenses when placed in the optic chamber of lentectomized eyes. In 1973, Connelly *et al.* succeeded in obtaining lens formation in normal dorsal iris cultured together with pituitary. Using the same culture condition but replacing the pituitary with frog larval retina, Yamada *et al.* (1973) demonstrated the formation of lens tissue from normal dorsal iris (Table II, Fig. 1). The experiment was expanded by testing iris epithelium and iris stroma separately in the presence of frog larval retina (Yamada and McDevitt, 1974). The results showed that dorsal iris epithelium was able to produce lens tissue in the presence of frog larval retina at a high frequency. The choice of frog retina was based on its easier cultivation compared with newt adult retina. Although the frog larval retina was sometimes contaminated by a fragment of lens epithelium, which gave rise to lens tissue during culture, the clear difference in stainability and the size of the nuclei allowed a quick distinction to be made between newt and frog lens tissue in sections. Immunofluorescence for amphibian γ-crystallins and total crystallins was used to demonstrate the authenticity of the lens tissue produced (Fig. 2). The possible contamination of cultured normal iris epithelium by iris stromal cells, and of iris stroma

TABLE II

Lens Formation by Iris Tissues of Adult Newts in Organ Culture[a]

Newt tissues cultured with frog retina	Cultures producing newt lens per total cultures		References
	n	Percentage	
Dorsal iris	57/61	93.4	Yamada *et al.*
Dorsal iris epithelium	36/41	87.8	Yamada and McDevitt
Dorsal iris stroma	1/45	2.2	Yamada and McDevitt
Ventral iris	0/21	0	Yamada *et al.*
Ventral iris epithelium	9/49	18.3	Yamada and McDevitt

[a] Yamada *et al.*, 1973; Yamada and McDevitt, 1974.

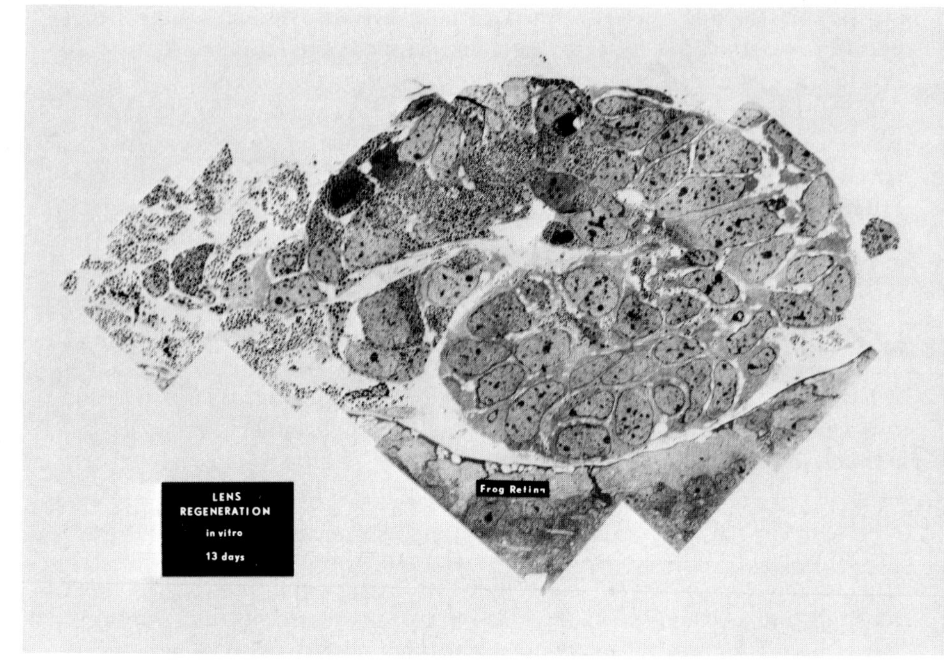

Fig. 1. A montage of electron micrographs of a culture of newt dorsal iris epithelium combined with larval frog retina. The depigmented cells in the center are in the process of converting to lens cells (Dumont and Yamada, original).

by IECs, was controlled by fixing tissue samples after their surgical preparation and by examining them in sections. The results showed that iris epithelium samples were completely free of contaminating cells in two-thirds of the cases and that in the remaining cases the average number of contaminating cells per sample was 3.5. On the other hand, it was not possible to sufficiently remove IECs sticking to the sample of iris stroma. All stroma samples were contaminated by IECs and the average number of IECs per sample was 24. Dorsal stromata produced a lens in only 1 out of 45 cases, probably as a result of this contamination.

These data on organ culture experiments support the classical notion that IECs are the progenitor cells for the regenerated lens and that stroma cells do not participate in lens regeneration. Another important point emerging from this investigation is that ventral iris epithelium, which does not participate in lens regeneration *in situ,* has a low level of lens-forming ability, in spite of the fact that the intact ventral iris under comparable culture conditions does not show any sign of lens

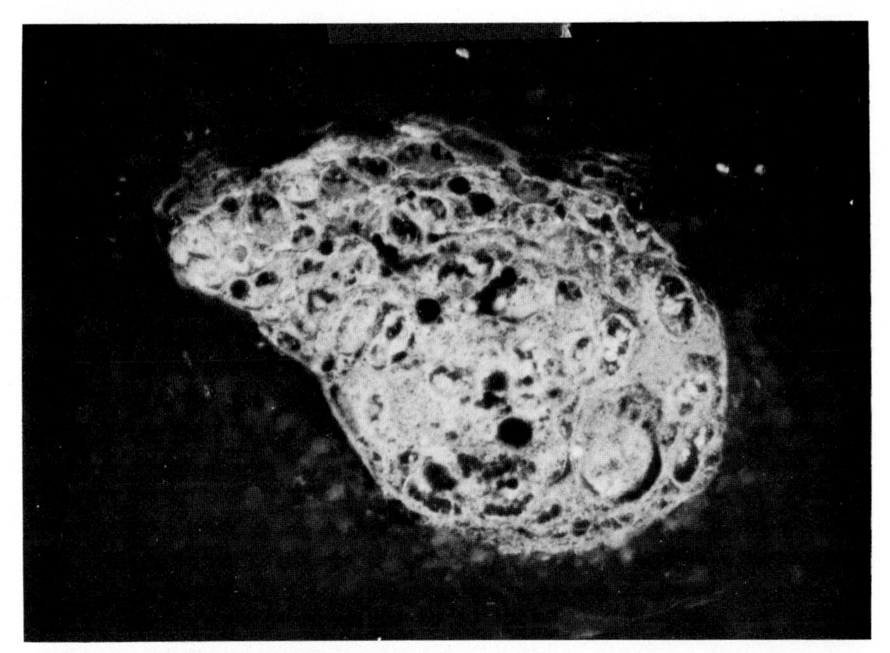

Fig. 2. Indirect immunofluorescence for γ-crystallins of newt lens tissue formed by newt dorsal iris epithelium cultured in the presence of frog retina, demonstrated in a tissue section (Yamada and McDevitt, 1974).

formation. This could suggest that ventral iris epithelial cells have an inherent capacity for conversion to lens cells, but that the expression of this capacity is suppressed by the tissue organization. The experiments of Ciaccio (1933) are interesting; he inserted a piece of celluloid into the pupillary opening after lentectomy and obtained supernumerary lenses produced all over the pupillary margin, suggesting that the lens-forming ability is present in all sectors of the iris ring including the ventral ones. One possible interpretation of this phenomenon may be that insertion of solid material in the pupillary opening causes localized disruption of tissue organization in the iris. Such a disruption may remove the alleged inhibitory effect of intact organization locally within the iris ring and lead to the expression of lens differentiation.

When one examines the results of the organ culture experiments reported by Connelly *et al.* (1973), Yamada *et al.* (1973), and Yamada and McDevitt (1974), it is clear that dorsal iris, dorsal iris epithelium, or a complex of whole iris and cornea cultured in the absence of other tissues can give rise to depigmented IECs and a limited number of lens fibers. This ability is, however, strongly enhanced in the presence of retina or pituitary. The addition of muscle or spinal ganglion showed a weaker enhancement of fiber formation (Connelly *et al.* 1973), while the addition of lung and spleen produced an enhancement of depigmentation (Yamada *et al.* 1973).

Cuny and Zalik (1981) recently tested a number of bovine pituitary hormone preparations for their effect on organ culture of adult newt dorsal iris, and found that thyrotropin added at different concentrations supported the appearance of lens cells which were identified by anti-γ crystallin in immunofluorescence. The mitotic index of cultures was significantly elevated by the addition of thyrotropin.

Although newt IECs were studied in cell cultures by Reese in 1967 and such cultures were used for various purposes (Ortiz, 1973; Ortiz *et al.*, 1973; Horstman and Zalik, 1974; Michel and Yamada, 1974), Eguchi *et al.* (1974) were the first to demonstrate, with immunochemical methods, the appearance of lens cells in primary cultures derived from the iris and mainly composed of pigmented cells. During the logarithmic phase, depigmented cells appeared and increased in number. Within the monolayer of nonpigmented cells, discrete cell aggregates appeared and increased in number. These aggregates showed a positive immunofluorescence reaction for total-lens antigens and an ultrastructure resembling that of lens fibers. Such lens fiber aggregates were produced not only in cultures derived from whole irises and dorsal irises but also in those from ventral irises. Immunodiffusion tests involving antibodies against whole-lens antigens, α-crystallins, and γ-crystallins, and immunofluorescence tests on smeared cells involving antibodies against whole-lens antigens revealed the appearance of lens cells after day 30 and their subsequent increase in primary cultures. Again, in these tests no difference was found among cultures from the three different sources. The ability of ventral IECs to produce lens cells was in line with the observation that in organ cultures (Yamada and McDevitt, 1974) limited lens differentiation occurs in ventral iris epithelium co-cultured with frog retina and supports the notion that tissue organization inhibits the expression of lens differentiation in ventral iris.

Horstman and Zalik (1974) estimated the cell cycle parameters of primary cultures of IECs derived from dissociated normal iris and from iris 10 days after lentectomy of adult newts. According to a later publication by the same group (Zalik and Dimitrov, 1980), the same culture contained cells positive in immunofluorescence for lens proteins during 42 days to 84 days of culture. These cells formed prominent aggregates. However, from 84 days and later, cell aggregates negative for the test appeared the nature of which remained obscure. Yamada and

Beauchamp (1978) estimated the cell cycle parameters of primary outgrowth cultures of dorsal and ventral IECs of adult newts. The ability of these cultures to give rise to lens fibers had been confirmed earlier by positive immunofluorescence using antibodies against purified γ-crystallins (Yamada, 1977) (Fig. 3). The estimated cell cycle parameters from both these papers are summarized in Table III. The shorter total cell cycle time (or shorter G_1 duration) of the 10-day series as compared with that estimated by Yamada and Beauchamp can be interpreted as resulting from the earlier activation of proliferation of the 10-day series by lentectomy *in situ*. This interpretation is supported by the growth curves estimated from the first group. The data of the second group show that there are no significant differences in the cell cycle parameters between dorsal and ventral IECs in culture. The estimated total cell cycle time of primary cultures of normal IECs is longer by a factor of 1.88 when compared with that of the conversion pathway *in situ* and slightly but not significantly longer than that of the retrieval pathway *in situ* (Table III).

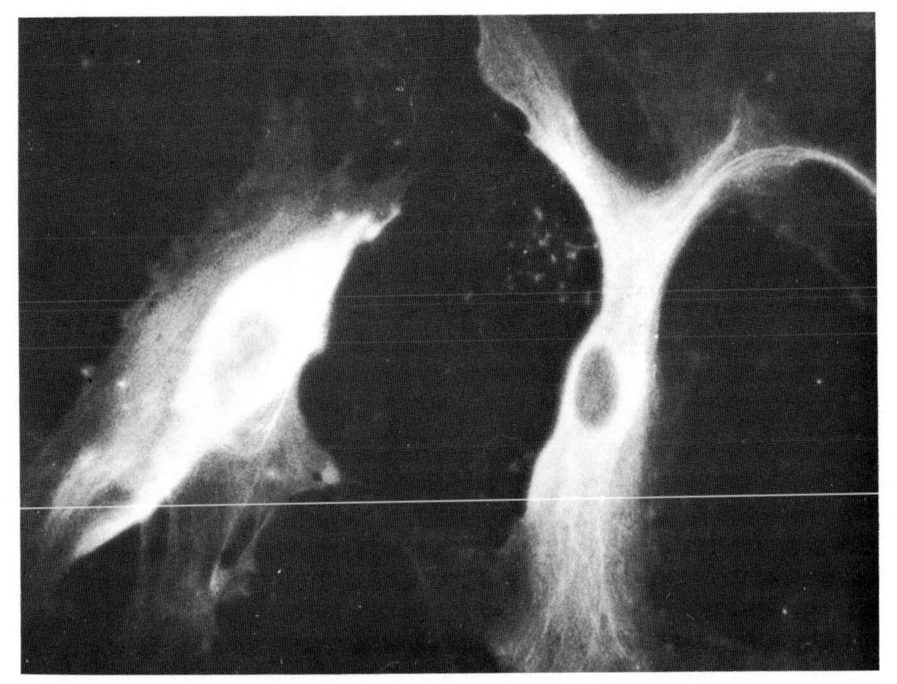

Fig. 3. Indirect immunofluorescence for γ-crystallins of cells in a 32-day outgrowth culture of newt iris epithelial cells (Yamada and McDevitt, original).

TABLE III

Estimates of the Duration of the Cell Cycle Phases of Iris Epithelial Cells of Adult Newts in Primary Cultures[a]

Source of iris epithelial cells	Duration of cell cycle phases (hr)[b]				Total duration of cell cycle (hr)
	G_1	S	G_2	M[c]	
Horstman and Zalik					
Dorsal, normal	—	36	6	2.1	—
Dorsal, 10-day regenerates	25	36	6	1.8	68.8
Yamada and Beauchamp					
Dorsal, normal	46.65 ± 14.41	34.58 ± 6.57	7.42 ± 5.42	—	88.65
Ventral, normal	41.99 ± 13.05	34.72 ± 5.80	6.93 ± 5.14	—	83.64
Dorsal plus ventral, normal	43.99 ± 14.44	35.08 ± 6.32	7.26 ± 5.69	—	86.33

[a] Horstman and Zalik, 1974; Yamada and Beauchamp, 1978.
[b] Duration in hours plus or minus the standard error.
[c] In the data of Yamada and Beauchamp, M is divided into G_1 and G_2.

Since primary cultures of iris epithelial cells cannot be completely free of contaminating cell types, it is desirable to have a cloned culture of IECs for the identification of progenitor cells for lens cells. Because they can be visually identified, even if the original cell suspension is somewhat contaminated, one should be able to obtain true clones of IECs. Abe and Eguchi (1977) followed the fate of spread single IECs seeded at the clonal concentration. The cells started to divide on day 10 and formed colonies of 20 and 40 weakly pigmented cells on day 20. When ca. 150–200 cells were reached, in some colonies lens cell aggregates were formed, which showed positive immunofluorescence for total crystallins. From 1 to 6% of the inoculated single cells formed colonies, and lens cell aggregates were formed in 20–50% of the examined colonies. No significant differences between dorsal and ventral IECs were found in these data.

In our culture of iris epithelial cells, cell motility was found to be too high to use colony formation as a measure of the clonal origin of cells. Utilizing the fact that epithelial cells cannot traverse the uneven surface of a plastic substrate produced with the tip of a steel blade, we surrounded the cell to be cloned with a continuous barrier of this type before culturing. The TVI cell line (Reese *et al.*, 1976) was used as a feeder layer. Coating the substrate with fibronectin was found useful for supporting the attachment and growth of cells. Such "segregated" cell populations originated from single or a small number of IECs produced lens cells when proliferation was optimum (see later). Lens cells thus produced were identified by indirect immunofluorescence for crystallins. No difference was found in the mode of appearance of lens cells between cell populations derived from the dorsal and ventral iris.

Thus, the results of cell culture work as a whole fully support the conversion of IECs into lens cells. It should be added, however, that they do not exclude the possibility that IECs are also converted into cell types other than lens cells.

There exist a number of apparent discrepancies in developmental parameters of the conversion between the system *in situ* and in cell culture: (1) The time required for the appearance of lens cells *in situ* is roughly a half of that required in cell culture. (2) *In situ* only the dorsal marginal IECs become converted, whereas in cell culture both dorsal and ventral IEC culture give rise to lens cells. (3) *In situ* the presence of retina is needed for the conversion. In cell culture no addition of the retinal factor is needed.

On the basis of the already discussed data on proliferation of IECs under various conditions, a working hypothesis was proposed, which can resolve these apparent discrepancies (Yamada, 1977). It assumes that there exists a critical number of cell cycles, the passage of which is basically required for the conversion, and that the retinal factor operates through the acceleration of proliferation, and not through direct induction of lens differentiation.

The observations discussed earlier that IECs have a longer cell cycle time in cell culture than *in situ* can explain point (1), since the time interval needed for the appearance of lens cells should be the function of the cell cycle time. Point (2) is interpreted by the local differentials in the proliferation of IECs *in situ,* and the elimination of these differentials reported above in cell culture. Point (3) may be due to the temporal limitation of proliferation *in situ* after lentectomy as opposed to its lack in cell culture. In order to complete the critical cell cycle number *in situ,* IECs need shortening of the cell cycle time by the retinal factor.

Since it was estimated earlier that cells in the conversion pathway *in situ* traversed at least six cell cycles before starting to express lens antigens, we assume that this is the critical number of cell cycles. In a previous publication, a preliminary estimation of the critical cell cycle number for cultured IECs indicating values close to six was reported (Yamada, 1977). However, at that time the estimation was based on parameters measured for different cultures (time of first mitosis, total cell cycle time, time of appearance of lens cells). We now have parameter values for the same culture, and the critical number calculated is 5.5–7.0 for a primary culture of *N. viridescens* IECs. No significant difference exists between the values for dorsal and ventral cells. Abe and Eguchi (1977) found in their clones that the number of cells of colony-producing lens cell aggregates distributed over a wide range, but that the minimum number was 100.

The segregated IEC cultures described above have been used for estimation of the critical cell cycle number (Yamada and McDevitt, in preparation). The number of cell cycles passed by a segregated culture at a given time was estimated by cell count and the stage of depigmentation. After the passage of a various number of cell cycles, the cultures were studied with immunoflourescence for γ crystallin or for total crystallins. A study involving 231 cultures showed that the percentage of positive cultures was 0 in series 1–4 cell cycles, increased from 31 to 95 in series 5–7 cell cycles, and attained 100 in series 8–11 cell cycles. The cell cycle number of each series denotes the maximum cell cycle number already traversed by any of the cells in the culture. The positive culture means a culture containing any number of cells positive for the immunoflourescence test. When the curves of positive cultures as the function of cell cycle number are prepared separately for cultures derived from dorsal and ventral irises, there was no significant difference between the two. The data as a whole are in good conformity with the concept of critical cell cycle number.

Direct support for the idea that neural retina functions in this system as a promoter of proliferation has been provided by investigations carried out in the laboratory of Y. Courtois (Arruti-Mizrali, 1979; Barritault *et al.*, 1981), which demonstrate the growth-promoting effect of a macromolecule of calf retina on mammalian cells in culture. In a preliminary test made at this institute (T. Yamada, unpublished), an extract of calf retina containing the molecule revealed a strong growth-promoting effect on cells of the TVI line (derived from newt iris, Reese *et al.*, 1976; McDevitt and Yamada, 1980). Now it is necessary to find out whether newt neural retina has a similar growth factor which enhances the proliferation of IECs.

The fact that in organ cultures of dorsal iris the supporting effect of retina or pituitary on lens formation requires contact of both tissues with iris, reported by Connelly (1977), apparently contradicts the idea that retina functions via its growth factor. Whether under these experimental conditions the secretion of the growth facter into media is inhibited remains to be analyzed.

The following observations are also relevant for judging the working hypothesis in question: When *in situ* or in organ culture lens regeneration was experimentally stimulated or supported, the progeny of IECs forming lens tissue was completely depigmented (Ciaccio, 1933, Eguchi and Watanabe, 1973; Reyer *et al.*, 1973; Connelly *et al.*, 1973; Yamada *et al.*, 1973; Yamada and McDevitt, 1974; Cuny and Zalik, 1981). In view of the correlation between depigmentation and proliferation demonstrated in segregated cultures (p. 201), this situation suggests that in those cases IECs should have traversed more than 5 cell cycles before expressing the lens phenotype. On the other hand, when lens regeneration *in situ* was completely suppressed by X-ray, the proliferation of IECs was inhibited, and the depigmentation of IECs was significantly reduced. In all those cases, one can argue that proliferation of IECs but not their differentiation was the direct target of the experimental control.

If the extent of proliferation is really decisive in this system of conversion, the next question to be asked is, What occurs while cells are proliferating? The most obvious event which occurs when iris epithelial cells are proliferating is the progressive loss of melanosomes. In the conversion pathway *in situ,* the cells completely lose melanosomes, express the lens phenotype, and do not resynthesize melanosomes. In the retrieval pathway *in situ,* a partial loss of melanosomes is followed by a recovery of the original level of melanosomes after withdrawal from the cell cycle. As we have discussed, there is evidence indicating that the difference in the pattern of depigmentation of the two pathways, i.e. whether melanosomes are eliminated completely or partially, is simply a reflection of the difference in the number of cell cycles traversed by cells in the two pathways. Transmission electron micrographs of adult IECs in the early stages of proliferation *in situ* revealed that premelanosomes were present only in exceptional cases and that the overwhelming majority of melanosomes were mature (Dumont and Yamada, 1972). This suggests that the synthesis of melanosomes is minimal in these cells. Hence, upon each cell division, the original population of melansomes should be diluted. However, this dilution alone cannot explain the rapid reduction in the number of melanosomes observed in proliferating IECs.

Transmission electron micrographs of adult newt IECs in the early stages of proliferation suggest that melanosomes are discharged from cells (Eguchi, 1963; Dumont and Yamada, 1972) and taken up by phagocytes. Eguchi expressed the view that phagocytes play an active role in this discharge. Later the discharge of melanosomes from cultured IECs was observed or inferred by workers involved in culture of this cell type (Ortiz, 1973; Horstman and Zalik, 1974; Eguchi *et al.*, 1974). The discharge of melanosomes can also be observed in segregated IECs. Thus one can conclude that this event can occur without the participation of phagocytes, even though it might be enhanced by phagocytes *in situ*.

Why melanosomes are removed from IECs before they differentiate into lens cells is an unanswered question. Although melanin is usually supposed to be inert, it may have some biophysical role in cell physiology. It may act as an electron acceptor and help control the free-radical concentration in the cells (Dain *et al.*, 1964), and its ability to act as an oxidation–reduction polymer has also been studied (Horak and Gillette, 1971). Thus it is possible that melanin must be removed before a specific biosynthesic event essential for the conversion is initiated. The observation that among the progeny of IECs the average length of the total cell cycle time is shorter in depigmented cells than in pigmented cells (Yamada *et al.*, 1975) can be interpreted to suggest that melanosomes have a retarding effect on progression through the cell cycle.

As for the mechanism of melanosome discharge, two processes have been suggested which may occur in sequence. The first event involves lysosomes. A transmission electron microscopic study coupled with cytochemical tests for acid phosphates has suggested that lysosomes are activated when IECs reenter the cell cycle (Yamada *et al.*, 1978). In organ cultures and in cell cultures, melanosomes are seen entrapped in the lysosomal membrane singly or in groups. Some of the melanosomes within the lysosomal membrane are found to be in the process of being discharged. Other organelles, e.g., multivesicular bodies and ribosomes, are also found sequestered in this membrane. These data have been interpreted as indicating that the reentry of IECs into the cell cycle leads to an autophagic condition which affects a number of preexisting organelles including melanosomes. It has further been suggested that the sequestered melanosomes are not degraded in the cell but are discharged. It was shown earlier that the activity of glucosaminidase, a lysosomal enzyme, increased during the depigmentation phase in iris after lentectomy *in situ* (Idoyaga-Vargas and Yamada, 1974).

Since it is rather unlikely that the lysosomal activity itself is involved in the final step of melanosome discharge, we investigated the role of Ca^{2+} in this process (Patmore and Yamada, in press). The principal role of the intracellular level of Ca^{2+} in various exocytotic events has been well established. The following methods were employed to increase the intracellular level of Ca^{2+} in cultured IECs: (1) Microinjection of $10^{-3} M$ Ca^{2+} into cells; (2) fusion of single lamellar phospholipid vesicles containing $10^{-3} M$ Ca^{2+} with the cell membrane; (3) incorporation of ionophore A23187 in the cell membrane. All methods produced a significant increase in the number of melanosomes discharged per unit time, suggesting that the intracellular level of Ca^{2+} controls the exocytosis. Each method was controlled by a sham treatment which did not produce a significant increase in the discharge. We speculate that melanosomes affected by lysosomes become sensitive for the discharge controlled by Ca^{2+}.

The suggestion that cyclic AMP (cAMP) may be involved in the discharge of melanosomes comes from an observation made by Ortiz (1973) that treatment of cultured IECs with agents which elevate the intracellular levels of cAMP altered the cell configuration from lamellar to stellate and that this reversible alteration was associated with the discharge of melanosomes (Ortiz *et al.*, 1973; Ortiz and Yamada, 1975; Yamada, 1977). The cell configuration similar to the stellate one induced *in vitro* by cAMP agents had earlier been found to occur *in situ* during the depigmentation phase in IECs by electron microscopic observations (Dumont and Yamada, 1972). According to the estimation of cAMP in iris *in situ* after lentectomy made by Velázquez and Ortiz (1980), the highest levels were in the depigmentation phase. This is in accord with the notion that cAMP is involved in melanosome discharge. However, these authors also showed, in agreement with Thorpe *et al.* (1974), that levels of cAMP in normal iris, where no

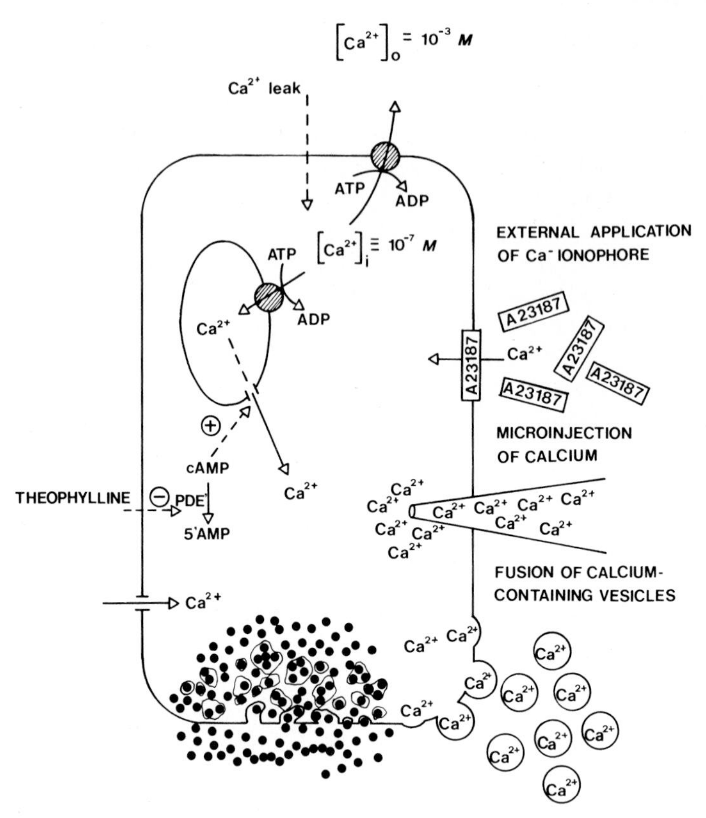

Fig. 4. A diagram indicating the principles of experimental elevation of intracellular levels of Ca^{2+} and the possible mechanism of its regulation (Patmore and Yamada, original).

depigmentation occurs, were relatively high, and during the transmission from G_0 to G_1 a rapid reduction in these levels occurred. These findings probably mean that cAMP is involved not only in depigmentation but also in regulation of the cell cycle. This is probably the reason why the addition of certain cAMP agents to an iridocorneal complex culture did not produce the expected enhancement of depigmentation (Connelly, 1980). It is possible that these agents inhibited the proliferation of IECs (Steinberg and Whittaker, 1975).

The observations that the intracellular level of Ca^{2+} can be controlled by the cAMP level in some mammalian cells (Berridge, 1975) seem to suggest that the above discussed role of cAMP in depigmenting

IECs is connected with an elevation of the Ca^{2+} level, which is necessary for the malanosome release. In a work to be published elsewhere (Patmore and Yamada, in press), discharge of melanosomes from cultured IECs was studied during a treatment with dibutyryl cAMP plus theophylline, which should increase the intracellular level of cAMP. A significant increase of melanosome discharge was observed during the treatment in the absence, as well as in the presence, of Ca^{2+} in the medium. The result was interpreted to suggest that the cAMP level controls the liberation of Ca^{2+} from the intracellular source and affects melanosome discharge. On the other hand, another possibility that the influx of Ca^{2+} from media through the Ca^{2+} channels which should open upon depolarization of membrane potential (Baker *et al.*, 1971; Trautwein, 1975) plays an essential role in melanosome discharge was not supported by the experiment involving a high potassium treatment of IECs. On the basis of these results, it appears that the high cAMP level in the dorsal iris during depigmentation of IECs (Velázquez and Ortiz, 1980) is associated with the liberation of intracellular Ca^{2+} in IECs, which in its turn triggers the release of melanosomes.

It is still possible that such an overt morphological process as depigmentation is simply a superficial indication of more fundamental but less visible changes. One such possible change was investigated by the Zalik's group. They measured electrophoretic mobilities of samples of newt iris cells (containing IECs as the major component) engaged in lens regeneration *in situ* (Zalik and Scott, 1972). In dorsal iris a slight early increase was followed by a progressive decrease from day 3 to day 10. Subsequently this value increased. The phase of the decreases coincided with the depigmentation of iris epithelial cells. In ventral iris the decreases in values was less than in dorsal iris. In subsequent experiments they studied the effects of neuraminidase, RNase, and chondroitinase on the mobility of iris cells at various stages of lens regeneration (Zalik and Scott, 1972, 1973). In normal and day-3 iris all three enzymes decreased in mobility. From day 5 on, RNase and chondroitinase failed to alter cell mobility. Neuraminidase, on the other hand, retained its effect on mobility until day 10. The effects of neuraminidase and RNase were additive, suggesting that they removed different groups from the cell periphery. But the effects of RNase and chondroitinase were not additive, suggesting that they removed an identical or similar group. The results implied a sequential disappearance of at least two groups from the cell surface during dedifferentiation *in situ*. However, a similar study with IECs in primary culture showed that the effect of RNase on the mobility disappeared on day 5, while the effect of neuraminidase persisted up to day 35 (Zalik *et al.,* 1976). The reason for this difference between cells *in situ* and those in primary cultures is not understood.

The traditional attitude has been to interpret the depigmentation of IECs as a dedifferentiation event by which cells are released from the original determination for cell-type differentiation (Yamada, 1972, 1976). At the present stage of research it is appropriate to interpret all events associated with depigmentation and the cell surface alteration discussed above as integral parts of dedifferentiation, which are as a whole essential for the cell-type conversion.

In summary, these results suggest a regulative role for cell proliferation, which is coupled with dedifferentiation. Thus dedifferentiation can be controlled by altering the extent of proliferation. The missing link in this whole approach seems to be a connection between the dedifferentiated condition and the expression of new cell-type differentiation. It has been speculated that the ultimate issue of cell-type conversion is the specificity of chromatin and that the role of dedifferentiation is to produce a signal for an alteration in the synthesis of chromatin proteins leading to the appearance of a lens-specific chromatin configuration. Following the completion of dedifferentiation there may be a decisive cell cycle in which a new chromatin configuration is established. This implies that a feedback control on nuclear

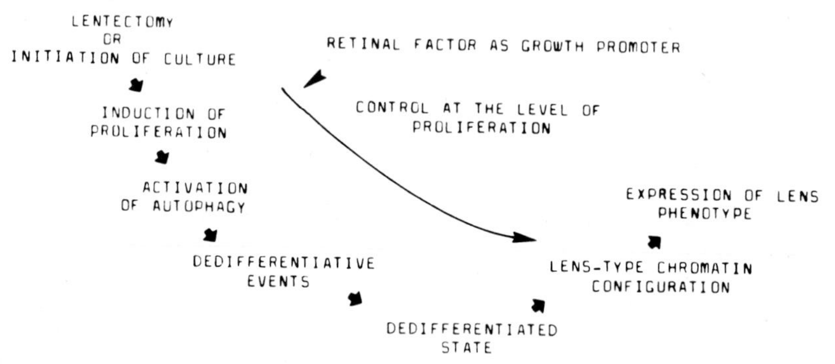

Fig. 5. A diagram of the proposed chain of events controlling the conversion of iris epithelial cells to lens cells. The progression of events toward the right is presumably coupled with the traversal of cell cycles. Whether the conversion occurs (the pathway of conversion) or fails (the pathway of retrieval) is regulated by the number of cell cycles traversed by cells in sequence. The retinal factor, which is essential only for the conversion *in situ* or in organ culture, influences the events through the proliferative activity of cells.

activities is exerted by the conditions in the cytoplasm and at the cell surface, which is decisive for cell-type conversion (Fig. 5).

IV. PIGMENTED RETINA AS A SOURCE OF LENS CELLS

System B: Conversion of Pigmented Retina to Lens in Adult Newts

It has been known for a long time that in amphibians degeneration or surgical removal of neural retina is followed by its regeneration from pigmented retina (pigmented retinal epithelium) and associated tissues—ora serrata and ciliary epithelium (Wachs, 1920; Stone, 1950; Hasegawa, 1965; Mitashov, 1968; Reyer, 1971; Keefe, 1973a,b,c,d; for review, see Reyer, 1977). Autoradiographic evidence suggests that cell-type conversion occurs here also, following a period of activated proliferation associated with melanosome discharge (Reyer, 1971; Keefe, 1973a,d). The conversion of pigmented retina to neural retina has also been demonstrated in larval and adult frogs (Lopashov and Sologub, 1972). This transformation of pigmented retina does not strictly belong to the subject of this chapter, however, because of its close relationship to conversion of the same tissue to lens, the existence of this phenomenon should be kept in mind.

That pigmented retina of the adult newt has the capacity to develop into lens tissue was shown by Sato (1951), who grafted test tissue into the posterior optic chamber of lentectomized eyes. The middorsal zone of pigmented retina, when isolated and tested by this method, produced lens tissue at a frequency of 22–24%, while dorsal iris produced lens tissue at a frequency of 93% under the same experimental conditions. Dorsal ciliary epithelium (pars ciliaris retinae), which is located between dorsal iris and retina, formed lens tissue in 54% of the cases. As stated at the beginning of this section, pigmented retina *in situ* produces neural retina if the latter tissue is removed. The lens-forming capacity of pigmented retina is expressed when it is separated and cultured in the optic chamber. Sato (1951) argues that the nutritional levels of the tissue may determine whether this alternative pathway is followed. *In situ,* pigmented retina may have optimal nutritional levels, but when grafted to the optic chamber, it may have lower levels of nutrition. Another possible mechanism determining the pathway of pigmented retina is the extent of proliferation. From what we have discussed in Section III, the separation of pigmented retina and its grafting into the optic chamber may provide better conditions for proliferation than leaving it *in situ.*

System C: Conversion of Pigmented Retinal Cells to Lens Cells in Chick Embryos and Human Fetuses

Eguchi and Okada (1973) cultured cells derived from pigmented retina of 8.5-day chick embryos. In primary cultures the originally pigmented cells became less pigmented or nonpigmented during active growth, as previously observed by Whittaker (1963). When cultures reached confluence, many foci of pigmented cells reappeared. Subsequently the number and size of pigmented foci increased. When primary cultures were subcultured, the same sequence of events occurred in the secondary cultures. But in one secondary culture, in addition to pigmented foci, piles of elongated nonpigmented cells developed, which had the appearance of the lens fiber aggregates observed earlier in cultures of chick embryo lens epithelium (Okada *et al.,* 1971). The aggregates showed an ultrastructure comparable to that of chick lens fibers, and positive immunofluorescence for chick lens fiber proteins. For the purpose of obtaining clonal cultures of pigmented cells, the pigmented foci of a secondary culture were separated and pooled; the cell suspension was then plated at the clonal concentration and cultured in conditioned medium. When some of these pigmented colonies orginating from single cells were subcultured, in a portion of cultures lens fiber aggregates appeared after 100–130 days of culture. An ultrastructural and immunofluorescence study confirmed the lens fiber nature of these aggregates. In a later work (Yasuda *et al.,* 1981) the appearance of lens cells in clonal cultures of chick pigmented retinal cells was studied as a function of the age of the donor. The capacity of transdifferentiation was found to decrease with the developmental stage and to be lost completely after 15–16 days of incubation.

According to Yasuda (1979), coating the culture substrate with collagen increases the spreading of cells in the initial phase of primary culture of pigmented retinal cells but inhibits their conversion into lens cells. That collagen does not select cells lacking the potency for lens differentiation was suggested by the observation that cultures reared on a collagen-coated substrate could develop lens fiber aggregates when subcultured on a noncoated substrate.

Yasuda *et al.* (1978) reported that, when pigmented retina of human fetuses at 12 weeks' gestation was put in primary culture, depigmentation of cells was followed by repigmentation on about day 30. The repigmented cells were then collected and innoculated at clonal density. Some of the colonies developed lens fiber aggregates by day 30. Immunofluorescence with antibodies against rat lens, and immunoelectrophoresis with the same antibodies, were used to demonstrate crystallins in lens fibers. Furthermore, the same authors observed the development of lens fiber aggregates in primary cultures of iris epithelial cells from the same fetuses, starting from day 30.

V. NEURAL RETINA AS A SOURCE OF LENS CELLS

System D: Conversion of Neural Retinal Cells to Lens and Pigment Cells in Chick Embryos

There exist a considerable number of reports which suggest the formation of lens tissue from neural retina under divergent experimental conditions in the amphibian and chick embryos (Amprino, 1949; Filatow, 1925; Fischel, 1900, 1903; Moscona, 1957, Rübsaamen, 1950, etc.). The observation that primary culture of chick embryo neural retina gives rise to lens fibers and pigment cells was first reported by Okada *et al.* and Itoh *et al.* in 1975. Since then a considerable number of publications have described this system of transdifferentiation, and the system is now the most vital issue in the area covered by this chapter. In the above-cited work, in primary cultures of neural retina of 8- to 9-day chick embryos, Eagle's minimum essential medium (MEM) supplemented with fetal calf serum, glutamine, and sodium bicarbonate was used. Small bipolar cells interpreted to be neuroblasts appeared on about day 10 but completely disappeared before day 20. Then foci of densely packed small epithelial cells formed and produced populations of pigment cells at their centers. The development of pigment cells was soon followed by the appearance of irregular aggregates of transparent, elongated cells. These aggregates were called "lentoid bodies." However, since this term was used earlier for various kinds of lenslike structures including some that seemed to lack crystallins, in this chapter it will be replaced by "lens fiber aggregates." Thereafter the number of aggregates and the size of the pigmented area increased up to days 40–50. In subcultures, lens fiber aggregates and pigment cells appeared from about day 10 onward. In immunodiffusion tests using anti-α- and anti-δ-crystallins, freshly dissociated neural retinal cells of primary cultures were negative up to day 12. All cultures of days 38–40 tested with immunodiffusion were positive for both crystallins. In indirect immunofluorescence, lens fiber aggregates and nonpigmented epithelial cells located immediately around the aggregates were found to be positive. Further tests measuring the immunofluorescence of smear preparations showed that cells positive for δ-crystallins increased from 2% on day 20 to 38% on day 45, while the inoculum was negative.

The factors that affect the expression of lens and pigment cell differentiation in primary cultures of embryonic neural retina have been studied to some extent. According to Okada (1976), when chick embryo neural retina was cultured in Ham's F-12 medium supplemented with 6% fetal calf serum, as opposed to Eagle's MEM medium supplemented with 6% fetal calf serum, the differentiation of both lens fibers and pigment cells was completely suppressed. However, in F-12 medium with the same supplement, neural retina of quail embryos differentiated lens fibers. Clayton *et al.* (1977) compared the effect of MEM with Earle's salt formulation (EMEM) and that of MEM with Hank's salt formulation (HMEM), adjusting the pH of the latter with sodium bicarbonate. In EMEM, 8- to 9-day chick embryo neural retina gave rise to a larger area covered with pigment cells and a larger number of pigmented colonies than in HMEM. But there was no obvious difference in the frequency or size of the lens fiber complex under the two culture conditions. Cell count data suggested a higher proliferation in HMEM than in EMEM. A study on the effect of initial cell density, utilizing the standard culture medium, showed that pigment cells appeared in cultures of the inoculum at a density of 10^5–10^7, but no pigment cells differentiated in cultures inoculated at 10^8. The time of appearance of pigment cells was earlier, and the number of pigment cells was larger in the 10^6 series than in the 10^7 series. On the other hand, lens fiber cells appeared in the 10^6, 10^7, and 10^8 series. Inoculation at 10^7 appeared to be optimal, giving rise to lens fibers at an earlier time than in the 10^6 series and in a larger number than in the 10^6 and 10^8 series. How cell proliferation was affected by inoculum size was not reported. The requirement of cell proliferation for the appearance of lens fibers and pigment cells was impressively demonstrated by the following experiment of Itoh (1976). Primary cultures of 9-day chick embryo neural retina grew and produced lens fibers and pigment cells in MEM supplemented with 6% undialyzed fetal calf serum. However, in MEM supplemented with 6% dialyzed fetal calf serum, proliferation of the same tissue was inhibited and no differentiation of lens fibers and pigment cells was observed. The addition of ascorbic acid to an inhibited culture restored proliferation which was followed by differentiation of lens fibers and pigment cells. Even if the presence of ascorbic acid was limited to 5 days, a short period of active proliferation was followed by a lower level of lens fibers and pigment cells after a lag of ca. 20 days in the absence of ascorbic acid. As a whole, the studies cited above seem to suggest that the differentiation of lens fibers and pigment cells in primary cultures of neural retina requires a certain level of proliferation.

Subsequently de Pomerai and Clayton (1980) studied the effects of dialyzed and undialyzed sera, the absence and presence of ascorbic acid on the formation of lens fiber aggregates, and the amounts of crystallins in cultures of 8-day chick embryo neural retina. Culture growth under different conditions was assessed by [^{3}H]thymidine incorporation into cultured cells, as well as by cell counts. The visual data suggested that the appearance and amount of lens fiber aggregates and pigment cells roughly correlated with the extent of proliferation. The amounts of the three crystallins estimated with hemagglutination inhibition assays also showed a similar trend. The ratio of δ-crystallins to α- plus β-crystallins was found to correlate with proliferation. They concluded that conditions permitting the greatest proliferation produced earlier and more extensive development of lens fiber aggregates and a higher proportion of δ-crystallins relative to α- plus β-crystallins, and vice versa. The addition of insulin promoted proliferation of neural retina cultures and increased the amount of δ-crystallins as well as the ratio of δ- to α- plus β-crystallins.

Araki *et al.* (1979) studied crystallin synthesis in cultures of 3.5-day chick embryo neural retina. In the first series, they estimated the amount of α- and δ-crystallins present in the cultures at different time intervals from inoculation by quantitative immunoelectrophoresis. δ-Crystallins started to appear on day 10, increased rapidly in amount, and on day 26 reached the maximum value which corresponded to 43% of the total soluble protein. α-Crystallins were first detected on day 10 and on day 20 reached a maximum value corresponding to 2.5% of the total soluble protein. On sodium dodecyl sulfate (SDS)-polyacrylamide gel electrophoresis of culture homogenates, the bands corresponding to δ-crystallins and those corresponding to the A and B subunits of α-crystallins could be recognized from day 12 and increased later. The incorporation of [^{14}C]leucine into the trichloroacetic acid (TCA)-insoluble fraction of culture homogenate was determined at different stages of culture, and the incorporation into δ- and α-crystallins during the course of culture was estimated by specific immunoprecipitation. The incorporation into δ-crystallins became detectable from day 9, increased to day 26, and then decreased. The incorporation into δ- and α-crystallins was also studied with autoradiographs of the SDS-gel electrophoretic pattern. The data obtained were in general agreement with the results of the last series. Although in this work the effort was concentrated upon quantitation of δ- and α-crystallins, the data leave no doubt that β-crystallins are also synthesized as important components of lens differentiation.

de Pomerai *et al.* (1977) studied crystallins accumulated in cultures of 8-day strain N (normal) chick embryo neural retina with SDS-polyacrylamide gel electrophoresis, which gave information on the majority of α- and β-crystallins in the molecular weight range 18,000–28,000, but no information on δ-crystallins and the high-molecular-weight β-crystallins (35,000), for technical reasons. The results showed that α- and β-crystallins in this range were not detectable in fresh embryonic neural retina and 16-day cultures but were detectable in 42-day cultures. The difference in the banding pattern in this range between 42-day cultures of neural retina and 1-day-old chick lens showed that the relative quantities of different crystallins were not identical in the two cell populations.

A study was reported by de Pomerai and Clayton (1978) on the relationship between the donor stage of neural retina and the appearance of crystallins in cultures derived from the tissue, utilizing SDS–polyacrylamide gel electrophoresis and hemagglutination inhibition assays. The amount of total crystallins relative to the total amount of soluble protein in the cultures decreased from 60 to 9% between the day-6 and day-15 donor stages. The proportion of δ-crystallins decreased from 80 to 30% during the same period. This decrease was compensated for by the increase in the proportion of α- and β-crystallins. A similar trend had been demonstrated earlier in cultures of chick embryo lens epithelium obtained at different developmental stages (de Pomerai *et al.*, 1978) and is known to occur *in situ* during chick lens ontogenesis (Rabaey, 1962; Genis-Galvez *et al.*, 1968; Truman *et al.*, 1972). This suggests that the redifferentiation phase of this transdifferentiation process follows the sequence of genetic expression characteristic of ontogenesis and that a common regulatory mechanism may exist.

Nomura and Okada (1979) examined the dependence of transdifferentiation of this system on the developmental stage of the donor over a span of 3 days of incubation and 3 days after hatching. The variation in cell attachment at different stages was adjusted by the inoculum size. The rate of transdifferentiation decreased steadily with the donor stage. The latest donor stage capable of producing lens fibers was day 18 of incubation, and that capable of producing pigment cells was day 15 of incubation. The amount of α-crystallins present in day-50 cultures as estimated by quantitative immunoelectrophoresis decreased from 39 to 6 in arbitrary units, and that of δ-crystallins from 300 to 60 units during the days 3–16 donor stage. The data did not show the relative decrease in δ-crystallins with the progression of donor stage suggested by de Pomerai and Clayton (1978). When conditioned medium was prepared from a 8.5-day embryonic neural retina culture containing a large number of neuronlike cells, and 3.5- and 8.5-day embryonic neural retinas were cultured in this medium, the first appearance of pigment cells was delayed but the appearance of lens fibers was not affected. This was interpreted as indicating the inhibitory effects of neuronlike cells on pigment cell formation.

Selected biochemical parameters of neurons were studied in primary cultures of 3.5- and 8.5-day chick embryo neural retina, and these results were related to the amount of δ-crystallins (Nomura *et al.*, 1980). Acetylcholine (ACh) and γ-aminobutyric acid (GABA) were estimated by electrophoresis combined with chromatography after labeling with methyl[^{3}H]choline chloride and [^{14}C(U)]glutamic acid, respectively. Identification of ACh and GABA was made with specific enzymes. The choline acetyltransferase (CAT) activity was measured according to Fonnum (1969). Quantitative immunoelectrophoresis was used for the estimation of δ-crystallins. The results with ACh and GABA indicated that, in cultures of 3.5-day embryonic neural retina, appreciable amounts of both neurotransmitters are synthesized on day 12 but not about day 50. Neurons were detected on day 12 but not on day 50. The results of measurements of CAT as a function of culture duration in primary cultures of 3.5- and 8.5-day embryonic neural retina are reproduced in Fig. 6, together with δ-crystallin data. In both series, enzyme activity was high only in the earlier phase, in general correlation with the presence of neuronlike cells. In the 3.5-day series, maximum activity was attained on about day 7.5, and the subsequent decrease was relatively fast. In the 8.5-day series, the maximum was already reached by day 2, and a swift decrease was followed by a slow decrease. In this series the percentage of small neuron-type cells was estimated to be 64, 41, and 7% on days 4, 7, and 14. Thus there is a parallelism between CAT activity and the number of neurons in the culture. Although not mentioned in the paper, it is interesting to note that, since the donor stages of the two series differed by 5 days, the maximum activity on day 7.5 of the 3.5-day series roughly coincided with day 2 of the 8.5-day series. Furthermore, the appearance of CAT activity *in ovo* in chick embryo neural retina (Crisanti-Combes *et al.*, 1978) roughly coincided in time with the beginning of the same activity observed *in vitro*. This very probably means that the first population of neuroblasts to differentiate in neural retina *in ovo* express neuronal differentiation in cultures in both series. Thus in both series of cultures the same population of neurons should be present. The authors are now attempting to determine how well the idea of direct conversion of neurons into lens cells, proposed by the same group (see p. 226), fits in with these results by comparing the relation between CAT activity and the amount of δ-crystallins. They suggest that the direct conversion of neuronal cells into lens fibers occurred in the 3.5-day series but not

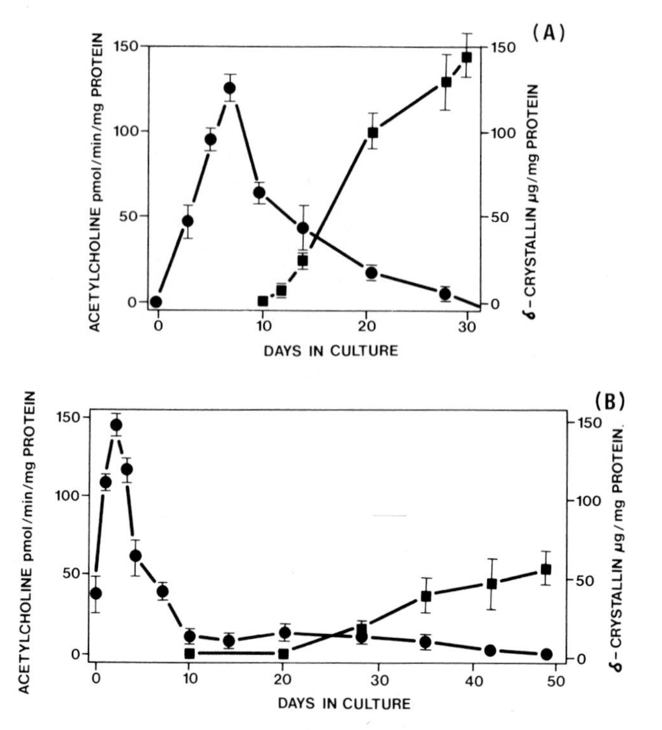

Fig. 6. CAT activity as shown by the synthesis of acetylcholine, and the amount of δ-crystallin as functions of the culture time in 3.5-day chick embryo neural retina (A) and 8.5-day chick embryo neural retina (B). Each value represents three or four independent inocula, and the bars indicate the SEM (Nomura *et al.*, 1980).

in the 8.5-day series. This interpretation does not appear to conform with the above-mentioned situation that the same group of neuronal cells should be present in both series. One could argue that, if direct conversion does not occur in the 8.5-day series, it is unlikely to occur in the 3.5-day series, and that the reciprocal relation of the CAT activity and the δ-crystallin amount in the latter series may be simply a coincidence.

The system in question was studied in cell clones by Okada (1977) and Okada *et al.* (1979a). In the former study, cells harvested from primary cultures of 8-day chick embryo neural retina 7–9 days after incubation were seeded at clonal density. Lens fiber aggregates were formed in 10% of the colonies obtained in F-12 medium and in 33.5% of the colonies cultured in MEM. Pigment cells appeared in only 1.4% of all the clones studied. When primary cultures were prepared separately from anterior and posterior parts of neural retina and colonies produced, lens fiber aggregates were found only in colonies derived from anterior neural retina. This was explained by the regional difference in the timing of the differentiation of neural retinal cells, the anterior region being ahead in withdrawing from predifferentiation proliferation compared with the posterior region (Kahn, 1974). One question to be asked is whether the possibility is excluded that the source of lens cells is ciliary cells which could be included in the anterior sample of neural retina. In adult newts, ciliary epithelium is known to produce lens tissue at a relatively high frequency when grafted into the optic chamber (Sato, 1951).

In the other cloning experiment published in 1979 (Okada *et al.*, 1979b), 3.5-day instead of 8-day chick embryos were used as donors of neural retina. About 1500 cells were inoculated per 60-mm dish to obtain primary clonal cultures. Secondary clonal cultures were prepared from a high-density culture. Selected colonies of primary clonal cultures were used for recloning. Most cultures were grown in F-12 medium supplemented with ascorbic acid, fetal calf serum, and glutamine, and the culture surface was coated with collagen. In primary clonal cultures some colonies produced lens fibers and pigment cells (allotypic differentiation). According to the type of allotypic differentiation expressed, colonies were classified as L type (lens fibers), P type (pigment cells), M type (both lens fibers and pigment cells mixed), or U type (allotypic cells absent). The frequencies of L, P, and M types in the primary clonal cultures were 32, 29, and 20%. "Neuronal" cells appeared in some colonies in early stages and disappeared later, while allotypic differentiation started to appear 14–16 days after inoculation. In secondary clonal cultures, 44% of the colonies were L type, 16% P type, and 8% M type. It was reported that 7 colonies which developed neuronal cells at an earlier stage later produced lens fibers. However, since there was 13 colonies with lens fibers, 6 colonies should have developed lens fibers without neuronal cells. This is an important point in a later discussion. In the recloning experiment the colonies again showed L type, P type, and M type differentiation. In one series of recloning experiments, recloning was done on day 30 of the original colonies, and the relationship between the different types of allotypic differentiation of original and derived colonies was studied. Although not pointed out in the paper, the reported data show that, when the original colony was L type, the derived colonies were either L type or U type, but when the original colony was P type or M type, the derived colonies could be L, P, M, or U type. Such a relationship may give some clue as to the mechanism involved in the selective expression of allotypic differentiation.

A time-lapse cinematographic study by Okada *et al.* (1979b), suggests that neuronal cells are directly converted to lens fibers. Although no final publication is as yet available, because of its importance a preliminary description of this work will be briefly cited. It is claimed that neuronal cells swell to form bottle cells which are assumed to be lens fibers, and that aggregates of neuronal cells are transformed directly into lens fiber aggregates. These authors have concluded that "it is highly probable that cells partially differentiated into neuronal specificity can transdifferentiate into lens cells." At this moment, without the cinematographic evidence, I cannot judge this proposal but would like to raise a few points which should be considered. (1) In reported observations of clones, the correlation of appearance of lens fibers and the presence of "neuronal" cells is poor. In many clones from 8-day embryos, which produced lens fibers, "neuronal" cells did not appear, and in a number of clones from 3.5-day embryos lens fibers developed in the absence of "neuronal" cells. The appearance of lens cells in the absence of "neuronal" cells can be explained by assuming other sources of lens cells. However, the lack of the observation of lens cells associated with "neuronal" cells in clones is a problem for the proposal. (2) Phase-contrast observations of three-dimensional structures have optic limitations, especially when the alleged transformation occurs on the monolayer which may contain the other precursor of lens fibers. (3) The brilliant appearance of lens fiber aggregates in phase contrast is due to the presence of blebs on the cell surface, which is not specific for lens fibers.

In a report made before the Basel Congress of the International Society of Developmental Biologists in 1981 (Nomura, K., Medaka, E., Takagi, S. and Okada, T. S.), an attempt to separate neuronal cells from a primary culture of 3.5-day embryo neural retina by labelling them with merocyanine 540 and sorting the labeled cells with the fluorescence-activated cell sorter was published. The separated cell population showed high specific activity of CAT, and produced lens fiber aggregates on further culture. In interpreting the results, it is necessary to know (1) whether the merocyanine staining marks the neuronal cell lineage or specifically indicates the differentiated state of neuronal cells, and (2) whether one can exclude cell-type contamination in this sorting procedure. Agata *et al.* (1980) studied the appearance of lens cells in 8-day chick embryo neural retina in different culture media, and observed that the conditions favorable for maintaining neuronal specificity are also favorable for transdifferentiation into lens cells. The authors interpret the result as supporting direct transdifferentiation of differentiated neuronal cells into lens cells. Obviously there are many other possible interpretations of the result.

In normogenesis embryonic neural retina, composed of cells without overt heterogeneity, produces a large number of cell types. The absence of precise information on the normal program of differentiation of neural retina at the cellular level makes understanding the mechanism of transdifferentiation difficult. Okada *et al.* (1979a) proposed the presence of a number of multipotential progenitor cells with slightly different specificities, which give rise to allotypic as well as autotypic cell types in culture. It may be that these progenitor cells function as precursors of a specific autotypic cell type in normogenesis. In a typical case of the clonal experiments, the progenitor cells give rise to neuronal cells in the early phase and later give rise to lens fibers and pigment cells during culture. Autotypic differentiation is not expressed in the progenitor cells themselves, hence the appearance of allotypic differentiation in this system is not based on cell-type conversion. However, the ability of the progenitor cells to produce cells expressing the autotypic differentiation in an early phase of proliferation *in vitro* can be interpreted as their determination. In this case we are dealing with transdetermination.

According to the general interpretation of Okada's group (Okada *et al.* 1979b; Okada, 1980), multipotent progenitor cells present in the neuroepithelium give rise in normogenesis to two cell lineages, one leading to Müller cells, and the other to neurones. Then cultured *in vitro* the neuronal cell lineage gives rise to lens fibers at its different phases, whereas the Müller cell lineage is able to produce pigment cells as well as lens fibers. Furthermore, pigment cells thus produced have the ability to become lens fibers. This means that the allotypic formation of lens fibers may be based on transdetermination, cell-type conversion, or any other intermediate mode of transdifferentiation. It appears that evidence for the derivation of lens fibers from the Müller cell lineage, probably through transdetermination, is rather convincing. On the other hand, formation of lens fibers from the neuronal cell lineage, in which cell-type conversion has been claimed, needs further clarification. (See also addendum, p. 234.)

The mRNA of this transdifferentiating system is being studied by R. M. Clayton's group. A DNA fraction complementary to the most abundant class of lens mRNA was used as a probe for crystallin sequences in different tissues (Thomson *et al.*, 1978; Jackson *et al.*, 1978). Cytoplasmic mRNA from pigmented retina of 8-day chick embryos saturated the probe with kinetics indicating the presence of putative crystallin sequences at a concentration 1/25,000 of that of the lens. Hybridization of the probe with neural retina cytoplasmic mRNA occurred more slowly than that of pigmented retina and was not completed at the highest R_0t values obtained. Polysomal mRNA from headless chick embryos and chick embryonic muscles studied as controls did not indicate any of the sequence (Fig. 7).

In another paper (Clayton *et al.*, 1979), eye cups were isolated from 3.5-day chick embryos and a poly(A)-containing RNA fraction was prepared. The abundant class of cDNA against 1-day chick lens mRNA was used to detect crystallin mRNA in mRNA samples of eye cup by hybridization. The data showed that 3.5-day embryo eye cups expressed crystallin mRNA at about 10 times the level found in 8-day

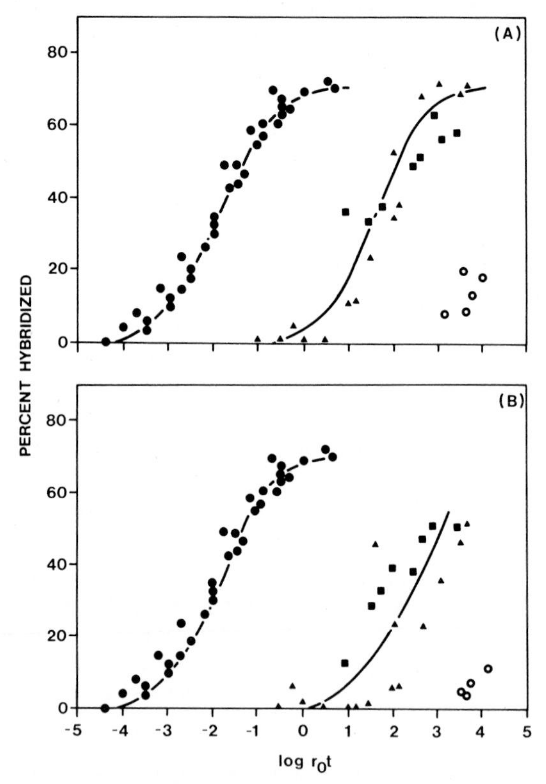

Fig. 7. Putative crystallin mRNA sequences in nonlenticular tissues of 8-day chick embryos. A cDNA fraction complementary to the most abundant class of lens mRNA was used to assay the putative crystallin sequences. (A) Two preparations of cytoplasmic mRNA of embryonic pigmented retina (■ and ▲) and headless embryos (○). (B) Two preparations of polysomal mRNA of embryonic neural retina (■ and ▲) and embryonic muscle (○) compared with the reaction of lens mRNA (●). (Jackson *et al.*, 1978.)

embryo neural retina. On the other hand, in the retina of 1-day-old chicks no crystallin mRNA could be detected. As already stated, the rate and frequency of transdifferentiation from neural retina of 3.5-day embryo is two to three times higher than that from 8-day embryo neural retina, and that from 1-day-old chick retina is nil.

Since neural retina and pigmented retina are closely associated with the lens, and the separation of tissues should be the more difficult the earlier the stage, the question of contamination of test tissue by fragments of transparent lens can be raised. The authors are confident that such contamination was avoided by their surgical method. Because of the importance attached to the results, it might be worthwhile to confirm this by a histological study on prepared tissue samples combined with immunofluorescence for crystallins.

Combining the available data on crystallin mRNA and on the capacity of neural retina and pigmented retina of chick embryo and newly hatched chick to transdifferentiate into lens, there seems to be a positive correlation. Based on this, Clayton *et al.* (1979) and Clayton (1979) propose that low levels of crystallin mRNA are the precondition for transdifferentiation into lens. This proposal inevitably raises the question, How can these sequences be involved in the mechanism of transdifferentiation? So far as I know, no clear description has been presented of the possible mechanism by which low levels of crystallin mRNA can direct cells into the pathway of lens differentiation. Another interpretation of the situation might be that the presence of crystallin mRNA at low levels indicates a special property of the chromatin configuration, which allows its transformation into the lens cell configuration. Such an idea appears compatible with the speculation presented elsewhere in this chapter that the appearance of a lens-specific chromatin configuration is the central issue in transdifferentiation.

An extensive series of studies report the presence of crystallins in low concentrations in iris, cornea, retina, and vitreous body of the chick after hatching (Clayton *et al.*, 1968; Brahma *et al.*, 1971; Bours and Doorenmaalen, 1972; for earlier literature, see Bours and Doorenmaalen). Although some efforts were made to rule out postmortem leaking of the crystallins of lens into the surrounding tissues (Brahma *et al.*, 1971), as suggested by Zwaan (1968), the site of synthesis of these proteins remains unknown. The possibility that a small fraction of crystallins synthesized in lens leak into the surrounding eye tissue under normal living conditions has not been excluded. In the immunofluorescence of cultured lens cells a large number of small vesicles positive for crystallins are found scattered around lens cells, suggesting exocytosis of crystallins. Such a mechanism could be operating in the leaking of crystallins from lens cells *in situ*. Another possibility is that these extralenticular crystallins are synthesized in adult nonlenticular tissue where they are found. This is improbable, at least for neural retina which loses crystallin mRNA before hatching. The last possibility is that all the nonlenticular tissues transcribe crystallin sequences in early embryonic stages and synthesize crystallins. Crystallins having a long half-life may remain intact at later stages. Although the presence of crystallins in various ocular tissues has been implicated in transdifferentiation into lens cells, because of the uncertainty concerning their origin, it is not possible to evaluate these observations on the mechanism of transdifferentiation. One recent development in this area is related to the finding of Barabanov (1977) that the adenohypophysis of chick embryos contains δ-crystallins. According to a short paper of Clayton (1979), primary culture of 4-day chick embryo adenohypophysis produced lens fiber aggregates positive for immunofluorescence for δ-crystallins.

VI. CORNEA AS A SOURCE OF LENS CELLS

System E: Transdifferentiation of Cornea into Lens in Larval Amphibians

The first observation of lens formation in the cornea was made by Ikeda (1939) in embryos and larvae of *Hynobius unangso,* a Japanese urodele. In order to confirm his earlier result suggesting lens regeneration from the cornea in this species, he grafted late embryonic and early larval corneas into the lentectomized optic chamber and demonstrated lens formation from grafts of all stages. The author interpreted this lens formation as an expression of a residual competence for embryonic induction of the lens. The cornea is produced from the part of the epidermis which participated in formation of the primary lens under local inductive influences. Hence the ectodermal cells remaining in this area could have retained the competence for lens induction.

In 1963 Freeman published a study on lens regeneration in the larvae of *X. laevis* showing that the lens was derived from the inner layer of the outer cornea, which thickened and produced a spherical compact aggregate protruding into the anterior optic chamber. The aggregate soon formed the primary lens fibers, separated from the corneal layers, and grew into a lens. Removal of the outer cornea together with the surrounding epidermis, followed by lentectomy, suppressed lens regeneration. When the cornea labeled with [^{3}H]thymidine was grafted into the eye cup of unlabeled larvae, the graft produced a labeled lens. Further, in the absence of the eye cup, grafted cornea failed to form a lens, suggesting an influence of the eye cup in lens formation.

According to Waggoner (1973) grafting of the cornea of *Xenopus* larvae onto an amputated hind limb gave rise to a lens, as well as grafting the same tissue into a lentectomized optic chamber. However, the cornea graft failed to produce a lens in the dorsal fin and under the skin. On the other hand, Campbell and Jones (1968) and Campbell (1970) reported the development of a lens from the cornea in an organ culture experiment, which showed positive immunofluorescence for total lens proteins. A systematic immunofluorescence study on lens regeneration from the cornea in *Xenopus* larvae after lentectomy was carried out by Brahma and McDevitt (1974) in which indirect immunofluorescence involving antibodies against total lens proteins critically diluted to remove nonspecific staining and antibodies against γ-crystallins was used. In Freeman stage IV (late compact aggregates) some cells started to show positive immunofluorescence with both antibodies. Later only the lens fibers showed staining for γ-crystallins, while all the cells of the later lens vesicle and young lens tissue were stained with anti-total lens proteins. Contrary to Campbell (1965) Brahma and McDevitt failed to find positive immunofluorescence in the corneal epithelium and in early regenerates up to Freeman stage III. That the proliferation of progenitor cells in the cornea precedes the appearance of lens cells has been shown in an autoradiographic study by Waggoner and Reyer (1975).

VII. GENERAL DISCUSSION

We have discussed five systems which transdifferentiate into lens from four different eye tissues. One group of systems which stands out from the others is the one starting with melanocytes (systems A, B, and C). In this group, the original tissue is composed of a single cell type, and good evidence exists for the formation of lens cells by cell-type conversion from these melanocytes. In all systems of the group, melanosomes are lost from the cells before the lens phenotype is expressed. It appears likely that the concept developed in system A that conversion is regulated by the extent of proliferation, which is coupled with dedifferentiative processes, can be applied to other systems of the group.

In many respects, system D is in contrast to the group mentioned above. Embryonic neural retina is heterogeneous in cellular composition, and the cells are in the process of completing their predifferentiative proliferation at the stage in question. The system gives at least two allotypic cell types as the result of transdifferentiation. The bulk of data accumulated suggest that allotypic differentiation of this system depends on extensive proliferation induced in culture. This is an aspect shared with the first group. Cloning experiments of this system support the idea that the source tissue contains multipotential progenitor cells which give rise, when cloned, first to autotypic cell types and later to allotypic cell types. The situation can be interpreted as transdifferentiation at the tissue level. However, a recent cinematographic study has suggested that neuronal cells produced from embryonic neural retina in culture are transformed directly into lens fibers. This should be a cell-type conversion without intervening cell divisions. This mode of cell-type conversion has been claimed to occur in some invertebrate cells (Burnett, 1968; Schmidt, 1974, 1975) but never in vertebrate cells. Such a conversion, if proved to occur in this system, should present a new challenge to the developmental biologist and cell biologist alike.

In system E, in which cornea gives rise to lens cells, we have a unique situation. In all the other systems discussed above, the source tissue does not have a direct connection with the ontogenetic source of lenses. The cornea is derived from the region of the epidermis from which the lens was formed in ontogenesis. The fact that the cornea retains the capacity for lens formation and expresses this capacity when the lens is removed is not entirely unexpected. Since the differentiated state of corneal epithelium is not precisely known, it is not possible to determine the mode of reprogramming in this system. As already stated, the mechanism of Wolffian lens regeneration (system A) was earlier supposed to be similar to that of ontogenetic induction of lens. But recent studies on system A indicate that the role of neural retina in this system is not induction of lens specificity but growth promotion. From what has been said, system E should occupy an intermediate position between Wolffian lens regeneration and embryonic lens induction. To know the mechanism by which a lens is produced in system E would be interesting. Some of the results reviewed above seem to suggest that the presence of an optic cup can be replaced by a certain level of proliferation. This would mean that the situation is like that in system A.

Two basic aspects of the reprogramming of differentiation are the factor responsible for realization of reprogramming and the competence of cells to be reprogrammed. With regard to the first aspect, the working hypothesis proposed for system A was that the occurrence of cell-type conversion is decided by the extent of proliferation. It is likely that this hypothesis can be applied to some other systems discussed here. But so far no serious efforts have been made in this respect. It should be recalled that cell proliferation plays a similar role in the transdetermination of *Drosophila* imaginal disks (for review, see Gehring and Nöthiger, 1973). The role of new cell contact in the formation of lens fibers in system D has been emphasized by Clayton *et al.* (1977) and (1979). It appears probable that cell contact plays the role of realization factor in different systems of transdifferentiation. However, under given experimental conditions, it is often difficult to distinguish whether cell contact contributes to transdifferentiation per se or to the appearance of lens fibers in a lens cell population.

The second aspect of reprogramming, competence, presents a complex problem. One should expect different types of competence in different systems. In system A, proliferation is coupled with sequential changes in the cells. The earlier changes are considered dedifferentiation because they are associated with the disappearance of overt differentiation of the original cell type. During this phase depigmentation is associated with alteration of cell surface properties. Between completion of dedifferentiation and the appearance of lens crystallins, there appears to be an interval of about two cell cycles. Since the synthesis of crystallins should follow the establishment of lens-specific chromatin configuration, which determines the new transcription pattern, it is speculated that the completion of dedifferentiation presents the necessary condition for the appearance of new chromatin configuration on which lens cell-type differentiation depends. Alteration of the configuration may occur during the above-mentioned interval.

Another entirely different aspect of competence is proposed in Section III—that the precondition for transdifferentiation into lens cells is the presence of low levels of crystallin mRNA. Although the parallelism between the ability to transdifferentiate into lens cells and the presence of low levels of crystallin mRNA has been demonstrated in a few chick tissues, it is still premature to judge this proposal. It may be interesting to find out whether a similar situation exists in the transdifferentiation of adult amphibian cells.

As a corollary to the previous hypothesis, one could speculate that the central issue common to all the systems discussed here is the emergence of a lens-specific chromatin configuration as a result of interaction of the realizing factor with the competence of cells. Once the configuration has emerged, the cells automatically differentiate into lens cells, and the stability of the differentiated state depends on the stability of the chromatin configuration. Then the relatively frequent occurrence of experimental reprogramming into lens cells may mean that the lens-specific chromatin configuration can be rather easily realized, especially when one starts with a cell type of eye tissues. Viewed from this aspect, the above-discussed proposal that low levels of crystallin mRNA are the precondition for transdifferentiation into lens cells could suggest that the chromatin configuration producing a leaky expression of crystallin sequences possesses the ability to transform into the configuration specific for lens. In this view crystallin mRNA is not involved in the transdifferentiation process but denotes the competence of the chromatin for transdifferentiation into lens cells.

One recent contribution to the mechanism of transdifferentiation is the working hypothesis of Pritchard (1981), which states that stimulation of the tricarboxylic acid cycle is involved in conversion of neural retinal cells into pigment cells. Although the hypothesis does not directly concern the transdifferentiation into lens cells, it is worth mentioning in this review, because it is the only theory which proposes a biochemical pathway as essential for transdifferentiation.

ADDENDUM

In a recent paper (Moscona and Degenstein, 1981) it is reported that rotation of cultures of chick embryo neural retina produces a large number of lens cell clusters. On the basis of an earlier study (Linser and Moscona 1979) the authors suggest that those lens cells are derived from the precursors of Müller cells (retinal glia cells), and that disruption of normal contact and interactions of these cells with neuroblasts is responsible for the shift of the differentiation pathway.

ACKNOWLEDGMENTS

Support of the work of the author reported in this chapter by Fond National Suisse de la Recherche Scientifique is greatly appreciated. Thanks are also due to Drs. Y. Courtois and D. Barritault for providing us with samples of the retinal growth factor, Drs. R. M. Clayton and T. S. Okada for allowing the use of their figures in the present article, and to Dr. D. S. McDevitt for critically reading the manuscript.

REFERENCES

Abe, S., and Eguchi, G. (1977). An analysis of differentiative capacity of pigmented epithelial cells of adult newt iris in clonal cell culture. *Dev. Growth Differ.* **19**, 309–317.

Agata, K., Kondoh, H., Takagi, S., Nomura, K. and Okada, T. S. (1980). Comparison of neuronal and lens phenotype expression in the transdifferentiating culture of neural retina with different culture media. *Dev. Growth Differ.* **22**, 571–577.

Amano, U., and Sato, J. (1940). Über die xenoplastische Implantation der larvalen Iris des *Triturus pyrrhogaster* in das entliste Auge der Larven des *Hynobius nebulosus*. *Jpn. Med. Sci.* (I Anat.) **8,** 75-81.

Amprino, R. (1949). Ricerche sperimentali sulla morfogenesi del cristallino nell'embrione di pollo. Induzione e rigenerazione. *Arch. Entwicklungsmech. Org.* **144,** 71-80.

Araki, M., Yanagita, M., and Okada, T. S. (1979). Crystallin synthesis in lens differentiation in cultures of neural retinal cells of chick embryos. *Dev. Biol.* **69,** 170-181.

Arruti-Mizrali, C. (1979). Effet d'un facteur rétinien sur la prolifération et le phénotype *in vitro* des cellules épithéliales du cristallin et endothéliales de la cornée. Thèse, Université René Descartes, Paris.

Baker, P. F., Hodgkin, A. L. and Ridgeway, E. B. (1971). Depolarization and calcium entry in squid giant axons. *J. Physiol.* (*London*) **218,** 709-755.

Barabanov, V. M. (1977). Expression of delta-crystallin in the adenohypophysis of chick embryos. *Dokl. Akad. Nauk* SSSR **234,** 195-198.

Barritault, D., Arruti, C., and Courtois, Y. (1981). Is there a ubiquitous growth factor in the eye? Proliferation induced in different cell types by eye derived growth factors. *Differentiation* **18,** 29-42.

Berridge, M. J. (1975). The interaction of cyclic nucleotides and calcium in the control of cellular activity. *Adv. Cyclic Nucleotides Res.* **6,** 2-98.

Bours, J., and van Doorenmaalen, W. J. (1972). The presence of lens antigens in the intra-ocular tissues of the chick eye. *Exp. Eye Res.* **13,** 236-247.

Brahma, S. K., and McDevitt, D. S. (1974). Ontogeny and localization of the lens crystallins in *Xenopus laevis* lens regeneration. *J. Embryol. Exp. Morphol.* **32,** 783-794.

Brahma, S. K., Bours, J., and van Doorenmaalen, W. J. (1971). Immunochemical studies of chick iris. *Exp. Eye Res.* **12,** 194-197.

Burnett, A. L. (1968). The aquisition, maintenance, and lability of the differentiated state in *Hydra*. *In* "The Stability of the Differentiated State" (W. Beermann and H. Ursprung, eds.), pp. 109-127. Springer-Verlag, Berlin and New York.

Campbell, J. C. (1965). An immunofluorescent study of lens regeneration in larval *Xenopus laevis*. *J. Embryol. Exp. Morphol.* **13,** 171-179.

Campbell, J. C. (1970). Cellular interactions in vertebrate lens regeneration. *In* "Intercellular Interactions in Differentiation and Growth" (G. V. Lopashov, N. N. Rott, and G. D. Tumanishvili, eds.), pp. 65-77. "Nauka," Moscow.

Campbell, J. C., and Jones, K. W. (1968). The *in vitro* development of lens from cornea of larval *Xenopus laevis*. *Dev. Biol.* **17,** 1-15.

Ciaccio, G. (1933). Ricerche di morfologia causale sulla rigenerazione multipla del cristallino dei tritoni adulti. *Arch. Biol.* (*Paris*) **44,** 179-249.

Clayton, R. M. (1970). Problems of differentiation in the vertebrate lens. *Curr. Top. Develop. Biol.* **3,** 115-180.

Clayton, R. M. (1978). Convergence and divergence in lens cell differentiation: Regulation of the formation and specific content of lens fibre cells. *In* "Stem Cells and Tissue Homeostasis" (B. I. Lord, C. S. Potten, and R. J. Cole, eds.), pp. 115-138. Cambridge Univ. Press, London and New York.

Clayton, R. M. (1979). Regulatory factors for lens fibre formation in cell culture. I. Possible requirement for pre-existing levels of crystallin mRNA. II. The role of growth condition and factors affecting cell cycle duration. *Ophthalmic Res.* **11,** 324-334.

Clayton, R. M., and Campbell, J. C., and Truman, D. E. S. (1968). A re-examination of the organ specificity of lens antigens. *Exp. Eye Res.* **7,** 11-29.

Clayton, R. M., de Pomerai, D., and Pritchard, D. J. (1977). Experimental manipulation of alternative pathways of differentiation in cultures of embryonic chick neural retina. *Dev. Growth Differ.* **19**, 319–328.

Clayton, R. M., Thomson, I., and de Pomerai, D. I. (1979). Relationship between crystallin mRNA expression in retina cells and their capacity to re-differentiate into lens cells. *Nature (London)* **282**, 628–629.

Colucci, V. (1891). Sulla rigenerazione parziale dell'occhio nei tritoni: Istogenesi e svilluppo—Studio sperimentale. *Mem. Accad. Sci. Ist, Bologna, Ser. 5* **1**, 593–629.

Connelly, T. G. (1977). Pituitary enhancement of Wolffian lens regeneration *in vitro:* Spatial and temporal requirements. *J. Exp. Zool.* **200**, 359–364.

Connelly, T. G. (1980). The influence of hormones and other substances on lens regeneration *in vitro. Differentiation* **16**, 85–91.

Connelly, T. G., Ortiz, J. R., and Yamada, T. (1973). Influence of the pituitary on Wolffian lens regeneration. *Dev. Biol.* **31**, 301–315.

Crisanti-Combes, P., Pessac, B., and Calothy, G. (1978). Choline acetyl transferase activity in chick embryo neuroretinas during development *in ovo* and in monolayer cultures. *Dev. Biol.* **65**, 228–232.

Cuney, R. and Zalik, S. E. (1981). Effect of bovine pituitary hormone preparation on newt lens regeneration *in vitro:* Stimulation by thyrotropin. *Develop. Biol.* **81**, 23–35.

Dain, A., Kerkut, G. A., Smith, R. C., Munday, K. A., and Wilmshurst, T. H. (1964). The interaction of free radicals in protein and melanin. *Experientia* **20**, 76–78.

de Pomerai, D. I., and Clayton, R. M. (1978). Influence of embryonic stage on the transdifferentiation of chick neural retina cells in culture. *J. Embryol. Exp. Morphol.* **47**, 179–193.

de Pomerai, D. I., and Clayton, R. M. (1980). The influence of growth-inhibiting and growth-promoting medium conditions on crystallin accumulation in transdifferentiating cultures of embryonic chick neural retina. *Dev. Growth Differ.* **22**, 49–60.

de Pomerai, D. I., Pritchard, D. J., and Clayton, R. M. (1977). Biochemical and immunological studies of lentoid formation in cultures of embryonic chick neural retina and day-old chick lens epithelium. *Dev. Biol.* **60**, 416–427.

de Pomerai, D. I., Clayton, R. M., and Pritchard, D. J. (1978). Delta crystallin accumulation in chick lens epithelial cultures: Dependence on age and genotype *Exp. Eye Res.* **27**, 365–375.

Dumont, J. N., and Yamada, T. (1972). Dedifferentiation of iris epithelial cells. *Dev. Biol.* **29**, 385–401.

Eguchi, G. (1963). Electron microscopic studies on lens regeneration. I. Mechanism of depigmentation of the iris. *Embryologia* **8**, 47–62.

Eguchi, G. (1967). *In vitro* analysis of Wolffian lens regeneration: Differentiation of the regenerating lens rudiment of the newt, *Triturus pyrrhogaster. Embryologia* **9**, 246–266.

Eguchi, G., and Okada, T. S. (1973). Differentiation of lens tissue from the progeny of chick retinal pigment cells cultured *in vitro:* A demonstration of a switch of cell types in clonal cell culture. *Proc. Natl. Acad. Sci. U.S.A.* **70**, 1495–1499.

Eguchi, G., and Shingai, R. (1971). Cellular analysis on localization of lens forming potency in the newt iris epithelium. *Dev. Growth Differ.* **13**, 337–349.

Eguchi, G. and Watanabe, K. (1973). Elicitation of lens from the ventral iris epithelium of the newt by a carcinogen, *N*-methyl-*N*-nitro-*N*-nitrosoguanidine. *J. Embryol. Exp. Morph.* **30**, 63–71.

Eguchi, G., Abe, S., and Watanabe, K. (1974). Differentiation of lens-like structure from newt iris epithelial cells *in vitro. Proc. Natl. Acad. Sci. U.S.A.* **71**, 5052–5056.

Eisenberg, S., and Yamada, T. (1966). A study of DNA synthesis during the transformation of the iris into lens in the lentectomized newt. *J. Exp. Zool.* **162,** 353–368.

Fell, H. B., and Mellanby, F. (1953). Metaplasia produced in cultures of chick ectoderm by high vitamin A. *J. Physiol. (London)* **119,** 470–488.

Filatow, D. (1925). Über die unabhängige Entstehung (Selbstdifferenzierung) der Linse bei *Rana esculenta. Arch. Entwicklungsmech. Org.* **104,** 50–71.

Fischel, A. (1900). Über die Regeneration der Linse. *Anat. Hefte* **14,** 1–256.

Fischel, A. (1903). Weitere Mittheilungen über die Regeneration der Linse. *Arch. Entwicklungsmech. Org.* **15,** 1–138.

Fonnum, F. (1969). Radiochemical micro-assays for the determination of choline acetyltransferase and acetylcholinesterase activities. *Biochem. J.* **115,** 465–472.

Freeman, G. (1963). Lens regeneration from cornea in *Xenopus laevis. J. Exp. Zool.* **154,** 39–66.

Gehring, W. J., and Nöthiger, R. (1973). The imaginal discs of *Drosophila. In* "Developmental Systems: Insects" (S. J. Counce and C. H. Waddington, eds.), Vol. 2, pp. 212–290. Academic Press, New York.

Genis-Galvez, J. M., Maisel, H., and Castro, J. (1968). Changes in chick lens proteins with aging. *Exp. Eye Res.* **7,** 593–602.

Grobstein, C. (1959). Differentiation of vertebrate cells. *In* "The Cell. I. Biochemistry, Physiology, Morphology" (J. Brachet, A. E. Mirsky, eds.), pp. 437–496. Academic Press, New York.

Hadorn, E. (1965). Problems of determination and transdetermination. *Brookhaven Symp. Biol.* **18,** 148–161.

Hadorn, E. (1966). Dynamics of determination. *In* "Major Problems in Developmental Biology" (M. Locke, ed.), pp. 85–104. Academic Press, New York.

Hasegawa, M. (1965). Restitution of the eye from the iris after removal of the retina and lens together with the eye coats in the newt, *Triturus pyrrhogaster. Embryologia* **8,** 362–386.

Horak, V., and Gillette, J. R. (1971). A study of the oxidation-reduction state of synthetic 3,4-dihydroxy-DL-phenylalanine melanin. *Mol. Pharmacol.* **7,** 429–433.

Horstman, L. P., and Zalik, S. E. (1974). Growth of newt iris epithelial cells *in vitro:* A study of the cell cycle . *Exp. Cell Res.* **84,** 1–14.

Idoyaga-Vargas, V., and Yamada, T. (1974). Glucosaminidase and dedifferation of newt iris epithelium. *Differentiation* **2,** 91–98.

Ikeda, Y. (1934). Beiträge zur Analyse der Wolffschen Linsenregeneration durch xenoplastische Implantation der Iris in das entlinste Auge bei *Triton* und *Hynobius. Arb. Anat. Inst. Kais. Jpn. Univ. Sendai* **16,** 69–82.

Ikeda, Y. (1936). Beiträge zur Frage der Fähigkeit zur Linsenregeneration bei einer Art von *Hynobius* (*Hynobius unnangso* Tago). *Arb. Anat. Inst. Kais. Jpn. Univ. Sendai* **18,** 17–50.

Ikeda, Y. (1939). Zur Frage der Linsenpotenz der Hornhaut in spätemryonalen und larvalen Stadien bei einer Art von *Hynobius* (*Hynobius unnangso* Tago). *Arb. Anat. Inst. Kais. Jpn. Univ. Sendai* **22,** 27–52.

Itoh, Y. (1976). Enhancement of differentiation of lens and pigment cells by ascorbic acid in cultures of neural retinal cells of chick embryos. *Dev. Biol.* **54,** 157–162.

Itoh, Y., Okada, T. S., Ide, H., and Eguchi, G. (1975). The differentiation of pigment cells in cultures of chick embryonic neural retina. *Dev. Growth Differ.* **17,** 39–50.

Jackson, J. F., Clayton, R. M., Williamson, R., Thomson, I., Truman, D. E. S., and de Pomerai, D. I. (1978). Sequence complexity and tissue distribution of chick lens crystallin mRNAs. *Dev. Biol.* **65,** 383–395.

Kahn, A. J. (1974). An autoradiographic analysis of the time of appearance of neurons in the developing chick neural retina. *Dev. Biol.* **38,** 30–40.
Keefe, J. R. (1973a). An analysis of urodelian retinal regeneration. I. Studies of the cellular source of retinal regeneration in *Notophthalmus viridescens* utilizing ^{3}H-thymidine and colchicine. *J. Exp. Zool.* **184,** 185–206.
Keefe, J. R. (1973b). An analysis of urodelian retinal regeneration. II. Ultrastructural features of retinal regeneration in *Notophthalmus viridescens. J. Exp. Zool.* **184,** 207–232.
Keefe, J. R. (1973c). An analysis of urodelian retinal regeneration. III. Degradation of extruded melanin granules in *Notophthalmus viridescens. J. Exp. Zool.* **184,** 233–238.
Keefe, J. R. (1973d). An analysis of urodelian retinal regeneration. IV. Studies of the cellular source of retinal regeneration in *Triturus cristatus carnifex* using ^{3}H-thymidine. *J. Exp. Zool.* **184,** 239–258.
Linser, P., and Moscona, A. A. (1979). Induction of glutamine synthetase in embryonic neural retina: Location in Müller fibers and dependence on cell interactions. *Proc. Nat. Acad. Sci. U.S.A.* **76,** 6476–6480.
Lopashov, G. V. (1977). Levels in stabilization of cell differentiation and its experimental transformation. *Differentiation* **9,** 131–137.
Lopashov, G. V., and Sologub, O. G. (1972). Artificial metaplasia of pigmented epithelium into retina in tadpoles and adult frogs. *J. Embryol. Exp. Morphol.* **28,** 521–546.
McAvoy, J. W. (1980). Induction of the eye lens. *Differentiation* **17,** 137–150.
McDevitt, D. S., and Yamada, T. (1980). Crystallin expression in the TVI cell line. *Differentiation* **16,** 125–128.
Mangold, O. (1931). Das Determinationsproblem. III. Das Wirbeltierauge in der Entwicklung und Regeneration. *Ergeb. Biol.* **7,** 193–403.
Michel, C., and Yamada, T. (1974). Cellular studies of X-ray induced inhibition of lens regeneration. *Differentiation* **2,** 193–201.
Mikami, Y. (1941). Experimental analysis of the Wolffian lens-regeneration in adult newt, *Triturus pyrrhogaster. Jpn. J. Zool.* **9,** 269–302.
Mitashov, V. I. (1968). Autoradiographic investigations into the regeneration of the retina in pectinate newts (*Triturus cristatus*). *Dokl. Akad. Nauk SSSR* **181,** 1510–1513 (411–415).
Moscona, A. A. (1957). Formation of lentoids by dissociated retinal cells of the chick embryo. *Science* **125,** 598–599.
Moscona, A. A., and Degenstein, L. (1981). Lentoids in aggregates of embryonic neural retina cells. *Cell Differ.* **10,** 39–46.
Nikitenko, M. F. (1939). Mehrmalige Restitution der Linse im Tritonauge. *Bull. Biol. Med. Exp. URSS* **8,** 136–139.
Nomura, K., and Okada, T. S. (1979). Age-dependent change in the transdifferentiation ability of chick neural retina in cell culture. *Dev. Growth Differ.* **21,** 161–168.
Nomura, K., Takagi, S., and Okada, T. S. (1980). Expression of neuronal specificities in "transdifferentiating" cultures of neural retina. *Differentiation* **16,** 141–147.
Ogawa, T. (1963). Appearance and localization of lens antigens during the lens regeneration of the adult newt. *Embryologia* **7,** 279–284.
Ogawa, T. (1964). The influence of lens antibody on lens regeneration in the larval newt. *Embryologia* **8,** 146–157.
Ogawa, T. (1967). The similarity between antigens in the embryonic lens and in the lens regenerate of the newt. *Embryologia* **9,** 295–305.
Okada, T. S. (1976). "Transdifferentiation" of cells of specialized eye tissue in cell cul-

ture. *In* "Tests of Teratogenicity *in Vitro*" (J. D. Ebert and M. Marrois, eds.), pp. 91-105. North-Holland, Amsterdam.

Okada, T. S. (1977). A demonstration of lens forming cells in neural retina in clonal cell culture. *Dev. Growth Differ.* **19,** 47-105.

Okada, T. S. (1980). Cellular metaplasia or transdifferentiation as a model for retinal cell differentiation. *Curr. Top. Develop. Biol.* **16,** 349-380.

Okada, T. S., Eguchi, G., and Takeichi, M. (1971). The expression of differentiation by chicken lens epithelium in *in vitro* cell culture. *Dev. Growth Differ.* **13,** 323-336.

Okada, T. S., Itoh, Y., Watanabe, K., and Eguchi, G. (1975). Differentiation of lens in cultures of neural retinal cells of chick embryos. *Dev. Biol.* **45,** 318-319.

Okada, T. S., Yasuda, K., Araki, M., and Eguchi, G. (1979a). Possible demonstration of multipotential nature of embryonic neural retina by clonal cell culture. *Dev. Biol.* **68,** 600-617.

Okada, T. S., Yasuda, K., and Nomura, K. (1979b). The presence of multipotential progenitor cells in embryonic neural retina as revealed by clonal cell culture. *INSERM Symp. No. 10,* pp. 335-346.

Ortiz, J. R. (1973). Dedifferentiation of iris epithelial cells *in vitro,* Dissertation, University of Tennessee, Knoxville, Tennessee.

Ortiz, J. R., and Yamada, T. (1975). Synergistic effects of adenosine and compounds related to adenosine 3′:5′-cyclic monophosphate on dedifferentiating iris epithelial cells in culture. *Differentiation* **4,** 135-142.

Ortiz, J. R., Yamada, T., and Hsie, A. W. (1973). Induction of the stellate configuration in cultured iris epithelial cells by adenosine and compounds related to adenosine 3′:5′-cyclic monophosphate. *Proc. Natl. Acad. Sci. U.S.A.* **70,** 2286-2290.

Patmore, L. and Yamada, T. (in press). The role of calcium in depigmentation of iris epithelial cells during cell-type conversion. *Develop. Biol.*

Piatigorsky, J. (1981). Lens differentiation in vertebrates. A review of cellular and molecular features. *Differentiation* **19,** 134-153.

Pritchard, D. J. (1981). Transdifferentiation of chick embryo neural retina into pigment epithelium: indications of its biochemical basis. *J. Embryol. Exp. Morph.* **62,** 47-62.

Rabaey, M. (1962). Electrophoretic and immunoelectrophoretic studies on the soluble proteins in the developing lens of birds. *Exp. Eye Res.* **1,** 310-316.

Reese, D. H. (1973). *In vitro* initiation in the newt iris of some early molecular events of lens regeneration. *Exp. Eye Res.* **17,** 435-444.

Reese, D. H., Puccia, E., and Yamada, T. (1969). Activation of ribosomal RNA synthesis in initiation of Wolffian lens regeneration. *J. Exp. Zool.* **170,** 259-268.

Reese, D. H., Yamada, T., and Moret, R. (1976). An established cell line from the newt, *Notophthalmus viridescens. Differentiation* **6,** 75-82.

Reyer, R. W. (1954). Regeneration of the lens in the amphibian eye. *Q. Rev. Biol.* **29,** 1-46.

Reyer, R. W. (1956). Lens regeneration from homoplastic and heteroplastic implants of dorsal iris into the eye chamber of *Triturus viridescens* and *Amblystoma punctatum. J. Exp. Zool.* **133,** 145-190.

Reyer, R. W. (1966). DNA synthesis and cell movement during lens regeneration in adult *Triturus viridescens. Am. Zool.* **6,** 329.

Reyer, R. W. (1971). DNA synthesis and incorporation of labeled iris cells into the lens during lens regeneration in adult newts. *Dev. Biol.* **24,** 533-558.

Reyer, R. W. (1977). The amphibian eye: Development and regeneration. *In* "Handbook of Sensory Physiology. VII. Visual System in Evolution" (F. Crescitelli, ed.), pp. 309-390. Springer-Verlag, Berlin and New York.

Reyer, R. W., Woolfitt, R. A., and Withersty, L. T. (1973). Stimulation of lens regeneration from the newt dorsal iris when implanted into the blastema of the regenerating limb. *Dev. Biol.* **32**, 258–281.

Rübsaamen, H. (1950). Die Wirkung des experimentellen Sauerstoffmangels auf die Entwicklung von Tritonkeimen nach beendeter Gastrulation. *Arch. Entwicklungsmech. Org.* **144**, 301–321.

Satasurja, L. A., and Nieuwkoop, P. D. (1974). The induction of the primordial germ cells in the urodeles. *Arch. Entwicklungsmech. Org.* **175**, 199–220.

Sato, T. (1930). Beiträge zur Analyse der Wolffschen Linsenregeneration. I. *Arch. Entwicklungsmech. Org.* **122**, 451–493.

Sato, T. (1935). Beiträge zur Analyse der Wolffschen Linsenregeneration. III. *Arch. Entwicklungsmech. Org.* **133**, 19–78.

Sato, T. (1951). Über die linsenbildende Fähigkeit des Pigmentepithels bei *Diemyctylus pyrrhogaster.* I. Pigmentepithel aus dorsalem Augenbereich. *Embryologia* **1**, 21–57.

Scheib, D. (1965). Recherches récents sur la régénération du cristallin chez les vertébrés. *Ergebn. Anat. Entwicklungsgesch.* **38**, 45–114.

Schmidt, V. (1974). Structural alterations in cultivated striated muscle cells from anthomedusae (Hydrozoa): A metaplastic event. *Exp. Cell Res.* **86**, 193–198.

Schmidt, V. (1975). Cell transformation in isolated striated muscle of hydromedusae independent of DNA synthesis. *Exp. Cell Res.* **94**, 401–408.

Selman, K., and Kafatos, F. C. (1974). Transdifferentiation in the labial gland of silk moths: Is DNA synthesis required for cellular metamorphosis? *Cell Differ.* **3**, 81–94.

Spemann, H. (1905). Über Linsenbildung nach experimenteller Entfernung der primären Linsenbildungszellen. *Zool. Anz.* **28**, 419–166.

Steinberg, M. L., and Whittaker, J. R. (1975). Stimulation of melanotic expression in a melanoma cell line by theophylline. *J. Cell. Physiol.* **87**, 265–276.

Stone, L. S. (1950). The role of retinal pigment cells in regenerating neural retinae of adult salamander eyes. *J. Exp. Zool.* **113**, 9–31.

Stone, L. S. (1958). Inhibition of lens regeneration in newt eyes by isolating the dorsal iris from the neural retina. *Anat. Rec.* **131**, 151–172.

Stone, L. S. (1967). An investigation recording all salamanders which can and cannot regenerate a lens from the dorsal iris. *J. Exp. Zool.* **164**, 87–104.

Stone, L. S., and Gallagher, S. B. (1958). Lens regeneration restored to iris membranes when grafted to neural retina environment after cultivation *in vitro*. *J. Exp. Zool.* **139**, 247–262.

Stone, L. S., and Steinitz, H. (1953). The regeneration of lenses in eyes with intact and regenerating retina in adult *Triturus viridescens viridescens*. *J. Exp. Zool.* **124**, 435–468.

Takata, C., Albright, J. F., and Yamada, T. (1964). Lens antigens in a lens-regenerating system studied by the immunofluorescent technique. *Dev. Biol.* **9**, 385–397.

Takata, C., Albright, J. F., and Yamada, T. (1966). Gamma-crystallins in Wolffian lens regeneration demonstrated by immunofluorescence. *Dev. Biol.* **14**, 382–400.

Thomson, I., Wilkinson, C. E., Jackson, J. F., de Pomerai, D. I., Clayton, R. M., Truman, D. E. S., and Williamson, R. (1978). Isolation and cell-free translation of chick lens crystallin mRNA during normal development and transdifferentiation of neural retina. *Dev. Biol.* **65**, 372–382.

Thorpe, C. W., Bond, J. S., and Collins, J. M. (1974). Early events in lens regeneration: Changes in cyclic AMP concentrations during initiation of RNA and DNA synthesis. *Biochim. Biophys. Acta* **340**, 413–418.

Trautwein, W. (1975). Membrane currents in cardiac muscle fibres. *Physiol. Rev.* **53,** 793-835.

Truman, D. E. S., Brown, A. G., and Campbell, J. C. (1972). The relationship between ontogeny of antigens and the polypeptide chains of the crystallins during chick lens development. *Exp. Eye Res.* **13,** 58-69.

Velázquez, F. M., and Ortiz, J. R. (1980). Intercellular levels of adenosine 3′:5′-cyclic monophosphate in the dorsal iris of adult newt during lens regeneration. *Differentiation* **17,** 117-120.

Wachs, H. (1914). Neue Versuche zur Wolffschen Linsenregeneration. *Arch. Entwicklungsmech. Org.* **39,** 384-451.

Wachs, H. (1919). Zur Entwicklungsphysiologie des Auges der Wirbeltiere. *Naturwissenschaften* **7,** 322-327.

Wachs, H. (1920). Restitution des Auges nach Extirpation von Retina und Linse bei *Triton. Arch. Entwicklungsmech. Org.* **46,** 328-390.

Waggoner, P. R. (1973). Lens differentiation from the cornea following lens extirpation or cornea transplantation in *Xenopus laevis. J. Exp. Zool.* **186,** 97-100.

Waggoner, P. R., and Reyer, R. W. (1975). DNA synthesis during lens regeneration in larval *Xenopus laevis. J. Exp. Zool.* **192,** 65-72.

Whittaker, J. R. (1963). Changes in melanogenesis during the dedifferentiation of chick retinal pigment cells in cell culture. *Dev. Biol.* **8,** 99-127.

Wolff, G. (1895). Entwicklungsphysiologische Studien. I. Die Regeneration der Urodelenlinse. *Arch. Entwicklungsmech. Org.* **1,** 380-390.

Yamada, T. (1961). A chemical approach to the problem of the organizer. *Adv. Morphog.* **1,** 1-53.

Yamada, T. (1966). Control of tissue specificity: The pattern of cellular synthetic activities in tissue transformation. *Am. Zool.* **6,** 21-31.

Yamada, T. (1967). Cellular and subcellular events in Wolffian lens regeneration. *Curr. Top. Dev. Biol.* **2,** 247-283.

Yamada, T. (1972). Control mechanisms in cellular metaplasia. *In* "Proceedings 1st International Conference of Cell Differentiation" (R. Harris, P. Allen, and D. Viza, eds.), pp. 56-60. Munksgaard, Copenhagen.

Yamada, T. (1976). Dedifferentiation associated with cell-type conversion in the newt lens regenerating system. *In* "Progress in Differentiation Research" (N. Müller-Bérat, ed.), pp. 355-360. North-Holland, Amsterdam.

Yamada, T. (1977). Control mechanisms in cell-type conversion in newt lens regeneration. *Monogr. Dev. Biol.* **13,** 1-124.

Yamada, T., and Beauchamp, J. J. (1978). The cell cycle of cultured iris epithelial cells: Its possible role in cell-type conversion. *Dev. Biol.* **66,** 275-278.

Yamada, T., and Dumont, J. N. (1972). Macrophage activity in Wolffian lens regeneration. *J. Morphol.* **136,** 367-384.

Yamada, T., and McDevitt, D. S. (1974). Direct evidence for transformation of differentiated iris epithelial cells into lens cells. *Dev. Biol.* **38,** 104-118.

Yamada, T., and Roesel, M. E. (1969). Activation of DNA replication in the iris epithelium by lens removal. *J. Exp. Zool.* 171, 425-431.

Yamada, T., and Roesel, M. E. (1971). Control of mitotic activity in Wolffian lens regeneration. *J. Exp. Zool.* **177,** 119-128.

Yamada, T., Reese, D. H., and McDevitt, D. S. (1973). Transformation of iris into lens *in vitro* and its dependency on neural retina. *Differentiation* **1,** 65-82.

Yamada, T., Roesel, M. E., and Beauchamp, J. J. (1975). Cell cycle parameters in dedifferentiating iris epithelial cells. *J. Embryol. Exp. Morphol.* **34,** 497-510.

Yamada, T., Dumont, J. N., Moret, R., and Brun, J. P. (1978). Autophagy in dedifferentiating newt iris epithelial cells *in vitro*. *Differentiation* **11**, 133–147.

Yasuda, K. (1979). Transdifferentiation of "lentoid" structures in cultures derived from pigmented epithelium was inhibited by collagen. *Dev. Biol.* **68**, 618–623.

Yasuda, K., Eguchi, G. and Okada, T. S. (1981). Age-dependent changes in the capacity of transdifferentiation of retinal pigment cells as revealed in clonal cell culture. *Cell Differentiation* **10**, 3–11.

Yasuda, K., Okada, T. S., Eguchi, G., and Hayashi, M. (1978). A demonstration of a switch of cell type in human fetal eye tissue *in vitro*: Pigmented cells of the iris or the retina can transdifferentiate into lens. *Exp. Eye Res.* **26**, 591–595.

Zalik, S. E., and Dimitrov, E. (1980). Shift in cell type of newt iris epithelial cells in culture. *Can. J. Zool.* **58**, 400–403.

Zalik, S. E., and Scott, V. (1972). Cell surface changes during dedifferentiation in the metaplastic transformation of iris into lens. *J. Cell Biol.* **55**, 134–146.

Zalik, S. E., and Scott, V. (1973). Sequential disappearance of cell surface components during dedifferentiation in lens regeneration. *Nature (London) New Biol.* **244**, 212–214.

Zalik, S. E., Scott, V., and Dimitrov, E. (1976). Changes at the cell surface during *in vivo* and *in vitro* dedifferentiation in cellular metaplasia. *In* "Progress in Differentiation Research" (N. Müller-Berat, ed.), pp. 361–367. North-Holland, Amsterdam.

Zalokar, M. (1944). Contribution à l'étude de la régénération du cristallin chez le triton. *Rev. Suisse Zool.* **51**, 444–521.

Zwaan, J. (1968). Lens specific antigens and cytodifferentiation in the developing lens. *J. Cell. Physiol.* (*Suppl. 1*) **72**, 47–71.

6

Biological–Physical Basis of Lens Transparency

FREDERICK A. BETTELHEIM AND ERNEST L. SIEW

I. TRANSPARENCY OF THE LENS

A. Scattering of Light

Light scattering plays an important part in our understanding of the lens function. Knowing the basic principles of light scattering we can

Cell Biology of the Eye

ISBN 0-12-483180-X

explain the transparency of the normal lens and the physical causes that may give rise to cataract formation. Consequently light scattering has played an important role in eye research. This role has a dual nature: phenomenological and analytical. Phenomenological investigations try to answer such questions as: How much light reaches the retina? How much is scattered and absorbed? What is the spectrum of light reaching the fovea? Analytical studies try to answer such questions as: What are the structural or optical units that cause scattering? What are the dimensions, shape, etc., of these structural units? In answering these questions the theories of light scattering are applied to experimental data. Therefore, it is pertinent to understand the basic principles of light scattering provided in the following section.

When light interacts with a particle, be it an atom, a molecule, or an aggregate of molecules, a dipole moment is induced in the particle. This is due to the fact that the particle in the electric field is subjected to polarization; i.e., the nuclei and the electrons move in opposite directions. Since electromagnetic radiation such as light carries an oscillating field, the induced dipole in the particle will also oscillate. An oscillating dipole, however, is the source of electromagnetic radiation, i.e., the scattered light that is propagated in all directions.

If we are dealing with elastic scattering (i.e, none of the energy of the incident radiation is used to put the atom or molecule in an excited—translation, rotational, vibrational, or electronic—state), the frequency of the scattered light will be the same as that of the incident light. We call this Rayleigh scattering. It was Lord Rayleigh who first derived the equation for the intensity of scattered radiation in 1871:

$$I_\theta/I_0 = 16\,\pi^4\alpha^2 \sin^2\theta/\lambda^4 r^2 \tag{1}$$

where I_θ is the intensity of the scattered light at any angle θ, θ is the scattering angle (i.e., the angle between the incident beam and the scattered beam), I_0 is the intensity of the incident beam, λ is the wavelength of the incident beam in vacuum, r is the distance of the observer from the scattering center, and α is the polarizability, a molecular parameter stating how much dipole moment is induced by a unit electric field. Combination of the intensities with the distance of the observer from the scattering center,

$$(I_\theta/I_0)(r^2/V) = R_\theta \tag{2}$$

is known as the Rayleigh ratio, which describes the intensity of the scattered light per unit volume V.

Equation (1) predicts that the intensity of the scattered light will depend on the inverse of the fourth power of the wavelength that is scattered. Blue light will have a greater intensity than scattered red light, which accounts for the blue sky.

The polarizability α is related to the refractive index n by the equation

$$n^2-1 = 4\,\pi N\alpha \tag{3}$$

where N is the number of particles per cm^3. In dilute gases the refractive index in turn can be expressed as a result of the fluctuation in the refractive index as a result of local concentration fluctuations (dn/dc).

Equations (1) and (2) combined can be rewritten as

$$R_\theta = [4\pi^2 \sin^2\theta(dn/dc)^2\, \mathrm{Mc}]/N_{AV}\lambda^4 \tag{4}$$

where the new quantities are M, the molecular weight of the particles, and c, the concentration of the particles (g/cm^3).

In contrast to the random position of gas molecules within the scattering volume the atoms in a crystal are rigidly fixed in a geometric array. Since the wavelength of the light is much larger than the individual scatterers (atoms), we always can find a pair of atoms (equal scatterers) which scatters out of phase from the point of view of the observer (or the detector) at any particular scattering angle. Thus, the destructive interference of a pair of scatterers is complete, and no scattered light will be observed.

The expression "crystal clear," the historic name "crystalline lens," and the isolated protein fraction designated "crystallins" have more to do with the transparency of the lens than with a crystalline organization of molecules within the lens.

Scattering from pure liquids is intermediate between that from crystals and gases. We could select pairs of small volume elements (much smaller than the wavelength of light) which would be independent scatterers that are out of phase, and destructive interference from these pairs would follow. However, since liquids are not orderly, the packing of particles in each member of the pair of volume elements may not be the same; therefore, the intensity of the scattered light from volume element 1 may differ from that from volume element 2. As a consequence, the destructive interference is not complete, and some scattering will occur.

The scattering from a two-component solution can be understood on a similar basis. The solution is construed as a system of small volume elements, each of which acts as a single scattering source. Each of these volume elements contains a large number of solvent molecules and a few solute molecules.

Based on the theories of Einstein (1910) and Debye (1944) the whole system can be looked upon as a fluctuation in concentration; that is, each small volume element has a different concentration, $\bar{c} + dc$, where $\bar{c}$ is the average concentration of the solution and dc is the deviation from this average. The deviation can be both positive and negative. This fluctuation of concentration is completely random. The concentration fluctuation corresponds to a refractive index fluctuation, and the intensity of the scattered light or the corresponding Rayleigh ratio will be given by the equation

$$\frac{I_\theta}{I_0}\frac{r^2}{V}\frac{1}{1+\cos^2\theta} = R'_\theta = \frac{2\pi^2 n_0 (dn/dc)^2 c}{N_{av}\lambda^4(1/M) + 2Bc + 3Cc^2 + \cdots)} \tag{5}$$

where R'_θ is the Rayleigh ratio corrected for unpolarized light, c is the concentration of the solute (g/cm^3), n_0 is the refractive index of the solvent, and B and C are the second and third virial coefficients representing thermodynamic interaction parameters between solute and solvent and solute and solute, respectively.

If all the constants are lumped together in an optical constant K,

$$K = 2\pi^2 n_0^2 (dn/dc)^2 N_{AV}\lambda^4 \tag{6}$$

then Eq. (5) can be rewritten in the familiar form:

$$(Kc/R_\theta') = (1/M) + 2Bc + 3Cc^2 \tag{7}$$

This equation shows that the intensity of the scattered light increases with concentration and molecular weight. When such an equation is applied to very dilute (infinitely dilute) solutions, the second and third terms on the right-hand side approach zero and thus Kc/R_θ' will be equal to the reciprocal of the molecular weight (Debye, 1947).

If the solutions contain macromolecules with at least one dimension comparable to the wavelength of the light, i.e., greater than $\lambda/20$, one macromolecule cannot be contained in the small volume element but will be part of a number of volume elements. This means in essence that the macromolecule will contain more than one scattering source. The light scatterers within the macromolecules will be somewhat out of phase with each other and will cause destructive interference. How much interference occurs depends on the scattering angle and the size and shape of the particle. The effect of the large size of the macromolecule is expressed by the particle-scattering function $P(\theta)$:

$$P(\theta) = \frac{\text{scattered intensity for macromolecule}}{\text{scattered intensity without interference}} \tag{8}$$

Since the interference reduces the scattered light intensity, $P(\theta)$ will always be less than 1, but it approaches 1 at zero scattering angle where the scattered light from the different parts of the macromolecule will be in phase and no destructive interference will occur.

The analytical expression of $P(\theta)$ for different-shaped molecules has been worked out by a number of authors and can be found in textbooks on light scattering such as those by Stacey (1956), Tanford (1961), and Kerker (1969).

For a flexible random coil when θ approaches zero

$$\text{Limit}\,[1/P(\theta)] = 1 + (16\,\pi^2/3\lambda^2)\,R_G^2 \sin^2(\theta/2), \qquad \theta \to 0 \qquad (9)$$

where R_g is the size parameter or the radius of gyration, i.e., the average distance to a segment of the macromolecule from the center of gravity of the molecule.

$$Kc/R_\theta' = [1/MP(\theta)] + Bc + Cc^2 + \cdots \qquad (10)$$

can be used for simultaneous evaluation of molecular weight, virial coefficient (interaction parameter), and size of molecule (radius of gyration). This is done with the aid of the Zimm (1948) plot in which experimental data are doubly extrapolated to both zero concentration and zero angle.

In the Zimm plot Kc/R_θ' is plotted against $c + \sin^2(\theta/2)$ (Fig. 1). The intercept of the zero angle and zero concentration line on the Kc/R_θ' axis gives the reciprocal of the molecular weight. The slope of the zero angle line gives the virial coefficient. At $\theta = 0$, $P(\theta) = 1$ and thus Eq. (10) becomes Eq. (7). The slope of the zero concentration line can yield the radius of gyration from Eq. (9) and (10):

$$R_G = \left(\frac{3\lambda^2/n^2}{16\pi^2} \; \frac{\text{initial slope of zero concentration line}}{\text{intercept}} \right)^{1/2} \qquad (11)$$

The light scattering of dilute solutions can be used to evaluate molecular parameters, for example, those of α-crystallins (Kramps *et al.*, 1975).

When the size of the particles becomes very large,λ or greater, the scattering intensity decreases very rapidly with the scattering angle θ and $P(\theta)$ becomes an oscillating function. Particles of this size in dilute solution are treated by the Mie theory (1908).

Nonpolarized light scattering from concentrated solutions, gels, and inhomogeneous solids can also be treated by the fluctuation theory (Debye and Bueche, 1949). An extension of this is found in Stein *et al.* (1959) for polarized light.

There are two kinds of fluctuations that can cause light scattering: density fluctuations and fluctuations in the orientation of optically

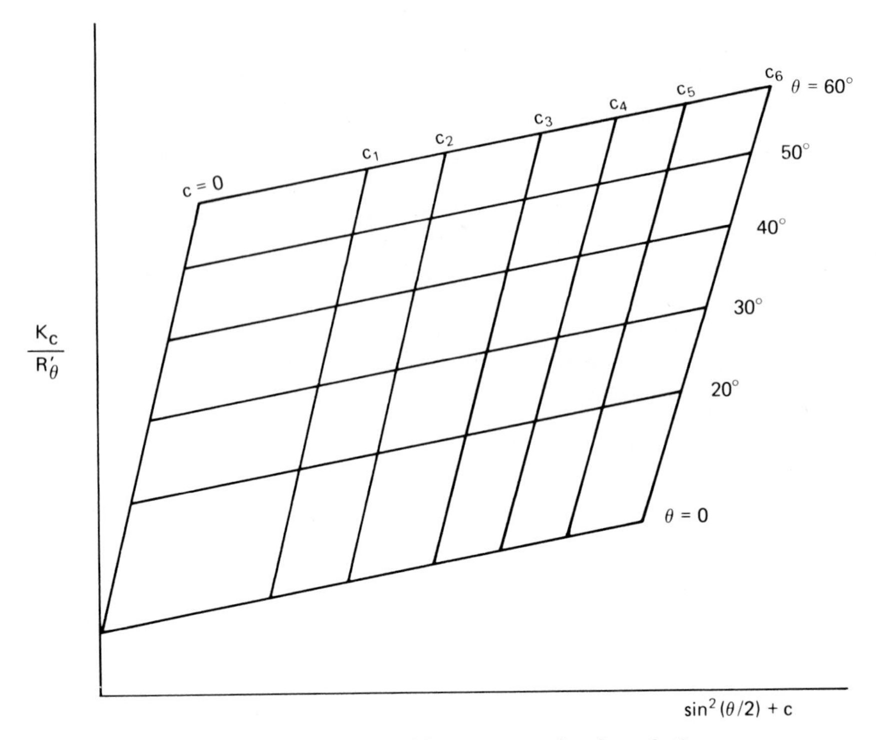

Fig. 1. Zimm plot of a dilute macromolecular solution.

anisotropic material. Both will result in variation in the local refractive index as the photon passes through the material (Fig. 2). In Fig. 2, the solid straight line represents the average polarizability (refractive index) of the medium. Here η shows the local deviation from this average, $\bar{\eta}^2$ is the average (mean squared) deviation from the average refractive index, and a is the size of the region that is correlated.

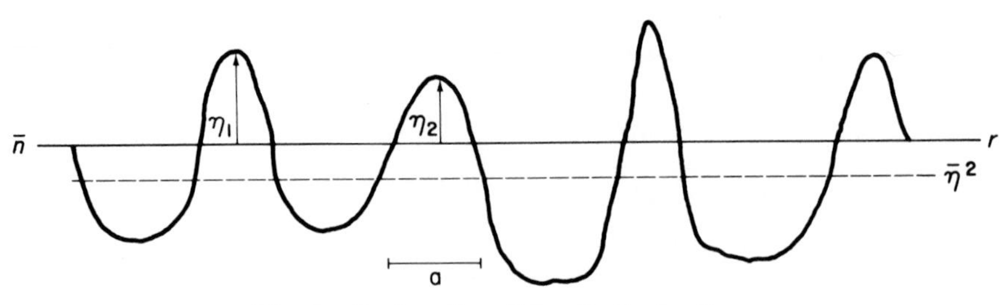

Fig. 2. Designation of density fluctuations.

The above symbols are used in the designation of density fluctuations. Similar symbols ($\bar{\delta}^2$ and b) are used in designating the orientation fluctuations. The reason for using two different designations for the same kind of refractive index fluctuation is that in most cases one can separate the contributions from the two fluctuations. This is done by the use of plane-polarized light as a source. If the scattering sample is placed between two polarizers whose axes of polarization are parallel, the measured scattered light intensity is designated $I_{\parallel}$. On the other hand, if the polarizer and analyzer are set 90° apart, the scattered light intensity is measured in the I_{+} mode.

The theory of light scattering from random density orientation fluctuation (Stein *et al.*, 1959) predicts that

$$I_{\parallel} = K\left\{ \bar{\eta}^2 \int \gamma(r) \frac{\sin hr}{hr} r^2\, dr + \frac{4}{45} \bar{\delta}^2 \int f(r) \left[\frac{\bar{\eta}^2}{\alpha^2} \gamma(r) + 1 \right] \frac{\sin hr}{hr} r^2\, dr \right\} \tag{12}$$

$$I_{+} = K\left\{ \frac{1}{15} \bar{\delta}^2 \int f(r) \left[\frac{\bar{\eta}^2}{\alpha^2} \gamma(r) + 1 \right] \frac{\sin hr}{hr} r^2\, dr \right\} \tag{13}$$

where $\bar{\eta}^2$ and $\bar{\delta}^2$ are the average deviations in density and orientation fluctuations, respectively, as explained before, $h = (4\pi/\lambda) \sin(\theta/2)$ is a function of the scattering angle θ, where λ is the wavelength of the light in the medium; and r is the distance of penetration of light in the medium. The correlation function of the density function $\gamma(r)$ is set up in such a manner that, when two volume elements are the same (i.e., their separation is zero), $r = 0$ and $\gamma(\mathrm{O}) = 1$. This is a perfect correlation. The form of the correlation function is

$$\gamma(r) = \langle \eta_1, \eta_2 \rangle_r / \bar{\eta}^2 \tag{14}$$

For all pairs of volume elements separated by a distance r, $\langle \eta_1, \eta_2 \rangle$ is the average of η_1 and η_2. On the other hand, η_1 and η_2 are the local deviations (of refractive indexes from the average) of two volume elements separated by a distance r.

At the other extreme, when the two volume elements are separated by $r = \infty$, no correlation exists and $\gamma(\infty) = 0$.

The experimental $\gamma(r)$ can be obtained by the numerical integration of a combined form of Eqs. (12) and (13). The experimental $\gamma(r)$ must be fitted to an analytical function obeying the boundary conditions. In some cases (Debye and Bueche, 1949; Gallagher and Bettelheim, 1962) a single exponential function,

$$\gamma(r) = e^{-r/a} \tag{15}$$

fits the experimental density correlation function, where a is the correlation length of the density fluctuation.

In other experiments it was found (Stein, 1969; Wun and Prins, 1974) that the scattering over the entire angular range could be described by a sum of Gaussians:

$$\gamma(r) = \Sigma X_i \exp(-r^2/a_i^2) \tag{16}$$

where $\Sigma X_i = 1$.

To interpret such an experimental correlation function, one can consider the inhomogeneous system a quasi-two-phase system (dispersed and dispersant; $i = 2$). Thus, two different correlation lengths can measure the size of the inhomogeneities under these conditions: a_1 and a_2 are the average sizes of the dispersed and the dispersant phases, respectively (Kahovec *et al.*, 1953; Kratky, 1966; Alexander, 1969).

When a proper analytical form of the experimental $\gamma(r)$ is obtained, the $\bar{\eta}^2$ values can be evaluated from Eq. (12) and (6) by using the Rayleigh ratio of intensities at a set angle.

Once this has been accomplished, the nature of the orientation correlation function $f(r)$ can be predicted, and thus the experimental orientation correlation function $f(r)$ can be obtained. The experimental orientation correlation function can be expressed as the result of two kinds of inhomogeneities and, similar to Eq. (16), the corresponding form of $f(r)$ will be

$$f(r) = \Sigma X_i \exp(-r^2/b_i^2) \tag{17}$$

where $i = 2$.

With the aid of the analytic form of $f(r)$, Eq. (13) is rearranged and solved for $\bar{\delta}^2$ at a set scattering angle. In this manner eight structural parameters can be obtained from the light-scattering measurements. In relation to lens transparency and cataract formation these parameters are size, volume fraction of the aggregates within the fiber cell, average distance of their separation, and refractive index difference between the scattering units and their environment. Similarly four parameters are obtained from the depolarized light-scattering envelope, giving information on the size, volume fraction, separation, and refractive index difference of the optically anisotropic units (cytoskeletal filaments, microtubles, etc.) within the lens (Bettelheim and Paunovic, 1979).

B. Absorption of Light

A young human lens in the first year of life transmits over 90% of visible light (400–800 nm), and it appears colorless. With aging the human lens takes on a yellowish appearance which can become brown in later stages. The amount of visible light transmitted may be reduced by as much as 30% by the age of 80 even in normal noncataractous lenses (Lerman and Borkman, 1976).

The reason for this absorption is the presence of chromophores, usually containing conjugated double bonds.

The chromophore in an excited state is metastable and usually loses absorbed energy by either radiative or nonradiative processes. Two kinds of radiative processes are important: fluorescence and phosphorescence.

In fluorescence (Fig. 3a) the transition from ground state to excited

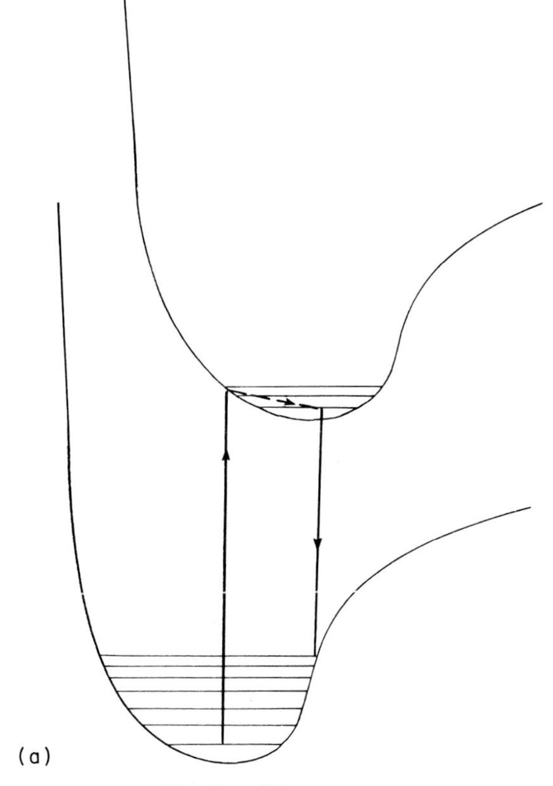

Fig. 3a. Fluorescence.

state (excitation) also involves a transition to higher vibrational energy levels. The excess vibrational energy will be dissipated by molecular collisions removing the excess energy by one vibrational step at a time. This energy will not appear as radiative energy but goes into thermal motion. Once the chromophore reaches the lowest vibrational state, a transition will occur to the ground state with the concomitant emission of energy (light). However, since the jump back to the ground state involves less energy than the excitation, the emitted fluorescent radiation will have a longer wavelength than the excitation radiation.

Similar considerations apply to phosphorescence (Fig. 3b) in which, after excitation to a singlet state, nonradiative processes through molecular collisions can cross over to a more stable neighboring triplet

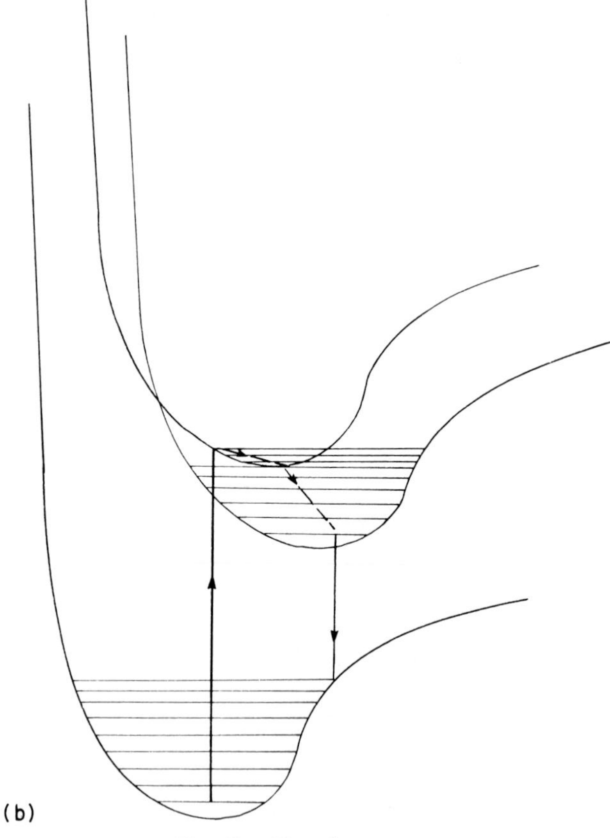

Fig. 3b. Phosphorescence.

state. The somewhat delayed emission is due to the transition from this stable triplet to the ground state. This transition has a low probability because of the change in spin state, and therefore phosphorescence will continue over a relatively long period. Since nonradiative processes dissipate a larger part of the excitation energy than in fluorescence, phosphorescence will occur at a longer wavelength than fluorescence.

Nonradiative processes can dissipate the excitation energy in two ways: vibrational deactivation and internal conversion. Vibrational deactivation, as mentioned previously, converts the excess vibrational energy of the excited state of the chromophore into thermal or vibrational motion of other molecules through molecular collisions. When a number of excited states cross over, as described in Fig. 4, the vibrational deactivation will be followed by an internal conversion in which the chromophore in the primary excited state crosses over to a neighboring lower excited state and eventually drops back to the ground state. Thus, in this case all the excitation energy will be dissipated into thermal or vibrational motions of molecules with which the chromophore collides.

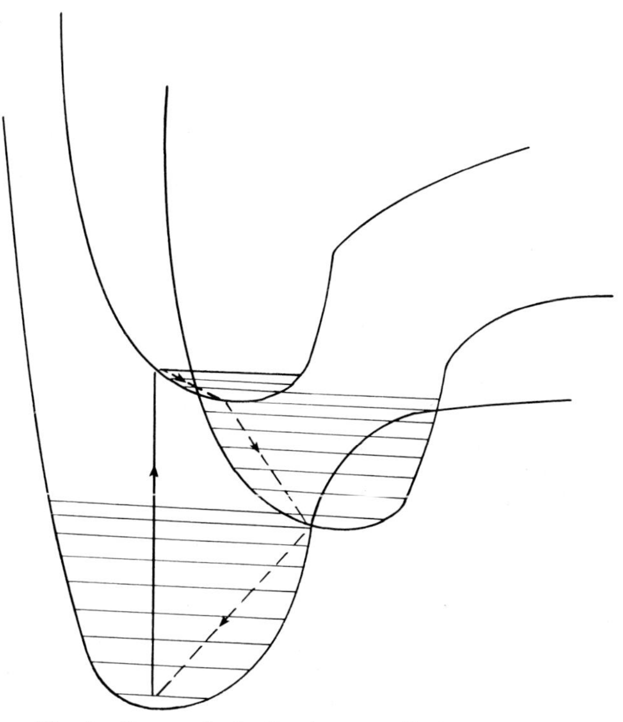

Fig. 4. Energy dissipation by nonradiative processes.

Chromophores that absorb in visible light are proteins and peptides containing photooxidation products of tryptophan and other aromatic residues.

Some schemes for the formation of these fluorescent pigments have been proposed, and they include the formation of kynurenine, formylkynurenine, 3-hydroxylkyurenine, and 3-hydroxyanthranilic acid (Mazur and Harrow, 1971).

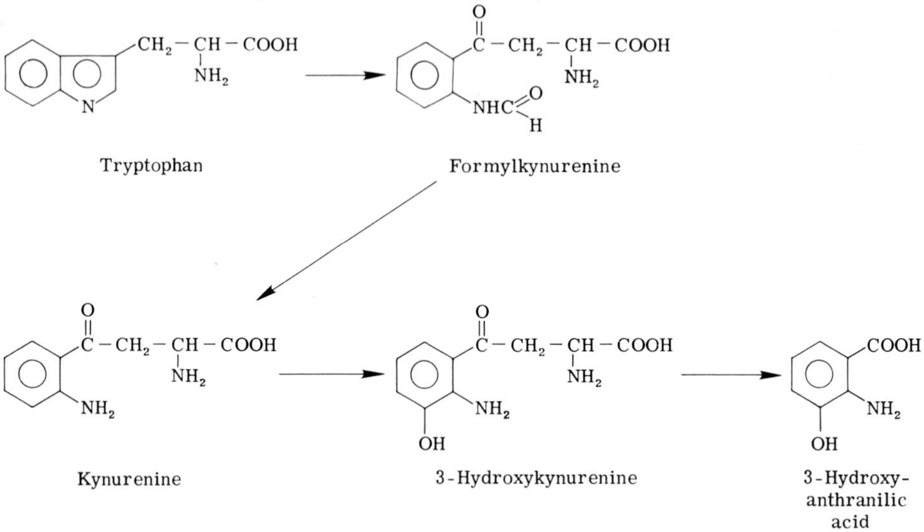

Tryptophan → Formylkynurenine → Kynurenine → 3-Hydroxykynurenine → 3-Hydroxyanthranilic acid

II. ULTRASTRUCTURE OF THE LENS

A. Cytoplasmic Structures

The lens system is an avascular tissue composed of cells of ectodermal origin. At different stages of development (for details, see the chapter by McDevitt and Brahma in this volume), it consists of one type of cells, namely, epithelial cells. In early embryonic life the lens vesicle closes off when the embryo reaches the 7-mm stage, and the lens is forever afterward an isolated organ. Toward the end of embryonic life and for all of adult life, the lens has no blood supply. The bulk of the lens is composed of long, tightly packed fibers derived from epithelial cells. The outer layer, the cortex, represents the newer part of the lens, and the central region, the nucleus, the older part. These fiber cells are very elongated (their length may be on the order of 1000 times their diameter) and have a hexagonal cross section. The lens has few organelles. However, the cells are filled with material of low density. This appears in electron micrographs either as filamentous structures about 10 nm thick without a well-defined length, or as a clump about 12.0-nm in diameter. Cliffe and Waley (1961) have found that nuclei, mitochondria, and microsomes acount for only 4% of the protein in calf lens compared to 50% in calf liver. The paucity of organelles can be understood as an aid toward transparency because of the size difference between a lens mitochondrion (500.0 nm) and a protein molecule (10.0 nm). Transparency is maintained during accommodation when the lens changes its shape. Moreover, the lens continued to grow at a decreasing rate throughout life, and this growth must be precisely correlated with the growth of the eye for correct refraction. The refractive index is different in different layers of the lens, being the highest toward the center. The changes in the refractive index are abrupt. There are layers where it increases sharply (Goldmann, 1964), its value being 1.38 in the cortex and rising to 1.40 in the nucleus (Huggert, 1948). The protein concentration rises from 27% in the calf cortex to 40% in the calf nucleus (Amoore *et al.*, 1959). This implies that there are "jumps" in the concentration of proteins in the different layers of the lens. (In dilute aqueous solution, the specific refraction increment is usually 0.185–0.195 ml/g for proteins and 0.14 ml/g for carbohydrates. For concentrated solutions, this value may not be constant.) An even greater refractive index gradient than this has been demonstrated along the optic axis, namely, cortex 1.38 and center of nucleus 1.50 in old rats (Philipson, 1969a). Furthermore, the refractive index along the surface of the lens is not uniform (Bettelheim and Wang, 1974). It is highest along the visual path, and moving vertically or horizontally from the central position along the surface its value diminishes. The refractive index reflects the distribution of protein in the fiber cells along the surfaces. These results indicate that the protein concentration somewhat counteracts the curvature of the lens, providing refractive power.

Morner (1894) first divided the protein from bovine lens into two main categories: soluble and insoluble (soluble referring to neutral solutions containing some salt). The soluble proteins were then divided into α-crystallin (precipitated at pH 5.00), β-crystallin (remaining in solution at pH 5.00) and γ-crystallin [a low-molecular-weight (LMW) basic protein]. (For a more detailed discussion see the chapter by McDevitt and Brahma in this volume.) The insoluble protein is called albuminoid and is closely related to α-crystallin. These crystallins, although morphologically they do not form organelles and physically appear to be evenly distributed, can therefore be considered a background matrix in the cytoplasm. One of the important factors accounting for the transparency of the normal lens is the relative absence of organelles or high-molecular-weight (HMW) aggregates, in addition to other factors such as degree of hydration and the opposite effects of birefringence (Section IV) (Phillipson, 1973; Griess and Zigman, 1973).

The cytoplasmic elements comprising the urea-soluble fraction of the lens water-insoluble protein include microtubules, intermediate filaments, actin. They show regional differences in their distribution. Filaments having a diameter of 10–14 nm have been obtained along with thin filaments in the pelleted fraction of disrupted lens fiber cells of various species by Maisel and Perry (1972) and Maisel (1977b). Intermediate filaments also occur in cells which contain actin filaments and exhibit contractility and/or motility. However, the function of the intermediate filaments in the lens is unknown.

Rafferty and Goossens, (1978) reported the finding of cytoplasmic filaments in the crystalline lens of mouse, rat, squirrel, human, and frog with an electron microscope. Two distinct patterns have emerged from their studies. One pattern, in which filaments are grouped in characteristic bundles around the nucleus, in processes, and throughout the subcortical cytoplasm of epithelial cells, is typical of the spherical, nonaccommodating lenses of mice and rats. The second pattern is associated with anteriorly flattened, accommodating lenses of infant humans, squirrels, and frogs. Filaments in both types of lenses range in diameter between 5 and 11 nm. The filaments are thought to be a mixture of thin and intermediate filaments which may have a dual function. The sheaths and thick bundles of filaments in nonaccommodating lenses may impart structural support to the epithelium to maintain the spherical shape of the lens. The loose arrangement of filaments in accommodating lenses may be compatible with the deformability of accommodation. Most of the lens filaments in all species studied are similar in diameter and fall into the range described for actin filaments. Actin has been found in all animal cells studied (Pollard and Weiling, 1974; Pollard, 1976) using specific labeling of the actin filament with heavy meromysin (Ishikawa *et al.*, 1969) or fluorescence-labeled antibodies against actin (Lazarides, 1975, 1976). A dual role for actin filaments has also been proposed by Pollard (1976). They are considered to possess not only a contractile but also a cytoskeletal function. Intermediate filaments, with diameters between 9 and 11 nm, shown by Cooke (1976) appeared to be dispersed among the thin filaments.

Bradley *et al.* (1979) found that chicken lens filaments were dispersed throughout the cytoplasm of the epithelial cells. They were particularly concentrated in a sheath near the plasma membrane junction between epithelial and fiber cells. Intermediate filaments were also identified in chicken epithelial and cortical fiber cells but were absent in nuclear fibers. Immunological studies have indicated that the 53K polypeptides are localized to the intermediate filaments (Maisel *et al.*, 1978). The presence of microtubules in lens epithelial cells in different species has also been well documented (Kuwabara, 1975; Longchampt *et al.*, 1976). In the chicken lens, microtubules are prevalent in the epithelial and annular pad cells. In the adult lens an occasional microtubule was seen only in the superficial cortical fiber cells.

Study reveals that a cytoskeleton of actin, microtubules, 10-nm filaments, and "chainlike" protein exists in lens cells and that the properties of this cytoskeleton change with differentiation and aging of the fiber cells. In particular the 10-nm filaments disappear with centralization of the cell toward the nucleus of the lens, and the protein chains become more clumped in the nucleus. The maturation of a fiber cell involves changes in the soluble protein, as well as in the cytoskeleton which represents the urea-soluble fraction of the lens water-insoluble protein (Maisel *et al.*, 1976a). The changes in the cytoskeleton occur in the normal developmental process and are reflected biochemically (Maisel *et al.*, 1977). Farnsworth *et al.* (1980) have reported observations on microtubules in stereo pair electron micrographs taken with a high-voltage electron microscope, and they appear to be the principal component of the cellular framework in both the cortical and nuclear fibers of the human lens. Superficial cortical fibers contain a network of filaments with 20 to 26-nm microtubules. They have no obvious geometrical pattern, extend for long distances and are relatively straight and do not curve with the undulations of the cell surface. Many microtubules enter the lateral interdigitations and appear to terminate in large numbers at their apexes. Others either span the width of the interdigitations or enter obliquely from the main cell body and terminate along the lateral borders of the interdigitations. The nuclear fibers are strikingly different. They contain densely packed microtubules with the same dimensions (20–26 nm) as those in the cortex. They are not extended but follow a winding path through the lens fiber cell. The microtubules have walls 9–10 nm thick with a continuous hollow core from 6.5 to 7.5 nm, which in some areas contains electron-dense particles about 5.5–6.5 nm in diameter sometimes spaced at regular intervals. Evidence for a microtubular complex in human cortical and nuclear lens fibers is therefore unequivocal. In general, cytoplasmic microtubules are known to be more labile than those found in the cilium, flagellum, or mitotic apparatus

(Burnside, 1975). Additionally, the convoluted configuration of the microtubules along with the uniformly distributed electron-dense particles in their core region may have obscured them in embedded tissue at lower powers of amplification.

Microtubules are thought to serve as a cytosketal component and maintain cell shape by remaining independent of one another with microfilaments serving as interconnections (Wolosewick and Porter, 1979).

B. Membrane Components

Generally speaking, the membrane components consist of intrinsic proteins, extrinsic proteins, carbohydrates, and lipids. The urea-insoluble fraction of the human lens contains a considerable amount of cell membrane (Maisel 1977a; Roy *et al.*, 1979; Horwitz *et al.*, 1979; Broekhuyse and Kuhlmann, 1979). Phospholipids as typical membrane components were exclusively present in the urea-insoluble fraction of clear animal and human lenses (Broekhuyse, 1969, 1973). Fiber cell membranes of adult human lens contain two major intrinsic polypeptides possessing properties similar to those of bovine lens membrane main intrinsic protein (MIP) (MW 26K) (Broekhuyse *et al.*, 1976; Alcala *et al.*, 1980) found the phospholipid-containing 25K and 27K polypeptides to consititute the main intrinsic protein of the human lens fiber membrane. The major age-dependent change in the protein composition of whole-lens human fiber membranes consisted of a gradual reversal in the preponderance of 27K polypeptide prenatally (and at birth) to a preponderance of 25K polypeptide in the membranes postnatally.

The fiber cell plasma membrane of the adult human lens comprises about 0.9% of the wet weight of the lens, close to the range previously noted for bovine (0.6%) (Broekhuyse and Kuhlmann, 1974; Alcala *et al.*, 1975) and chick (0.4%) lens. About 90–95% of the calf membrane consists of protein and lipid, and the remainder is carbohydrate. The protein/lipid ration for calf and bovine cortical and chick total fiber membranes is on the order of 1:0.9, while that of the adult human is 1:1.2. Carbohydrates comprise 2.7–8.2% of the dry weight of calf plasma membrane, depending on the method of analysis.

The main intrinsic component of bovine, rabbit, and chick lens membranes is a polypeptide of molecular weight 26K–27K (MP27K) (Broekhuyse and Kuhlmann, 1974; Alcala *et al.*, 1975; Maisel *et al.*, 1976b; Bagchi *et al.*, 1979), which is soluble only in detergents or strong organic solvents. Species differences in the composition of MP27 are also evident in bovine and chick MP27. In the human lens two major polypeptides have been identified with apparent molecular weights of about 27K (MP27) and 24K (MP24) (Alcala *et al.*, 1978; Roy *et al.*, 1979; Horwitz *et al.*, 1979). The main intrinsic polypeptides as percentages of membrane protein in the adult lens are 17–20% (MP27) and 25–36% (MP24) for human, 48% for bovine, and 54% for chick.

Chick MP27 has been localized to the plasma membrane of fiber cells by immunofluorescence (Waggoner and Maisel, 1978). The yield of chick and bovine MP27 is greatly enhanced in membrane preparations enriched in gap junctions, indicating that it is the transmembrane connection polypeptide (Alcala *et al.*, 1978). Under precise experimental conditions the main intrinsic polypeptides of bovine and human plasma membranes are heat-sensitive and upon heating in sodium dodecyl sulfate (SDS) aggregate into higher-molecular-weight components that do not enter polyacrylamide gels. (Horwitz *et al.*, 1979).

In contrast to intrinsic proteins, extrinsic proteins can be dissociated from the membrane without the use of detergents. They are free of lipids and soluble in neutral aqueous buffers. Two groups of extrinsic proteins can be identified in the lens. The first group consists of proteins that are generally extracted from lens water-insoluble material as urea-soluble protein (USP). Some of these polypeptides can be assigned to specific components of the cytoskeleton, such as the intermediate filaments (Maisel *et al.*, 1978), or to the chain filaments (Maisel *et al.*, 1977). They can be regarded as extrinsic proteins, since the lens cytoskeleton forms a network that criss-crosses the interior of the fiber cell and also interacts with or inserts into the membrane. The second group consists of proteins that are not released by urea but can be extracted from a urea-treated membrane by the action of water or EDTA (Broekhuyse *et al.*, 1978; Bouman *et al.*, 1979). In the first group, 54K polypeptide is the major component of the intermediate filaments. In the second, 47K polypeptide in the calf lens is extractable with water from the urea-insoluble membrane fraction (Broekhuyse *et al.*, 1978; Bouman *et al.*, 1979). Two components of 32K and 35K polypeptides can be further extracted with EDTA washing of water-treated, urea-insoluble membranes, suggesting that the latter polypeptides are extrinsic protein bound to the membrane by calcium ions. A portion of human lens MP24 is also bound to the membrane by divalent cations (Roy *et al.*, 1979), as is δ-crystallin of the chick lens membrane (Alcala *et al.*, 1978).

The main intrinsic polypeptides of the human lens (MP27 and MP24) show marked age-dependent changes in concentration. MP27 is the predominant component of the total fiber cell membrane in fetal life and in the newborn, and MP24 constitutes a minor fraction; further more, postnatally there is a rapid increase in the abundance of MP24. The age-dependent increase in MP24 is particularly evident in the nucleus. Even in 65-yr-old lenses the amount of MP27 exceeds that of MP24 in the cortical fiber cells, while the ratio is reversed in the nucleus. Early age-dependent changes in lipid metabolism of the human lens coincide with the time of rapid emergence of MP24 (Broekhuyse, 1973). This implies that M27 and M24 may differ only in the amount or nature of the associated lipid. It is unlikely that MP24 is a separate gene product, since little synthesis occurs in the nucleus (Wannemacher and Spector, 1968). The carbohydrates of the lens membrane were analyzed by Dische (1965, 1971) and more recently by Broekhuyse and Kuhlmann (1974, 1978). Sugar forms 2.7–8.4% of the dry weight of the calf lens membrane. The dominant sugars are galactose and *N*-acetylglucosamine. Only the 68K polypeptide of the bovine cortical fiber membrane was identified as the major glycoprotein of the plasma membrane (Alcala *et al.,* 1975). Almost all lipid in the lens is found in the fiber cell plasma membrane. Lipids make up from 47% (bovine and chick) to 55% (human) of the fiber plasma membrane dry weight. The lens largely synthesizes all its membrane lipids (Broekhuyse *et al.,* 1978). The cholesterol/phospholipid molar ratio rises progressively from equator to cortex to nucleus (Broekhuyse *et al.,* 1978). Aging of the fiber plasma membranes is associated with a higher cholesterol/phospholipid ratio in bovine and human lenses. A higher protein/lipid ratio is found upon aging of bovine lens fiber plasma membrane, but a lower ratio for aging human lens fiber plasma membrane.

III. BIOLOGICAL PROCESSES FOR THE MAINTENANCE OF TRANSPARENCY

A. Carbohydrate Metabolism

Generally speaking, a cell is characterized by its large molecules (proteins, nucleic acids, and polysaccharides). The small molecules (glucose, amino acids, coenzymes, salts) tend to be ubiquitous, and it is their movement, interconversions, degradation, and synthesis which are responsible for the maintenance, growth, and repair of the cell. The metabolism of the cell can be usefully divided into two groups:

1. Catabolic or degradative processes in which the raw material (glucose) is broken down step by step in an energy-yielding sequence of

reactions and chemical energy is built up in the form of ATP. These reactions also provide the carbon skeleton needed for biosynthetic processes.

2. An anabolic process or reaction in which the ATP is used in biosynthetic reactions to supply the special needs of the lens and also for the chemical work needed to maintain the structure and integrity of the lens.

The lens has no blood supply and relies on the aqueous humor and, to a lesser extent, the vitreous humor for its nutrients; it is by this route also that the waste produces of metabolism are removed. The raw materials needed for degradative and biosynthetic metabolism are also provided through the aqueous humor together with inorganic ions and such cofactors that the lens itself cannot make.

Energy for the lens is derived from the metabolism of glucose, most of which is converted by way of glycolysis to lactic acid which is formed in the aqueous humor at a considerably higher concentration than in the blood. Most of the lactic acid in the aqueous humor probably comes from the metabolism of the cornea, ciliary body, and iris, and the contribution of the lens is small.

The intermediary metabolism of carbohydrates in the lens consists of glycolysis and the pentose phosphate pathway, both of which are mediated by way of phosphorylated compounds, and the less universal sorbitol pathway. These pathways involve the participation of one or both of the coenzymes (NAD and NADP). These are hydrogen-transporting coenzymes which act in conjunction with the dehydrogenase group of enzymes; NAD is involved in glycolysis, NADP in the pentose phosphate pathway, and both in the sorbitol pathway. The relation between these pathways can be illustrated as

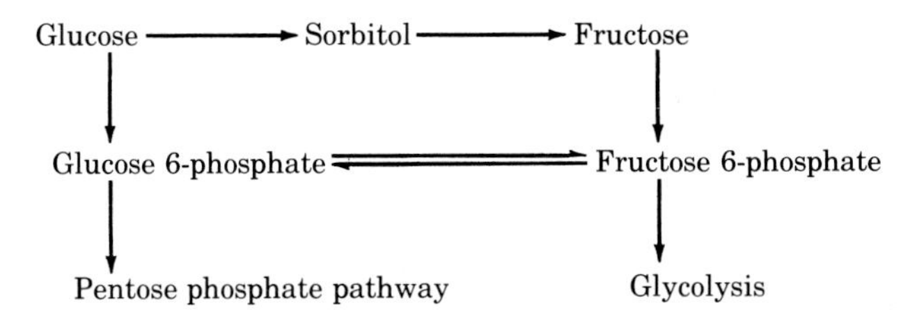

Glycolysis provides ATP (two molecules per molecule of glucose oxidized to lactate); the pentose phosphate pathway also provides ATP, but less per molecule of glucose used.

It has been calculated that 33% of the energy derived from glucose metabolism in the lens could come from oxidation by molecular oxygen

(Kleinfeld and Hockwin, 1960; Sippel 1962); while this estimate is probably too high, the reactions of oxidative phosphorylation yield more energy than those of glycolysis and could contribute appreciably to ATP formation.

The following is a summary of the possible carbohydrate metabolic pathways:

1. *Glycolysis:* Glucose is phosphorylated by means of ATP and the enzyme hexokinase to form glucose 6-phosphate from which triose phosphate and pyruvic acid are obtained. In the presence of an active citric acid cycle pyruvic acid is further oxidized, but in the lens it is almost exclusively reduced to lactic acid by means of an active lactate dehydrogenase and $NADH_2$.

$$\text{Glucose} + 2\ \text{phosphate} + 2\ \text{ADP} \longrightarrow 2\ \text{lactate} + 2\ \text{ATP} + H_2O$$

Addition of the enzyme hexokinase increases the rate of glycolysis, and this enzyme concentration limits the rate of glycolysis (Lou and Kinoshita, 1967).

2. *Sorbitol pathway:* The conversion of glucose to fructose involves two enzyme reactions. In the first, glucose is reduced to sorbitol, the coenzyme being $NADPH_2$; in the second, the sorbitol formed in the first reaction is oxidized to fructose by another enzyme, polyol dehydrogenase, the coenzyme being NAD;

$$\text{Glucose} + NADPH_2 \xrightleftharpoons{\text{Aldose reductase}} \text{sorbitol} + NADP$$

$$\text{Sorbitol} + NAD \xrightleftharpoons{\text{Polyol dehydrogenase}} \text{fructose} + NADH_2$$

$$\text{Glucose} + NADPH_2 + NAD \longrightarrow \text{fructose} + NAPD + NADH_2$$

The combined action of these enzymes in the normal lens has been discussed by Kuck (1961), Van Heyningen (1962), Lerman and Zigman (1965), and Kinoshita (1965).

3. *Pentose phosphate pathway:* The first step, the formation of glucose 6-phosphate, is common to both the glycolytic and the pentose phosphate pathways. This compound is then oxidized, in two stages, both reactions involving the reduction of coenzyme NADP. In the first stage 6-phosphogluconic acid is formed, and in the second stage this compound undergoes oxidative decarboxylation with the formation of ribulose 5-phosphate and the liberation of carbon dioxide. The ribulose 5-phosphate enters into a complicated series of reactions. Some of the intermediate phosphorylated compounds are drained off for biosynthetic reactions, and some of the D-glyceraldehyde phosphate is converted

to lactic acid by glycolysis. It has been shown that the pentose phosphate pathway accounts for a greater proportion of the glucose breakdown in tissues. In the lens, this route has been shown to account for only a small proportion of the total glucose metabolized but almost all of the carbon dioxide produced (Kinoshita and Wachtl, 1958).

There are probably many shuttles in the metabolic pathways. We will mention one that is worth noting. Nearly all mitochondria behave as if they are impermeable to $NADH_2$ *in vivo* and *in vitro* (Greville, 1966). For oxidative phosphorylation to take place, a mechanism must exist for transferring to the mitochondria the "reducing equivalents" of the $NADH_2$ in the cytochrome. $NADH_2$ is oxidized by the enzyme α-glycerolphosphate (α-GP) dehydrogenase, its reducing equivalent being passed to dihydroxyacetone phosphate (a glycolytic intermediate) with the formation of α-GP. The α-GP enters the mitochondria and is there reoxidized by a mitochondrial α-GP dehydrogenase and the respiratory chain to dihydroxyacetone phosphate, which returns to the cytoplasm. This mechanism operates in the lens (Pirie, 1962; Griffins, 1966).

B. Active Pumps

The lens, in common with other mamalian cellular structures, constitutes an integrated system in which there is a close interplay between release of energy in metabolism and utilization of energy in active transport (Whittam, 1964; Csaky, 1965). Active transport across a membrane can be defined as translocation, for which metabolic energy has to be provided and which may proceed against an electrochemical gradient. There is also a close interdependence between the water content of the lens and the transport of cations and amino acids. Swelling interfers with these transport systems and, conversely, interruption of the transport system (the use of poison) results in osmotic swelling.

The process by which a quantity of substance entering or leaving actively across the epithelium is balanced by that diffuses in the opposite direction has been termed the "pump-leak" system of transport. Kinsey (1965, 1966), using radioactive potassium, and rubidium, α-aminoisobutyric acid (AIB), has calculated various coefficients of active transport and parameters involved in the movement of ions. Sodium ions are actively transported across the membrane against a concentration gradient. The extrusion of sodium ions is a major energy-requiring function of most animal cells; sodium transport subserves volume regulation (Albers, 1967). The most common observation depicts an extrusion process which requires potassium in the external medium and is inhibited by a cardiac glycoside such as oubain. The activity of the pump is controlled by the internal sodium concentration or by electrochemical gradient of sodium. The immediate source of energy for sodium transport is ATP; there is much evidence that the energy is released by the action of an enzyme, Na^+,K^+-activated ATPase, ATP phosphorylase. This enzyme appears to be oriented across the cell membrane; the directional feature of the membrane is shown by the fact that, while both ions are required for optimal enzyme activity, the sodium and the ATP, in the case of the reconstructed erythrocyte membrane, must be on the inside and potassium on the outside of the cell; ATP outside the cell is not hydrolyzed (Whittam, 1962). The influx of potassium across the membrane is usually referred to in the literature as active transport, but its movement is considered to be linked to that of sodium. The ion transport in the lens fits the general pattern. The sodium and potassium contents of the lens are maintained by a transport system sensitive to oubain and dependent upon a supply of energy chiefly derived from glycolysis. The high potassium content (119–128 μmol/kg water) and low sodium content (19–23 μmol/kg water) are maintained, although the lens is surrounded by aqueous and vitreous humor with a reverse cation ratio (143–148 μmol sodium and 4.7–5.6 μmol potassium/kg water) (Bonting, 1965).

The calculated ratios of cation transported per mole of ATP hydrolyzed by the epithelial Na^+,K^+-ATPase system, ranging from 2.9 to 2.4, are similar to those found for many other tissues (Glynn, 1962; Bonting *et al.*, 1963; Bonting 1965, 1966a).

Mammalian and amphibian lenses are asymmetrical structures, having a simple layer of epithelial cells between the capsule and fiber cell mass on the anterior surface, while the outermost fiber cells on the posterior face make direct contact with the capsule. Kinsey (1965) has shown that ^{42}K penetrates the anterior surface of the lens more than three times as fast as it penetrates the posterior surface, whereas ^{24}Na penetrates the posterior surface more rapidly. Removal of the capsule and epithelium tends to abolish these conditions. Sodium and potassium gradients within the lens exist, both ions being more concentrated in the outermost layers (Amoore *et al.*, 1959).

The pump–leak model envisions that ion movement occurs by passive diffusion and/or active transport and takes into account the effect of the electric as well as the chemical gradient on the rate of diffusion across the outer fiber membranes of the lens (Kinsey and Hightower,

1976). Hightower and Kinsey (1977) studied the effects of lowered temperatures and ouabain on rabbit lens. Their results indicated that active transport of potassium into the lens and sodium out of it was depressed, but that there was no effect on the diffusion fluxes of these cations. These observations are thought to indicate an impairment of the electrogenic cation pump. On the contrary, dinitrophenol (DNP) did not affect active transport of sodium or potassium. Thus the reduction in potential by this metabolic poison may have resulted indirectly from a decrease in the diffusion potential. Kinsey and Hightower (1978) formulated an equation which expresses the electrogenic and diffusional components of the potential difference in terms of active and passive transport coefficients and the cation concentration gradients existing in lenses maintained in rabbit lens cultures on the basis of the pump–leak model of ion transport, with an assumption of the equality of influx and efflux of sodium and potassium ions at a steady state.

In the lens, there is much evidence that the active transport of amino acids into cells is closely related to the Na^+,K^+-ATPase system. In fact, it appears to be largely dependent upon sodium transport. However, there are many different amino acids, and they are not all transported into cells by identical mechanisms. The amino acids in the lens of the rabbit (Reddy and Kinsey, 1962), cow (Dardenne *et al.,* 1962), rat, cat, and monkey (Reddy, 1967) are higher than those in the surrounding fluids. The concentration ratio is particularly high for acidic amino acids. For example, glutamic acid is at a concentration of 5.8 mol/g in the water of the rabbit lens, a value 17 times that in the aqueous humor and 35 times that in the vitreous humor. It is believed that there are, in general, three main transport systems for neutral, acidic, and basic amino acids, respectively (Csaky, 1965). Neutral and acidic amino acids, which at physiological pH carry a net negative charge, enter the lens and other cells against an electric potential, whereas basic amino acids move in the direction of the potential gradient.

The following evidence suggests that the transport of amino acids in the lens is closely related to that of cations:

1. Both processes are inhibited by oubain, which also inhibits Na^+,K^+-/ATPase, the enzyme believed to be responsible for sodium transport.
2. Uptake of AIB by a rabbit lens is proportional to the concentration of sodium in the medium, provided the normal osmolarity of the lens is maintained (Cotlier and Beaty, 1967).

3. Both processes depend upon ATP, as shown by the finding that they are inhibited by procedures which inhibit glycolysis and probably proceed equally well in the absence of oxygen.
4. Both systems are impaired by osmotic swelling induced by the accumulation of sugar alcohols in the lens. Amino acid transport is more strongly inhibited by a small degree of swelling, and in fact loss of free amino acids is one of the early signs of sugar cataract.
5. Removal of the capsule and epithelium abolishes both AIB and cation transport.

Observations 1 and 2 are strong evidence of the close relationship between cation and amino acid transport, while observations 3–5 are evidence that both processes require energy and take place in the epithelium, but not necessarily that they are coupled. The ocular lens is capable of transporting amino acids by carrier-mediated processes which are energy- and temperature-dependent. The energy of this active transport is derived chiefly from the anaerobic metabolism of glucose, while the site of active transport is apparently located in the epithelial layer.

Unlike that of amino acids, the passage of sugar into most cells does not necessarily require metabolic energy, since they are uncharged and they do not move against an electrochemical gradient. Sugars enter cells by carrier-mediated transport, free diffusion, or both. D-Glucose and D-xylose enter the lens by facilitated transport, D-galactose probably enters the lens both by diffusion and by facilitated transport, and D-fructose and D-arabinose enter the cell water only by simple diffusion. Mediated transfer of glucose therefore presumably occurs both at the epithelium and at the membrane of the individual fibers.

Thoft and Kinoshita (1965) have found that, in the absence of calcium, the rat lens becomes more permeable to mannitol and sucrose, two inert substances used to estimate the size of the extracellular space. The effect of a lack of calcium on the leakiness of the lens to sucrose is almost completely reversible.

Recently, in investigating the role of Ca^{2+}, Bentley and Cruz (1978) (by electrochemical methods) found that the amphibian lens *in vitro,* like the mammaliam lens, accumulated sodium and lost potassium more rapidly when Ca^{2+} was excluded from the bathing solution. Other alkaline earth metal ions, Mg^{2+}, Be^{2+}, and Sr^{2+}, could substitute for the Ca^{2+} and maintain lenticular sodium and potassium. The sodium gain was greater from the anterior than from the posterior bathing solution. Their results suggest that Ca^{2+} exerts a relatively specific type of membrane effect in controlling lenticular permeability of sodium and potassium, but that the nature of the required divalent ion is not especially unique. Fagerholm (1979) studied the influence of different concentrations of calcium on isolated lens fiber bundles of rat and rabbit lenses *in vitro* and concluded that calcium induced lens fiber swelling by increasing lens fiber permeability to sodium and water.

Baldwin and Bentley (1980), in their studies on calcium metabolism of the amphibian lens, concluded that in the lens there was little free calcium in the cell water, a large proportion of the divalent ions being in a bound but exchangeable form, much of which was associated with the cell membrane. Relatively low levels of free calcium are probably maintained as a result of the action of an energy-dependent calcium pump which extrudes calcium. Calcium–sodium exchange does not appear to be involved. Clark *et al.* (1980b) studied the dependence of calf lens transparency on the concentration of calcium in bathing solutions using turbidimetric methods and scanning electron microscopy. The calf lens is transparent in concentrations of Ca^{2+} between 0.5 and 1.3 m*M*, which corresponds to the level of Ca^{2+} found inside the normal transparent lens. The calf lens cortex opacifies in concentrations of calcium either above or below the level of Ca^{2+} found in the normal lens. The lens fiber membranes are disrupted at both high and low calcium concentrations.

Hightower *et al.* (1980) suggested the presence of Ca^{2+},Mg^{2+}-ATPase in the rat, bovine, and rabbit lens based on the following observation: Calcium levels in the lens are lower than those measured in the aqueous humor and are reduced by inhibitors of calcium transport. There exists a calcium-activated, magnesium-dependent ATPase. To determine the nature of the mechanism responsible for maintaining these low levels of calcium in the lens, calcium fluxes were measured. The rate of ^{45}Ca efflux at 4°C was reduced by 85% compared with that found at 37°C. The efflux was not altered in the absence of external Na^+. Calcium efflux was reduced by lanthanum and propranolol, inhibitors of Ca^{2+},Mg^{2+}-ATPase.

IV. NORMAL LENSES

A. Light Scattering from Membranes

We can separate the contributions of the different morphologies to the total amount of light scattered. Ocular halos and corneas have been attributed by Simpson (1953) to the diffraction of light by the fiber cells which are aligned geometrically.

When lasers became available, we were the first to produce sharp diffraction patterns from thin sections of bovine lenses which showed even second- and third-order diffractions (Bettelheim and Vinciguerra, 1971; Vinciguerra and Bettelheim, 1971). The repeating distances calculated from the laser diffraction patterns corresponded well to the membrane-to-membrane distances in fiber cells observable in electron micrographs. These repeating distances changed as one proceeded from the cortical region to the nucleus, becoming progressively smaller. The diffraction patterns could be observed only with unpolarized light or polarized light in the $I_{\parallel}$ mode. This indicates that the membranes causing the diffraction have only periodic density fluctuations and no optical anisotropy. Similar diffraction patterns were found in thin sections of human lenses by Philipson (1973), and the calculated spacings from such patterns corresponded well to the thickness of cortical fiber cells, 7 μm.

Using the laser diffraction patterns from thin lens sections, we studied (Bettelheim *et al.*, 1973) the behavior of the fiber cells and their orientation under stress. Thin slices of lens were clamped between the chucks of a dynamic viscoelastometer. Periodic stresses were induced with different cycles and amplitudes; at the same time a laser beam was passed through the sample, and the light-scattering pattern was photographed with a high-speed movie camera.

The most important aspect of the dynamic laser diffraction experiment was that, in each case investigated, the interplanar spacing remained constant during the cyclic vibrations. This means that the width of the fiber cells and their packing remains virtually constant during a dynamic stretching cycle. Cortical and nuclear regions of the lens behaved the same way in this respect. Furthermore, the nuclear region gave the same interplanar spacings whether the direction of the stretch was parallel or perpendicular to the long axis of the fiber cells. In the cortical region, when the direction of the stretch was perpendicular to the long axis of the fiber cells, the diffraction spot split into a V-form doublet at maximum extension. One could observe a scissoring motion of this V, merging and splitting again. Thus, only in perpendicular stretching of the cortical fiber cells is there any change in their packing. It was observed that the cortical fiber cells responded to the stretching direction by becoming oriented more parallel to each other, and that upon contraction this forced alignment was relaxed reversibly to the original position. The resting cortical fiber cells have less than perfect alignment, and this alignment is improved upon stretching.

The same can be said of the nuclear region; when we see that the intensity of a diffraction spot has increased, we must assume that the refractive index differences between the cytoplasm and membrane have increased. This can be visualized as water migrating from the membrane region into the cytoplasmic region, making the refractive index of the membrane higher and that of the cytoplasm lower. The opposite process may be assumed when the intensity of a diffraction spot decreases. In the cortical region, the highest intensity is observed when the vibrating cycle is at its maximum contraction. When the fiber cells are pulled away from each other laterally, water may flow into the membrane regions, thereby decreasing the refractive index difference.

In the nuclear region of the bovine lens the opposite is true. Maximum intensity is obtained at maximum extension (perpendicular to the fiber cells). This implies that the nuclear and cortical fiber cells behave differently with respect to water migration. Thus the cortical fiber cells have a capacity to store water instantaneously, possibly between two adjacent cell membranes. In the nuclear region, no such instantaneous accumulation occurs, and it seems that nuclear fiber cells have degenerated cell membranes that allow a more-or-less hindrance-free migration of water from fiber cell to fiber cell. This occurs in spite of the fact that laser diffraction proves the presence of aligned membranes of high refractive index throughout the nucleus. Recent studies by Benedek *et al.* (1979) and Clark *et al.* (1980a) have shown that the intensity of the diffraction pattern also increases upon cold cataract and salt cataract formation in calf lenses, although the positions of the diffraction spots do not change. Scanning electron microscopic observation showed that in these cataracts patches occurred on the surface of the cell membrane, which increased the refractive index fluctuation and therefore could account for the increase in intensity of the diffracted laser beam. These observations further reinforce our belief that the diffraction in the lens is a membrane-caused phenomenon.

B. Light Scattering from the Cytoplasm

Trokel (1962) proposed that, in order to account for the transparency of the lens, the proteins must be uniformly distributed in the lens fibers. Schachar and Solin (1975) carried the uniformity idea to extreme by suggesting that the pleated-sheet conformation of the crystalline lens detectable by Raman spectroscopy as well as by circular dichroism (CD) (Jones and Lerman, 1971; Li and Spector, 1974; Horwitz *et al.*, 1977), provides a hexagonal packing of the lens proteins into quasi-crystalline lattices that underlie the hexagonal shape of the fiber cell. The desire to account for the transparency of the lens by destructive interference in a simple crystalline molecular organization is obviously unattainable. The teleological question often asked is why the lens needs so many distinct structural proteins to accomplish uniformity (Kuck, 1970).

From the principles of light scattering (Section I) it emerges that turbidity is due to fluctuations in the refractive index over spatial domains comparable to the wavelength of the light. There are two contributions to the refractive index fluctuations: (1) fluctuations in density and (2) fluctuations in optical anisotropy.

Fluctuations in density arise from the fact that solvent and solute molecules have different densities (refractive indexes). The thermal motions of the molecules make these fluctuations random. If the solvated molecules are small compared to the wavelength of the light, no light scattering will occur and the lens will be transparent. Benedek (1971) has pointed out that, using a dilute solution theory of light scattering (Section I,A), the calculated molecular weight below which little scattering can occur is $\simeq 10^6$.

A dilute solution theory predicts an increase in turbidity with an increase in the concentration and in the molecular weight of aggregates of molecular weight 50×10^6 and above. However, it is difficult to see how dilute solution theory can apply, except as a first approximation, to the lens cytoplasm which has a 33–45% protein content. In spite of this, Benedek's calculation initiated a wide search for HMW protein aggregates, which will be reviewed in Section V. The prediction, qualitative as it is, has been confirmed many times by showing that, as lens turbidity increases, the number and size of the HMW aggregates increase.

The same calculations can be made for condensed-phase models using the Debye and Bueche (1949) theory.

These calculations show that in gels or concentrated solutions not only the size of the particles is important in predicting turbidity, but also their interparticle separation (Bettelheim and Siew, 1980, 1982).

Thus, in addition, because of an increase in the size and number of the HMW protein aggregates, the lens may become turbid also because of dilution if the interparticle distances reach the order of the wavelength of the light. For example, we calculated that a 14% swelling of the lens caused by an influx of water may increase the turbidity 20–40%. In these calculations no other effects except the dilution was taken into consideration.

A second condition for transparency of the lens is that the amplitude of the density fluctuations be small. This means that, even if HMW aggregates are present, if their hydration is such that there is very little difference between the refractive index of the aggregates and their surroundings, turbidity will not develop. As a consequence of this condition, any process that increases the refractive index difference may lead to cataractogenesis. Syneresis (Bettelheim, 1979), in which the internal collapse of highly hydrated molecules and aggregates results in the exudation of water into the immediate environment, is such a process. The collapsed dehydrated molecule (aggregate) will acquire a higher refractive index than before, while the surroundings will have a lower refractive index. The refractive index difference or, more correctly, a statistical measure of it (i.e., the mean square deviation from the average refractive index) is proportional to the amount of light scattered.

Fluctuations in optical anistropy will also lead to fluctuations in the refractive index. Optical anisotropy means that the refractive index in a medium is a function of orientation. The optical anisotropy of a body can be measured by its birefringence Δ:

$$\Delta = n_{\parallel} - n_{\perp} = \Gamma/d$$

where $n_{\parallel}$ and $n_{\perp}$ refer to the refractive indexes of the sample in two mutually perpendicular directions, both of which are perpendicular to the direction of the propagation of the light in the sample. The actual measurement is made by determining the number of waves retarded Γ and by the sample of thickness d.

The spatial domains of optical anisotropy fluctuations must be of the same order as that of the wavelength of the light in order to cause turbidity—the same requirement that exists for density fluctuations. In order for lens cytoplasmic elements of the proper size to be optically anisotropic, they must possess two properties: The molecules and molecular segments must be optically anisotropic, and their orientation in space must not be random. These two properties combined will give rise to intrinsic birefringence Δ_i, that is, optical anisotropy due to molecular structure. Another kind of birefringence, form birefringence Δ_f, develops when geometrically anistropic bodies such as needles or plates are embedded in a matrix with a different refractive index. The form birefringence increases with the preferentially parallel alignments of the needles or thin plates.

The sum total of form and intrinsic birefringence will give the total birefringence that will measure the optical anisotropy of the sample. Lenses under a polarizing microscope exhibit a Maltese cross pattern, a fact first reported by Brewster in 1816, which implies birefringence components. In spite of this, the total birefringence of the lens is very small, on the order 10^{-6}–10^{-7} (Bettelheim, 1975).

Because of the particular morphology of the lens—thin membranes (plates or tubes) surrounded by a cytoplasmic medium of different refractive index, the lens must possess form birefringence. The estimated form birefringence is on the order of 10^{-4} (Bettelheim, 1975). The fact that the total birefringence is 100 or 1000 times less implies that an intrinsic birefringence of the opposite sign must cancel out the total birefringence. This theory (Bettelheim, 1977) has been proved also by light scattering. The normal lens scatters light mainly in the $I_{\parallel}$ mode (see Section I). This implies the absence or the relative small value of optical anisotropy. The first sign of cataract formation is the large increase in scattering in the I_{+} mode compared to the small increase in the $I_{\parallel}$ mode, leding to a drastic increase in the $I_{+}/I_{\parallel}$ ratio (Bettelheim and Bettelheim, 1978). In the normal lens, the intrinsic birefringence is provided by the alignment of supramolecular structures such as cytoskeletal thin and intermediate filaments (Maisel and Perry, 1972; Rafferty *et al.*, 1974; Maisel, 1977b; Rafferty and Goossens, 1978; Ireland *et al.*, 1978) made of optically anisotropic actin molecules or microtubules (Kuwabara, 1975; Longchampt *et al.*, 1976; Fransworth *et al.*, 1980). Other supramolecular structures that could provide intrinsic birefringence may be the α- and β-crystallins aligned in semicrystalline matrices by LMW phosphopeptides (Bettelheim, 1972; Bettelheim and Mehrotra, 1972; Bettelheim and Wang, 1977).

The theory of minimal total birefringence of the lens (Bettelheim, 1977) through cancelation of the intrinsic birefringence by the form birefringence has a certain teleological significance. The lens is a cellular body composed of onionlike layers of fiber cells. In order to minimize the light scattering that is inevitable from such a morphology (i.e., resulting from form birefringence) some supramolecular organization within the fiber cells must provide intrinsic birefringence to minimize the optical anisotropy. A corollary statement to this is that any disturbance that perturbs the delicate balance between form and intrinsic birefringence will inevitably lead to turbidity, be it the disintegration of membranes or the disorganization of cytoplasmic supramolecular structures.

A second requirement so that optical anisotropy fluctuations will not generate appreciable light scattering is that the amplitude of such fluctuations be small. This was found to be the case in normal human lenses (Bettelheim and Paunovic, 1979), where the statistical measure of the refractive index differences due to optical anisotropy fluctuations $\bar{\delta}^2$ was 10^3–10^5 times smaller than that due to density fluctuations $\bar{\eta}^2$. This further demonstrates that intrinsic birefringence or the alignment of supramolecular structures plays an important role in minimizing light scattering in normal lenses. In summary the transparency of the normal lens is accounted for by three factors: (1) the relative absence of HMW aggregates or organelles, (2) the degree of hydration of supramolecular aggregates that minimizes the refractive index between them and their surroundings, and (3) the relative absence of optical anisotropy fluctuations due to the opposite effects of intrinsic and form birefringence.

Consequently, any process that leads to changes in the above conditions will initiate opacity and cataract formation.

V. AGE-DEPENDENT CHANGES IN LENS TRANSPARENCY AND CATARACTOGENESIS OF A DIFFERENT ETIOLOGY

A. Aggregation Process

An increase in light scattering by the lens with age *in vivo* was reported by Goldmann in 1964. This increase has been attributed to the aggregation of proteins about to form water-insoluble particles (Mach, 1963). This has been justified by the concomitant increase in the insoluble protein fraction with age. HMW proteins (mainly in the central region of the lens) of molecular weight greater than 10^6 (Spector, 1972; Hoenders and Van Kamp, 1972) and up to molecular weight 3×10^8 (Jedziniak *et al.*, 1973) have been reported. Such HMW protein aggregates increase with age in bovine and human lenses (Spector *et al.*, 1971; Jedziniak *et al.*, 1973; Spector *et al.*, 1973; Spector *et al.*, 1974; Jedziniak *et al.*, 1975). The insoluble protein fraction also increases with age (Pirie, 1968; Clark *et al.*, 1969; Spector *et al.*, 1975), and it has been proposed that HMW protein is the precursor of the insoluble protein aggregates (for review, see Harding and Dilley, 1976). HMW proteins were also found in increased amounts in human senile cataractous lenses (Jedziniak *et al.*, 1975) and in x-ray-induced cataracts (Liem-The *et al.*, 1975; Giblin *et al.*, 1978). The aggregates as viewed in the electron microscope showed irregular shapes up to 500 nm in diameter (Kramps *et al.*, 1975). Various causes for the aggregation process have been proposed. These include Ca^{2+}, disulfide, and nondisulfide covalent cross-links, as reviewed by Harding and Dilley (1976). The HMW aggregates usually contain subunits of α- as well as β-crystallin (Roy and Spector, 1976). That γ-crystallin is excluded from the reaggregation of α- and β-crystallins was found first by Spector (1965). Bloemendal *et al.* (1975a) also found that γ-crystallin did not participate in the molecular self-assembly.

Bettelheim (1972) has shown that α and β chains aggregate into a crystalline matrix in the presence of phosphopeptide and that the presence of γ-crystallin prevents the crystalline organization (Bettelheim and Wang, 1977). As the above cited literature indicates, the theory that cataract formation in essence is an aggregation process, progressing through HMW proteins to insoluble proteins, has been found to be very attractive by a large number of researchers and that such a process contributes to cataractogenesis has been substantiated.

Benedek and co-workers have advanced the idea that aggregation to HMW proteins eventually leads to phase separation which is the main cause of opacity. Their model is cold cataract formation. Tanaka *et al.* (1977) noted that the cold cataract formation that can be seen in young mammalian lenses when the temperature of the lens is lowered resembles very much in its physicochemical properties the "phase separation" occurring in a protein–water binary mixture. "Phase separation" in this case does not mean the observation of two visible phases, rather it refers to the interpretation of the appearance of opacity. Opacity, the argument goes, is due to the aggregation of protein molecules forming microphases of a size comparable to the wavelength of the light. When one studies the spread of cold cataracts in young rat lenses, one finds that at first an annular cataract is formed approximately 0.5 mm from the cortical surface. When the temperature is lowered, this opacity spreads toward the surface of the lens as well as toward the nucleus. Plotting the positions in the lens where the edge of turbidity becomes a function of temperature, they obtained a bell-shaped coexistence curve which resembled very much the coexistence curves for a protein–water mixture in which the protein concentration was varied and the temperature at which turbidity appeared was plotted against the protein concentration.

The same similarity was observed between the intact lens and a protein–water mixture when the rate of decay of the concentration fluctuation was measured by quasi-elastic light scattering as a function of temperature at different positions within the lens on the one hand, and at different protein concentrations on the other.

Therefore, these authors have proposed that cold cataract formation can be explained on the basis of phase separation. Ishimoto *et al.* (1979) have shown that with aging the coexistence curves for rat lenses shift their maxima progressively toward the radial positions and that the temperature at which the maxima occur (critical point) decreases with age.

Further experiments of Benedek *et al.* (1979) and Clark and Benedek (1980) demonstrated that, when bovine lenses were soaked in glycerol and glycols, cold cataract formation at 5°C could be prevented, i.e., the critical temperature T_c at which opacity appeared was lowered to below 5°C. Once the glycols, etc., were removed by soaking in a buffered saline solution, the critical temperature returned to 5°C. These authors believe that the reversibility of the coexistence curves by agents such as glycols demonstrates the phase separation nature of cold cataract formation.

Quasi-elastic light scattering by the same lenses with or without soaking in glycols has shown that these agents increase the decay time of the autocorrelation function. This implies that the diffusivity of proteins, which is proportional to the decay rate, increases in glycol-soaked lenses. If we interpret these observations in equilibrium rather than in kinetic terms, it implies that glycols, etc., prevents the aggregation of proteins to large sizes and therefore that one needs a lower temperature to decrease the solubility of proteins and to form opacities in the presence of these agents than in normal lenses. Tanaka and Benedek (1975) have applied quasi-elastic light-scattering techniques in studying the intact lens. The spectrum of the scattered laser light or the intensity of the autocorrelation function is measured. Assuming that some energy of the incident beam has been dissipated in thermal (Brownian) motion of the protein molecules, one can calculate from the measured autocorrelation a mean decay rate which can be related to the diffusivity of the particles. As the temperature of the lens decreases, the diffusivity D decreases linearly. At the temperature at which opacity (cold cataract) appears T_c, the diffusivity is still nonzero. Further extrapolation of the diffusivity–temperature lines to zero diffusivity provides the value of T_s, or the spinodal temperature. This temperature represents the limit of supercooling below which phase separation will occur. Again "phase separation" refers to microphases and it is not visible macroscopically.

The diffusivity of normal human and bovine lenses approached the diffusivity of α-crystallins in solution, and the diffusivity of one cataractous lens had a value similar to that of the soluble HMW component of the lens. Therefore, the authors proposed that quasi-elastic light scattering could be applied to lenses *in vivo* in studying cataractogenesis. The intensity of the laser beam necessary to make such measurements is low enough that it will not cause any damage to eye tissues (Tanaka and Ishimoto, 1977).

It has been suggested that cold cataract formation is initiated by a conformational change in γ-crystallins (Zigman and Lerman, 1965). However, Horwitz *et al.* (1977) have shown that no such conformational change can be seen in the CD spectra of γ-crystallins. They have suggested that γ-crystallin precipitation from high-concentration solutions within a narrow pH is simply isoelectric precipitation and that cold cataract formation involves more than just γ-crystallin precipitation (microphase separation).

Although in mammalian lenses cold cataract formation may serve as a model for reversible cataract formation that can be understood on the basis of microphase separation, it is not by any means a universal phenomenon. In fishes cold cataract formation results in irreversible annular opacity. Therefore, in fish lenses cold cataract formation (Loewenstein and Bettelheim, 1979) must involve changes other than reversible microphase separation of protein aggregates. Since the same annular irreversible ring appears in pressure cataracts of fish lenses as well (Loewenstein and Bettelheim, 1980), and no changes can be observed in the protein subunit composition due to such opacity, it may be assumed that the irreversible process involves supramolecular organization and/or possible membrane-related changes. Irreversible cataracts are produced in fish lenses also by cutting, in contrast to the situation in mammalian lenses (Bettelheim and Bettelheim, 1980).

The main criticism of the argument that cold cataract formation and consequently microphase separation are general models for aging and all kinds of cataract formations is as follows: The concept of phase separation is a thermodynamic concept, and it is based on reversibility. Mammalian cold cataract formation satisfies this condition. Most aging and natural cataract formation are irreversible processes, and therefore microphase separation is too simple a model to explain the underlying mechanism.

B. Syneresis

Besides the increase in the molecular weight of protein aggregates, another cause of turbidity is an increase in the amplitude of density fluctuations. This amplitude reflects the refractive index difference between aggregates and their surroundings.

A change in refractive index fluctuations can be caused by syneresis, that is, a change in the hydration of the supramolecular structure of the protein aggregates. Changes in the primary, secondary, tertiary, and quaternary structures may result in a decrease in the hydration of proteins. Exuding a dilute aqueous solution into the medium surrounding the protein aggregates will result in a lowering of the refractive index of the medium and an increase in the refractive index of the aggregates (Bettelheim, 1979).

Evidence that dehydration plays a role in cataractogenesis was established quite early. Cataracts can be caused by external conditions leading to dehydration. It was observed early (Goldmann and Rabinovich, 1928) that the rat lens developed cataracts when eyelid closure was prevented. Similar data are available on guinea pig lens (Fraunfelder and Burns, 1962). In both cases the dehydration was caused by exposure of the cornea. Other experiments (Fraunfelder and Burns, 1966; Weinstock and Scott, 1967) showed that drug-induced cataract formation due to analgesics, morphine, or epinephrine could be prevented if the eyelids were kept closed.

One must consider the experimental evidence that supports such a hypothesis. The focus must be on the question, what kind of change in macromolecular suprastructures would lead to the gradual dehydration of protein and the gradual sequestering of water in the surrounding medium?

Four types of experimental evidence will be reviewed in support of the syneretic process leading to cataractogenesis.

1. *Changes in Primary Structure*

Age-related changes in the primary structure of crystallins are well documented. Schoenmakers *et al.* (1969b) and Palmer and Papaconstantinou (1969) have observed that the αA_1 subunit is absent in embryonic calf lenses and appears gradually with an increase in the age of the embryo. Palmer and Papaconstantinou (1969) have suggested that αA_1 is formed from αA_2, and Schoenmakers *et al.* (1969a,b,c) have demonstrated that this is a posttranslational process involving the deamination of glutamine residues. A similar mechanism was suggested for the age-dependent $\alpha B_2 \rightarrow \alpha B_1$ transformation (Stauffer *et al.*, 1974). In the calf lens the $A_2 \rightarrow A_1$ conversion seemed to be faster than the $B_2 \rightarrow B_1$ conversion. In human lenses Harding and Dilley (1976) found this conversion to be much slower. In the bovine lens these conversions proceed until two-thirds of the original A_2 and B_2 remains (Van Kamp, and Hoenders, 1973; Li and Spector, 1974), but in the human lens the process never goes to completion, even in old lenses (Harding and Dilley, 1976).

Other major postsynthetic changes in αA and αB polypeptides are the degradative processes that cleave four to seven amino acids from the carboxy terminal of the polypeptide chains (de Jong *et al.*, 1975; Van Kleef, 1975).

A further indication of primary structure alteration of lens proteins in aging and cataractogenesis was the observation that new nontryptophan fluorescent compounds, covalently attached to the lens protein, appeared in the process (Dilley and Pirie, 1974; Spector *et al.,* 1975; Augusteyn, 1975). It was suggested that this was the result of tyrosine oxidation (Pirie, 1968), an alternation of tryptophan in the polypeptide chain (Zigman *et al.,* 1973; Kurzel *et al.,* 1973; Lerman, 1976), or the involvement of other compounds originally not bound to the protein (Zigman, 1971; Obara *et al.,* 1976). A number of fluorescent components were found in an enzymatic digest of the insoluble yellow protein of human cataractous lens. Among these, 3,3′-bityrosine and anthranilic acid were identified by Garcia-Castineras *et al.* (1978). McNamara and Augusteyn (1980) contend that 3,3′-bytyrosine is probably a capsule component and therefore does not participate in the cross-linking of lens proteins.

2. Secondary Structure

There is a vast literature concerning the transition of soluble lens proteins to aggregates and then to insoluble lens protein as a function of aging and cataractogenesis. This literature has been reviewed by Harding and Dilley (1976).

The majority of the literature indicates that cataractogenesis, especially nuclear cataract formation, may occur via unfolding of protein chains, cross-linking by disulfide or other covalent bonds (especially in brown cataracts), and formation of HMW aggregates and water-insoluble proteins.

Increased amounts of insoluble protein in nuclear cataracts were correlated with the severity of the cataract (Truscott and Augusteyn, 1977a; Anderson and Spector, 1978). Also, increased amounts of protein disulfide groups were found with an increase in the severity of the cataracts (Truscott and Augusteyn, 1977b). Anderson and Spector (1978) suggested that, for the soluble-to-insoluble protein transformation, two alternative pathways may be available: (1) Soluble protein with an increasingly higher level of disulfides may precede the transition to insoluble protein, and (2) such a transformation occurs without the oxidation of protein sulfhydryl groups. They suggest that pathway 1 operates to a greater extent.

Evidence indicating that a mechanism other than disulfide cross-linking caused a change in protein conformation and may lead to cataract formation has been presented by Masters *et al.* (1977, 1978) and Garner and Spector (1978). Aspartic acid racemization was greater in cross-linked or apparently degraded lens proteins from the urea-soluble fraction than in normal-molecular-weight lens proteins. The greater D/L ratio was found in older lenses, and the increase in water-insoluble protein content was correlated with the D/L ratio of aspartic acid.

3. Tertiary and Higher Structures

Another indication of age-dependent changes that also occur in nuclear cataract formation involves a LMW polypeptide (MW 11,000–9600). Spector *et al.* (1975) and Kramps *et al.* (1976) reported that this polypeptide was present only in lenses 20 yr or older.

This polypeptide is present in both water-soluble and -insoluble fractions (Roy and Spector, 1978a). It does not self-aggregate to large particles (Roy and Spector, 1978b). Therefore, it has been suggested that it interacts with other polypeptides to form HMW protein or water-insoluble protein. The same conclusion has been reached by Jedziniak *et al.* (1978).

It is well known (Kendrew *et al.*, 1958) that globular proteins tend to project their hydrophilic side chains toward the surroundings and achieve an equilibrium stable structure by maximizing the hydrophobic interactions inside the molecule and the hydrophilic interactions at the surface. The fact that most acidic and basic sites of globular proteins are titratable is proof that this thermodynamic requirement is met (Lehninger, 1970).

Specifically in α-crystallin of bovine lenses, Bettelheim and Finkel (1976) have shown, using high-vacuum water vapor sorption, that aggregates of the four subunits (αA_1, αA_2, αB_1, and αB_2) exhibit greater affinity for water uptake than aggregates of individual subunits or of any binary or tertiary combination of subunits. This implies that any change in the thermodynamically stable stoichiometric interaction of subunits, even without a change in the size of the aggregates, leads to diminishing hydration of the protein.

The state of water in concentrated solutions and gels is occasionally referred to as bound or free. Bound water cannot be frozen at the melting point or slightly below, while free water solidifies between 0° and −10°C. Nuclear magnetic resonance (NMR) studies have often been used to estimate the amount of bound and free water from relaxation times. Neville *et al.* (1974) have shown that in the lens part of the water is bound or is in an ordered state. Racz *et al.* (1979) measured the spin-spin and spin-lattice relaxation times in normal and senile cataractous human lenses as a function of temperature. They found that below −9°C there was no difference between the relaxation times of normal and cataractous lenses. Above −9°C both the spin-spin and the spin-lattice relaxation times were higher in the cataractous lenses than in the normal lenses. This was interpreted as an increase in the free water versus the bound water content upon cataractogenesis. While the overall hydration of the lens was the same in normal and cataractous lenses, only 76% of the total water was free in normal lenses but 87% was free in cataractous lenses.

This strongly supports a syneretic process, i.e., the release of strongly bound water from protein aggregates into the surroundings.

A similar syneretic exudation of water from protein aggregates was found in ultraviolet (UV)-induced cataracts by studying the Raman spectra of lenses (Thomas and Schepler, 1980).

4. *Other Processes Contributing to Syneresis*

A different and more complex form of syneresis that affects the amplitude of refractive index fluctuations is present in sugar cataracts. Kinoshita (1974) described the effect of a galactose diet on cataract formation by attributing an increase in water uptake in the cortical region to an accumulation of galactitol in the cortex. At first, this hydration of the cortex does not have an visible effect on ultrastructure, but later small vacuoles appear in the cortex and eventually lake formation results, causing turbidity. Although sugar cataracts largely involve imbibition of water from the surroundings, due to the development of an osmotic pressure gradient across the membranes, they are also partly the result of a syneretic effect. An increase in the galactitol or xylitol concentration in the lens fibers results in an effective competition for water between the lens proteins and the smaller molecules. Thus, an increase in sugar alcohol may decrease the hydration of the lens proteins. Li has shown recently (1978) that the gel permeation chromatography of lens proteins is different in water than in sucrose solutions. Part of the difference can be explained by a less hydrated form, hence a smaller hydrodynamic radius of lens proteins in the sucrose medium than in the aqueous medium.

Philipson (1969a,b) devised a quantitative microradiographic procedure in which freeze-sectioned, freeze-dried, 10-μm thick sections were exposed to soft x-rays and the resulting radiation measured photometrically. He detected the dry weight or densities in volumes on the order 10 μm^3. Among such volume elements Philipson (1973) found a uniform distribution of protein masses in normal lenses, but sudden changes in density and correspondingly in the refractive index at points where opacities occurred in the lens in cortical cataracts. Such changes corresponded to a refractive index fluctuation of 0.02–0.04. These fluctuations occurred over domains larger than 1 μm.

On the other hand, no refractive index fluctuations were found in nucelar cataracts over domains larger than 1 μm (Philipson and Fagerholm, 1981). Philipson's measurements using this restricted scale of 1 μm reinforce the arguments advanced on the basis of electron microscopic observations that cortical opacities in the final stage are due to large (<1 μm) lake formations, while nuclear opacities are due to density fluctuations over a smaller domain, 0.1–0.6 μm (Siew *et al.*, 1981b).

Philipson (1969a) has proposed that in sugar cataracts the water in the cortex comes from the surrounding fluids and includes water from the nucleus. Bettelheim (1978) and Bettelheim and Bettelheim (1978) showed that this, in effect, was the case in incipient cataract formation in galactose and xylose cataracts, respectively. While the cortex showed an increase in form birefringence, hence in the relative contribution of the optical anisotropy as compared to density fluctuations, the nucleus showed a decrease. If such a dehydration process can occur across lens membranes, it can occur with greater ease within the same fiber cell.

Finally, in any type of cataract formation, the light scattering increases at wide angles as well as at low angles. In some cases, at very low angles the scattering even decreases (Philipson, 1969b; Bettelheim and Bettelheim, 1978). This implies that the change in the refractive index amplitude is an important contributor to the opacity formation. If the size of the aggregates were the only contributor to the turbidity, one would expect that, as the size increased, more and more scattering would occur at low angles. On the other hand, the amplitude of the refractive index fluctuation contributes to the scattering, independently of the scattering angle, at least when the Debye–Bueche (1949) theory of scattering is involved.

C. Morphologically Degradative Processes

Drastic morphological changes involving organelles can be seen in many cataracts. Only a few examples will be summarized here. Upon physical injury to the lens, capsular and epithelial rifts have been observed (Wiedenthal and Schepens, 1966). Further progression of a subcapsular cataract causes a totally opaque lens with a swollen cortex. Finally the lens ruptures. In younger individuals the major part of the lens is resorbed (Fagerholm and Phillipson, 1979). The posterior lens capsule and the remaining intact part of the anterior capsule and cortex often form a ring-shaped structure called Soemmering's ring.

Epithelial regeneration and the formation of new capsular material tend to seal the lens and stop the free migration of ions inward and outward. The success of this regenerative process is the most important prognostic factor (Rafferty *et al.,* 1974; Fagerholm and Philipson, 1979). Morphological changes that accompany sugar cataracts, in particular galactose cataracts, have been well documented (Kuwabara *et al.,* 1969; Beyer-Mears *et al.,* 1978; Unakar *et al.,* 1978). These changes include disalignment of lens fibers, membrane-bound vesicles randomly distributed throughout the fiberplasm, variable degrees of swelling, membrane disruption, enlargement of intercellular spaces, vacuole and lake formation, electron-dense aggregates, decreased amounts of interdigitation between adjacent fiber cells, and granulation of the surface of fiber cells. Galactose cataracts are reversible, and both of the morphological changes disappear in reverse, parallelling the biochemical changes.

Other cataracts, such as those formed in myotonia dystrophica, show numerous filamentous inclusions, vacuoles containing multilaminar whorls, multilayer plasma lamellar membranes, lipid droplets, and crystalloid proteins (Dark and Streeten, 1977). Lenses from patients with myotonic dystrophy also show the development of myelin bodies from groups of membrane globular processes. (Erhegian *et al.,* 1978).

Globular bodies have been observed in senile and diabetic posterior cortical subcapsular cataracts (Creighton *et al.,* 1978). Gorty (1977, 1978), studying senile supranuclear and posterior subcapsular cataracts, proposed that the swelling and the degeneration of cortical fibers was caused by intercellularly located hydrolases. Further degeneration resulted in the formation of globular bodies and multilameller inclusions.

Cataract formation is enhanced by exposure to sunlight. Studies show that people living near the equator have significantly more cataracts than those living in a temperate climate (Van Heyningen, 1976; Hiller *et al.,* 1977; Zigman *et al.,* 1979). Ionizing radiative damage such as x-ray cataracts show disalignment of fiber cells, disintegration of membranes, rounding out of swollen fiber cells, intracytoplasmic vacuoles, and lakelike intercellular spaces (Lambert and Kinoshita, 1967; Liem-The *et al.,* 1975; Palva and Palkama, 1978). Similar vacuolization and intercellar spaces are also found in cataracts due to nonionizing radiation (Zaret *et al.,* 1976).

D. Combined Effects of Different Processes in Aging and Cataractogenesis

Age-dependent changes in human lens were reported first by Goldman (1964) with drawings of slit-lamp images. The nuclear region was shown to become less transparent with age. The backscatter from human lenses with age has been studied by Siegelman *et al.* (1974), using a slit-lamp camera and tracing the photographs with a microdensitometer. The cortical backscatter is quite appreciable and is independent of age. This does not seem to impair visual acuity. On the other hand, the backscatter from nuclear regions is insignificant in young lenses and increases with age. There is a direct correlation between the visual acuity and the nuclear backscatter above the threshold value. In patients with nuclear sclerosis and senile cataracts the backscatter from the nuclear region increases considerably.

Similar studies with similar results have been performed (Ben-Sira *et al.,* 1980) with equipment resembling Goldman's (1964) apparatus. Recently a more quantitative clinical characterization of lens opacity was attempted by stereoscopic photographing of lenses and their classification (Chylack and Kinoshita, 1972; Chylack, 1978a). Further improvements in clinical description and characterization are accomplished by rotating slit-image photography employing *Scheimpflug* principles (Topcon Instrument) (Dragomirescu *et al.,* 1978; Lerman *et al.,* 1981; Chylack *et al.,* 1981).

An important aspect of the clinical characterization is that a correspondence must exist between this classification and those of the physicochemical and biological processes elucidated with different measurements. Such an attempt was initially made with nuclear cataracts. Light-scattering parameters and classifications were compared in an empirical study (Siew *et al.,* 1981b).

In studying the light-scattering patterns and obtaining structural parameters responsible for turbidity from analysis of the light-scattering envelope there is one requirement to be met for a correct interpretation. This is that one should analyze only primary scattering data. Thus, one needs to study very thin sections of the lens to avoid secondary and multiple scattering which otherwise would mask the primary information. In normal human lenses (Bettelheim and Paunovic, 1979) we found that sectioning with a refrigerated microtome at −10°C did not alter the optical properties of the lens during freeze–thaw cycles. This condition is also met in cataractous lenses (Chylack *et al.,* 1981; Siew *et al.,* 1981a). A freeze–thaw cycle (15 min at −20°C and thawing) alters neither the appearance of the cataractous lens nor the optical properties.

The description of the human cataractous lens using structural parameters obtained from the analysis of light-scattering intensities agreed quite well with the clinical description obtained from stereoscopic photographs (Siew *et al.,* 1981b). The variations in parameters such as the size of aggregates, their concentrations, and the refractive index difference between aggregates and their surroundings, etc., in different regions of lenses with nuclear cataracts corresponded well with the observable turbidities in these regions.

In order to ascertain that the measured sizes of aggregates were correct, a one-to-one correspondence was established between light scattering and electron microscopy. Previously it had been shown that the membrane-to-membrane distance in the lens agreed well in these two measurements (Bettelheim and Vinciguerra, 1971; Vinciguerra and Bettelheim, 1971).

Bettelheim *et al.* (1981b) compared the protein aggregates based on the analysis of light-scattering intensities to those based on scanning electron micrograph measurements of the same lens sections. They found good agreement not only between the size parameters (~100 nm in the cortex and ~300 nm in the nucleus of young human lens) but also between the concentrations of aggregates obtained by the two techniques.

Delaye *et al.* (1982b) measured light scattering from the onset of cold cataract formation in intact calf lenses both by dissymmetry and quasi-elastic light-scattering spectroscopy. Partly on the basis of the quasi-elastic scattering data, which could not be fitted with a simple exponential decay curve, these authors proposed a bimodal distribution of scatterers. Small scatterers on the order of 10 nm and large scatterers with an average size of 130 nm were calculated from both the dissymetry and quasi-elastic spectroscopy data. With an increase in temperature the number of large scattering units decreased, implying that cold cataract formation induced aggregates having microphase dimensions of only 130 nm and disaggregating above the cold cataract temperature. They also found that, in the nucleus, although the aggregates were only slightly larger, they disaggregated much faster upon an increase in temperature.

These values seem to be too small compared to our measurements on human lenses (Siew *et al.,* 1981a,b; Bettelheim *et al.,* 1981a). The different lenses (calf versus human) may be one explanation. More likely, however, the fitting of the autocorrelation curve with different relaxation times (one, two, or a broad distribution of relaxation times) may yield different average-sized data. Further critical examinations of both our two-phase model and Delaye *et al.*'s (1981a) two kinds of scatterer models are in order. The latest report from Benedek's laboratory (Delaye *et al.,* 1982a) indicates that agreement maybe at hand. In this work on the temperature dependence of the aggregate sizes of calf lenses, they report that in the intact nucleus when cold cataract occurs the size of the aggregates reach the value of 300–500 mm. These are the values we found in human nuclear cataracts as well as in aging human lenses as the size of the scatterers causing turbidity. A more complex picture emerges from an age-dependent study of the structural units responsible for light scattering in the human lens. Siew *et al.* (1981a) have analyzed the light-scattering intensities of thin sections of human lenses in the $I_{\parallel}$ and I_{+} modes. In such a study two aging processes are actually compared. Within one lens the transition from cortex to nucleus indicates the aging occurring within the lens. A comparison of normal lenses from donors of different ages from 2½ to 81 yr old provided another aging gradient. The center sections of each human lens and sections at set distances from the centers were compared to each other. In this manner a true aging factor among different lenses could be observed which was not obscured by internal age differences in different parts of the lens.

One must clearly indicate what parameters are studied to avoid confusion. It is generally acknowledged that aggregation is an age-dependent phenomenon and therefore that it leads to HMW proteins in increasing concentrations. In our study we found that the size of the aggregate particles with age increased in the cortex but decreased deeper into the nucleus. At the same time we also found that the interparticle distance increased and the volume fraction of the aggregates decreased with aging.

The only interpretation that allows for an increase in molecular weight and at the same time a decrease in the size and volume fractions and a concomitant increase in interparticle separation is that the aggregation process is followed by a severe collapse of the hydrated volume. In this way heavier particles are actually smaller in size, occupy a smaller volume fraction, and are further separated from other collapsed protein aggregates. That such a syneretic process is operative in aging can be also seen by the age dependence of the amplitude factor of the density fluctuations $\bar{\eta}^2$. This increases with age in every part of the lens, indicating that the refractive index difference between protein aggregates and their surroundings steadily increases with age. Such an increase in the refractive index difference can be due to a loss of the water of hydration from the aggregates to the surroundings because of the collapse of the protein network.

Aging is also associated with a decrease in the size of cytoskeletal elements and their more random dispersion in older lenses than in younger ones—a sort of degradative process.

Bettelheim *et al.* (1981a) have calculated the correlation coefficients between light-scattering intensity and different structural parameters in human lenses with nuclear cataracts.

The most important contributor to cataractogenesis is the syneretic process. The highest correlation was found between turbidity and an increase in refractive index differences due to network collapse and dehydration of the protein aggregates (syneresis). Turbidity was also highly correlated with an increase in the concentration (volume fraction) of the HMW aggregates but not in the size of the aggregates. A third major contributor is the optically anisotropic units (birefringent bodies such as cytoskeletal elements entangled with protein aggregates). As cataractogenesis progresses, the size of these birefringent bodies increases. Their alignment decreases, in contrast to the situation in the aging process in the lens (Siew *et al.*, 1981a) where the size of the birefringent bodies decreases with age. The randomization of the birefrigent bodies is common in aging as well as in cataractogenesis. Another major difference between aging and nuclear cataractogenesis is that in aging the volume fraction of HMW protein aggregates decreases as a result of syneresis, but in cataractogenesis, in spite of collapsing networks (syneresis), the volume fraction of HMW aggregates increases.

REFERENCES

Albers, R. W. (1967). Biochemical aspects of active transport. *Annu. Rev. Biochem.* **36,** 727-756.

Alcala, J., Lieska, N., and Maisel, H. (1975). Protein composition of bovine lens cortical fiber cell membranes. *Exp. Eye Res.* **21,** 581-596.

Alcala, J., Bradley, R. H., Kusak, J., Waggoner, P., and Maisel, H. (1978). Biochemical and structural features of chick lens gap junctions. *J. Cell Biol.* **79,** 219a.

Alcala, J., Valentine, J., and Maisel, H. (1980). Human lens fiber cell plasma membranes. I. Isolation, polypeptide composition and changes associated with aging. *Exp. Eye Res.* **30,** 659-677.

Alexander, L. E. (1969). "X-ray Diffraction Methods in Polymer Science," p. 296. Wiley, New York.

Amoore, J. E., Bartley, W., and van Heyningen, R. (1959). Distribution of sodium and potassium within cattle lens. *Biochem. J.* **72,** 126-133.

Anderson, E. I., and Spector, A. (1978). The state of sulfhydryl groups in normal and cataractous human lens proteins. I. Nuclear region. *Exp. Eye Res.* **26,** 407-417.

Augusteyn, R. C. (1975). Distribution of fluorescence in the human cataractous lens. *Ophthalmic Res.* **7,** 217-224.

Bagchi, M., Gordon, P. A., Alcala, J. R., and Maisel, H. (1979). The plasma membrane of the rabbit lens cortical fiber. *Invest. Ophthalmol.* **18,** 562-569.

Baldwin, G., and Bentley, P. J. (1980). The calcium metabolism of the amphibian lens. *Exp. Eye Res.* **30,** 379-389.

Benedek, G. G. (1971). Theory of transparency of the eye. *Appl. Opt.* **10**, 459–473.

Benedek, G. B., Clark, J. I., Serralach, E. N., Young, C. Y., Mengel, L., Sanke, T., Bagg, A., and Benedek, K. (1979). Light scattering and reversible cataracts in the calf and human lens. *Philos. Trans. R. Soc. London, Ser. A* 293–329.

Ben-Sira I., Weinberger, D., Bodenheiner, J., and Yassur, Y. (1980). Clinical method for measurement of light backscatterings from the *in vivo* human lens. *Invest. Ophthalmol. Vis. Sci.* **19**, 435–437.

Bentley, P. J., and Cruz, E. (1978). The role of Ca^{2+} in maintaining the Na and K content of the amphibian lens. *Exp. Eye Res.* **77**, 335–341.

Bettelheim, F. A. (1972). Nature of crystallinity of α-crystallin preparation of bovine lenses. *Exp. Eye Res.* **14**, 251–258.

Bettelheim, F. A. (1975). On the optical anisotropy of lens fiber cells. *Exp. Eye Res.* **21**, 231–234.

Bettelheim, F. A. (1977). Organization of lens proteins in the fiber cell. *Atti Fond. Giorgio Ronchi* **32**, 331–336.

Bettelheim, F. A. (1978). Induced optical anisotropy fluctuation in the lens of the eye. *J. Colloid Interface Sci.* **63**, 251–258.

Bettelheim, F. A. (1979). Syneresis and its possible role in cataractogenesis. *Exp. Eye Res.* **28**, 189–197.

Bettelheim, F. A., and Bettelheim, A. (1978). Small angle light scattering studies on xylose cataract formation in bovine lenses. *Invest. Ophthalmol. Visual Sci.* **17**, 896–904.

Bettelheim, A. A., and Bettelheim, F. A. (1980). Cut induced turbidity in fish lenses. *Ophthalmic Res.* **12**, 199–202.

Bettelheim, F. A., and Finkel, J. (1976). Water vapor sorption of the subunits of α-crystallin and their complexes. *Colloid Interface Sci. Proc. Int. Conf., 50th, 1976* **5**, 203–16.

Bettelheim, F. A., and Mehrotra, K. N. (1972). On the hydration of α-crystallin of bovine lenses. *Exp. Eye Res.* **14**, 251–258.

Bettelheim, F. A., and Paunovic, M. (1979). Light scattering of normal human lens I. Application of random density and orientation fluctuation theory. *Biophys. J.* **26**, 85–100.

Bettelheim, F. A., and Siew, E. L. (1980). Light scattering and lens morphology in red blood cell and lens metabolism. *Dev. Biochem.* **9**, 443–446.

Bettelheim, F. A., and Siew, E. L. (1982). The effect of change in concentration upon lens turbidity as predicted by the random fluctuation theory. *Biophys. J.* (in press).

Bettelheim, F. A., and Vinciguerra, M. J. (1971). Laser diffraction patterns of highly ordered superstructures in the lenses of bovine eyes. *Ann. N. Y. Acad. Sci.* **172**, 429–439.

Bettelheim, F. A., and Wang, T. J. Y. (1974). Topographic distribution of refractive indices in bovine lenses. *Exp. Eye Res.* **18**, 351–356.

Bettelheim, F. A., and Wang, T. J. Y. (1977). X-ray diffraction studies of macromolecular aggregates of bovine lens. *Exp. Eye Res.* **25**, 613–620.

Bettelheim, F. A., Vinciguerra, M. J., and Kaplan, D. (1973). Dynamic laser diffraction of bovine lenses. *Exp. Eye Res.* **15**, 149–155.

Bettelheim, F. A., Siew, E. L., and Chylack, T. L., Jr. (1981a). Studies on human cataracts III. Structural elements in nuclear cataracts and their contribution to the turbidity. *Invest. Ophthalmol. Vis. Sci.* **20**, 348–354.

Bettelheim, F. A., Siew, E. L., Shyne, S., Farnsworth, P., Burke, P. (1981b). A comparative study of human lens by light scattering and scanning electron microscopy. *Exp. Eye Res.*, **32**, 125–130.

Beyer-Mears, A., Farnsworth, P. N., Fu, J. S. C., and Yeh, C. K. (1978). Regional analyses of the reversal process in the neonatal galactose cataract. *Exp. Eye Res.* **27**, 627–635.

Bloemendal, H., Zweers, A., and Walters, H. (1975a). Self-assembly of lens crystallins *in vitro*. *Nature (London)* **255**, 426–427.

Bloemendal, H., Zweers, A., Benedetti, E. L., and Walters, H. (1975b). Selective reassociation of the crystallins. *Exp. Eye Res.* **20**, 463–478.

Bonting, S. J. (1965). Na-K activated ATP and active cation transport in the lens. *Invest. Ophthalmol.* **4**, 723–738.

Bonting, S. J. (1966a). Physiological chemistry of the eye. *AMA Arch. Ophthalmol.* **74**, 561–578.

Bonting, S. J. (1966b). Physiological chemistry of the eye. *AMA Arch. Ophthalmol.* **76**, 607–622.

Bonting, S. L., Caravaggio, L. L., and Hawkins, N. M. (1963). Studies on sodium-potassium-activated ATP. V. Correlation of enzyme activity and cation flux in six tissues. *Arch. Biochem. Biophys.* **101**, 37–46.

Bouman, A. A., deLeeuw, A. L. W., Tolhuyzen, E. F. J., and Broekhuyse, R. M. (1979). Lens membranes. VI. Some characteristics of the EDTA-extractable protein (EEP) from bovine lens fiber membranes. *Exp. Eye Res.* **29**, 83–93.

Bradley, R. H., Ireland, M., and Maisel, H. (1979). The cytoskeleton of chick lens cells. *Exp. Eye Res.* **28**, 441–453.

Brewster, O. (1816). On the structure of crystalline lenses in fish and quadrupeds as ascertained by its action on polarized light. *Philos. Trans. R. Soc. London.* (1816), 311–317.

Broekhuyse, R. M. (1969). Phospholipids in tissues of the eye. III. Composition and metabolism of phospholipids in human lens in relation to age and cataract formation. *Biochim. Biophy. Acta* **187**, 354–365.

Broekhuyse, R. M. (1973). Membrane lipids and proteins in aging lens and cataract in the human lens: In relation to cataract. *Ciba Found. Symp.* **19**, 135–149.

Broekhuyse, R. M., and Bögermann, B. (1978). Lipids in the tissues of the eye. XVI. Uptake of lipids by the rabbit lens *in vitro*. *Exp. Eye Res.* **26**, 567–572.

Broekhuyse, R. M. and Kuhlmann, E. D. (1974). Lens membranes. I. Composition of urea-treated plasma membranes from calf lens. *Exp. Eye Res.* **19**, 297–302.

Broekhuyse, R. M., and Kuhlmann, E. D. (1978). Lens membranes. IV. Preperative isolation and characterization of membranes and various membrane proteins from calf lens. *Exp. Eye Res.* **26**, 305–320.

Broekhuyse, R. M., and Kuhlmann, E. D. (1979). Lens membranes. V. The influence of reduction and heating on the electrophoretical polypeptide pattern of lens fiber membrane. *Exp. Eye Res.* **28**, 615–618.

Broekhuyse, R. M., and Soeting, W. J. (1976). Lipids in tissues of the eye. XV. Essential fatty acids in lens lipids. *Exp. Eye Res.* **22**, 653–657.

Broekhuyse, R. M., Kuhlmann, E. D., and Stols, A. L. H. (1976). Lens membranes. II. Isolation and characterization of the main intrinsic polypeptide (MIP) of bovine lens fiber membranes. *Exp. Eye Res.* **23**, 365–367.

Broekhuyse, R. M., Kuhlmann, E. D., Bijvelt, J., Verkleij, A. J., and Vergrrt, P. H. J. T. (1978). Lens membranes. III. Freeze fracture morphology and composition of bovine lens fibre membranes in relation to aging. *Exp. Eye Res.* **26**, 147–156.

Burnside, B. (1975). The form and arrangement of microtubules: An historical primarily morphological review. *Ann. N. Y. Acad. Sci.* **253**, 14–26.

Chylack, L. T., Jr. (1978a). Classification of human cataracts. *Arch. Ophthalmol. (Chicago)* **96**, 888–892.

Chylack, L. T., Jr. (1978b). The cooperative cataract research group. *Invest. Ophthalmol. Vis. Sci.* **17,** 1131–1134.

Chylack, L. T., Jr., and Kinoshita, J. H. (1972). The high glucose cataract in a lens-vitreous preparation. *Exp. Eye Res.* **14,** 58–64.

Chylack, L. T., Jr., Bettelheim, F. A., and Tung, W. H. (1981). Studies on human cataracts. I. Evaluation of techniques of human cataract preservation after extraction. *Invest. Ophthalmol. Vis. Sci.* **20,** 326–334.

Clark, J. I., and Benedek, G. B. (1980). The effect of glycols, aldehydes and acrylamide on phase separation on opacification in the calf lens. *Invest. Ophthalmol. Vis. Sci.* **19,** 771–776.

Clark, J. I., Mengel, L., and Benedek, G. B. (1980a). Scanning electron microscopy of opaque and transparent states in reversible calf lens cataracts. *Ophthalmic Res.* **12,** 16–33.

Clark, J. I., Mengel, L., Baggi, A., and Benedek, G. (1980b). Cortical opacity, calcium concentration and fiber membrane structure in the calf lens. *Exp. Eye Res.* **31,** 399–410.

Clark, R., Zigman, S., and Lerman, S. (1969). Studies on the structural proteins of the human lens. *Exp. Eye Res.* **8,** 172–182.

Cliffe, E. E., and Waley, S. G. (1961). Acidic peptide of the lens. 6. Metabolism of gamma-glutamyl peptide in subcellular fractions of rabbit liver. *Biochem. J.* **79,** 118–128.

Cooke, P. (1976). A filamentous cytoskeleton in vertebrate smooth muscle fibers. *J. Cell Biol.* **68,** 539–556.

Cotlier, E., and Beaty, C. (1967). The role of Na ion in the transport of alpha-aminoisobutyric acid and other amino acids into the lens. *Invest. Ophthalmol.* **6,** 64–75.

Creighton, M. O., Trevithick, J. R., Mousa, G. Y., Percy, D. H., McKinna, A. J., Dyson, C., Maisel, H., and Bradley, R. (1978). Globular bodies: A primary cause of opacity in senile and diabetic posterior cortical and subcapsular cataracts. *Can. J. Ophthalmol.* **13,** 166–181.

Csaky, T. Z. (1965). Transport through biological membranes. *Annu. Rev. Physiol.* **27,** 415–450.

Dardenne, V., and Kirsten, G. (1962). Presence and metabolism of amino acids in young and old lenses. *Exp. Eye Res.* **1,** 415–421.

Dark, A. J., and Streeten, B. W. (1977). Ultrastructural study of cataract in myotonia distrophia. *Am. J. Ophthalmol.* **84,** 666–674.

De Berardinis, E., Tieri, O., Inglio, N., and Polzella, A. (1965a). The concentration of lactic acid in the human aqueous humour is not determined by the metabolism of the lens. *Experientia* **21,** 589–590.

De Berardinis, E., Tieri, O., Inglio, N., and Polzella, A. (1965b). The chemical composition of human aqueous humour in normal and pathological conditions. *Exp. Eye Res.* **4,** 179–186.

Debye, P. (1944). Light scattering in solutions. *J. Appl. Phys.* **15,** 338–342.

Debye, P. (1947). Molecular weight determination by light scattering. *J. Phys. Colloid Chem.* **51,** 18–32.

Debye, P., and Bueche, A. M. (1949). Scattering by an inhomogeneous solid. *J. Appl. Phys.* **20,** 518–526.

de Jong, W. W., Terwindt, E. C., and Bloemendal, H. (1975). The amino acid sequences of the A chain of human alpha-crystallin. *FEBS Lett.* **58,** 310–313.

Delaye, M., Clark, J. I. and Benedek, G. B. (1982a). Identification of the scattering elements responsible for lens opacification in cold cataracts. *Biophys. J.* **37,** 647–656.

Delaye, M., Clark, J. I., Benedek, G. B. (1982b). Light Scattering Studies of Cataracts, Proc. NATO Int. Conf. Light Scattering (in press).

Dilley, K. J. (1975). The proportion of protein from the normal and cataractous human lens which exists as higher molecular weight aggregates *in vitro*. *Exp. Eye Res.* **20,** 73-78.

Dilley, K. J., and Pirie, A. (1974). Changes to the proteins of the human lens nucleus in cataract. *Exp. Eye Res.* **19,** 59-72.

Dische, Z. (1971). The effect of sulfhydryl compounds on the stability of cell membranes of the lens and red cells of lens capsule. *Exp. Eye Res.* **11,** 338-350.

Dragomirescu, V., Hockuwin, O., Koch, H. R., and Sasaki, K. (1978). Development of a new equipment for rotating slit image photography according to Scheimpflug's principle. *Interdiscip. Top. Gerontol.* **13,** 1-13.

Einstein, A. (1910). Theory of the opalescence of homogeneous liquids and liquid mixtures in the neighborhood of the critical state. *Ann. Phys.* **33,** 1275-1298.

Erhegian, J., March, W. F., Goossens, W., and Rafferty, N. (1978). Ultrastructure of cataract in myotonic distrophy. *Invest. Ophthalmol. Vis. Sci.* **17,** 289-294.

Fagerholm, P. P. (1979). The influence of calcium on lens fibers. *Exp. Eye Res.* **28,** 111-122.

Fagerholm, P. P., and Philipson, B. (1979). Human traumatic cataract. *Acta Ophthalmol.* **57,** 20-32.

Farnsworth, P. N., Shayne, S. E., Caputo, S. J., Fasano, A. V., and Spector, A. (1980). Microtubules: A major cytoskeletal component of the human lens. *Exp. Eye Res.* **30,** 611-615.

Fraunfelder, F. T., and Burns, R. P. (1962). Production of unilateral reversible cataracts in the hamster. *Proc. Soc. Exp. Biol. Med.* **110,** 72-74.

Fraunfelder, F. T., and Burns, R. P. (1966). Effect of lid closure in drug-induced experimental cataracts. *AMA Arch. Ophthalmol.* **76,** 599-601.

Gallagher, L., and Bettelheim, F. A. (1962). Light scattering studies of cross-linking unsaturated polyesters with methylacrylate. *J. Polym. Sci.* **58,** 697-714.

Garcia-Castineiras, S., Dillon, J., and Spector, A. (1978). Non-tryptophane fluorescence associated with human lens protein: Apparent complexity and isolation of bityrosine and anthranilic acid. *Exp. Eye Res.* **26,** 461-476.

Garner, W. H., and Spector, A. (1978). Racemization in human lens: Evidence of rapid insolubilization of particular polypeptides in cataract formation. *Proc. Natl. Acad. Sci. U.S.A.* **75,** 3618-3620.

Giblin, F. J., Chakrapani, B., and Reddy, V. N. (1978). High molecular weight protein aggregates in X-ray induced cataract. *Exp. Eye Res.* **26,** 507-519.

Glynn, I. M. (1962). Activation of adenosine triphosphatase activity in a cell membrane by external potassium and internal sodium. *J. Physiol.* (*London*) **160,** 181-191.

Goldmann, H. (1964). Senile changes of the lens and the vitreous. *Am. J. Ophthalmol.* **57,** 1-13.

Goldmann, H., and Rabinowitz, G. (1928). Ueber eine unbekannte, reversibel Kataraktform bei jungen Ratten. *Klin. Monatsbl. Augenheilkd.* **81,** 771-785.

Gorthy, W. L. (1977). Cataracts in the aging rat lens. *Ophthalmic Res.* **9,** 329-342.

Gorthy, W. C. (1978). Cataracts in the aging rat lens: Morphology and acid phosphates histochemistry of incipient forms. *Exp. Eye Res.* **27,** 301-322.

Greville, G. O. (1966). Factors affecting the utilization of substrates by mitochondria. *In* "Regulation of Metabolic Processes in Mitochondria" (J. M. Tager, S. Quaglierielle, E. Papa, and E. C. Slater, eds.), Pp. 86-107. Elsevier, Amsterdam.

Griffins, M. H. (1966). The components of an α-glycerophosphate cycle and their relation to oxidative metabolism in the lens. *Biochem. J.* **99,** 12-21.

Griess, G., and Zigman, S. (1973). Lens protein interaction and transparency. *Exp. Eye Res.* **15,** 143–148.

Harding, J. J., and Dilley V. J. (1976). Structural proteins of the mammalian lens: A review with emphasis on changes in development, aging and cataract. *Exp. Eye Res.* **22,** 1–73.

Hightower, K. R., and Kinsey, E. (1977). Studies on the crystalline lens. XXIII. Electrogenic potential and cation transport. *Exp. Eye Res.* **24,** 587–593.

Hightower, K. R., Leverenz, V., and Reddy, V. N. (1980). Calcium transport in the lens. *Invest. Ophthalmol.* **19,** 1059–1066.

Hiller, R. L., Giacometti, L., and Yuen, V. S. (1977). Sunlight and cataract: An epidemiological study. *Am. J. Epidemol.* **105,** 450–462.

Hoenders, H. J. (1976). Changes in the subunit structure of alpha-crystallin from bovine and rabbit eye lenses. *Doc. Ophthalmol. Proc. Ser.* **8,** 105–112.

Hoenders, H. J., and Van Kamp, G. J. (1972). Eye lens development and aging processes of crystallins. *Acta Morphol. Neerl. Scand.* **10,** 215–221.

Hoenders, H. J., van Tol, J., and Bloemendal, H. (1968a). Release of an N-terminal tetrapeptide from α-crystallin. *Biochim. Biophy. Acta* **160,** 283–285.

Hoenders, H. J., Shoenmaker, J. G. B., Gerding, J. J. J., Tesser, G. I., and Bloemendal, H. (1968b). N-Terminus of α-crystalline. *Exp. Eye Res.* **7,** 291–300.

Horwitz, J., Kabasawa, I., and Kinoshita, J. H. (1977). Conformation of γ-crystallins of the calf lens: Effects of temperature and denaturing agents. *Exp. Eye Res.* **25,** 199–208.

Horwitz, J., Robertson, N. P., Wong, M. M., Zigler, J. S., and Kinoshita, J. H. (1979). Some properties of lens plasma membrane polypeptides isolated from normal human lenses. *Exp. Eye Res.* **28,** 359–365.

Huggert, A. (1948). On the iso-indicial surfaces of the human crystalline lens. *Acta Ophthalmol. Suppl.* **30,** 1–126.

Ireland, M., Maisel, H., and Bradley, R. (1978). The rabbit lens cytoskeleton: An untrastructural analysis. *Ophthalmic Res.* **10,** 231–236.

Ishikawa, H., Bischoff, R., and Holtzer, H. (1969). Formation of arrowhead complexes with heavy meromysin in a variety of cell types. *J. Cell Biol.* **43,** 312–328.

Ishimoto C., Goalwin, P. W., Sun, S. T., Nishio, I., and Tanaka, T. (1979). Cytoplasmic phase separation information of galactosemic cataract in lenses of young rats. *Proc. Natl. Acad. Sci U.S.A.* **76,** 4414–4416.

Jedziniak, J. A., Kinoshita, J. H., Yates, E. M., Hocker, L. O., and Benedek, G. B. (1973). On the presence and mechanism of formation of heavy molecular weight aggregates in human normal and cataractous lenses. *Exp. Eye Res.* **15,** 185–192.

Jedziniak, J., Kinoshita, J. H., Yates, E. M., and Benedek, G. B. (1975). The concentration and localization of heavy molecular weight aggregates in aging normal and cataractous human lenses. *Exp. Eye Res.* **20,** 367–369.

Jedziniak, J. A., Nicoli, D. F., Baram, H., and Benedek, G. B. (1978). Quantitative verification of the existence of high molecular weight protein aggregates in the intact normal human lens by light scattering spectroscopy. *Invest. Ophthalmol. Vis. Sci.* **17,** 51–57.

Jones, H. A., and Lerman, J. (1971). Optical rotatory dispersion and circular dichroism on ocular lens proteins. *Can. J. Biochem.* **49,** 426–430.

Kahovec, L. G., Porod, R., and Rick, H. (1953). X-ray small-angle investigations on close packed colloid systems. *Kolloid Z.* **133,** 16.

Kendrew, J. C., Bodo, G., Dintzis, H. M., Parrish, R. G., Wyckoff, H., and Philips, D. C. (1958). A three dimensional model of the myoglobin molecule obtained by X-ray analysis. *Nature (London)* **181,** 662–666.

Kerker, M., (1969). "The Scattering of Light and Other Electromagnetic Radiation." Academic Press, New York.

Kinoshita, J. H. (1965). Cataracts in galactosemia (Friedenwald lecture). *Invest. Ophthalmol.* **4**, 786-799.

Kinoshita, J. H. (1974). Mechanism initiating cataract formation. *Invest. Ophthalmol.* **13**, 713-724.

Kinoshita, J. H., and Wachtl, C. (1958). A study of C^{14}-glucose metabolism of rabbit lens *J. Biol. Chem.* **233**, 5-7.

Kinsey, V. E. (1965). The pump-leak concept of transport in ocular lens. *In* "Eye Structure," II. Symposium (J. W. Rohen, ed.), pp. 383-394. Schattauer, Stuttgart.

Kinsey, V. E. (1966). Studies on the crystalline lens. XIV. Kinetics of alpha-aminoisobutyric acid transport. *Doc. Ophthalmol.* **20**, 30-43.

Kinsey, V. E., and Hightower, K. R. (1976). Studies on the crystalline lens. XXII. Characterization of chloride movement based on the pump leak method. *Exp. Eye Res.* **23**, 425-433.

Kinsey, V. E., and Hightower, K. R. (1978). Studies on crystalline lens. XXV. An analysis of the dependence of the components of the potential on sodium-potassium fluxes based on the pump-leak model. *Exp. Eye Res.* **26**, 157-164.

Kleinfeld, O., and Hockwin, O. (1960). The importance of oxygen to the metabolism of the lens. *Graefes Arch. Ophthalmol.* **162**, 346-349.

Kramps, H. A., Stols, A. L. H., Hoenders, H. J., and deGroot, K. (1975). On the quaternary structure of high molecular weight proteins from the bovine eye lens. *Eur. J. Biochem.* **50**, 503-509.

Kramps, H. A., Hoenders, H. J., and Wollensak, J. (1976). Soluble and insoluble lens proteins in human nuclear cataract. *Colloq. Inst. Natl. Sante Rech. Med.* **60**, 237-247.

Kratky, O. (1966). Possibilities of X-ray small-angle analysis in the investigation of dissolved and solid high-polymer substances. *Pure Appl. Chem.* **12**, 483.

Kuck, J. R. R., Jr. (1961). The formation of fructose in the ocular lens. *AMA Arch. Ophthamol.* (*Chicago*) **65**, 840-846.

Kuck, J. F. R., Jr. (1970). Chemical constituents of the lens. *In* "Biochemistry of the Eye" (C. N. Graymore, ed.), p. 187. Academic Press, New York.

Kurzel, R., Wolbarsht, M. L., Yamanashi, B. S., Staton, G. W., and Borkman, R. F. (1973). Tryptophane excited states and cataracts in the human lens. *Nature* (*London*) **241**, 132-133.

Kuwabara, T. (1975). The maturation of the lens cell: A morphological study. *Exp. Eye Res.* **20**, 427-443.

Kuwabara, T., Kinoshita, J., and Cogan, D. (1969). Electron microscopic study of galactose induced cataract. *Invest. Ophthalmol.* **8**, 133-149.

Lambert, B. W., and Kinoshita, J. H. (1967). The effects of ionizing radiation on lens cation permeability transport and hydration. *Invest. Ophthalmol.* **6**, 624-634.

Lazarides, E. (1975). Immunofluorescence studies on the structure of actin filaments in tissue culture cells. *J. Histochem. Cytochem.* **23**, 507-528.

Lazarides, E. (1976). Actin, alpha-actinin, and tropomysin interaction in the structural organization of actin filament in nonmuscle cells. *J. Cell Biol.* **68**, 202-219.

Lehninger, A. L. (1970). "Biochemistry." Worth, New York.

Lerman, S. (1976). Lens fluorescence in aging and cataract formation. *Doc. Ophthalmol. Proc. Ser.* **8**, 241-260.

Lerman, S., and Borkman, R. F. (1976). Spectroscopic evaluation and classification of the normal, aging and cataractous lens. *Ophthalmic Res.* **8**, 335-353.

Lerman, S., and Zigman, S. (1965). The metabolism of the lens as related to aging and experimental cataractogenesis. *Invest. Ophthalmol.* **4,** 643-660.
Lerman, S., Hockwin, O., and Dragomirescu, V. (1981). UV-visible slit lamp densitography of the human eye. *Exp. Eye Res.* **33,** 587-596.
Li, L. K. (1978). Effects of sucrose on interactions of calf lens soluble proteins. *Exp. Eye Res.* **27,** 553-556.
Li, L. K., and Spector, A. (1974). Circular dichroism and optical rotatory dispersion of the aggregates of purified polypeptides of α-crystallin. *Exp. Eye Res.* **19,** 49-57.
Liem-The, K. N., Stols, A. L. H., Jap, P. H. K., and Hoenders, H. J. (1975). X-ray induced cataract in rabbit lens. *Exp. Eye Res.* **20,** 317-328.
Longchampt, M. D., Laurent, M., Courtos, Y., Trenchso, P., and Hughes, R. S. (1976). Microtubules and microfilaments of bovine lens epithelial cells: Electron microscopy and immunofluorescence staining with specific antibodies. *Exp. Eye Res.* **23,** 505-518.
Lou, M. F., and Kinoshita, J. H. (1967). Control of lens glycolysis. *Biochim. Biophys. Acta* **141,** 547-559.
Lowenstein, M. A., and Bettelheim, F. A., (1979). Cold cataract formation in fish lenses. *Exp. Eye Res.* **28,** 651-663.
Lowenstein, M. A., and Bettelheim, F. A. (1980). Pressure induced turbidity in fish lenses. *Exp. Eye Res.* **30,** 315-317.
McNamara, M. K., and Augusteyn, R. C. (1980). 33′-Dityrosine in the proteins of senile nuclear cataracts. *Exp. Eye Res.* **30,** 319-321.
Mach, H. (1963). Untersuchungen von Linseneiweiss und Mitroelektrophorese von wasser löslichem Eiweiss im Altersstar. *Klin Monatsol. Augenheilkd.* **143,** 689-710.
Maisel, H. (1977a). The nature of the urea-insoluble material of the human lens *Exp. Eye Res.* **24,** 417-419.
Maisel, H. (1977b). Filaments of the vertebrate lens. *Experientia* **33,** 525.
Maisel, H., and Alcala, J. R. (1980). The plasma membrane of the normal lens and cataract. *In* "Red Blood Cell and Lens Metabolism" (Srivastava, ed.), pp. 213-228. Elsevier, Amsterdam.
Maisel, H., and Perry, M. M. (1972). Electron microscope observations on some structural proteins of chicken lens. *Exp. Eye Res.* **14,** 7-12.
Maisel, H., Alcala, J., and Lieska, N. (1976a). Protein composition of bovine lens cortical fiber cell membrane. *Doc. Ophthalmol.* **8,** 121-133.
Maisel, H., Perry, M., Alcala, J., and Waggoner, P. (1976b). The structure of chick lens water-insoluble material. *Ophthalmic Res.* **8,** 55-63.
Maisel, H., Alcala, J., Lieska, N., and Rafferty, N. (1977). Regional differences in the polypeptide composition of chick lens intracellular matrix. *Ophthalmic Res.* **9,** 147-154.
Maisel, H., Lieska, N., and Bradley, R. (1978). Isolation of filaments of the chick lens. *Experientia* **34,** 352-353.
Masters, P. M., Bada, J. L., and Zigler, J. S., Jr. (1977). Aspartic acid racemization in human lens during aging and in cataract formation. *Nature (London)* **268,** 71-73.
Masters, P. M., Bada, J. L., and Zigler, J. S. (1978). Aspartic acid racemization in heavy molecular weight crystallins and water insoluble protein from normal human lenses and cataracts. *Proc. Natl. Acad. Sci. U.S.A.* **75,** 1204-1208.
Mazur, A., and Harrow, B. (1971). "Textbook of Biochemistry," 10th ed. Saunders, Philadelphia, Pennsylvania.
Meyer, D. (1977). The avian eye and its adaptations. *Hand. Sens. Physiol.* **7,** No. 5, 550-611.

Mie, G. (1908). Optics of trubid media. *Ann. Phys.* **25**, 377-445.

Morner, C. T. (1894). Untersuchungen der Proteinsubstanzen in lichtbrechaunches Mieden des Auger. *Hoppe-Seyler's Z. Physiol Chem.* **18**, 61-101.

Neville, M. C., Paterson, L. A., Rae, J. L., and Woessner, D. F. (1974). Nuclear magnetic resonance studies and water "ordering" in the crystalline lens. *Science* **184**, 1072-1077.

Obara, Y., Cotlier, E., Lindberg, R., and Horn, J. (1976). Cholesterol, cholesterol ester sphingomyelin complexed to protein of normal human lens and senile cataracts. *Doc. Ophthalmol. Proc. Ser.* **8**, 193-203.

Palva, M., and Palkama, A. (1978). ultrastructural lens changes in X-ray induced cataract of the rat. *Acta Ophthalmol.* **56**, 587-598.

Palmer, W. G., and Papaconstantinou, J. (1969). Aging of α-crystalline during development of the lens. *Proc. Natl. Acad. Sci. U.S.A.* **64**, 404.

Philipson, B. (1969a). Galactose cataract: Changes in protein distribution during development. *Invest Ophthalmol.* **8**, 3-11.

Philipson, B. (1969b). Light scattering in lenses with experimental cataract. *Acta Ophthalmol.* **47**, 1089-1101.

Philipson, B. (1969c). Distribution of protein within the normal rat lens. *Invest. Ophthalmol.* **8**, 258-270.

Philipson, B. (1973). Changes in the lens related to the reduction of transparency. *Exp. Eye Res.* **16**, 29-39.

Philipson, B., and Fagerholm, P. P. (1981). Refractive properties and light scattering in normal and cataractous lenses. *Exp. Eye Res.* **33**, 621-630.

Pirie, A. (1962). Metabolism of glycerophosphate in the lens. *Exp. Eye Res.* **1**, 427-435.

Pirie, A. (1968). Color and solubility of the proteins of human cataracts. *Invest. Ophthalmol.* **7**, 634-650.

Pollard, T. D. (1976). Cytoskeletal functions of cytoplasmic contractile proteins. *J. Supramol. Struct.* **5**, 317-334.

Pollard, T. D., and Weiling, R. R. (1974). Actin and myosin and cell movement. *CRC Crit. Rev. Biochem.* **2**, 1-65.

Racz, P., Tompa, V., and Pocsik, T. (1979). The state of water in normal and senile cataractous lenses studied by nuclear magnetic resonance. *Exp. Eye Res.* **28**, 129-135.

Rafferty, N., and Goossens, W. (1978). Cytoplasmic filaments in the crystalline lens of various species: Functional correlations. *Exp. Eye Res.* **26**, 177-190.

Rafferty, N. S., Goossens, W., and March, W. F. (1974). Ultrastructure of human traumatic cataract. *Am. J. Ophthalmol.* **78**, 985-995.

Rayleigh, Lord (J. W. Strutt) (1871). On the light from the sky, its polarization and color on the scattering of light by small particles. *Philos. Mag.* **41**, 107-120.

Reddy, D. V. M. (1967). Distribution of free amino acids and related compounds in ocular fluids, lens and plasma of various mammalian species. *Invest. Ophthalmol.* **6**, 478-483.

Reddy, D. V. M., and Kinsey, V. E. (1962). Studies on the crystalline lens: Quantitative analysis of free amino acids and related compounds. *Invest. Ophthalmol.* **1**, 635-641.

Roy, D., and Spector, A. (1976). High molecular weight protein from human lenses. *Exp. Eye Res.* **22**, 273-279.

Roy, D., and Spector, A. (1978a). Human insoluble lens protein. I. Separation and partial characterization of polypeptides. *Exp. Eye Res.* **26**, 429-443.

Roy, D., and Spector, A. (1978b). Human insoluble lens protein. II. Isolation and characterization of a 9000 dalton polypeptide. *Exp. Eye Res.* **26**, 445-459.

Roy, D., Spector, A., and Farnsworth, N. P. (1979). Human lens membrane: Comparison of major intrinsic polypeptides from young and old lenses isolated by a new methodology. *Exp. Eye Res.* **28**, 353-358.

Schachar, R. A., and Solin, S. A. (1975). The microscopic protein structure of the lens with a theory for cataract formation as determined by Raman spectroscopy of intact bovine lenses. *Invest. Ophthalmol.* **14**, 380-396.

Shoenmakers, J. G. G., Matze, R., van Poppel, M., and Bloemendal, H. (1969a). The isolation of non-identical polypeptide chains of α-crystallin. *Int. J. Prot. Res.* **1**, 19-27.

Shoenmakers, J. G., Gerding, J. J., and Bloemendal, H. (1969b). The subunit structure of α-crystalline. Isolation and characterization of the *S*-carboxy-urethylated acidic subunits from adult and embryonic origin. *Eur. J. Biochem.* **11**, 472-481.

Shoenmakers, J. G., Matze, R., van Pappel, M., and Bloemendal, H. (1969c). The isolation of non-identical polypeptide chain of α-crystallin. *Inst. J. Post. Res.* **1**, 19-27.

Siew, E. L., Opalecky, D., and Bettelheim, F. A. (1981a). Light scattering of normal lens II. Age dependence of the light scattering parameters. *Exp. Eye Res.* **33**, 603-614.

Siew, R. L., Bettelheim, F. A., Chylack, L. T., Jr., and Tung, W. H. (1981b). Studies in human cataracts. II. Correlation between the clinical description and the light scattering parameters of human cataracts. *Invest. Ophthalmol. Visual Sci.* **20**, 334-348.

Siegelman, J., Trokel, S. L., and Spector, A. (1974). Quantitative biomicroscopy of lens light back scatter. *Arch. Ophthalmol.* (*Chicago*) **92**, 437-442.

Simpson, G. C. (1953). Ocular haloes and coronas. *Br. J. Ophthalmol.* **37**, 450-486.

Sippel, T. O. (1962). Pyridine nucleotides in normal and diabetic rat lens. *Exp. Eye Res.* **1**, 368-371.

Spector, A. (1965). The soluble proteins of the lens. *Invest. Ophthalmol.* **4**, 579-591.

Spector, A. (1972). Aggregation of alpha-crystallin and its possible relationship to cataract formation. *Isr. J. Med. Sci.* **8**, 1577-1582.

Spector, A., and Roy, D. (1978). Disulfide-linked with high molecular weight protein associated with human cataracts. *Proc. Natl. Acad. Sci. U.S.A.* **75**, 3244-3248.

Spector, A., Freund, T., Li, L. K., and Augusteyn, R. C. (1971). Age dependent changes in the structure of alpha-crystallin. *Invest. Ophthalmol.* **10**, 677-686.

Spector, A., Stauffer, J., and Sigelman, J. (1973). Preliminary observations upon the proteins of human lens. *Ciba Found. Symp.* **19** (New Series), 187-206.

Spector, A., Li, L. K., and Sigelman, J. (1974). Age-dependent changes in the molecular size of human lens protein and their relationship to light scatter. *Invest. Ophthalmol.* **13**, 795-798.

Spector, A., Roy, D., and Stauffer, J. (1975). Isolation and characterization of an age dependent polypeptide from human lens with non-tryptophane fluorescence. *Exp. Eye Res.* **21**, 9-24.

Stacey, K. A. (1956). "Light Scattering in Physical Chemistry." Academic Press, New York.

Stauffer, J., Rothschild, G., Wandel, T., and Spector, A. (1974). Transformation of α-crystallin polypeptide chains with aging. *Invest. Ophthalmol.* **13**, 135-146.

Stein, R. S. (1969). The determination of the inhomogeneity of crosslinking of a rubber by light scattering. *Polym. Lett.* **1**, 657-659.

Stein, R. S. J., Keane, J. J., Norris, F. H., Bettelheim, F. A., and Wilson, P. R. (1959). Some light scattering studies of the texture of crystalline polymers. *Ann. N. Y. Acad. Sci.* **83**, 37-59.

Tanaka, T., and Benedek, G. B. (1975). Observation of protein diffusivity in intact human and bovine lenses with application to cataract. *Invest. Ophthalmol.* **14**, 449-456.

Tanaka, T., and Ishimoto, C. (1977). *In vivo* observation of protein diffusivity in rabbit lenses. *Invest. Ophthalmol. Vis. Sci.* **16,** 135–140.

Tanaka, T., Ishimoto, C., and Chylack, L. T., Jr. (1977). Phase separation of a protein-water mixture in cold cataract in the young rat lens. *Science* **197,** 1010–1012.

Tanford, C. (1961). "Physical Chemistry of Macromolecules," p. 307. Wiley, New York.

Thoft, R. A., and Kinoshita, J. H. (1965). The effect of calcium on rat lens permeability. *Invest. Ophthalmol.* **4,** 122–128.

Thomas, D. M., and Schepler, K. L. (1980). Raman spectra of normal and ultraviolet induced cataractous rabbit lens. *Invest. Ophthalmol. Vis. Sci.* **19,** 904–912.

Trokel, S. L. (1962). The physical basis for transparency of the crystalline lens. *Invest. Ophthalmol.* **1,** 493–501.

Truscott, R. J. W., and Augusteyn, R. C. (1977a). Changes in human lens protein during nuclear cataract formation. *Exp. Eye Res.* **24,** 159–170.

Truscott, R. J. W., and Augusteyn, R. C. (1977b). Oxidative changes in human lens proteins during senile nuclear cataract formation. *Biochim. Biophys. Acta* **492,** 43–52.

Unakar, N. J., Genyea, C., Reddan, J. R., and Reddy, V. N. (1978). Ultrastructural changes during the development and reversal of galactose cataract. *Exp. Eye Res.* **26,** 123–133.

Van Heyningen, R. (1962). The sorbitol pathway in the lens. *Exp. Eye Res.* **1,** 396–404.

Van Heyningen, R. (1976). What happens to the human lens in cataract? *Sci. Am.* **233,** 70–92.

Van Kamp, G. J., and Hoenders, H. J. (1973). The distribution of the soluble proteins in the calf lens. *Exp. Eye Res.* **17,** 417–426.

Van Kleef, F. S. M. (1975). Post-synthetic modification of bovine alpha-crystallin. Thesis, University of Nijimegen, The Netherlands.

Vinciguerra, M. J., and Bettelheim, F. A. (1971). Packing and orientation of fiber cells. *Exp. Eye Res.* **11,** 214–219.

Waggoner, P. R., and Maisel, H. (1978). Immunofluorescent study of a chick lens fiber cell membrane polypeptide. *Exp. Eye Res.* **27,** 151–157.

Waley, J. G. (1969). The lens: Function and macromolecular composition. *In* The Eye. (H. Davson, ed.), Vol. 1, 2nd ed., p. 303. Academic Press, New York.

Wannemacher, C. F., and Spector, A. (1968). Protein synthesis in the core of calf lens. *Exp. Eye Res.* **7,** 623–625.

Weinstock, M., and Scott, J. D. (1967). Effect of various agents on drug-induced opacities of the lens. *Exp. Eye Res.* **6,** 368–375.

Whittam, R. (1962). The assymmetrical stimulation of a membrane ATP in relation to active cation transport. *Biochem. J.* **84,** 110–118.

Whittam, R. (1964). *In* "Transport and Diffusion in Red Blood Cells" Arnold, London.

Wiedenthal, D. T., and Schepens, C. L. (1966). Peripheral fundus changes associated with ocular contusion. *Am. J. Ophthalmol.* **62,** 465–477.

Wolosewick, J., and Porter, K. (1979). Microtubular lattice of the cytoplasmic ground substance. *J. Cell Biol.* **82,** 114–139.

Wun, K. L., and Prins, W. (1974). Assessment of non-random crosslinking in polymer networks by small angle light scattering. *J. Polym. Sci. Part A-2* **12,** 533–543.

Zaret, M. M., Snyder, N. Z., and Birenbaum, L. (1976). Cataract after exposure to nonionizing radiant energy. *Brit. J. Ophthalmol.* **60,** 632–637.

Zigman, S. (1971). Eye lens color: Formation and function. *Science* **171,** 807–809.

Zigman, S., and Lerman, S. (1965). Properties of cold precipitable protein fraction in the lens. *Exp. Eye Res.* **4,** 24–30.

Zigman, S., Schultz, J., Yulo, T., and Griess, G. (1973). The binding of photo-oxidized tryptophane to a lens gamma crystallin. *Exp. Eye Res.* **17**, 209–217.

Zigman, S., Datiles, M., and Torczyznski, E. (1979). Sunlight and human cataracts. *Invest. Ophthalmol.* **18**, 462–467.

Zimm, B. H. (1948). Apparatus and methods for measurement and interpretation of the angular variation of light scattering: Preliminary results on polystyrene solutions. *J. Chem. Phys.* **16**, 1099–1116.

7

Control of Cell Division in the Ocular Lens, Retina, and Vitreous Humor

JOHN R. REDDAN

Cell Biology of the Eye

ISBN 0-12-483180-X

I. REGULATION OF CELL DIVISION IN THE OCULAR LENS

A. Introduction

Understanding the extracellular factors and intracellular events regulating cell division is one of the most fundamental problems in cell biology. Investigations on the process of mitosis indicate that environmental factors play a role in determining whether or not cells enter mitosis (Harding *et al.,* 1971; de Asua *et al.,* 1980). Of the numerous factors that may regulate mitosis (Clarkson and Baserga, 1974; Holley, 1980), attention has recently focused on a group of polypeptides many of which are structurally and functionally homologous to insulin (Rinderknecht and Humbel, 1978a). In view of the complexity of the *in vivo* environment, attempts have been made to define the conditions and factors required for the growth of cells in tissue and/or organ culture (Ham and McKeehan, 1978; Gospodarowicz *et al.,* 1978a; Barnes and Sato, 1980; Reddan *et al.,* 1981a). It is the general assumption in such studies that a delineation of the environmental factors required for the growth of a particular cell type *in vitro* may provide insight into the factors required for the division of these cells *in vivo.* A detailed review of the effect of the environmental factors on the growth of corneal cells is presented by Gospodarowicz in this volume.

The ocular lens has several properties which make it amenable to studies on cell division. The adult lens is avascular, lacks innervation, is derived solely from ectoderm, and is enclosed within a limiting basement membrane, i.e., the lens capsule. These properties make it possible to place the lens in organ culture without interruption of either the nerve or blood supply. Moreover, under appropriate culture conditions, the isolated lens retains the overall tissue organization and mitotic pattern characteristic of the organ *in vivo* (Kinsey *et al.,* 1955; Constant, 1958; Wachtl and Kinsey, 1958; Harding *et al.,* 1962; Rothstein *et al.,* 1965; Reddan *et al.,* 1970a, 1972, 1975). In addition, encapsulation of the lens effectively isolates the epithelium from other cell types and, with appropriate precautions, ensures the availability of a pure population of epithelial cells for the initiation of tissue culture (Reddan *et al.,* 1981a,b). It is thus possible to investigate the environmental factors that control cell division in the lens *in vivo* and in a pure population of epithelial cells in both organ and tissue culture. In this regard, insulin (Reddan *et al.,* 1972, 1975) and insulin-like growth factors (Reddan and Wilson, 1978) trigger cell division in mammalian lens epithelial cells maintained in organ or tissue culture. Moreover, studies from Rothstein's laboratory (Rothstein *et al.,* 1980) indicate that mitosis in the amphibian lens *in vivo* is abolished by procedures which decrease the level of insulin-like growth factors and is reinstated by the systemic administration of such factors. The lens epithelium represents one of the very few investigational systems wherein a specific type of growth factor is known to initiate a series of events which trigger cell division both *in vivo* and *in vitro*.

The initial portion of this chapter concentrates on the environmental factors and conditions that regulate cell division in lens epithelial cells, details the conditions required for the growth of these cells in a completely defined serum-free medium, and suggests a stratagem for the establishment of lines of human lens epithelia. Emphasis in Section II of this chapter is focused upon retinal neovascularization and on the consequences of cell proliferation in the vitreous chamber. Although the role of mitogens is emphasized, it is recognized that cell proliferation may be subject to control by natural inhibitors (Holley *et al.*, 1980), including interferon (Sreevalsan *et al.*, 1980). It is assumed that information gathered on the control of cell division in the lens will be applicable, at least in part, to understanding the mechanisms regulating mitotis in other tissues.

B. Cell Division and Differentiation in the Embryonic Lens

After adhesion of the optic vesicle to the overlying surface ectoderm, the latter is induced to form a lens placode. Subsequent morphogenetic movements convert the placode into a vesicle and eventually into an isolated lens. Initially the cells engaged in DNA synthesis and mitosis may be found throughout the very early lens placode. Eventually the cells comprising the central posterior region of the invaginating lens placode show a decrease in proliferative activity relative to the peripheral regions of the vesicle. The region that exhibits the decrease in proliferative activity is in juxtaposition to the underlying presumptive neural retina. Indeed, as development proceeds, the region of the lens facing the neural retina shows a progressive decrease in DNA synthesis and exhibits a concomitant increase in cellular elongation which heralds the formation of the prospective primary lens fibers (see Reddan, 1974; McAvoy, 1980; and McDevitt and Brahma, this volume, for reviews).

Schaper (1897) reported the localization of mitotic figures on the luminal side of the lens placode. Zwaan *et al.* (1969) have documented the relationship between DNA synthesis, mitosis, and cell shape in the developing lens placode (Fig. 1). Cells located on the basal side of the lens placode undergo DNA synthesis. Subsequent to DNA synthesis,

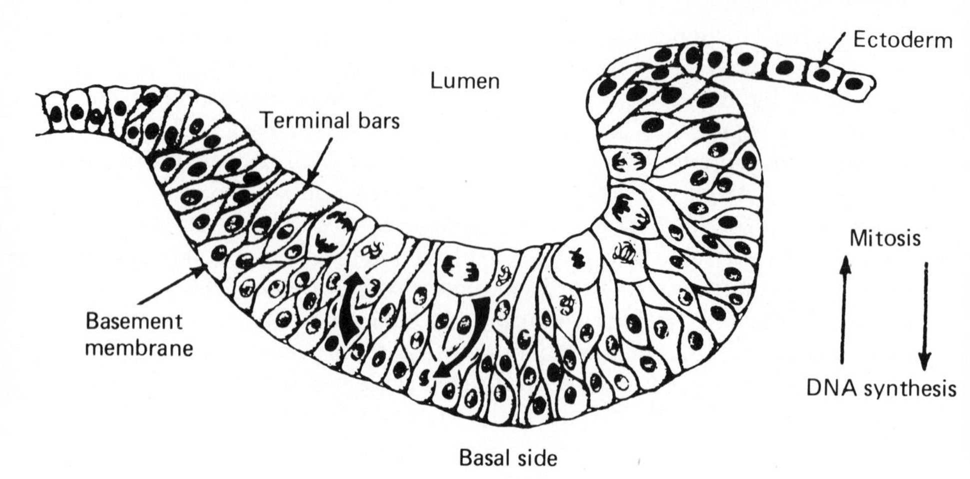

Fig. 1. Drawing of a stage-14.5 chick lens rudiment. Note that mitotic activity is confined to the cells near the lumen, whereas DNA synthesis occurs solely on the basal side of the rudiment. (From Zwaan *et al.*, 1969.)

the nuclei migrate to the apical end of the cell and the cell undergoes mitosis. After mitosis the cells elongate and the daughter nuclei return again to the basal region of the cell. As a result of nuclear migration, the genetic material is in close proximity to the underlying presumptive neural retina during the S phase of the cell cycle. This positioning might be most amenable to genetic induction or depression.

In an investigation of the process of nuclear migration in the lens placode, Pearce and Zwaan (1970) showed that antimicrotubular agents brought about an accumulation of C mitoses on the luminal side of the placode. However, Colcemid does not bring about a perceptible change in the overall architecture of the lens placode. Beebe *et al.* (1979) also report that the elongation of 6-day-old chick lens epithelia is not disrupted by treatment with antimicrotubule agents. A morphometric analysis of cell shape relative to the cell cycle has led Hendrix and Zwaan (1974) to suggest that cell volume regulation may play a role in determining the shape of embryonic lens cells. Cell death may also play a role in the morphogenesis of the eye (Silver and Hughes, 1973). The presence of junctional complexes on the luminal surface of the lens placode suggests the possible role of these structures in the process of morphogenesis (Cohen, 1961; Wrenn and Wessels, 1968).

The extracellular matrix material, known to connect the lens placode to the optic vesicle (Mann, 1950; McAvoy, 1980), has been suggested to exert an effect on the shape of the lens vesicle (Pearce and Zwaan, 1970; Hendrix and Zwaan, 1975) and to play a primary role in the morphogenesis of other tissues (Grobstein, 1954, 1962; Hay, 1978; Kefalides, 1978). In this respect, basement membranes are known to be among the first of the extracellular matrices in the developing embryo. Detailed studies on the role of basement membranes (see e.g., Foidart and Reddi, 1980) are possible, since several components have recently been purified (Timpl *et al.*, 1979; Hassell *et al.*, 1980) including type IV collagen and laminin. Thesleff *et al.* (1981) have suggested that "specific receptors on the epithelial cells react with components of the basement membrane. While these interactions endure the cells divide; when they are lost, the cells differentiate." Recent studies indicate that a laminin substrate permits the growth of rabbit lens epithelial cells in a serum-free medium (Reddan *et al.*, 1981c).

Yamada (1967) and McDevitt *et al.* (1969) studied the pattern of crystallin synthesis during amphibian lens development. A temporal relationship between DNA synthesis and the beginning of γ-crystallin synthesis was noted (Yamada, 1967). The primary lens fibers do not give a positive reaction for γ-crystallin until approximately 36 hr after the completion of DNA synthesis. Hence, in the amphibian there is a temporal relationship between termination of DNA synthesis and the synthesis of γ-crystallin. This relationship between DNA synthesis and differentiation does not appear to be ubiquitous (see Cameron and Jeter, 1971). Modak *et al.* (1968) and Zwaan and Pearce (1970) have described the pattern of cell proliferation in the developing chick lens. They have concluded that "cessation of cell replication and the onset of crystallin production are not directly related."

C. Influence of the Intraocular Environment on Cell Division and Differentiation in the Embryonic Lens

Coulombre (1969) demonstrated that the position of the lens cells in the intraocular milieu determined the locus of dividing and differentiation cells. The lens was removed from a 5-day chick embryo (Coulombre and Coulombre, 1963), reversed, and reimplanted in the eye so that the side of the lens originally facing the retina faced the cornea. Within 48 hr, the lens fibers originally adjacent to the vitreous humor stopped elongating upon exposure to the milieu of the anterior chamber. The cells of the former anterior epithelium elongated upon exposure to the vitreous humor and were activated to initiate the synthesis of proteins characteristic of the differentiated state (Genis-Galvez and Castro, 1971). Similiar results were obtained with intraocular implants of lens epithelium in *Ambystoma maculatum* (Reyer, 1977a), with intraocular implants of dorsal iris in adult *Notophthalmus viridescens* (Gulati and Reyer, 1980), in reversed regenerating lenses in adult newt (Reyer, 1977b), and in the mouse lens (Yamamoto, 1976). In all instances, the epithelium was oriented toward the cornea and the lens fibers faced the vitreous humor. In addition, the most favorable transplantation sites for the transdifferentiation of dorsal iris implants into lens were the pupillary space and vitreous chamber (Gulati and Reyer, 1980). Implants placed in the anterior chamber had a well-developed epithelium but exhibited poorly developed lens fibers.

Rotation of the chick lens also influences the location of the germinative zone, which shifts to occupy a location characteristic of the zone of proliferation in normal lenses. The cells comprising the new germinative zone form lens fibers if they migrate toward the neural retina. If they travel toward the cornea, the cells form a characteristic lens epithelium. The changes in the morphology of the reversed lens can be seen in Fig. 2.

These studies indicate that the majority of the epithelial cells of the embryonic lens possess the inherent capacity to form lens fibers if they are exposed to a suitable microenvironment. Moreover, it is equally evident that certain factors present in the posterior segment permit the initiation and continued elongation of lens fibers *in vivo*. It is also possible that factors in the aqueous humor inhibit lens fiber formation. Although it is known that other tissues can act as inducers of the lens (Jacobson, 1966), the presumptive neural retina is generally believed to exert the predominant influence. The importance of the neural retina stems from the original observation of Spemann (1905) and has been buttressed by observations from several laboratories (e.g., Wachs, 1920; Zalokar, 1944; Stone, 1957; Hasegawa, 1965; Reyer, 1971; Okada, 1977; Clayton *et al.*, 1979; Barritault *et al.*, 1981). This question is treated in more detail by Yamada, in this volume.

In addition to the influence of the intraocular environment, systemic factors have been involved in the control of cell division (Van Buskirk *et al.*, 1975) and wound healing (Rothstein *et al.*, 1976) in the adult frog

Fig. 2. An axial section from an 11-day chick embryo in which the lens was reversed at 5 days of incubation. Note that new lens fibers have developed posteriorly from the original anterior lens epithelium and that a new anterior lens epithelium has been formed. The orientation, size, shape, and position of the lens are appropriate to an 11-day nonreversed chick lens. (From Coulombre and Coulombre, 1963. Copyright 1963, by the American Association for the Advancement of Science.)

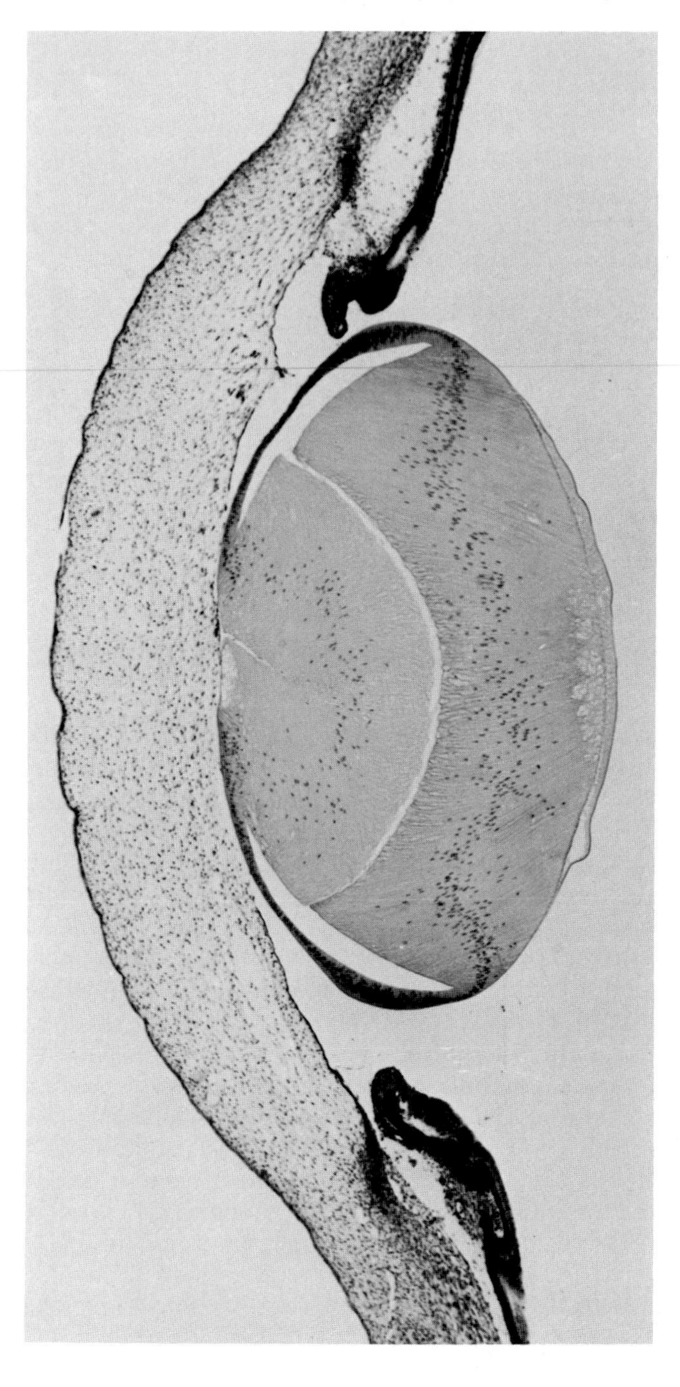

lens, and in Wolffian lens regeneration (Connelly *et al.*, 1973; reviewed by Yamada 1977; and Reyer, 1977c). Coculture of the dorsal iris epithelium of a newt with the pituitary enhances the ability of the iris epithelium to transdifferentiate into lens cells (Connelly, 1977). Cuny and Zalik (1981) have documented the effect of pituitary factors on the transdifferentiation of dorsal iris into lens. The dorsal iris was cultured in medium 199 containing 10% fetal bovine serum and various pituitary factors. Bovine thyrotropin at 30–3000 μg/ml stimulates lens formation and results in an increased mitotic index. The authors note that the response could be due to contaminating factors.

Some of the events associated with fiber formation *in vivo* occur in cultured embryonic lens epithelial cells (Mann, 1948; Philpott and Coulombre, 1965; McDevitt and Yamada, 1969; Philpott, 1970; Okada *et al.*, 1973). Recently a factor has been isolated from chick vitreous humor that stimulates elongation and synthesis of δ-crystallin in lens epithelia *in vitro* (Beebe *et al.*, 1980). Piatigorsky *et al.* (1973) described lens fiber formation in 6-day-old chick epithelial explants cultured in serum-containing medium. The morphological changes were consistent with those found *in vivo*. These investigators noted a correlation between the synthesis of δ-crystallin subunits in cultured embryonic chick lenses and the intracellular concentrations of sodium and potassium (Shinohara and Piatigorsky, 1980; Piatigorsky *et al.*, 1980; Piatigorsky, 1980). Changes in ionic fluxes have also been implicated as possible intracellular regulators of cell growth (Leffert and Koch, 1980; Rozengurt and Mendoza, 1980; Holley, 1980) and differentiation (Hennings *et al.*, 1980).

A detailed characterization of extracellular environmental factors and intracellular events associated with the expression of the crystallin genes (Zelenka and Piatigorsky, 1974, 1976; Bloemendal, 1977; Clayton *et al.*, 1979; Bhat *et al.*, 1980) should permit an understanding of the signaling mechanism involved in the regulation of cellular differentiation at the molecular level.

The amenability of the lens to studies aimed at gaining a better understanding of the control of cell differentiation has been admirably stated by Yamada (1967): "As a subject of developmental biology, the system obviously possesses an incomparable potentiality; if it were possible to gather complete information of cellular and molecular events occurring during the whole process of tissue transformation in Wolffian lens regeneration, we might be able to understand the control mechanisms of tissue specificity." The quotation, of course, can be extended to normal lens development. The ability to trigger cellular differentiation in a completely defined serum-free medium may assist in the realization of this goal.

D. Pattern of Cell Division in the Adult Lens *in Vivo*

In the early stages of lens development the epithelium is multilayered and cell proliferation occurs throughout the entire population of cells that face the cornea. As shown in Fig. 3, the adult lens epithelium consists of a monolayer lying immediately subjacent to the lens capsule. The capsule with the adhering epithelium can be removed from the lens, and all of the epithelial cells from an individual lens can be viewed on one whole-mount preparation (Howard, 1952). Analysis of the pattern of DNA synthesis (Harding *et al.*, 1959) and mitosis (von Sallmann, 1952) on whole-mount preparations indicates that cell proliferation in the adult lens is confined to a narrow band of cells in the preequatorial region of the lens referred to as the germinative zone (Fig. 3).

In the normal sequence of events, the cells in the germinative zone synthesize DNA, divide, line up in meridional rows, and subsequently differentiate into lens fibers (Hanna and O'Brien, 1961; Mikulicich and Young, 1963). The terminology used to describe various regions of the lens is shown in Fig. 3. The lens continues to grow throughout the life

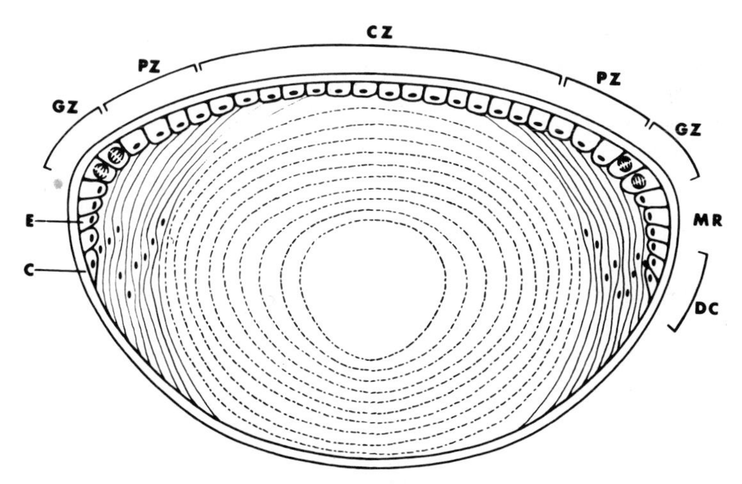

Fig. 3. Diagrammatic cross section of vertebrate lens. C, Capsule; E, epithelium; GZ, germinative zone; PZ, preequatorial zone; CZ, central zone or region; MR, meridional rows; DC, differentiating cells. (From Harding *et al.*, 1971, with permission.)

span of the animal, and the superimposition of older fibers on their younger counterparts results in a montage in which the entire life history of the organ is recorded. The cells that reside in the germinative zone, like those of the intestinal epithelium, constitute a renewal cell population (Messier and Leblond, 1960). In such a system, cells that are lost by differentiation are replaced by the process of cell division.

In marked contrast to the renewal population in the germinative zone, cells throughout the central region of the epithelium do not normally undergo mitosis but can be prompted to enter the cell cycle upon appropriate stimulation (see Harding *et al.,* 1971). These cells have withdrawn from the cell cycle and exist in a protracted G_1 period, a condition that has been called the G_0 state by some investigators (Lajtha, 1963). Pardee (1974) suggests that term "restriction point" or "R point" to describe cells that are arrested in G_1. The fact that the cells in the central epithelium are effectively noncycling and arrested in G_1 phase of the cell cycle was originally reported by Harding *et al.* (1959) and has been confirmed by numerous investigators, e.g., Reddan and Rothstein (1965), Rafferty and Smith (1976), and Treton and Courtois (1981).

The events associated with the release of cells from G_1 are of utmost significance, since an understanding of the mechanism that controls cell division is predicated on a knowledge of the cellular events that regulate transit into the cell cycle. The transit times for the various subphases of the cell cycle of the lens epithelium in several species are listed in Harding *et al.* (1971). The cell cycle times for $S + G_2 + M$ for the lens epithelium are comparable to those for other cells (Prescott, 1976), with G_1 exhibiting the greatest variability.

An understanding of the events regulating the entrance of cells into the cell cycle may also define the role of mitosis in cataractogenesis. In this respect certain hereditary and nutritional cataracts are known to exhibit hyperactive mitosis (Grimes and von Sallmann, 1968; Zwaan and Williams, 1968; Gorthy and Abdelbaki, 1974; Robb and Marchevsky, 1978; Gorthy, 1979). A disorientation of mitotic figures in the early lens rudiment of aphakic mice has also been reported (Zwaan and Kirkland, 1975). Moreover, changes in the normal pattern of cell division or in the orderly formation of lens fibers accompany x-ray cataracts (Worgul and Rothstein, 1975), are implicated in Marfan's syndrome (Farnsworth *et al.,* 1977), and typify anterior polar cataracts in humans (Ogata and Matsui, 1972; Font and Brownstein, 1974; Dilley *et al.,* 1976).

Mitotic activity in the lens *in vivo* varies as a function of age, undergoes seasonal and diurnal variation, and is subject to hormonal manipulation. The total number of mitotic figures in the rabbit (von Sallmann *et al.*, 1957) and rat lens (Cotlier, 1962; Treton and Courtois, 1981) decreases as a function of age. The number of cells undergoing DNA synthesis also diminishes with increasing age (von Sallmann and Grimes, 1966; Treton and Courtois, 1981). The influence of age on the mitotic index in lenses of male rats is shown in Fig. 4. Investigators have also noted a decrease in the total number of epithelial cells per unit area in the central region of the epithelium as a function of age (von Sallmann, 1952; Treton and Courtois, 1981). An increase in the number of cytoplasmic filaments has been reported in the rat lens with increasing age (Rafferty *et al.*, 1979). A reduction in cell height, karyopycnosis, elaborate endoplasmic reticula, and an increase in the frequency of vacuoles and multilamellar bodies are associated with aging in the rat lens epithelium (Gorthy, 1978).

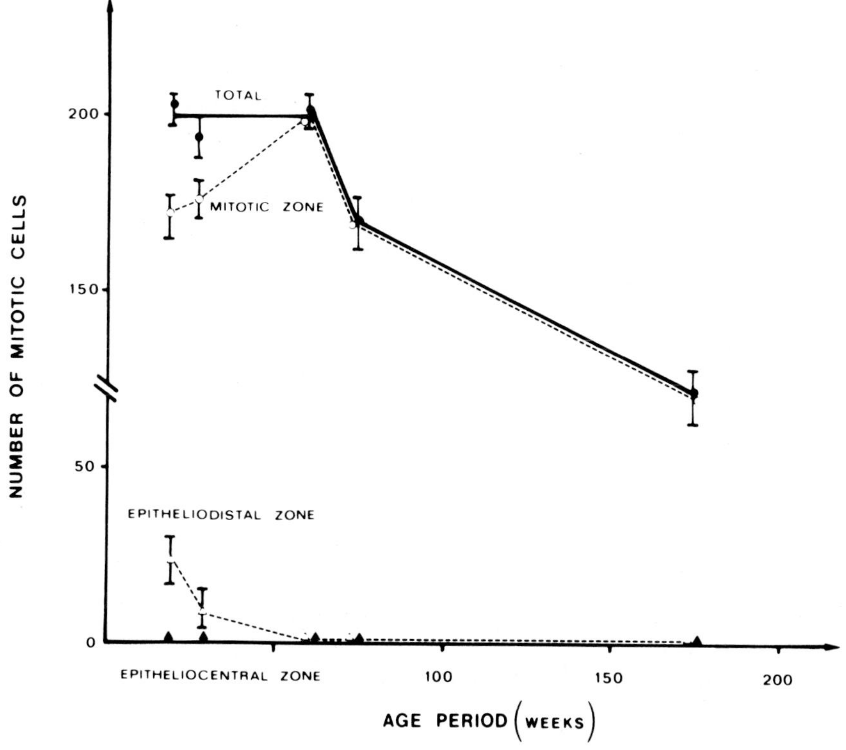

Fig. 4. Total number of mitoses in each zone of the rat lens as a function of age. (From Treton and Courtois, 1981, with permission.)

Mitosis in rat (von Sallmann and Grimes, 1966), rabbit (von Sallmann *et al.,* 1962), and mouse lens exhibits diurnal variation. The counts are highest in the early morning hours and lowest at midnight. Mitosis is also subject to seasonal variation. The mitotic index in the rabbit lens is highest in the summer and decreases throughout fall, winter, and spring (Harding *et al.,* 1971). Mitotic activity in the frog (Golichenkov *et al.,* 1966; Rosenbaum and Rothstein, 1972) and toadfish (Reddan *et al.,* 1976b) lens also exhibits seasonal variation. The toadfish lens epithelium is essentially amitotic from January to May. Mitosis commences in June and remains elevated from June through August. The seasonal variation in mitosis is thought to be regulated by factors produced by the pituitary gland (Van Buskirk *et al.,* 1975).

Response to Injury and Wound Healing in the Lens in Vivo

The cells in the central region of the epithelium, which have effectively withdrawn from the cell cycle, can be prompted to divide as a result of mechanical (Harding and Srinivasan, 1961; Rafferty, 1963; Reddan and Rothstein, 1966), chemical (Weinsieder *et al.,* 1975), or immunological (Worgul and Merriam, 1979) insult. The cells that enter mitosis subsequent to injury replace the damaged lens cells and play an integral role in the reestablishment of transparency. Repeated paracentesis of the anterior chamber (Harding *et al.,* 1971) or x-irradiation of the eye (von Sallmann *et al.,* 1955) also triggers a transient increase in mitosis in the normally amitotic region of the epithelium. Opacities associated with certain mechanical or chemical insults are either partially or completely reversible (Rafferty, 1973; Rafferty and Goossens, 1977; Unakar *et al.,* 1973, 1979) and are being used to probe the cellular events required for the restoration of transparency in the traumatized lens.

Injury to the anterior polar region of the rabbit (Harding *et al.,* 1959; Harding and Srinivasan, 1960), rat (Riley and Lindgren, cited in Harding *et al.,* 1971), mouse (Rafferty, 1973), sea bass (Rothstein and Harding, 1962), toadfish (Reddan *et al.,* 1975), or frog lens (Reddan and Rothstein, 1965; Rothstein *et al.,* 1965; Rafferty, 1965, 1967) initiates a sequence of events which triggers DNA synthesis and mitosis in the lens epithelium. In all instances, the cells must synthesize DNA before entering mitosis.

In the rabbit (Harding and Srinivasan, 1961), the proliferative response is propagated outward from the site of injury (Fig. 5), a situation not realized in the other animals. In mammalian lenses, DNA synthesis commences at approximately 14 hr after insult, a tenure that

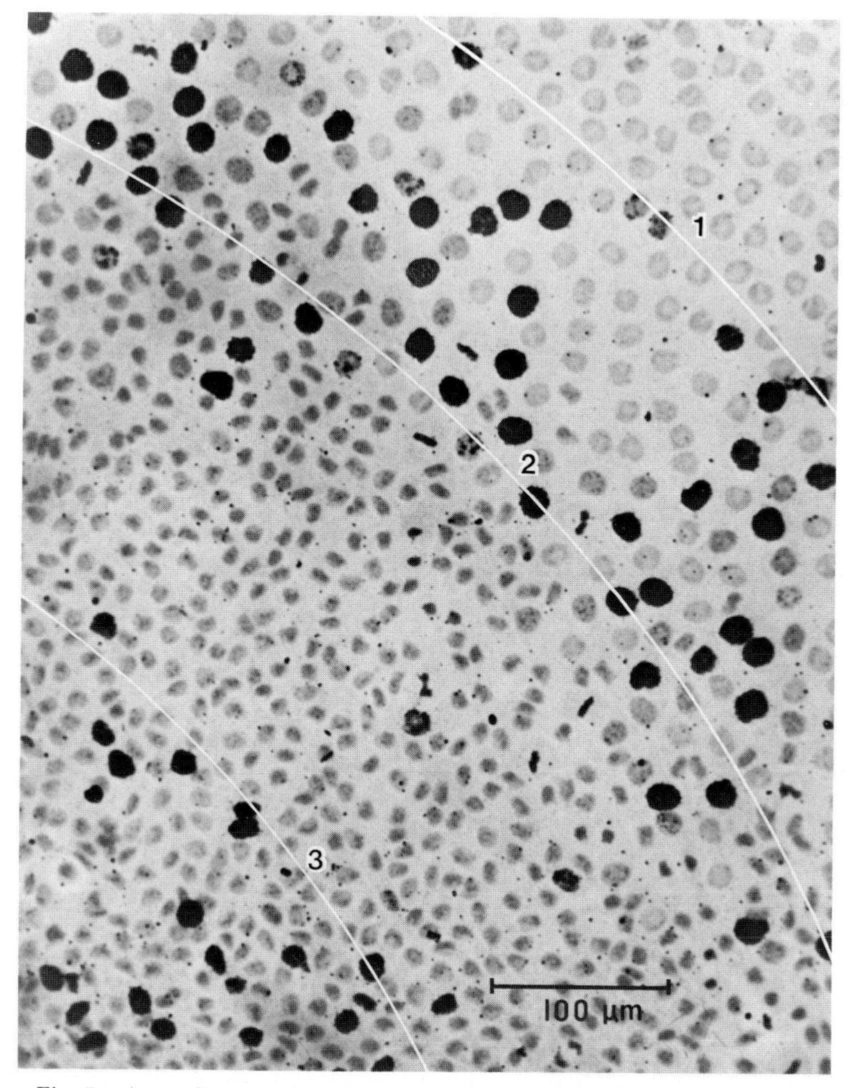

Fig. 5. Autoradiogram of a 44-hr injury. [^{3}H]thymidine was injected into the anterior chamber 2 hr before the lens was fixed. The site of injury is out of field toward the lower left. Note that the cells situated near the injury (left of line 2) appear relatively small and that their concentration is relatively high compared with the cells which lie further out (right of line 2). Mitotic figures are situated along the border between the large and small cells (line 2). The majority of radioactive nuclei lie external to the zone of mitotic figures (between lines 1 and 2). The beginning of a second wave of thymidine incorporation is evident at the lower left (left of line 3). (From Harding and Srinivasan, 1961, with permission.)

is significantly longer in poikilotherms. This is due in part to the fact that each of the subphases of the cell cycle is significantly longer in cold-blooded animals (Reddan and Rothstein, 1966). Injury-induced cell proliferation is preceded by increments in RNA and protein synthesis (Rothstein *et al.,* 1966), is curtailed by agents which inhibit genetic transcription and translation (Rothstein *et al.,* 1967), and is thought to be dependent on prior gene activation (Weinsieder *et al.,* 1973; Briggs *et al.,* 1977).

E. Initiation of Cell Division in the Lens in Organ Culture

1. Response of the Lens Epithelium to Serum and Plasmoid Aqueous Humor

Many injuries which trigger hyperplasia in the lens are associated with changes in the composition of the aqueous humor; such changes also accompany other forms of ocular insult, including x-irradiation and immunogenic uveitis (Worgul *et al.,* 1977). Thus the environment of the healing or traumatized lens differs from that of its noninjured counterpart, showing among other things an increase in the level of plasma proteins (von Sallmann and Moore, 1948; Reddy, 1968; Weinsieder *et al.,* 1975), prostaglandins (Eakins, 1973; Bito, 1974), and phospholipids (Varma and Reddy, 1972).

Another common feature of ocular injury is the influx of blood cells, including lysosome-rich leukocytes, into the anterior chamber (Unakar *et al.,* 1973; Rafferty and Goossens, 1977). Acid phosphatase is found at extracellular sites in the damaged lens and is thought to play a primary role in the removal of tissue debris (Unakar *et al.,* 1975). The reaction product for acid phosphates has been detected at extracellular sites in precataractous rat lenses (Gorthy, 1978). However, in this instance the hydrolases appear to degrade the plasma membranes, which leads to lens degeneration and opacification (Gorthy, 1978). The role of hydrolytic enzymes in the generation of senile cataract (Gorthy, 1978) and the entire mechanism responsible for the removal of debris from traumatized lenses requires extensive study.

Following injury, cellular debris and fibrin are found in the immediate vicinity of the wound and are part of the wound coagulum (Unakar *et al.,* 1973; Rafferty and Goossens, 1977). The occurrence of fibrin in the coagulum implies the presence of the entire cascade of enzymes involved in the blood-clotting process. Thrombin, one of the serine proteases can stimulate cell division in the cultured mammalian lens (Reddan *et al.,* 1976a). Lens epithelia also contain an endogenous serine protease, plasminogen activator (Reddan *et al.,* 1981).

In view of the numerous changes that occur in the environment of the injured lens *in vivo,* attempts were made to study the injury response under standardized conditions in organ culture. It was found that exposure of the cultured lens to a serum-containing medium engendered a highly reproducible stimulation of DNA synthesis and mitosis throughout the central region of the lens epithelium (Harding *et al.*, 1962; Harding and Thayer, 1964; Reddan and Rothstein, 1966). In the absence of serum, the mitotic pattern is ostensibly similar to that noted in the noninjured lens *in vivo.* The typical pattern of DNA synthesis and mitosis in rabbit lenses exposed to a serum-containing medium is shown in Fig. 6. Proliferation is preceded by heightened levels of RNA and protein synthesis, is accompanied by an increase in the number of ribosomes (Reddan *et al.*, 1970b), and is curtailed by

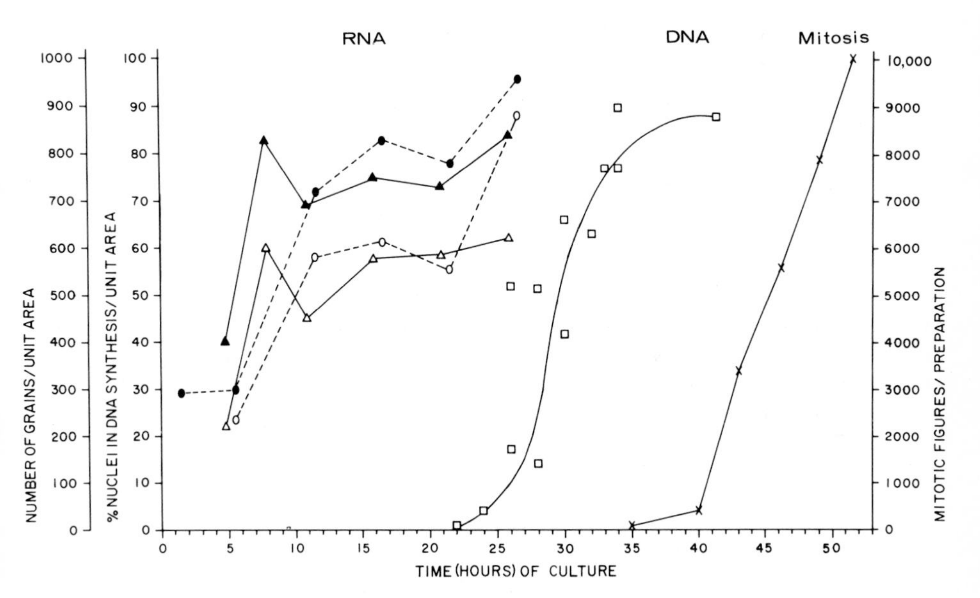

Fig. 6. Temporal relationships among RNA synthesis, DNA synthesis, and mitosis, and the effect of actinomycin D (2.4×10^{-8} *M*) on RNA synthesis in the epithelium of cultured rabbit lenses. Lenses were exposed to [^{3}H]uridine for the final 90 min (experiment 1) or 45 min (experiment 2) of culture, or incubated in [^{3}H]thymidine for periods between 21 and 40 hr, or assayed for mitosis at various intervals. Some lenses were exposed to actinomycin D from 0 to 4 hr and subsequently exposed to [^{3}H]uridine for the last 45-90 min of culture. Each point represents the mean value obtained from at least three lenses. ×, Mitotic figures per preparation: □, percentage of nuclei in DNA synthesis per unit area; ●, grain counts per unit area [^{3}H]uridine, experiment 1; ○, grain counts per unit area [^{3}H]uridine (actinomycin D,0–4 hr), experiment 1;▲, grain counts per unit area [^{3}H]uridine, experiment 2; △, grain counts per unit area [^{3}H]uridine (actinomycin D, 0–4 hr), experiment 2. (After Reddan *et al.*, 1970a, with permission.)

agents that block transcription and/or translation (Rothstein *et al.*, 1966; Reddan *et al.*, 1970a).

The importance of the aqueous environment in lens clarity, growth, and repair stems from the early studies of Boeve (1927) and Bakker (1936a,b). Despite the fact that the composition of the aqueous humor changes following injury to the eye, the role of the aqueous humor in repair of the injured lens has received little attention. In order to gain insight into the possible role of the aqueous humor in the regulation of mitosis in the injured lens, lenses from nontraumatized eyes were exposed to samples of aqueous humor collected after a variety of ocular insults (Weinsieder *et al.*, 1976; Reddan *et al.*, 1979).

The plasmoid aqueous humor which forms subsequent to chemical injury triggers mitosis in the cultured lens (Weinsieder *et al.*, 1976). This reflects a general phenomenon, since other types of ocular insult yield aqueous humors (Table I) which promote lenticular cell division (Reddan *et al.*, 1979). The increase in mitosis is primarily due to an activation of mitosis in the central region of the lens.

Since plasmoid aqueous humor contains all the major constituents of serum (Unger *et al.*, 1975) and since serum induces mitosis in the

TABLE I

Mitogenic Effect of Posttraumatic Aqueous Humors on Cultured Lenses[a]

Treatment preceding aqueous humor collection	Protein in aqueous humor (mg/ml)	Protein in culture medium (mg/ml)	Mitotic figures
Drainage of primary aqueous humor	33 ± 9 (86)	2	7720 ± 2248 (9)
Mechanical injury	27 ± 8 (18)	2	2890 ± 1090 (4)
X-irradiation	9 ± 4 (17)	2	5973 ± 964 (5)
Intravitreal injection of BSA (3 days)	6 ± 2 (14)	2	4898 ± 1958 (4)
Intravitreal injection of BSA (3 days)	14 ± 10 (14)	2	2970 ± 1223 (6)
None (control)	—	2 (RSA)	43 ± 28 (5)
None (control)	—	0	513 ± 69 (6)

[a] Lenses were cultured for 20 hr in medium KEI-4 containing samples of aqueous humor collected after each type of ocular injury and then maintained in medium KEI-4 alone and fixed at 52 hr; then the total number of mitotic figures per whole-mount preparation was counted. Lenses cultured in KEI-4 alone or in KEI-4 supplemented with rabbit serum albumin (RSA) served as controls. The values represent the mean plus or minus the standard deviation (*n* is shown in parentheses). (From Reddan *et al.*, 1979, with permission.)

organ-cultured lens, the relative proliferative potentials of plasmoid aqueous humor and serum were compared. The aqueous humor used in these experiments (Reddan *et al.,* 1979) was collected 15 min after initial drainage of the anterior chamber. Lenses were incubated in medium containing either serum or secondary aqueous humor at equivalent concentrations of protein. The mitogenicity of the aqueous humor exceeded that of serum. Lenses exposed to secondary aqueous humor also exhibited an elevated level of [^{3}H]thymidine incorporation relative to those exposed to serum.

The timing of the mitotic response brought about by plasmoid aqueous humor is in agreement with that engendered by serum or insulin (Reddan *et al.,* 1970a, 1975). The lens need be exposed to the plasmoid aqueous humor for only the initial hour of culture in order to trigger mitosis in regions of the lens epithelium that are normally amitotic. The magnitude of the mitotic response at 52 hr of culture appears to depend upon the time of exposure of the cultured lens to the plasmoid aqueous humor. Lenses exposed to the plasmoid aqueous humor for only the initial hour of culture had fewer mitotic figures than lenses exposed to plasmoid aqueous humor throughout the first 20 hr of culture. In other investigational systems, the number of cells prompted to enter the cell cycle has been shown to depend on the time of exposure of the cells to specific mitogens (Leffert and Koch, 1977). The stimulation of mitosis in the cultured lens in response to a brief exposure to a mitogen is not without precedent, since a short exposure of the lens to trypsin (Reddan *et al.,* 1976a) or human thrombin (Reddan *et al.,* 1981) can trigger cell division in the cultured lens.

These findings suggest that plasmoid aqueous humor may play an important role in inducing or sustaining the mitotic response which precedes wound healing in the traumatized lens *in vivo.*

2. *Stimulation of Cell Division by Insulin in Lenses Cultured in a Completely Defined Serum-Free Medium*

It is of obvious interest to determine the factors in plasmoid aqueous humor which are capable of triggering mitosis in the lens. However, the addition of aqueous humor or serum to the medium exposes the lens to a highly complex mixture of growth-promoting and -inhibiting factors. Attempts were made therefore to obtain a reproducible stimulation of mitosis in lenses cultured in a completely defined medium. The addition of crystalline insulin to medium 199 stimulates cell division in both mammalian and amphibian lenses (Reddan *et al.,* 1972). Insulin also initiates cell proliferation in rabbit lenses cultured in medium KEI-4, a serum-free medium which mimics rabbit aqueous humor (Kinsey and Reddy, 1965; see Reddan *et al.*, 1975, for formulation). Lenses cultured in KEI-4 alone exhibit mitosis in the peripheral region of the epithelium. The mitotic response evoked by insulin has a time course that parallels that elicited by serum or plasmoid aqueous humor (Reddan *et al.*, 1970a; Reddan and Wilson, 1978). Changes in cell membranes and an increase in ribosomes precede the insulin-induced mitotic response (Reddan and Unaker, 1976).

The response engendered by insulin is dependent on the integrity of the insulin molecule. Neither the A chain, the B chain, nor the combined A and B chains trigger mitosis. The requirement for the intact hormone suggests that lens cells possess an insulin receptor. Cultured mammalian lens epithelia have receptors for insulin, epidermal growth factor (EGF), and fibroblast growth factor (FGF) (Hollenberg, 1975; Gospodarowicz *et al.*, 1977). The concentration of insulin used in the above-noted studies exceeds the known values for the hormone in either primary or plasmoid aqueous humor (Coulter and Knebel, 1977; Coulter *et al.*, 1980). This suggests that molecules that are structurally or functionally homologous to insulin might represent the type of factor capable of regulating cell division in the lens *in vivo* and *in vitro*.

F. Stimulation of Cell Division in the Lens Epithelium by Insulin-like Growth Factors *in Vivo* and *in Vitro*

1. Mitogenicity of Insulin-like Growth Factors, Somatomedin C, and Epidermal Growth Factor on the Cultured Mammalian Lens

Certain polypeptides which exhibit insulin-like activity are potent mitogens (Van Wyk *et al.*, 1973; Bomboy and Salmon, 1975; Rinderknecht and Humbel, 1976). The insulin-like growth factors, which include insulin growth factors (IGF-I and IGF-II) and the somatomedins, are immunologically distinct from insulin, exhibit structural and functional homologies with insulin (Froesch *et al.*, 1979), and may bind to insulin receptors (Hintz *et al.*, 1972). The primary structures of IGF-I and IGF-II have been determined (Rinderknecht and Humbel, 1978a,b). The IGFs, which are the most highly purified of the insulin-like growth factors, show extensive cross-immunological reactivity with certain somatomedins (Phillips and Vassilopoulou-Sellin, 1980a,b). In addition to the possible involvement of IGFs in the control of mitotic activity in the lens, EGF is known to induce precocious eyelid opening in mice and to stimulate hypertrophy and hyperplasia

TABLE II

Comparison of the Effect of Insulin and IGF on Cell Division in the Cultured Rabbit Lens[a]

Lens pairs	Total number of mitotic figures[b]		IGF/ insulin
	Insulin	IGF	
1	2028	8816	4.4
2	1837	7936	4.3
3	2161	9724	4.5
4	1594	7916	4.5

[a] Lenses were cultured in KEI-4 containing insulin or IGF at 8.8×10^{-10} *M*.

[b] Lenses were fixed after 52 hr of culture, and the total number of mitotic figures per whole mount was counted. (From Reddan and Wilson-Dziedzic, 1982, with permission.)

of corneal cells in organ culture (Carpenter and Cohen, 1978). Recent studies indicate that IGFs and EGF initiate mitosis in lenses cultured in a serum-free medium (Reddan and Wilson, 1978, 1979; Reddan and Wilson-Dziedzic, 1982; and unpublished observations).

As shown in Table II, IGFs and/or somatomedin C (Fig. 7) are more mitogenic toward the rabbit lens than insulin. The addition of IGF to the medium is followed by a stimulation of DNA synthesis and mitosis throughout the central epithelium (Fig. 8), a situation not realized in lenses cultured in the absence of the mitogen. The mammalian lens responds to IGF-I, IGF-II (Reddan and Dziedzic, unpublished observations), and EGF in a dose-dependent manner (Figs. 9 and 10). Furthermore, the total number of mitotic figures present in lenses exposed to both EGF and IGFs exceeds that realized by equimolar quantities of either polypeptide (Table III).

2. *Regulation of Cell Division in the Adult Amphibian Lens by Insulin-like Growth Factors* in Vivo

Hypophysectomy totally curtails DNA synthesis and mitosis in the frog lens *in vivo* and prevents the mitotic activation normally associated with mechanical insult (Van Buskirk *et al.*, 1975; Rothstein *et al.*, 1976; Hayden and Rothstein, 1979). Figure 11 shows the decline in mitotic activity in the injured frog lens as a function of time after hypophysectomy. If lenses are injured 4 weeks after hypophysectomy, DNA synthesis and mitosis do not occur. However, cell proliferation can be restored in the hypophysectomized animal if replacement

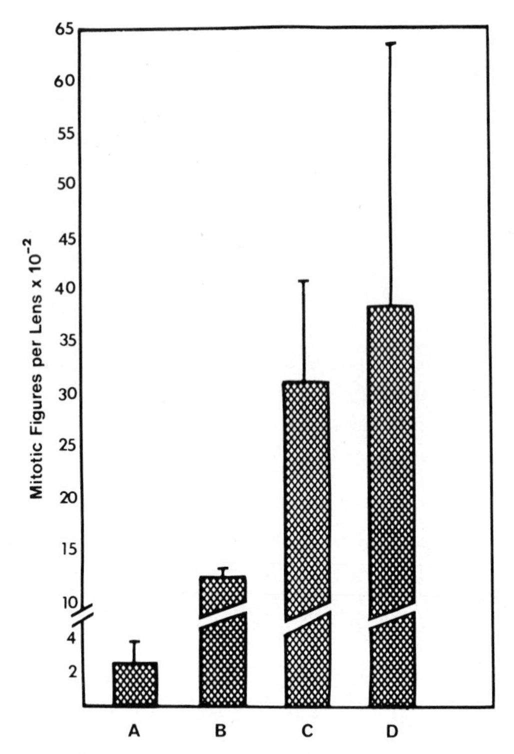

Fig. 7. Total number of mitotic figures in rabbit lenses cultured for 52 hr in medium KEI-4 (A) or in KEI-4 containing bovine insulin (B), IGF (C), or somatomedin C (D), all at 4.66 ng/ml. (Reddan and Dziedzic, to be published.)

therapy is initiated with either frog pituitary powder or bovine somatotropin. Hypophysectomy also exerts a radioprotective effect on the frog lens (Hayden *et al.,* 1980) and prevents x-ray cataracts (H. Rothstein, B. V. Worgul, C. Medvedovsky, and G. R. Merriam, personal communication). Worgul and Rothstein (1975) have proposed that "lens cell damage from ionizing radiation is transduced by mitosis and expressed as an opacity by subsequent errant fiber genesis."

The concept that the pituitary acts indirectly on various tissues stems from the original observation of Salmon and Daughaday (1957), who hypothesized that the enhancement of cartilage growth brought about by growth hormone (somatotropin) was mediated by a "circulating factor" which triggered cell proliferation. Mitogens known to be partially regulated by growth hormone include certain somatomedins and the insulin-like growth factors. The pituitary gland is thought to facilitate growth by mediating the synthesis and/or release of insulin-

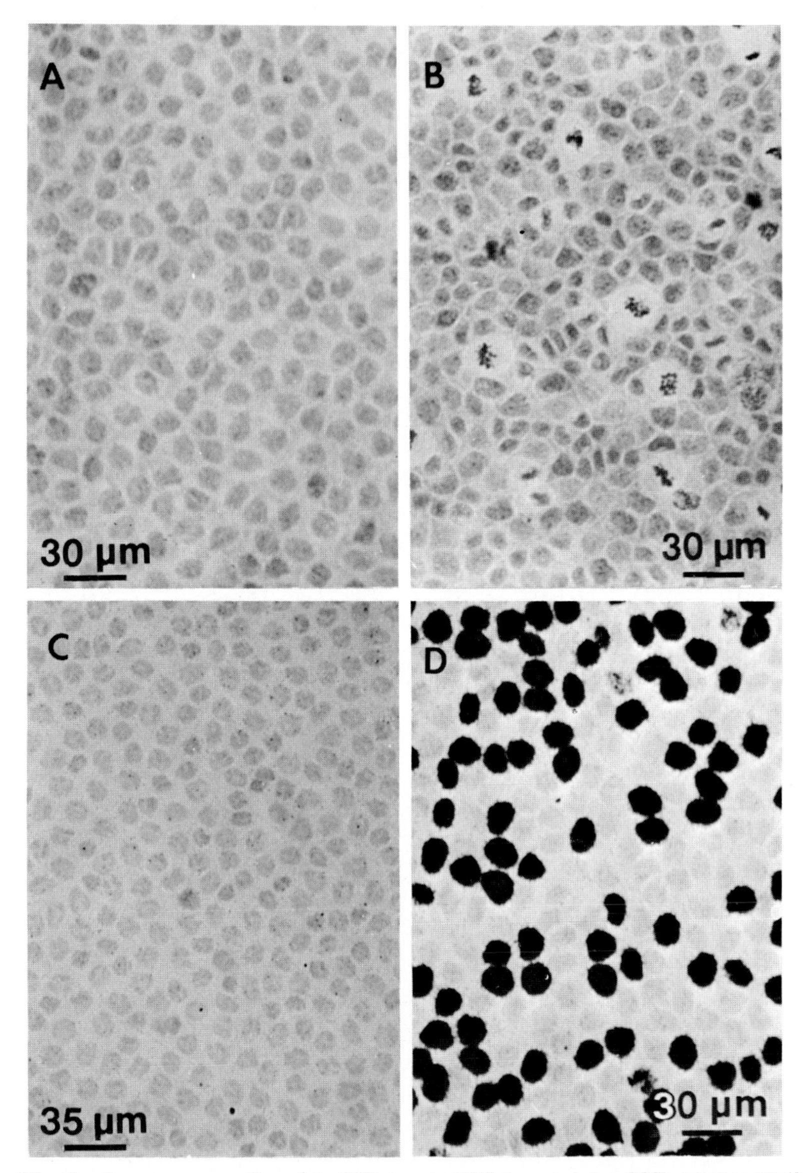

Fig. 8. Lenses were cultured in KEI-4 or in KEI-4 containing IGF at 8.8×10^{-10} *M*. Lenses were fixed at 52 hr and examined for mitosis or were exposed to [^{3}H]thymidine from 30–36 hr of culture and prepared for autoradiography. Central epithelium from a lens cultured in KEI-4 alone (A) or in KEI-4 plus IGF (B). (C) Autoradiogram of the central epithelium from a lens cultured in KEI-4 alone. (D) Autoradiogram of the central epithelium from a lens cultured in KEI-4 plus IGF.

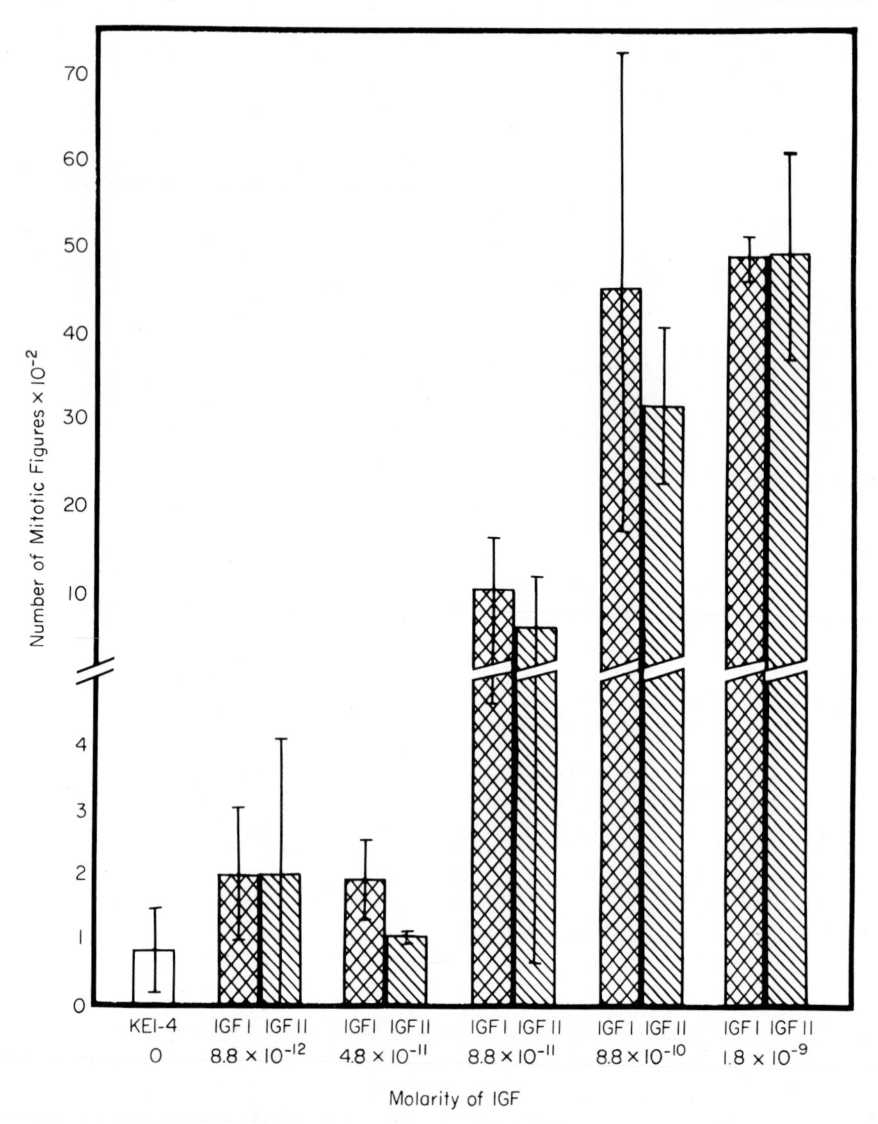

Fig. 9. Rabbit lenses were cultured in KEI-4 containing various concentrations of IGF-I or IGF-II. Lenses were fixed at 52 hr, and the total number of mitotic figures per lens was determined. (Reddan and Dziedzic, to be published.)

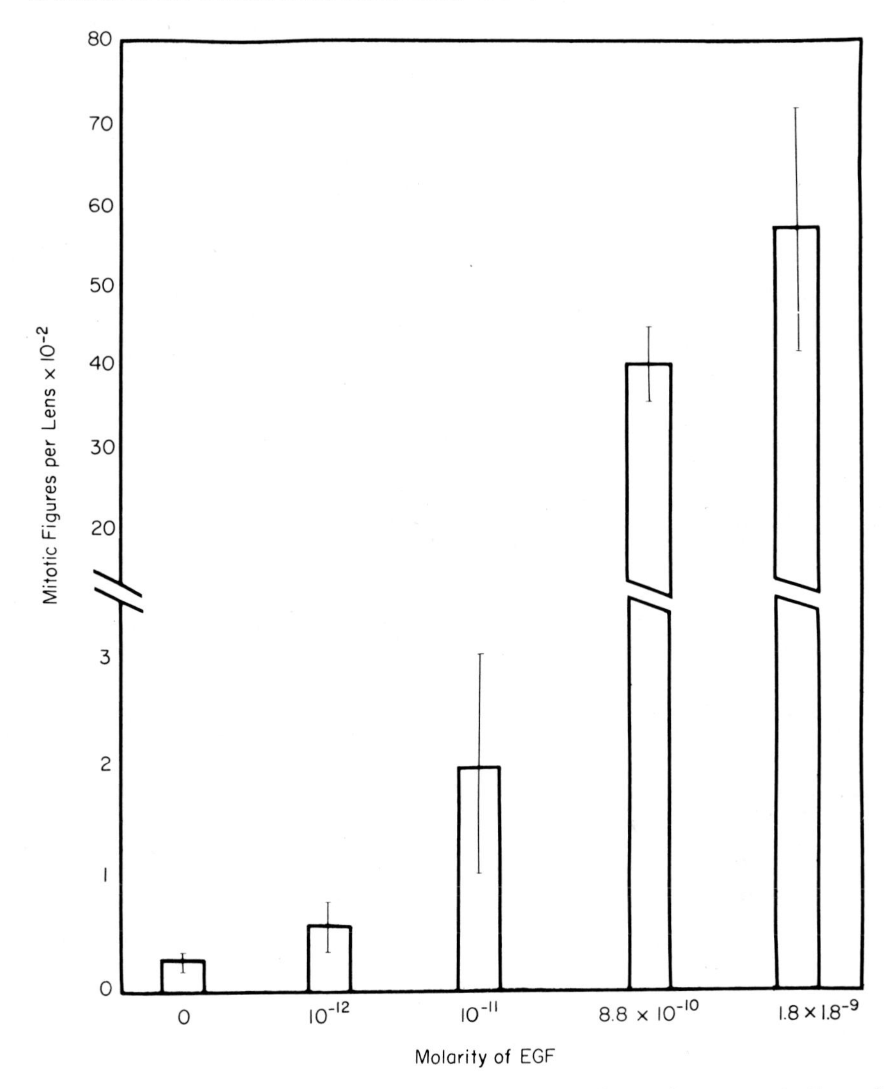

Fig. 10. Rabbit lenses were cultured in KEI-4 containing various concentrations of EGF. Lenses were fixed at 52 hr, whole mounts of the epithelium were prepared, and the total number of mitotic figures per lens was counted. (From Reddan and Wilson-Dziedzic, 1982, with permission.)

TABLE III

Synergistic Effect of IGF plus EGF on Cell Division in the Cultured Rabbit Lens

Lens pair	Total number of mitotic figures[a]		
	2 × EGF[b]	EGF + IGF[b]	(EGF + IGF)/(2 × EGF)
1	7,127	17,195	2.41
2	3,910	13,864	3.55
3	6,183	16,768	2.71

[a] Lenses were fixed after 52 hr of culture, and the total number of mitotic figures was counted.

[b] Lenses were cultured in KEI-4 containing EGF and IGF with both factors at 8.8×10^{-10} *M*, or in KEI-4 containing EGF at 1.8×10^{-9} *M* (2 × EGF). (From Reddan and Wilson-Dziedzic, 1982, with permission.)

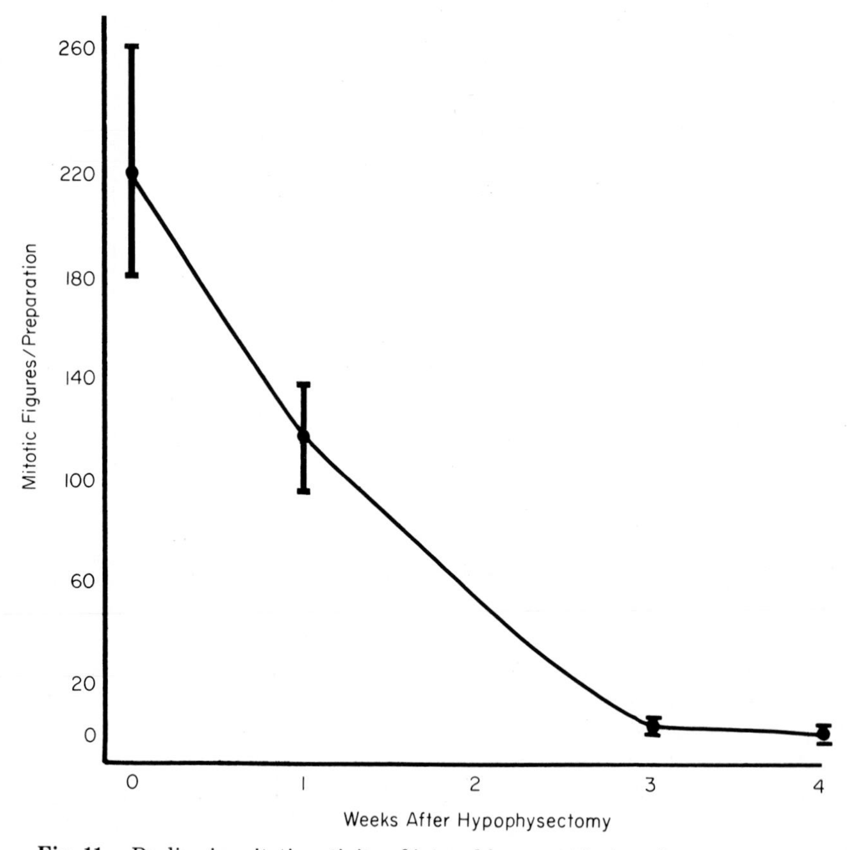

Fig. 11. Decline in mitotic activity of injured lenses with time. Lenses were injured and fixed 4 days later. Six *Rana catesbiana* were used at each time interval. (From Rothstein *et al.*, 1976, with permission.)

like growth factors from other tissues (see Phillips and Vassilopoulou-Sellin, 1980a,b, for a detailed discussion).

Relative to the ocular lens, the administration of bovine, human, or frog growth hormone, frog pituitary powder (Wainwright *et al.,* 1976), or thyroid hormones (thyrotropin, thyroxine, triiodothyronine) (Weinsieder and Roberts, 1979, 1980) reinitiates cell proliferation in the lens epithelium of the hypophysectomized frog. However, these hormones do not appear to stimulate cell division in the amphibian lens in organ culture. Moreover, while serum from hypophysectomized frogs is not mitogenic toward the frog lens in organ culture, mitogenicity is restored to the serum following the initiation of replacement therapy with bovine growth hormone (Wainwright *et al.,* 1978).

The question remained as to whether administration of an insulin-like growth factor could initiate cell division in the amphibian lens *in vivo.* As shown in Fig. 12, administration of somatomedin C to the hypophysectomized froglet restores DNA synthesis and mitosis in the lens (Rothstein *et al.,* 1980). Administration of somatomedin to Snell dwarf mice also results in a restoration of overall growth comparable to that elicited by growth hormone (van Buul-Offers *et al.,* 1979, cited in Philips and Vassilopoulou-Sellin, 1980a). These findings on the ocular lens constitute the first evidence that a growth hormone-dependent mitogen can reinstate growth in an organ that has been completely stopped in the animal. As noted in Section I,F,3, pituitary factors have also been implicated in amphibian lens regeneration and in the transdifferentiation of newt iris into lens.

3. *Do Insulin-like Growth Factors Regulate Cell Division in the Mammalian Lens* in Vivo?

Pituitary ablation brings about a slight decrease in the growth of the rat lens (Kuck, 1970). An increase in the number of dividing cells and a concomitant decrease in the number of C mitoses occurs in the lens of the hypophysectomized rat (Cotlier, 1962). These findings might indicate that insulin-like growth factors do not influence cell division in the mammalian lens. However, the level of insulin-like growth factors in mammals is not determined solely by the pituitary but is subject to regulation by both insulin and adequate levels of nutrition (Phillips and Vassilopoulou-Sellin, 1980b). The effect of growth hormone on somatomedin generation is abolished in hypophysectomized rats fed a low-protein diet (Takano *et al.,* 1980). Moreover, following removal of the pituitary, hyperphagic children grow and maintain normal levels of somatomedin. In the mammal, insulin and adequate nutrition are more important in the generation of insulin-like growth factors than

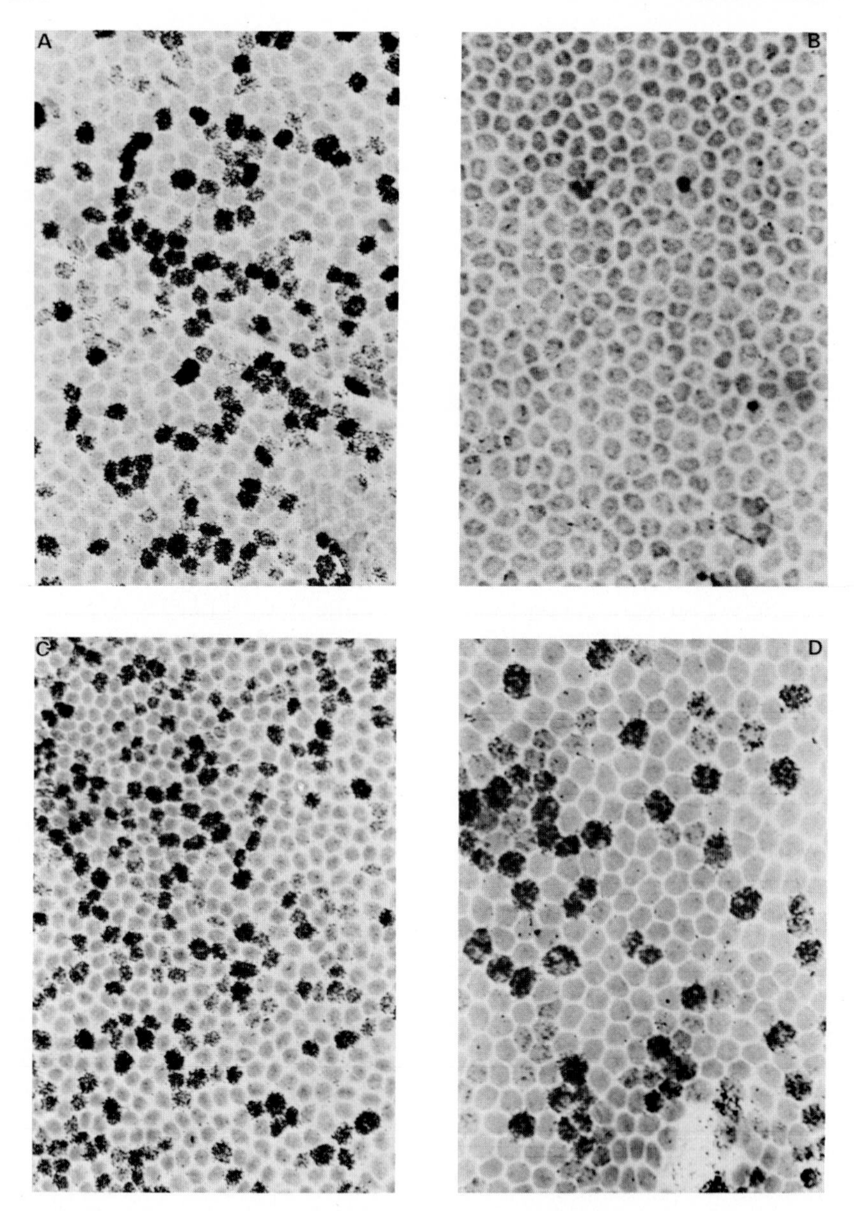

Fig. 12. Autoradiograms of lens epithelial whole mounts. (A) Lens epithelium from a normal, intact, postmetamorphic bullfrog that had been injected with Earle's balanced salt solution and [^{3}H]thymidine. (B) A similar preparation from a frog that had been hypophysectomized and had received injections of isotope for 2 weeks. None of the cells show evidence of DNA synthesis or mitosis. (C) Lens epithelium from an animal that was hypophysectomized and later injected with human growth hormone; labeling is evident.

growth hormone itself. A. Weinsieder (unpublished observations) found a 40% reduction in the number of cells in DNA synthesis in lenses from hypophysectomized monkeys relative to age-matched controls. This was accompanied by a 90% decrease in the circulating level of somatomedins. H. Rothstein and C. Nooriley (unpublished observations) noted a reduction in mitosis in lenses of rats that were hypophysectomized and starved. Although cultured mammalian lens epithelia and amphibian lens epithelia *in vivo* divide in response to insulin-like growth factors, the effect, if any, of these factors on the mammalian lens *in vivo* remains to be demonstrated.

G. Environmental Factors Regulating Cell Division in Lens Cells in Tissue Culture

Lens epithelia in organ culture, wherein the cells lie subjacent to their natural substrate, i.e., the lens capsule, undergo mitosis if a single growth factor is added to a serum-free medium (Reddan *et al.*, 1972, 1975, Reddan and Wilson, 1979). This suggests that it should be possible to establish lines of lens epithelia in a completely defined, serum-free medium. Furthermore, tissue culture has an added advantage since it permits investigation of the interaction among cells, substrate, and growth factors. In addition, the use of cell hybrids should provide insight into the specific chromosomes involved in the expression of lens proteins (Church, 1980).

The initial attempt to culture lens cells was made by Fischer in 1922. Kirby (1927; Kirby *et al.*, 1929, 1932) initiated a series of studies that delved into "the nature of the cause of senile cataract and the possibility of its prevention by non-surgical means." These studies were followed by those of Mann (1948) who maintained mouse cells *in vitro* for approximately 10 days.

The conditions required for the establishment of lens epithelial cell lines from older animals have been reported by several investigators (Tamura 1965; Shapiro *et al.*, 1969; Gospodarowicz *et al.*, 1977; Russell *et al.*, 1977; Courtois *et al.*, 1978; Taylor-Papadimitriou *et al.*, 1978;

(D) Lens epithelium from a hypophysectomized frog that received injections of purified somatomedin C starting on day 75 after hypophysectomy. This animal received a total of ten 700-ng injections and was killed on day 88 after the operation; note the mitotic figures and evidence of incorporation of [^{3}H]thymidine. Cells manifesting these phenomena were confined to the germinative zone in the tissue shown in (D). The percentages of labeled cells were: (A) 37.7, (B) 0.0, (C) 39.7, and (D) 24.4. (From Rothstein *et al.*, 1980. Copyright 1980, by the American Association for the Advancement of Science.)

Hamada *et al.*, 1979; Rink and Vornhagen, 1980; Masterson and Russell, 1981). However, lens epithelial cells from younger animals of several species exhibit very limited growth in tissue culture (Albert *et al.*, 1969; Philpott, 1970; Piatigorsky and Rothschild, 1972; Creighton *et al.*, 1976; Taylor-Papadimitriou *et al.*, 1978; Hamada *et al.*, 1979).

The following studies (Reddan *et al.*, 1981a) document the influence of donor age on the growth of rabbit lens cells *in vitro,* characterize the conditions that permit the growth of these cells in a serum-free medium, and demonstrate that a specific basement membrane component, laminin, modulates the response of lens epithelia to polypeptide growth factors.

Figure 13 documents the influence of donor age on the growth of lens epithelial cells cultured in minimal essential medium (MEM) containing 10% fetal bovine serum. There is a gradual increase in the number of cells as a function of donor age. In contrast to the cell populations

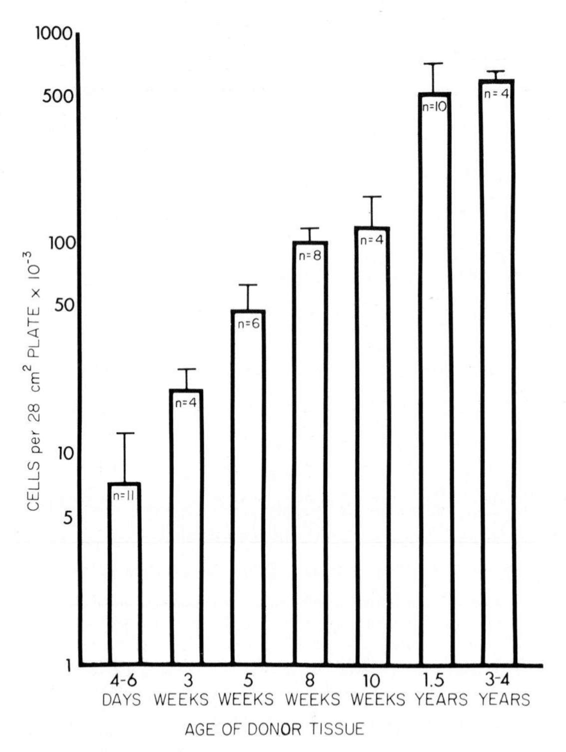

Fig. 13. Single-lens capsule epithelial preparations from rabbits of different ages were cultured in MEM containing 10% FBS. The total number of cells was counted 17 days after explantation. (From Reddan *et al.*, 1981a, with permission.)

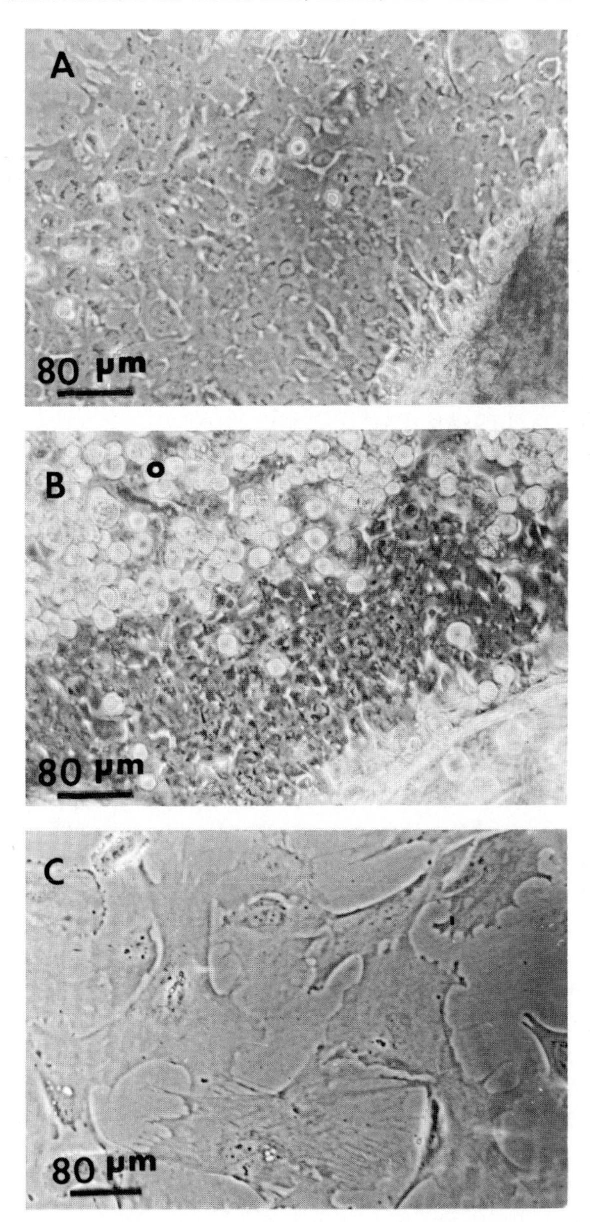

Fig. 14. Phase-contrast photomicrographs of lens epithelial cells from a 4- to 6-day-old rabbit, cultured in MEM plus 10% FBS. (A) Two days after attachment. (B) Ten days after attachment. Note numerous blebs. (C) Leading edge of the outgrowth 17 days after attachment. Note cellular enlargement, stress fibers, and pigment granules. (From Reddan *et al.*, 1981a, with permission.)

from the older animals, cells from 4- to 6-day-old or 3- or 5-week-old animals represent static populations since the number of cells does not increase over that shown in Fig. 13 even if cultured for 90 days. Figure 14 depicts the pattern of epithelial outgrowth in cells from 4- to 6-day-old donors. The outgrowth continues for approximately 11 days. However, few mitotic figures are noted on routine examination. Beyond 10 days of culture the epithelial cells at the leading edge of the outgrowth enlarge and do not exhibit cell-to-cell contact. If the cells are pulsed with [^{3}H]thymidine and prepared for autoradiography, labeled nuclei are found throughout the outgrowth at 1 and 3 days. However, by day 6 and thereafter, few nuclei are labeled. Analysis of time-lapse video tapes indicates that the outgrowth is primarily due to cellular migration. Cells that are enzymatically dissociated from 4- to 6-day-old lenses also have a limited potential for growth.

This pattern of limited growth typifies *in vitro* senescence in many culture systems (Hayflick, 1965; Martin *et al.*, 1970). This phenomenon in cultured lens cells is clearly attributable to the culture conditions (see below) and is not an inherent property of the cells. Whether this limited growth of lens epithelia cultured under standard conditions is indicative of precocious senescence or represents an intermediate stage toward terminal differentiation is not known (see Bell *et al.*, 1978, for discusion of these alternatives).

Since the lens in 4- to 6-day-old animals exhibits mitosis *in vivo,* it seems that cell division could be sustained *in vitro* under the appropriate environmental conditions. Cells from 4- to 6-day-old donors reach confluency and can be subcultured if placed in MEM containing non-heat-inactivated rabbit serum. The cells do not enlarge, contain lens-specific proteins (Fig. 15), have the diploid number of 44 chromosomes, and one cell line is currently at population doubling level 190. Typical growth curves are shown in Fig. 16. The cells can be maintained in a viable but nondividing state in medium supplemented with low concentrations of serum.

A procedure has been described which permits the routine establishment of normal cell lines from individual adult rabbit lenses (Reddan *et al.*, 1980). The morphological appearance of cultured cells from an 18-month-old rabbit is depicted in Fig. 17. The cells have a modal number of 44 chromosomes, contain lens-specific proteins, and have currently been maintained through 30–40 passages. Typical growth curves are presented in Fig. 18. The initial cell number can be maintained in medium supplemented with low concentrations of fetal bovine serum. However, a decline in cell number occurs in serum-free medium.

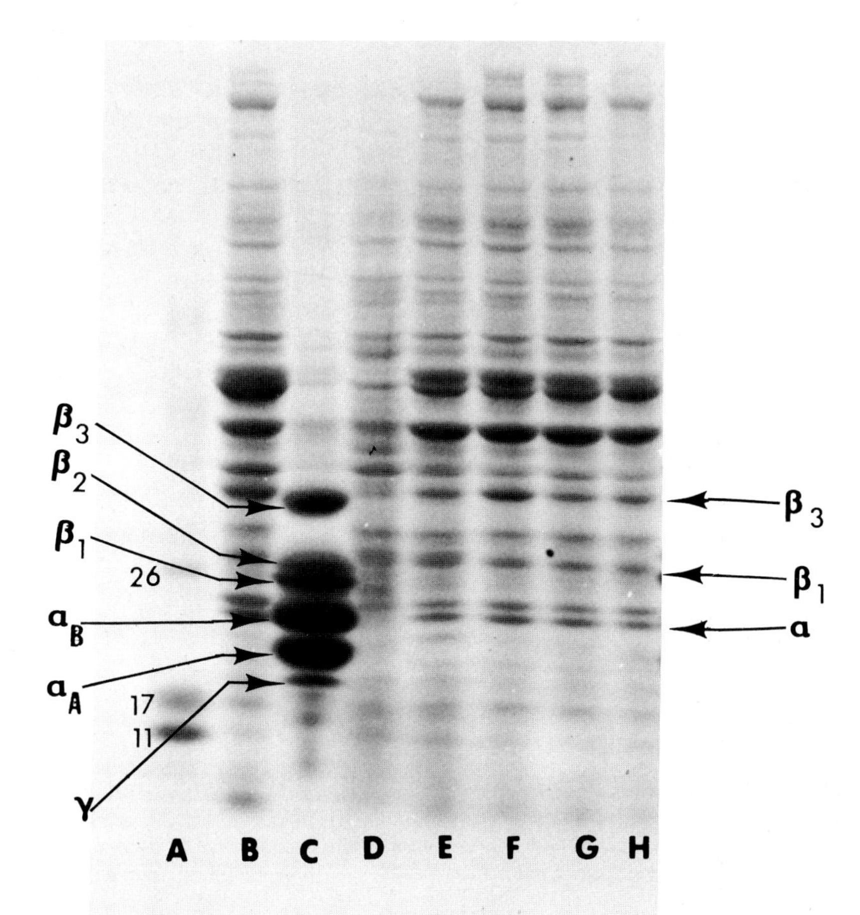

Fig. 15. Resolution of polypeptides from cultured lens cells. (A) Molecular-weight markers; (B) total water-soluble protein extract from epithelial cells; (C) polypeptides extracted from 2-year old rabbit cortex; (D) polypeptides extracted from cultured rabbit fibroblasts; (E) polypeptides extracted from cell line N/N 1135A at passage 10; (F) polypeptides extracted from cell line N/N 1120A at passage 10; (G and H) polypeptides extracted from cell line N/N 1003A at passages 10 and 37, respectively. Note that polypeptides extracted from rabbit lens epithelia have a migration pattern and apparent molecular weight characteristic of α- and β-crystallins.

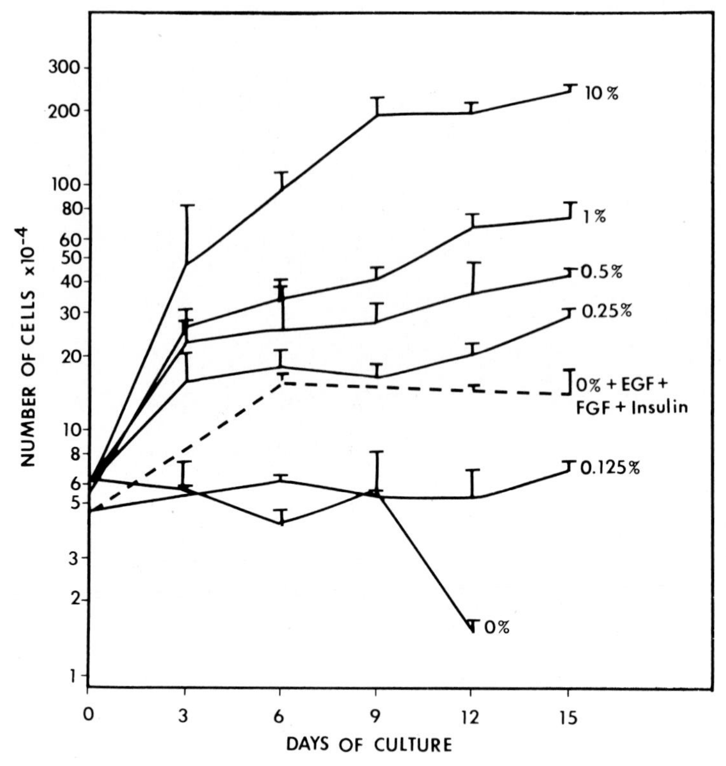

Fig. 16. Growth curves of newborn rabbit lens epithelial cells cultured in MEM or in MEM containing various concentrations of non-heat-inactivated rabbit serum. 5×10^4 cells were placed in each dish, and the total number of cells from at least three plates was counted for each time period and concentration. Note the substantial decrease in cell number in cultures maintained in MEM alone. The addition of EGF, insulin, and FGF to MEM alone permitted the cells to undergo one population doubling.

1. *Growth of Rabbit Lens Epithelial Cells in a Completely Defined, Serum-Free Medium*

Influence of Laminin and Lens Epithelial Basement Membrane on the Growth and Morphology of Lens Cells. Barnes and Sato (1980) have reviewed the conditions required for the growth of several cell types in serum-free media. However, little information is available concerning the conditions required for growth of epithelial cells in the absence of serum. This seems to be an essential prerequisite for understanding the mechanisms by which growth factors regulate mitosis. Attempts were made, therefore, to reduce or eliminate the serum requirement

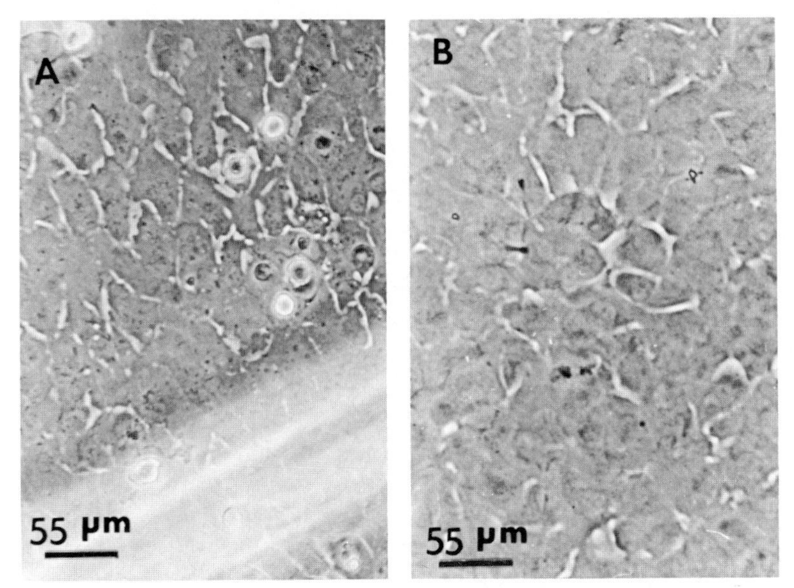

Fig. 17. Phase-content photomicrograph of rabbit lens epithelia from an 18-month-old donor cultured in MEM plus 10% fetal calf serum. (A) four days after capsule attachment; (B) 18 days after capsule attachment. Note the epithelial morphology.

for lens epithelial cells in tissue culture. Toward this end, the mitogenicity of various polypeptides was determined and optimal concentrations of the growth factors were used in an attempt to grow cells under serum-free conditions (J. R. Reddan, S. McGee, E. Goldenberg, and D. Dziedzic, unpublished).

The substrate upon which cells are cultured is known to influence their response to plasma, serum, or growth factors (Ham and McKeehan, 1978; Wolfe *et al.*, 1980; Gospodarowica and Ill, 1980a; Gospodarowicz *et al.*, 1981). Since the lens in organ culture can be stimulated to enter DNA synthesis and cell division by the addition of a single polypeptide growth factor to a serum-free medium (Reddan *et al.*, 1972), it should be possible to grow lens cells in tissue culture under similar conditions. In organ culture, the lens epithelial cells lie subjacent to the capsule, a basement membrane produced by the epithelium (Young and Ocumpaugh, 1966). This suggests that an appropriate substrate might permit the growth of lens epithelial cells in a serum-free medium. The conditions required for the production of basement membrane by cultured lens epithelia have recently been determined. Rabbit lens epithelia cultured on this membrane, in MEM supplemented with EGF, FGF, and insulin at 10^{-9}, 10^{-9}, and 10^{-6} *M*,

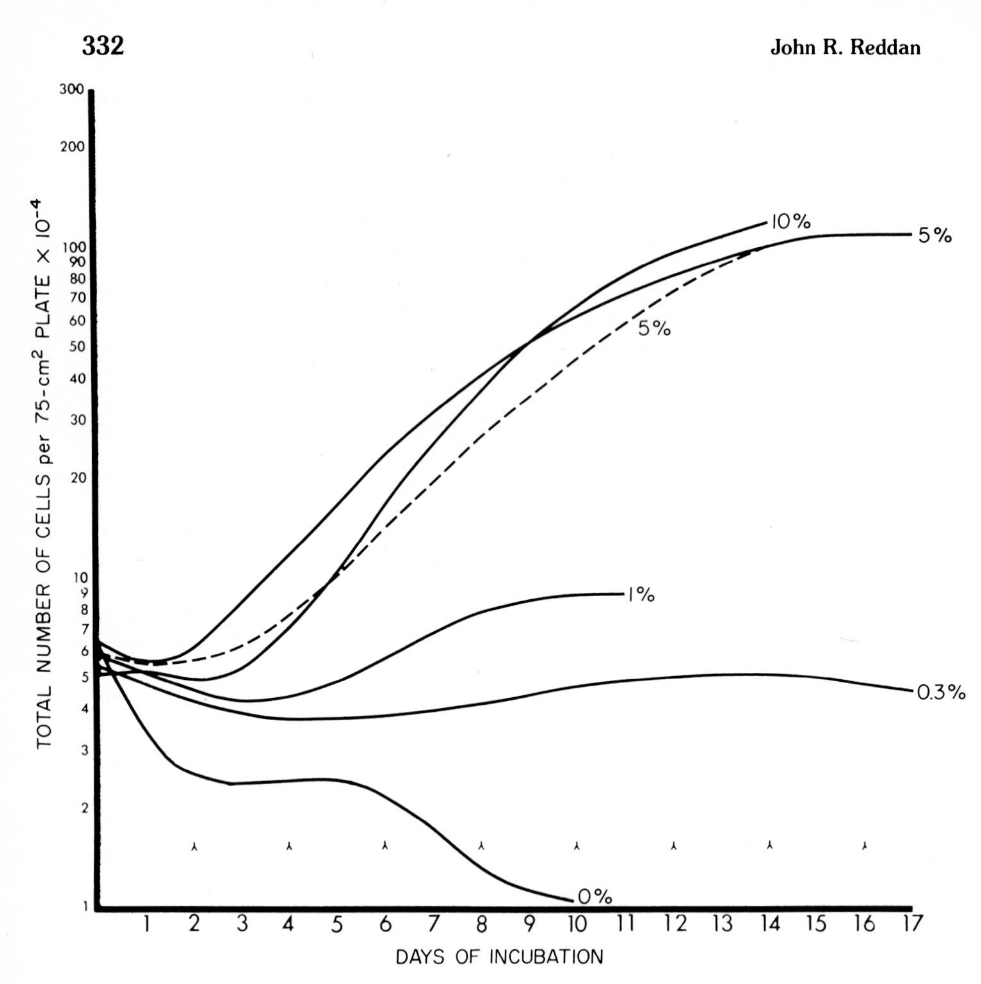

Fig. 18. Lens epithelial cells from an 18-month-old rabbit were cultured in MEM or MEM containing various concentrations of fetal bovine serum. 6×10^4 cells were placed in each dish, and the total number of cells from at least three plates was counted for each time period and concentration. (From Reddan *et al.*, 1981a, with permission.)

respectively, complete three population doublings within 7 days. IGF at 10^{-9} *M* also permits the growth of lens cells in a serum-free medium. Thus a substrate consisting of basement membrane produced by lens epithelium permits the growth of lens cells in a serum-free medium (Reddan, DeHart, and Sackman, 1981). Gospodarowicz and Ill (1980b) reported that bovine lens epithelia cultured on plastic dishes in medium containing 10% calf serum completed only one population doubling in 8 days. If bovine cells are cultured on a matrix produced by corneal endothelial cells, the lens cells enter logarithmic growth if cultured in medium supplemented with 10% calf serum. Lens cells cultured on plastic in 10% calf serum plus 100 ng/ml FGF also enter into logarithmic growth.

These results suggest that a specific basement membrane component may permit the growth of lens cells in a serum-free medium. Reddan *et al.* (1981d) have reported that laminin permits growth of lens epithelia in a serum-free medium. Laminin is a glycoprotein isolated from the basement membrane of a murine tumor (Timpl *et al.*, 1979) and is known to be present in the lens capsule (Drs. G. W. Laurie and C. P. Leblond, personal communication). Adult rabbit lens epithelial cells were placed in MEM containing 0.25% rabbit serum and cultured on plastic tissue culture dishes or on plastic dishes coated with rabbit serum, fibronectin, laminin (a gift of Drs. George R. Martin and Victor P. Terranova), or type-A and -B collagen (a gift of Dr. Mauricio Lande), all at 3 $\mu g/cm^2$. Although the cells attached to all the substrates, only laminin permitted, a marked increase in cell number at serum concentrations that are normally growth-limiting on plastic (Fig. 19). More recent studies (J. Reddan and D. Dziedzic, unpublished) indicate that a similar response can be obtained with fibro nectin.

On laminin, cells cultured in a serum-free medium (Fig. 20) complete approximately 3.5 population doublings in 7 days and can be subcultured. As indicated (Fig. 21), cells cultured on plastic in MEM supplemented with EGF alone do not increase in number over 7 days. However, if cells are cultured on laminin in the presence of EGF alone, they complete 3.3 population doublings in 7 days.

Thus lens cells cultured on an appropriate substrate in a serum-free medium can undergo mitosis in response to a single growth factor, either IGF or EGF. Although laminin appears to exert its effect by acting as a suitable substrate, the data do not preclude the possibility that growth enhancement is due to the presence of soluble laminin in the culture medium.

In the majority of tissue culture systems growth in a serum-free medium requires a set of four or five growth-promoting factors (see Barnes and Sato, 1980). In these systems, certain factors may regulate growth by stimulating the formation of basement membrane components. The effect of other basement membrane components and other polypeptides on the growth and/or differentiation of lens cells in tissue culture remains to be determined. It has not escaped our notice that basement membrane components may be involved in the regulation of cell division in the lens *in vivo*. Indeed, the early studies of Dische and Borenfreund (1954) and Dische and Zelmenis (1965) showed regional and age-related changes in the glycoproteins and mucopolysaccharides in the lens capsule.

The precise manner by which the basement membrane or its compo-

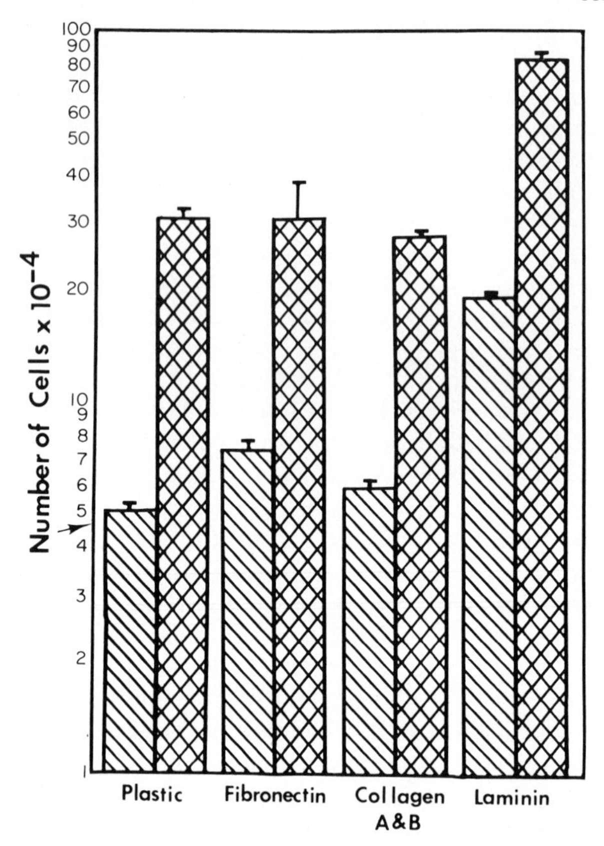

Fig. 19. Influence of substrate on cell growth. 5.0×10^4 cells from an 18-month-old rabbit were inoculated into MEM containing either 0.25% rabbit serum (cross-hatched areas) or 0.25% rabbit serum with EGF (10^{-9} *M*) and insulin (10^{-6} *M*) (double cross-hatched areas). Cells were grown on plastic dishes or on plastic dishes coated with fibronectin, collagens A and B, or laminin and counted on day 7. Values shown represent the mean and standard deviation ($n = 3$).

nents influence cell division remains to be elucidated. However, this culture stratagem, i.e., the use of specific basement membrane components as a cellular substrate, appears to be of general significance and may obviate the need for the feeder layers presently required for the growth of many types of epithelia (Rheinwald, 1980). It is quite possible that the behavior of cultured cells, including the apparent *in vitro* senescence of many epithelial cells, is a reflection of an inappropriate substrate.

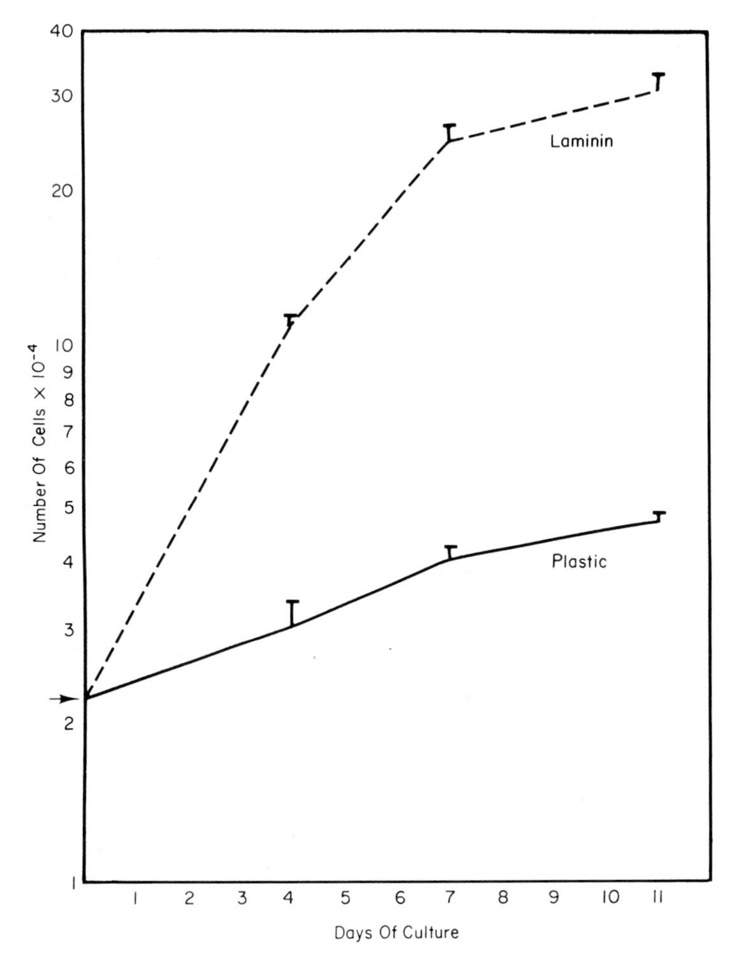

Fig. 20. Influence of laminin and plastic on the growth of rabbit lens epithelial cells in a serum-free medium. 2.0×10^4 cells were placed in MEM supplemented with EGF (10^{-9} *M*), FGF (10^{-9} *M*), and insulin (10^{-6} *M*). Cells were counted as indicated ($n = 3$).

2. *Tissue Culture of Human Lens Epithelium*

Although a limited number of human epithelial cell lines have been established from neoplastic tissues, attempts to establish long-term cultures of normal human epithelial cells have proven unsuccessful (Price *et al.*, 1980). The catalog of the American Type Tissue Collection lists 171 normal human fibroblast cell lines; however, no diploid human epithelial cell lines are listed (Rafferty, 1980).

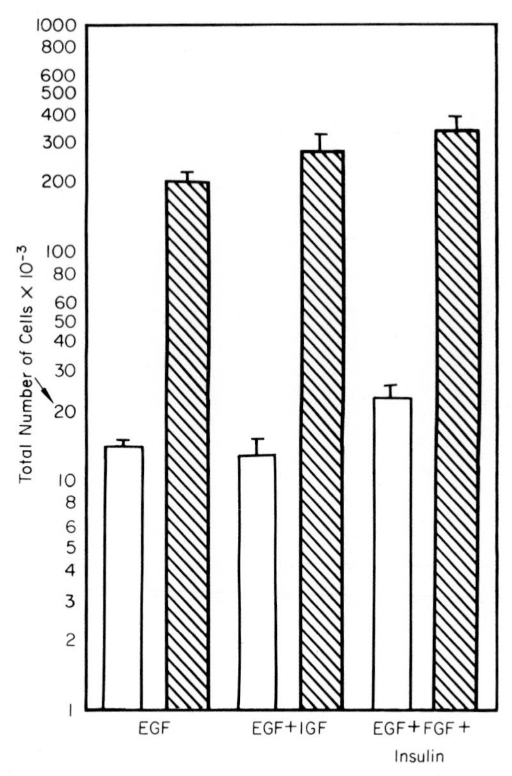

Fig. 21. Cells cultured on a laminin substrate respond to a single polypeptide growth factor in a serum-free medium. 2.0×10^4 cells from an adult rabbit were placed in MEM containing EGF (10^{-9} *M*), EGF and IGF (10^{-9} *M*), or EGF, FGF (10^{-9} *M*), and insulin (10^{-6} *M*) and cultured on plastic tissue culture dishes (open areas) or on plastic tissue culture dishes coated with laminin (cross-hatched areas). Cells were counted on day 7. Values represent the mean with standard deviation ($n = 3$).

Despite the interest in and potential use of human lens epithelial cells, little information is available on the culture conditions required for the long-term growth and survival of these cells. Such an accomplishment would be important not only in understanding human lens epithelium but could also provide a model for investigating the environmental factors involved in the transformation of human epithelial cells. This is of particular interest, since most human solid tumors are of epithelial origin (Price *et al.*, 1980). A few investigators have maintained human lens epithelial cells *in vitro* for relatively short periods (Mamo and Leinfelder, 1958; Bryan *et al.*, 1960; Tassin *et al.*, 1976; Hamada and Okada, 1978; Courtois *et al.*, 1978). Hamada and

Okada, (1978) report that fetal human lens epithelia are capable of only one subculture. Tassin *et al.* (1979) note that human lens cells show a progressive age-related decline in their *in vitro* growth capacity (Fig. 22). The human lens epithelial cells reported to yield continuous cell lines are either overtly aneuploid (Awasthi *et al.*, 1975) or appear to be aneuploid (Geeraets, 1972). Moreover, neither Awasthi *et al.* (1975) nor Geeraets (1972) has documented the presence of lens-specific proteins in their cultures.

Reddan *et al.* (1981a) reported that human lens epithelial cells cultured in a variety of sera, media, and growth factors showed limited growth and exhibited cellular enlargement. Cultures initiated from explants or from cells that were enzymatically dissociated from lens capsules could be subcultured a maximum of three times and thus exhibited a limited doubling potential, as described by other investigators. Figure 23 shows the typical pattern of outgrowth of human lens epithelial cells cultured in MEM with 10% fetal bovine serum. The cells lost cell-to-cell contact, entered a nondividing state and, in this respect, mimicked the behavior of cells from 4- to 6-day-old rabbits

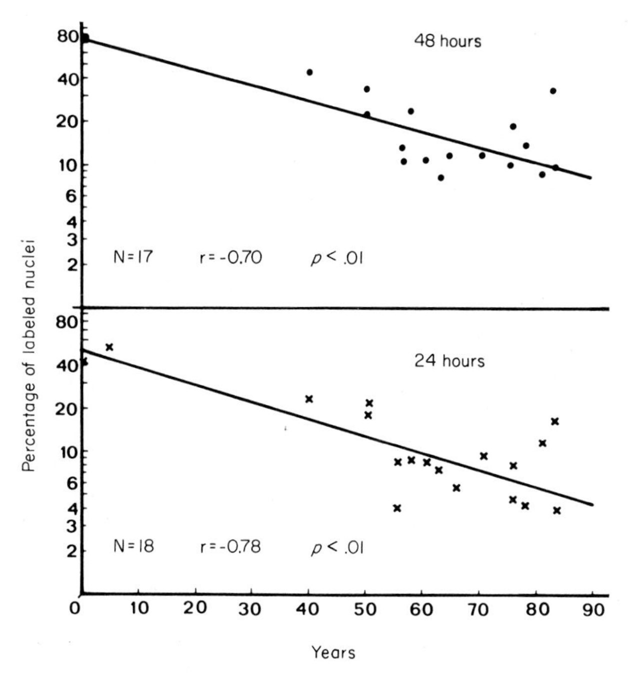

Fig. 22. Labeling index of human lens cells as a function of donor age. The cells were labeled for 24 and 48 hr with [^{3}H]thymidine. (From Tassin *et al.*, 1979, with permission.)

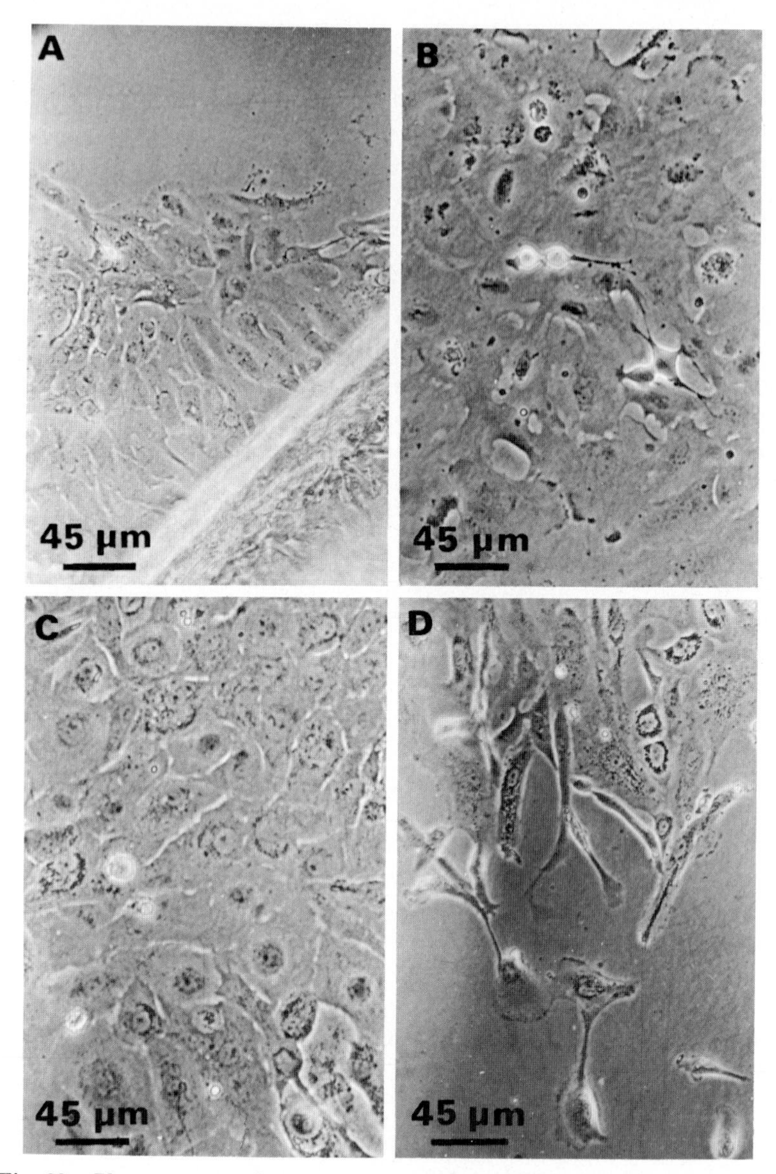

Fig. 23. Phase-contrast photomicrograph of human lens epithelial cells. (A) Four days after capsule attachment; (B) center of outgrowth 9 days after capsule attachment; (C) a mitotic figure in the center of the outgrowth 26 days after capsule attachment; (D) a loss of cell-to-cell contact at the leading edge of the outgrowth, day 32 of culture.

cultured under standard conditions. A few lentoid-like bodies were present 30 days after initiation of the culture.

Limited growth has also been reported for other human cell types, including mammary epithelium (Kirkland *et al.,* 1979; Taylor-Papadimitriou *et al.,* 1980), keratinocytes (Rheinwald and Green, 1975; Peehl and Ham, 1980a,b; Rheinwald, 1979), prostrate epithelium (Kaighn, 1980), and bronchial epithelium (Stoner, 1980). It is recognized that human epithelial cells *in vitro* may require tissue-specific nutrients or hormonal factors not provided by growth media in common use (Owens, 1976).

H. Epilogue

Cell division in the ocular lens appears to be regulated by polypeptide growth factors. The lens is one of the few investigational systems wherein a specific type of growth factor appears to be operative *in vivo* and *in vitro*. The addition of insulin, IGF, EGF, or somatomedin C to a completely defined serum-free medium triggers a highly reproducible stimulation of cell division in the rabbit lens in organ culture. Cell division in the frog lens *in vivo* is totally abolished by hypophysectomy, a procedure which brings about a decrease in the level of insulin-like growth factors. Furthermore, mitosis is reinstated in the frog lens by the systemic administration of the insulin-like growth factor somatomedin C. Preliminary experiments suggest that mitosis in the mammalian lens *in vivo* is reduced under conditions known to bring about a decrease in the level of insulin-like growth factors.

Studies on the control of cell division in cultured cells are usually hindered by the presence of serum or by the addition of elaborate "sets" of growth factors. Rabbit lens epithelia can be grown in tissue culture in a completely defined serum-free medium if the cells are placed on an appropriate substrate. Cells grown on an extracellular matrix produced by lens epithelia, or on the highly purified basement component laminin, increase in number in the presence of a single growth factor. The use of a "natural" substrate appears to be a general stratagem for tissue culture studies and may facilitate the growth of other types of epithelia in a serum-free medium. Since growth factors tend to be cell-type-specific, information gleaned from various animal models should be invaluable in planning experiments on human lens epithelia.

An understanding of the control of cell division in the human lens seems to be predicated on a delineation of the conditions required for the establishment of continuous lines of these cells. Through a judicious selection of growth factors and substrates, it should be possible to establish continuous lines of normal human lens epithelial cells. Lines of lens epithelial cells that remain diploid and produce lens-specific proteins will provide an investigational system that may aid in understanding the mechanisms and/or factors regulating the growth, development, repair, and senescence of these cells. Tissue culture studies are particularly well suited for characterizing age-related changes in the lens epithelium. Moreover, the availability of cell lines from humans subject to hereditary cataracts may permit a dissection of the underlying genetic factors.

II. REGULATION OF CELL PROLIFERATION IN THE RETINA, RETINAL VESSELS, AND VITREOUS HUMOR

A. Introduction

Vascularization of tissues is a fundamental biological process that is a normal adjunct of embryonic development, wound healing, and neo plasia. Ocular neovascularization, a major cause of blindness (Lynn *et al.,* 1974; Henkind, 1978; Patz, 1980), implies the growth of new blood vessels and can involve the cornea, iris, lens, choroid, and retina (Henkind, 1978). The cells of the retinal capillaries which normally exhibit an extremely low index of proliferation in the adult animal posess the latent ability to form new blood vessels. This section focuses on the normal development of the retinal vascular system, on retinal neovascularization which is defined as "the growth of retinal vessels in a configuration different from normal vessels with the abnormal configuration lying in a plane anterior to that of the normal retina vessels, often extending into the vitreous cavity" (Patz, 1980), and on the consequences of cell proliferation in the vitreous chamber. In many instances, glial cells, retinal pigment epithelia or other cell types enter the vitreous humor as a normal adjunct of retinal neovascularization or subsequent to ocular injury (van Horn *et al.,* 1977; Machemer *et al.,* 1978; Topping *et al.,* 1979). These cells proliferate and may attach to the retina. Subsequent traction exerted by the cells on the retina leads to retinal detachment (Radtke *et al.,* 1981).

Many ocular pathologies, including retrolental fibroplasia (Patz, 1980), proliferative diabetic retinopathy (L'Esperance, 1978), and massive periretinal proliferation (Machemer and Laqua, 1975), reflect a breakdown in the mechanism that regulates cell proliferation. These pathologies are associated with heightened mitotic activity in the retinal capillaries and/or vitreous humor. Although blood vessels are noted in the vitreous humor of embryonic and postnatal eyes, the adult vitreous humor is normally acellular save for the presence of hyalocytes (Bloom and Balazs, 1965). Indeed the absence of dividing cells in the normal vitreous humor has prompted the hypothesis that vitreous humor contains amitotic inhibitor (Brem *et al.*, 1976, 1977; Finkelstein *et al.*, 1977). As noted in Section I,C, exposure of the lens epithelium to vitreous humor results in a cessation of cell proliferation and a concomitant induction of cell differentiation.

In order to understand the factors that regulate neovascularization and the proliferation of cells in the vitreous humor, attempts have been made to isolate growth-promoting and -inhibiting agents from the retina and vitreous humor (Brem *et al.*, 1977; Chen and Chen, 1980; Glaser *et al.*, 1980a). Isolation and characterization of such factors may help define the basic mechanism that leads to or permits aberrant mitosis in retinal capillaries and vitreous humor and may suggest new nonsurgical techniques for inhibiting neovascularization, curtailing cell division in the vitreous humor, and interfering with the mechanism responsible for the attachment of "foreign cells" to the retina.

Techniques have been developed which permit the culture of endothelial and mural cells from retinal capillaries (Buzney *et al.*, 1975; Buzney and Massicotte, 1979; Frank *et al.*, 1979; Frank, 1980; Cohen *et al.*, 1980). Thus it should be possible to evaluate putative retinal angiogenesis factors on retinal capillary endothelial cells under defined and controlled conditions. It is known that solid tumors produce an angiogenesis factor (Folkman, 1974). In this regard, Finkelstein *et al.* (1977) reported that the presence of cancerous cells in the retina stimulated neovascularization. Although many investigators feel that retinal neovascularization is due to the release of diffusible growth factor(s) from the retina, systemic factors may also be involved. It is of great interest that pituitary ablation, a procedure that curtails mitosis in the ocular lens, causes a regression of the neovascularization associated with diabetic retinopathy (Davis, 1974; Valone and McMeel, 1978). However, at this time there is no compelling evidence that specific growth factor(s) initiate mitosis in retinal capillaries *in vivo* or stimulate cell division in cultured retinal endothelial cells.

B. Development of the Vascular System of the Retina

The developing retina receives its initial nutrient supply from the choroidal circulation (Wise, 1956; Wise *et al.*, 1971). As development proceeds, the retina thickens (Kuwabara and Weidman, 1974) and its innermost layers become further removed from the choroidal circulation. Concomitant with the increase in the thickness and the subsequent displacement of the tissue from an adequate blood supply a swelling is noted on the hyaloid artery where it passes through the optic disk. The development of the retinal circulation in the human, rat, and cat is similar and involves the growth of blood vessels from the region of the disk toward the periphery of the retina (Michaelson, 1948). Many of the vascular channels in the eye, e.g., the tunica vasculosa lentis, hyaloid systems, and some of the retinal vessels, are prominent in the embryonic stages but are diminished or undergo atrophy as a function of age (Henkind, 1978). It appears that certain of the ocular vessels are subject to programmed cell death. In addition, degeneration of mature blood vessels is a normal adjunct of the aging process and is thought to lead to or contribute to senescence.

Michaelson (1948) documented the spatiotemporal pattern of the development of the retinal circulation in the cat and human. The pattern of retinal circulation in the cat embryo at 35, 45, 51, and 56 days and in kittens is as follows: At day 35 the hyaloid artery is present at the disk. No retinal vessels are present at this time. By day 45, vessels are noted at the edge of the disk and progress for a slight distance into the retina. The temporal vessels reportedly begin growth earlier than the nasal ones. The vessels continue to progress toward the periphery of the retina and are 0.36–0.72 mm from the optic disk at day 51. By day 56, three major vessel complexes have developed and are reportedly 5.6, 4.8, and 4.6 mm from the optic disk. Arteries and veins can be distinguished at this stage. A capillary system develops on a side of the vein distal to the neighboring artery. As development proceeds, the embryonic capillary net, which emanates from the veins, grows toward, but does not reach, the neighboring artery. A capillary-free area remains around the artery and is a prominent feature in the definitive eye. The vessels continue to grow toward the periphery of the retina and are relatively complete in the 22-day-old kitten. Beyond this stage, the growth of the capillaries essentially ceases. Indeed, the vascular endothelial cells of older animals rarely divide under normal circumstances (Engerman *et al.*, 1976; Schwartz and Benditt, 1973). Constant infusion of mice with [^{3}H]thymidine for 48 hr labeled 0.2% of the retinal endothelial cells and gave a calculated turnover time of 3 yr or more (Engerman *et al.*, 1967). A single injection of [^{3}H]thymidine labeled 0.01% of the retinal capillary cells in the adult mouse. The retinal capillary endothelial cells, which normally exhibit a very low rate of turnover, can be induced to proliferate by a variety of pathological stimuli.

In all animals studied the retinal vessels commence growth at a comparatively late stage in eye development. The pattern of retinal vascularization in the human is similar to that in the cat. When the vessels reach the ora serrata, at 8 months of development, the embryonic vasculature is considered to be complete. However, fetal weight may be a better indicator of the time that the vascularity has reached the ora serrata. This corresponds to a weight of ca. 2000 g. Infants with a lower birth weight show a less extensively vascularized retina (Patz *et al.*, 1953). Again, as is in the cat, capillary growth is initiated from the side of the vein that is distant from the neighboring artery. A capillary-free space around the arteries is characteristic of the mature retina.

The pattern of growth of the retinal capillaries from the side of the vein distal to the neighboring artery, the existence of a capillary-free space around the artery, and the fact that capillaries develop from the veins prompted Michaelson (1948) to hypothesize the presence of a factor(s) in the extravascular tissue of the retina capable of triggering the growth of new blood vessels. Within the framework of these findings, it was surmised that capillary growth prevailed in areas where the oxygen concentration was low and did not occur in proximity to the arteries where the oxygen concentration was higher. These early findings suggested that tissue hypoxia might be an important permissive and/or causative factor in regulation of the growth of retinal capillaries. Moreover, Michaelson (1948) noted that the abnormal proliferation of retinal vessels associated with certain retinal pathologies was preceded by venous occlusion and attendant retinal ischemia.

C. Retinal Neovascularization

New blood vessel formation and attendant proliferative retinopathies are a major cause of blindness. Retrolental fibroplasia (RLF) (Ashton, 1957; Kinsey *et al.,* 1977; Patz, 1980), diabetic retinopathy (McMeel, 1971; L'Esperance, 1980; James, 1980), branch or central vein occlusion (Shilling and Kohner, 1976), sickle cell disease (Condon and Sergeant, 1972), and other disorders that lead to the occlusion of retinal vessels are characterized by subsequent new blood vessel growth. RLF and diabetic retinopathy constitute two well-defined pathologies, and attention will focus on the common features of these disorders. A detailed description of the clinical and cellular changes associated with RLF can be gleaned from the studies of Wise (1956), Ashton (1957), Henkind (1978), and Patz (1980).

Terry (1942) reported RLF in premature infants. High concentrations of oxygen were suggested to be the etiological agent (Kinsey and Zacharias, 1949; Kinsey *et al.,* 1956). RLF is characterized by the growth of newly formed retinal capillaries into the vitreous body and commences following the transfer of animals to a normal atmosphere. Proliferative changes similar to those noted in human RLF have been observed in mice, kittens, puppies, and ratlings (Patz *et al.,* 1953; Gyllensten and Hellstrom, 1954; Ashton, 1957). High concentrations of oxygen lead to a partial or total obliteration of retinal blood vessels in the immature eye. In the presence of continuously high concentrations of oxygen there is a constriction of arteries and arterioles, which is followed by the disappearance of the entire capillary bed. The vaso-obliteration brought about by oxygen is inversely proportional to the maturity of the retinal vessels and, once maturation is realized, high concentrations of oxygen no longer evoke neovascularization. The newly formed vessels in the peripheral region of the immature retina are most susceptible to oxygen insult, whereas their counterparts at the optic disk are most resistant (Ashton, 1957). In all instances, the growth of new vessels is preceded by arteriolar obstruction, venous stasis, retinal ischemia and local hypoxia, and retinal capillary destruction.

Although the natural history, time course, and incidence of the neovascularization associated with proliferative diabetic retinopathy (Dobree, 1964; McMeel, 1971; Adler *et al.,* 1975; Shabo and Maxwell, 1976; Henkind, 1978; L'Esperance, 1978; Wolbarsht and Landers, 1980; Davis and Engerman, 1980) is decidedly different from that of RLF, certain common features emerge. Some of the characteristics of proliferative diabetic retinopathy include obliteration of retinal vessels, proliferation of vascular endothelial cells and other cell types, increased permeability of vascular walls, growth of cells in the vitreous body, and eventual retinal detachment (Davis, 1974; L'Esperance, 1980). A selected loss of mural cells, "pericytes" (Cogan and Kuwabara, 1967), has also been reported (Cogan *et al.,* 1961; Speiser *et al.,* 1968). As in RLF, areas of capillary nonperfusion and presumed retinal ischemia foreshadow the onset of the neovascularization of proliferative diabetic retinopathy (Patz, 1980).

The pattern of capillary growth during development is thought to be mediated by factor(s) released from hypoxic retinal tissue (Michaelson, 1948; Ashton and Cook, 1954; Campbell, 1957). In a like manner, neovascularization is thought to be due to a vasoproliferative factor

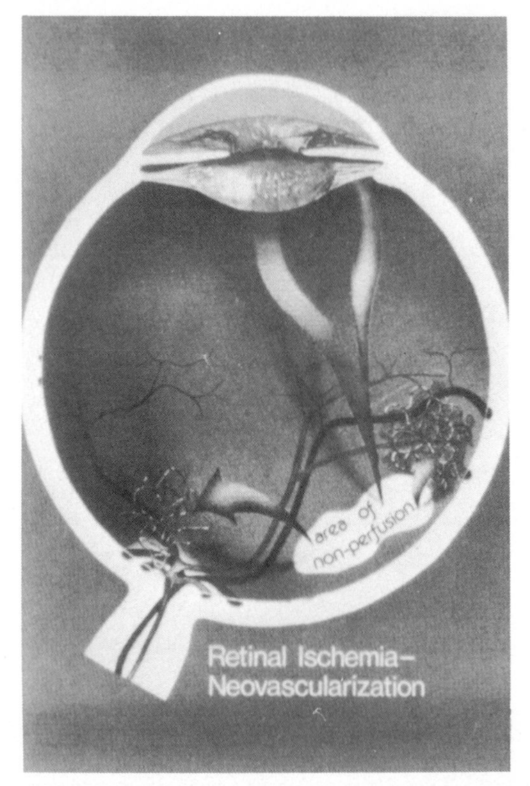

Fig. 24. Area of nonperfusion from retinal capillary closure. That resultant ischemia is presumed to liberate a diffusible angiogenesis substance that stimulates neovascularization in the adjacent area, at the optic disk, or on the surface of the iris. (From Patz, 1980, with permission.)

which is produced in and diffuses from ischemic retinal tissue (Fig. 24). Indeed capillary nonperfusion and resultant ischemia are characteristic events which usually precede retinal neovascularization (Figs. 25 and 26). However, support for the existence of a diffusible angiogenic factor is circumstantial and is based on the pattern of development of the retinal blood vessels (Wise, 1956), on the regression of retinal and iridial neovascularization following photocoagulation treatment, and on the finding that retinal neovascularization is often accompanied by hyperplasia of other ocular blood vessels (Fig. 24). Photocoagulation treatment is not aimed at the newly growing vessels but is focused upon the surrounding tissue and is thought to limit or inhibit the release and/or production of angiogenic factors. Exposure of mice to

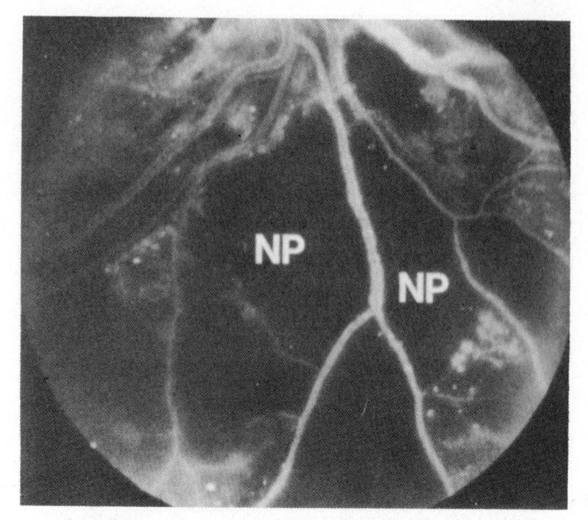

Fig. 25. Fluorescein angiogram of a young diabetic patient with preproliferative retinopathy. Note the large areas of capillary nonperfusion (NP). Four months later the patient developed neovascularization at the optic disk. (From Patz, 1980, with permission.)

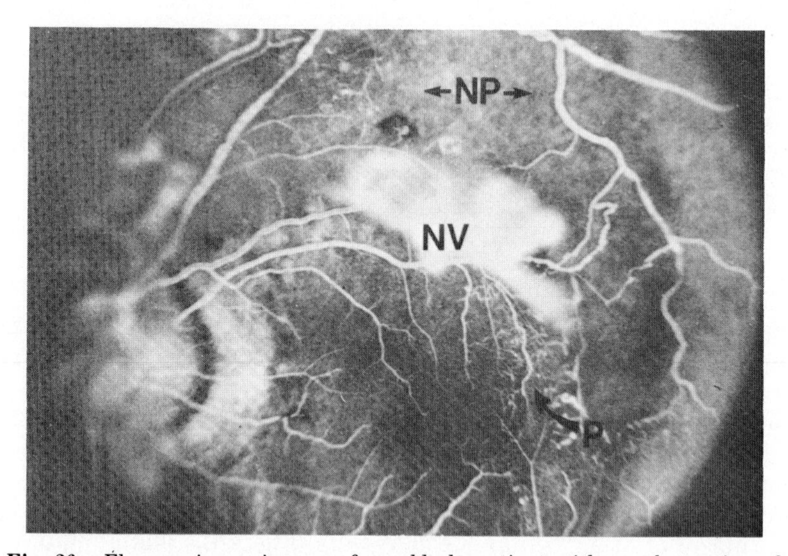

Fig. 26. Fluorescein angiogram of an elderly patient with an obstruction of the superior temporal vein. Note the area of capillary nonperfusion (NP) and area of perfusion (P). Neovascularization (NV) developed at the border of nonperfused and perfused retina. (From Patz, 1980, with permission.)

high concentrations of oxygen results in neovascularization of the retina and iris and hyperplasia of the tunica vasculosa lentis (Gyllensten and Hellstrom, 1954). Iris neovascularization may also accompany RLF (Wise, 1956; Ashton, 1957; Henkind, 1978); central retinal vein occlusion (Laatikainen and Blach, 1977) and retinoblastoma (Walton and Grant, 1968). These findings have prompted investigators to look for diffusible angiogenic factors in the retina and vitreous humor.

Influence of Pituitary Ablation on Neovascularization Associated with Diabetic Retinopathy

Pituitary ablation often results in regression of the new blood vessel proliferation associated with diabetic retinopathy (Urrets-Zavalia, 1977, a review). This suggests the intervention of systemic factors in the regulation of neovascularization. The involvement of the pituitary in the control of retinal neovascularization has been gleaned from studies on humans. Experiments aimed at evaluating the role of the pituitary in new blood vessel formation in the retina in various animal models have not been forthcoming. In terms of animal models of diabetic retinopathy, rhesus monkeys exhibit an antigen-induced retinopathy which resembles diabetic retinopathy (Shabo and Maxwell, 1976, 1977; Shabo *et al.*, 1980). Moreover, dogs made diabetic with alloxan or growth hormone exhibit retinal abnormalities reported to be morphologically indistinguishable from those of human diabetic retinopathy (Engerman and Bloodworth, 1965). Additional animal models seem to be highly desirable.

Luft *et al.* (1955) first introduced hypophysectomy as a therapeutic modality for diabetic retinopathy. An actual regression of newly formed blood vessels resultant to pituitary ablation has been noted by several investigators (Bradley *et al.*, 1965; Wright *et al.*, 1965; Lee *et al.*, 1966; Deckert *et al.*, 1967; Kohner *et al.*, 1972; 1976a,b; Kohner, 1974; Valone and McMeel, 1978). Indeed, hypophysectomy appears to be an effective treatment of rapid, progressive, severe, proliferative retinopathies (Kohner *et al.*, 1976a,b). Hypophysectomy provides a powerful experimental approach for investigating the environmental factors and/or conditions regulating retinal neovascularization *in vivo*. The regression of retinal neovascularization noted subsequent to pituitary ablation suggests that the release, production, or binding of yet unidentified retinal angiogenic factors may be controlled by agents(s) produced at sites distant from the retina itself. In addition, as noted in Urrets-Zavalia (1977), diabetic subjects with proliferative diabetic retinopathy whose lesions remained stable or exhibited a relatively slow deterioration had circulating levels of growth hormone decidedly lower than those found in patients with marked retinal deterioration. Thus, growth hormone or substances that may act in a manner similar to growth hormone (see Section I,F) may play a major role in the development of proliferative diabetic retinopathy. The data do not preclude the direct effect of other types of growth factors on the cells of the retinal vessels, e.g., IGFs or somatomedins. Suitable animal models are required to test these alternatives. As noted in Section I,F, hypophysectomy, which can curtail retinal neovascularization, also lowers or abolishes mitosis in the ocular lens.

D. Isolation and Characterization of Possible Angiogenic Factors from Ocular Sites

It has been known for some time that newly formed blood vessels invade solid tumors (Virchow, 1863). Algire and Chalkey (1945) reported that tumors induced new capillary formation from preformed vessels, and it was hypothesized that factors from the tumor initiated neovascularization (Greenblatt and Shubic, 1968; Ehrmann and Knoth, 1968). In living animals, vascular endothelial and mural cells undergo DNA synthesis and mitosis (Cavallo *et al.*, 1972, 1973) in response to a diffusible substance released from tumor cells (Folkman, 1974; Folkman *et al.*, 1971). In addition, implants or rabbit V-2 carcinoma placed 2.5 mm from the rabbit corneoscleral limbus elicit corneal neovascularization within 4 days (Gimbrone *et al.*, 1974). Regional infusion of an angiogenesis inhibitor isolated from cartilage retards the neovascular response normally elicited by corneal implants of V-2 carcinoma cells (Langer *et al.*, 1980). As pointed out by Langer *et al.* (1980), the results suggest, but do not prove, that the angiogenesis inhibitor acts on the blood vessels and not on the tumor itself. Medroxyprogesterone, dexamethasone, or cortisone applied in a sustained-release polymer, blocked the corneal neovascularization normally induced by V-2 carcinoma cells (Gross *et al.*, 1981). Injection of V-2 carcinoma cells into the vitreous humor of an adult rabbit induced proliferation of retinal vessels (Finkelstein *et al.*, 1977). The growth of retinal vessels into the tumors occurred when the tumors were contiguous with the retina. These results coupled with the information gathered on apparent tumor angiogenic factors presaged the search for angiogenic factors in the retina and vitreous humor. Isolation and purification of potent nontoxic angiogenesis inhibitors would be of obvious import to the field of ophthalmology and to all investigators seeking to understand cell growth.

Chen and Patz (1976) noted a correlation between the presence of vitreous humor-soluble proteins and the rate of retinal vessel growth in the newborn dog. Exposure of newborn dogs to high concentrations of oxygen, a procedure which induces RLF, was associated with heightened levels of vitreous humor-soluble proteins. Extracts of the retina (Chen and Chen, 1980; Glaser *et al.*, 1980a,b) and vitreous humor (Chen and Chen, 1980) of several species possess vasoproliferative activity. The efficacy of retinal or vitreous extracts in inducing neovascularization *in vivo* has been demonstrated using the cornea (Gimbrone *et al.*, 1974) and chick chorioallantoic membrane as a bioassay system. The mitogenicity of the extracts has also been evaluated on fetal bovine aortic endothelia cultured in MEM containing 1.0 or 0.25% fetal bovine serum (Chen and Chen, 1980; Glaser *et al.*, 1980a,b).

In dogs, the retina is fully vascularized at approximately 4 weeks after birth. Vitreous humor from 3-day-old normal dogs or from 8-day-old oxygen-treated puppies induced corneal neovascularization (Fig. 27). In contrast, vitreous humor from 3-month-old or adult dogs did not elicit new vessel formation in the rabbit cornea (Fig. 27). The

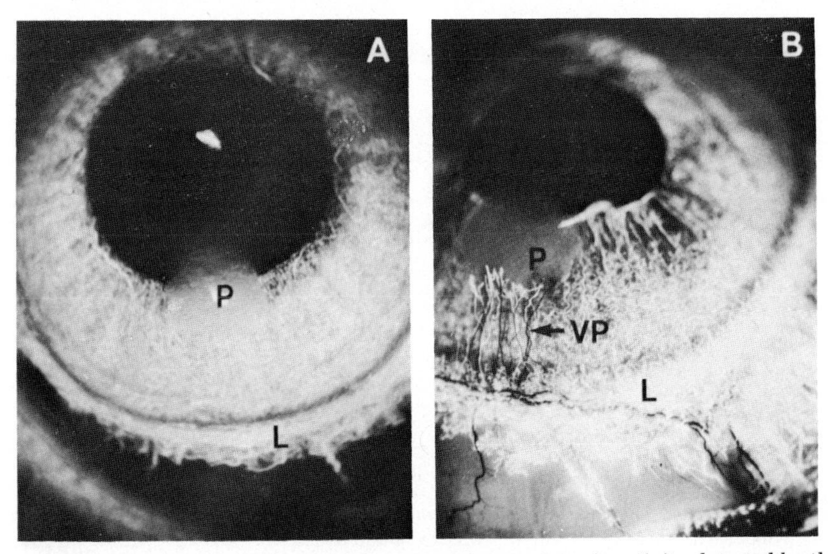

Fig. 27. Fluorescein angiograms demonstrating angiogenic activity detected by the corneal micropocket method. Angiograms were taken 12 days after a test substance had been added to the pocket. At this stage, the micropocket (P) was still seen. (A) Vitreous humor prepared from a 3-month-old normal puppy induced no apparent proliferating vessels from the limbus. (B) The vitreous humor of an 8-day-old oxygen-treated puppy induced significant vasoproliferation (VP) from the limbus (L) toward the corneal micropocket. (From Chen and Chen, 1980, with permission.)

influence of animal age and hyperoxia on the growth-promoting properties of vitreous humor on cultured aortic bovine endothelial cells is shown in Fig. 28. The specific activity of vitreous humor from younger animals is decidedly higher than that from older animals, however, an increase in the number of cultured endothelial cells was noted with vitreous humor from both the younger and older animals (Fig. 28). It should be noted that the vitreous humor from 3-month-old animals failed to elicit corneal neovascularization but was mitogenic toward cultured endothelial cells. Vitreous humor collected from dogs exposed

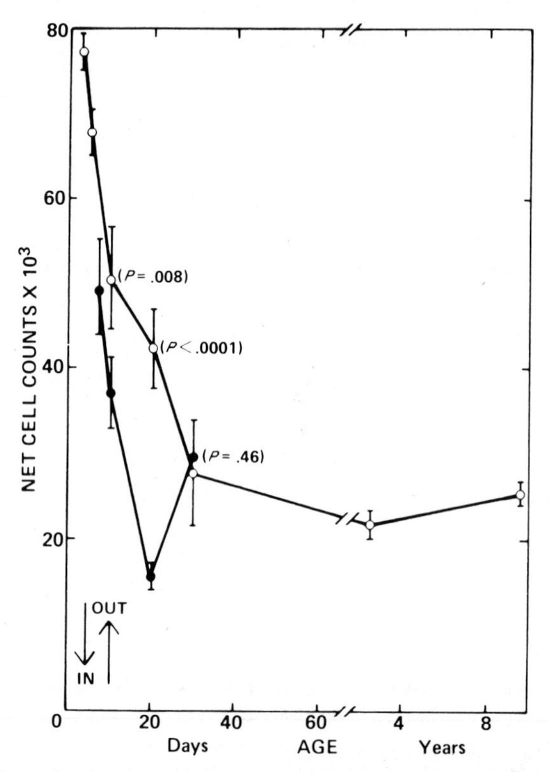

Fig. 28. Effects of animal age and hyperoxia on dog vitreous humor in growth stimulation of cultured endothelial cells. A volume of 100 μl vitreous sample (fourfold dilution) was used in each assay. Data as given have been subtracted by an assay control of 45.2×10^3 cell counts/2 days in which no test substance was added. Data were expressed as net cell counts/2 days/100 μl of vitreous. Each point is the mean of four preparations. Bars indicate the standard error. ○, Normal animals; ●, animals treated with 85% oxygen. The statistical significance of the difference in activity of test substances in paired treated and untreated dogs is shown in parentheses. Arrows indicate the period of oxygen treatment. (From Chen and Chen, 1980, with permission.)

to high oxygen concentrations did not show a substantial increase in mitogenicity when compared to controls. Moreover, retinal extracts prepared from normal animals or from dogs subjected to hyperoxia stimulated the growth of cultured vascular endothelial cells. If these findings are representative of the *in vivo* situation, they do not support the original hypothesis of Ashton and Cook (1954) and Michaelson (1948) which implied that the presence, release, or synthesis of angiogenic factors was predicated on prior retinal ischemia. Extracts from human, bovine, feline, and rabbit retinas (Glaser *et al.*, 1980a,b) are known to stimulate the growth of cultured bovine aortic endothelial cells and to induce neovascularization in the chick chorioallantoic membrane. Whether the retinal and/or vitreous humor extracts are mitogenic in and of themselves or modulate the response of the endothelial cells to serum, which is present in the culture medium, remains to be determined. Moreover, there is no evidence at this time that the retinal angiogenic factor is synthesized in the retina or that it is mitogenic toward retinal capillary endothelial cells. A determination of the tissue culture conditions required for the growth of retinal capillary endothelia and other retinal cells is needed in order to evaluate potential angiogenic factors and to determine the synthetic capacity of these cells.

E. Growth of Cells from Retinal Capillaries in Tissue Culture

Recent advances have permitted the long-term growth of endothelial cells from large vessels (Gimbrone *et al.*, 1974; Gospodarowicz *et al.*, 1978a; Striker *et al.*, 1980; Castellot *et al.*, 1980) and capillaries (Folkman *et al.*, 1979). The availability of cultured endothelial cells, pericytes, and vascular smooth muscle cells permits study of the mechanism(s) regulating the growth, repair, regeneration, and senescence of these cells. It should also permit an understanding of the synthetic and functional properties of normal vascular cells (Cohen *et al.*, 1980) and of abnormalities associated with various vascular pathologies, e.g. diabetic retinopathy, RLF, and atherosclerosis (Buzney *et al.*, 1975; Gimbrone 1977).

The procedures used for the isolation of vascular cells (Meezan *et al.*, 1974; Gimbrone *et al.*, 1974) usually result in the admixture of cell types (Striker *et al.*, 1980). However, the identification of cultured endothelial cells has been greatly facilitated by the use of morphological and immunological markers. Electron microscopic examination of human umbilical vein endothelia and human capillaries (Folkman *et al.*, 1979) shows the presence of Weibel–Palade bodies

(Weibel and Palade, 1964). However, there is a wide variation in the frequency and ease of demonstration of Weibel-Palade bodies in various species (Gimbrone, 1977). Factor VIII, an antihemophilic factor, is synthesized by vascular endothelial cells (Jaffe *et al.*, 1973) and distinguishes endothelial cells from other cell types that might contaminate the culture. Another marker for endothelial cells is the angiotensin-converting enzyme, a cell surface enzyme found in all vascular endothelial cells examined (Caldwell *et al.*, 1976; Striker *et al.*, 1980).

The initial attempt to grow retinal blood vessels for extended periods was that of Tripathi *et al.*, (1973). They took advantage of the fact that the developing retinal vessels of the rabbit lie free on the retinal surface. The entire complex of growing retinal vessels was removed and placed in organ culture. Very limited growth was observed; the majority of the cultures survived for only 3 weeks. Cells of capillaries from bovine, monkey, and human retinas undergo limited proliferation *in vitro* (Buzney *et al.*, 1975; Frank, 1980). Exposure of the isolated capillaries to [^{3}H]thymidine revealed that the mural cells were undergoing DNA synthesis. The mural cells grew very slowly and required 2–4 months to reach confluence on a 60-mm culture plate and did not form cell lines. Capillary endothelial cells did not grow under these conditions. It should be noted that the retinal capillaries used in these studies were from adult animals. The pattern of cell growth observed in monkey retinal capillaries as a function of time in culture is shown in Fig. 29. In marked contrast to the lack of capillary endothelial cell proliferation realized from preparations obtained from the retinas of older animals, endothelial cells obtained from younger animals proliferated *in vitro* (Buzney and Massicotte, 1979; Frank *et al.*, 1979). Retinal vessels from fetal calf (Buzney and Massicotte, 1979) or kitten (Frank *et al.*, 1979) eyes yielded colonies consisting of endothelial cells, pericytes, and smooth muscle cells. Bovine endothelial cells displayed factor VIII immunofluorescence. The morphological appearance of kitten retinal capillary endothelial cells and of calf retinal capillary pericytes is shown in Fig. 30.

Under standard culture conditions endothelial cells from the younger animals did not reach confluence, were not subculturable, and became senescent within 3 months. However, when the endothelial cells were cocultured with viable mural cells, the endothelial cells reached confluence in primary culture. This suggests that conditioned medium (Puck and Marcus, 1955) might be required for serial cultivation of retinal capillary endothelial cells (Buzney and Massicotte, 1979). A determination of the mechanism by which retinal pericytes enhance the growth of retinal capillary endothelial cells might permit

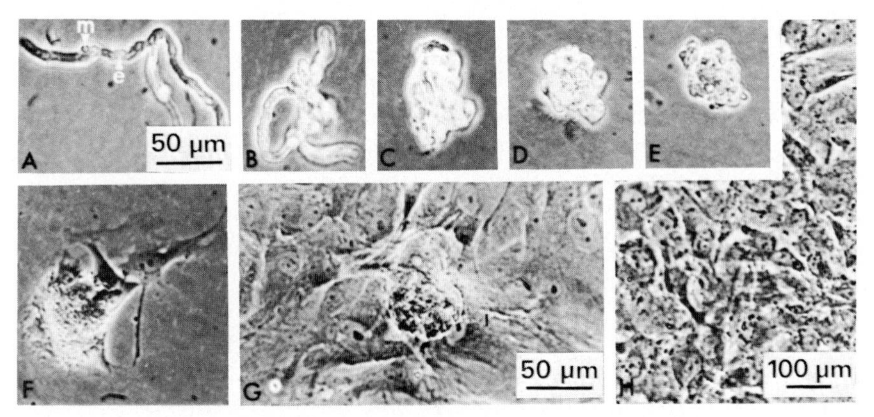

Fig. 29. Phase-contrast photomicrographs of a single rhesus monkey retinal capillary observed in culture for more than 30 days. (A) Day 0; (B) day 1; (C) day 2; (D) day 3; (E) day 6; (F) day 9; (G) day 17; (H) day 30. e, Endothelial cell; m, mural cell. (A to G) ×265; (H) ×81. (From Buzney *et al.*, 1975, with permission. Copyright 1975, American Association for the Advancement of Science.)

an understanding of the possible metabolic cooperation between these cells and lead to the identification of endothelial growth factors. It is somewhat of a paradox that there is a preferential loss of retinal pericytes in diabetic retinopathy *in vivo* (Kuwabara and Cogan, 1963), a situation associated with the growth of retinal endothelial cells, whereas growth of retinal capillary endothelial cells *in vitro* is enhanced by mural cells.

Although limited cell proliferation of endothelial cells was noted in cells from the immature retina (Buzney and Massicotte, 1979; Frank *et*

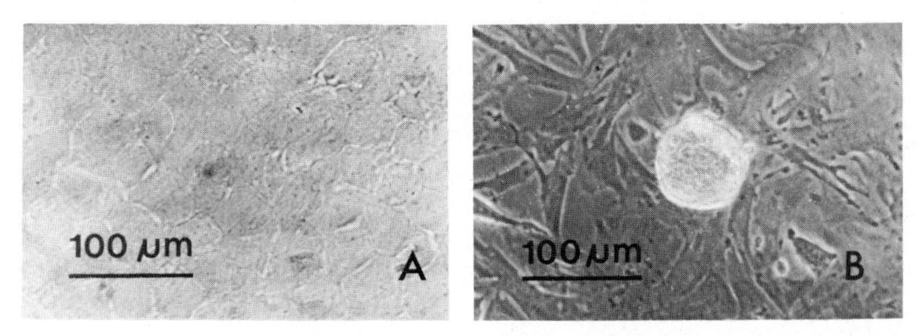

Fig. 30. (A) Kitten retinal capillary endothelial cells following 1 month in culture. Phase-contrast, ×200. (B) Calf retinal capillary pericytes following 2 weeks in culture. (From Frank *et al.*, 1979 with permission.)

al., 1979), endothelial cells from the retinas of older animals of several species, including bovine, monkey, human (Buzney *et al.*, 1975), calf, lamb, and dog (Frank *et al.*, 1979), did not grow when cultured under identical conditions. As is the case with lens epithelia (Reddan *et al.*, 1981a), the growth requirements of capillary endothelial cells and pericytes can vary as a function of donor age.

Techniques have recently been reported that permit serial culture of capillary endothelial cells from the adrenal gland of rats, calves, and humans (Folkman *et al.*, 1979; Zetter, 1980), a situation not yet realized for cells obtained from retinal capillaries. The specific innovations included the use of a nonplastic substrate and tumor-conditioned medium. It is known that certain transformed cells liberate factors into the culture medium that modulate their own growth (Todaro *et al.*, 1979) and stimulate the proliferation of other cell types (Rubin *et al.*, 1981). Bovine capillary endothelial cells have been cloned and maintained in serial passage for longer than 8 months (Folkman *et al.*, 1979) by using tumor-conditioned medium and gelatinized tissue culture plates coupled with the selective removal of nonendothelial cells. Bovine aortic endothelial cells, on the other hand, do not require tumor-conditioned medium for serial culture (Gospodarowicz *et al.*, 1978b; Folkman *et al.*, 1979). Human capillary endothelial cells cultured in microwell dishes could be passaged every 3 weeks for up to 5 months when cultured on an appropriate substrate in a medium supplemented with tumor-conditioned medium, FGF, human thrombin, and endothelial cell growth supplement. Whether the approach introduced by Folkman *et al.* (1979) will permit the serial culture of retinal capillary cells remains to be determined. The use of natural basement membrane components such as laminin (Reddan *et al.* 1981c) or various collagens produced by the retinal cells (Cohen *et al.*, 1980) as a tissue culture substrate in conjunction with the use of tumor-conditioned medium seems to be a suitable approach for enhancing the growth of cultured retinal capillary cells. Identification of the mitogenic factor(s) present in tumor-conditioned medium could lead to the growth of retinal capillary cells in serum-free medium.

F. The Role of Cell Proliferation in Retinal Detachment

"Despite its frequent occurrence, massive periretinal proliferation, the dreaded complication of retinal detachments, is a little understood disease" (Machemer *et al.*, 1978). Uncontrolled proliferation of retinal cells (Laqua and Machemer, 1975a; van Horn *et al.*, 1977) or of nonretinal cells introduced into the retinal environment subsequent to ocular injury (Topping *et al.*, 1979; Cleary and Ryan, 1979) can initiate a sequence of events leading to retinal detachment and blindness. The cells proliferate at epiretinal and/or subretinal sites, secrete basement membrane materials, and attach to the retina; subsequent contraction of the cells exerts a force which triggers retinal detachment (Kampik *et al.*, 1980). In view of important role of cell proliferation in retinal detachments, Laqua and Machemer (1975b) suggest the term "massive periretinal proliferation" (MPP) to describe the pathology.

In the nontraumatized eye the presence of cells, other than hyalocytes, near the inner aspect of the retina is thought to be due to a definite break in the internal limiting membrane of the retina (Clarkson *et al.*, 1977). The entire complex of cells and cell products which adhere to the inner aspect of the retina in pathological states are termed epiretinal, subretinal, or vitreoretinal membranes. Morphological examination of epiretinal membranes from humans reveals the presence of glial cells, pigment epithelial-like cells, fibrocyte-like cells, macrophage-like cells, and extracellular materials produced by the cells (van Horn *et al.*, 1977; Machemer, *et al.*, 1978; Kampik *et al.*, 1980). Although pigment epithelial cells and glial cells are the primary cells involved in epiretinal membrane formation, the identification and origin of specific cell types responsible for epiretinal membrane formation will require both immunological and morphological markers. One additional problem in cell identification is due to the fact that retinal pigment epithelial (RPE) cells become depigmented when they enter a program of active cell proliferation. Moreover, following photocoagulation of the rabbit retina, which is accompanied by heightened mitotic activity in the choroid, RPE, and Müller cells, an increase in cell proliferation has been reported to occur in the hyalocytes (Gloor, 1969). Thus another cell type, namely, the hyalocyte, could possibly play a role in MPP. In view of the dire consequences of MPP in humans, attempts have been made to develop animal models that mimic human MPP.

Studies on experimentally induced retinal detachment in the owl monkey (Machemer, 1968; Kroll and Machemer, 1968; Machemer and Norton, 1968) reveal a clinical and histological picture closely resembling human retinal detachment. In subretinal regions, the pigment epithelial cells migrate into the wound focus and undergo mitosis. The mitotic response is often of such magnitude that a multilayer of pigment epithelial cells and extracellular matrix material occupies the injured region. Subsequent detailed studies (Laqua and Machemer, 1975b) on monkey eyes show that proliferation of pigment epithelial and glial cells as evidenced by the presence of mitotic figures and the incorporation of [^{3}H]thymidine (Laqua and Machemer, 1975b; Machemer and Laqua, 1975) is a normal adjunct of induced retinal detachment. The cells grow and form "membranes" on the inner and outer retinal surfaces as well as in the vitreous humor. In addition to glial cells, three distinct cell types thought to be derived from retinal pigment epithelial cells (Machemer and Laqua, 1975), namely, pigment epithelial macrophages, fibrocyte-like pigment epithelial cells, and highly differentiated pigment epithelial cells, comprise the periretinal membranes. Contraction of the cells comprising the intravitreal, epiretinal (preretinal), or subretinal membrane results in retinal folds thus mimicking the situation in human MPP. In this regard, it is known that cloned chick RPE cells contain an array of microfilaments and undergo focal contractions (Crawford, 1979). It would be of interest to know the distribution of contractile proteins (Sanger and Sanger, 1976) in the cells comprising the epiretinal and subretinal membranes.

An additional experiment approach has been developed (Mandelcorn *et al.*, 1975; Radtke *et al.*, 1981) in an effort to understand the environmental factors and/or conditions that regulate cell division, epiretinal membrane formation, and retinal detachment. In these studies, freshly isolated RPE cells (Radtke *et al.*, 1981), RPE cells and intraocular fibroblasts (Trese *et al.*, 1980), cultured RPE cells (Radtke *et al.*, 1981), or nonocular fibroblasts (Sugita *et al.*, 1980) were injected into the vitreous humor of the rabbit. As the needle used for injection of specific cell types traverses the various layers of the eye, it undoubtedly carries with it certain cells, e.g., fibroblasts from the penetrated tissue. Autotransplantation of freshly isolated or cultured rabbit RPE cells into the midvitreal cavity resulted in the formation of intravitreal membranes (strands), traction detachment, and retinal puckers. In these experiments, cells and cell products were noted on the retinal surface and on the posterior surface of the lens. DNA synthesis was intense at 1 week following injection of the cells, and by day 28 elongated cells with prominent myofilaments were aligned on extracellular matrix material. Collagen, a normal product of RPE cells (Newsome and Kenyon, 1973), was interspersed between the cells (Radtke *et al.*, 1981). In an additional experiment, Mueller-Jensen *et al.*, (1974) placed a pure population of freshly isolated rabbit RPE cells in a tightly sealed diffusion chamber and inserted the chamber into the rabbit vitreous humor. The cells, which were effectively isolated from all other intraocular cell types, synthesized DNA, produced collagen, and resembled macrophages and fibrocytes. In agreement with the above-noted studies, intravitreal injection of freshly isolated RPE cells and fibroblasts (Trese *et al.*, 1980) or cultured rabbit fibroblasts (Sugita *et al.*, 1980; Tano *et al.*, 1980a) was followed by cell proliferation, membrane formation, and retinal detachment. Whether the injected cells release a bound mitogen from the vitreous humor, neutralize an inhibitor of cell growth (see Section II,D), or utilize the vitreous humor as a substrate which favors cell growth (see Section 1,G) remains to be determined. It is equally tenable that the vitreous humor enhances the production of mitogenic factor(s) from the injected cells.

The model systems are being used in an attempt to develop agents that will inhibit intravitreal cell proliferation. Recent experiments (Tano *et al.*, 1980a,b) suggest that steroids may be effective in reducing cell proliferation and retinal detachment following the intravitreal injection of rabbit fibroblasts into the rabbit vitreous humor. Whether the effect of the steroids is due to their antiinflammatory properties, to an action on some phase of the fibroblast cell cycle, or to the elimination of a portion of the fibroblast progenitor population cannot be ascertained from the present studies. It will be of interest to determine the efficacy of steroids and other pharmacological agents on the intravitreal growth of RPE cells and on glial cells following induced retinal detachment in the owl monkey, and to determine the effect of these compounds on the growth of RPE and glial cells following injection into the rabbit vitreous humor. A determination of tissue culture conditions required for the growth of human and rabbit RPE cells in a completely defined serum-free medium could provide a model system for testing the effects of steroids and other growth-inhibiting agents. It is equally important to understand the mechanisms responsible for production of the epiretinal membranes and to delineate the mechanisms responsible for the attachment of retinal or other cells to the retinal surface. Inhibition of cell division, prevention of the elaboration of extracellular matrix, or inhibition of the attachment of "membranes" to the retinal surface should curtail retinal detachment.

G. Epilogue

The environmental factors that regulate cell division in the developing, injured, or diseased retina reamin to be elucidated. Equally perplexing at present are the cellular mechanisms that control vascularization, neovascularization, and subretinal and epiretinal membrane formation. The challenge is not only to define the nature of the signals involved but to determine the mechanism by which they control cell division. What is clear, however, is that an understanding of the cell biology of specific retinal cells and the influence of local and systemic factors on these cells will be required to gain insight into the regulatory events. Extension of such studies to humans seems to be predicated on elucidation of the conditions required for the serial cultivation of human retinal cells.

ACKNOWLEDGMENTS

I thank Dorothy C. Dziedzic, Stephen J. McGee, Eric M. Goldenberg, and Del J. DeHart for their contributions to the research project. I am particularly indebted to Dr. Rene Humbel for his generous gift of IGF, to Dr. Judson Van Wyk for somatomedin C, to Drs. George R. Martin and Victor P. Terranova for laminin, and to Drs. Stanley Cohen and Denis Gospodarowicz for reference preparations of EGF and FGF, respectively.

The author's work was supported by grant EY-00362 from the National Eye Institute. The human lens work was part of the Cooperative Cataract Research Program.

REFERENCES

Adler, R., Freedman, J., Kukar, N., and Gombos, G. M. (1975). Various parameters of diabetic retinopathy among clinic populations. *Ann. Ophthalmol.* **7,** 1447-1454.

Albert, D. M., Rabson, A. S., Grimes, P. A., and von Sallmann, L. (1969). Neoplastic transformation *in vitro* of hamster lens epithelium by simian virus 40. *Science* **164,** 1077-1078.

Algire, G. H., and Chalkey, H. W. (1945). Vascular reactions of normal and malignant tissue *in vivo*. I. Vascular reactions of mice to wounds and to normal neoplastic transplants. *J. Natl. Cancer Inst.* **6,** 73-85.

Ashton, N. (1957). Retinal vascularization in health and disease. *Am. J. Ophthalmol.* **44,** 7-17.

Ashton, N., and Cook, C. (1954). Direct observation of the effect of oxygen on developing vessels: Preliminary report. *Br. J. Ophthalmol.* **38,** 433-440.

Awasthi, Y. C., Miller, S. P., Arya, D. V., and Srivastava, S. I. (1975). The effect of copper on human and bovine lens and on human cultured lens epithelium enzymes. *Exp. Eye Res.* **21,** 251-257.

Bakker, A. (1936a). Eine Methode die Linses erwachsener Kaninchen auszerhalb ders Korpers am Leben zur erhalten. *A. Graefes Arch. Ophthalmol.* **135,** 581-592.

Bakker, A. (1936b). Die Regeneration der verwandenten Linses-kapsel von Kaninchenlinsen in der Durchstrommungskultur. *A. Graefes Arch. Ophthalmol.* **136,** 333-340.

Barnes, D., and Sato, G. (1980). Methods for growth of cultured cells in serum-free medium. *Anal. Biochem.* **102,** 255-270.

Barritault, D., Arruti, C., and Courtois, Y. (1981). Is there a ubiquitous growth factor in the eye? Proliferation induced in different cell types by eye-derived growth factors. *Differentiation (Berlin)* **18,** 29-42.

Beebe, D. C., Feagans, D. E., Blanchette-Mackie, E. J., and Nau, M. E. (1979). Lens epithelial cell elongation in the absence of microtubules: Evidence for a new effect of colchicine. *Science* **206,** 836-839.

Beebe, D. C., Feagans, D. E., and Jebens, H. A. H. (1980). Lentropin: A factor in vitreous humor which promotes lens fiber cell differentiation. *Proc. Natl. Acad. Sci. U.S.A.* **77**, 490-493.

Bell, E., Marek, L. J., Levinstone, D. S., Merrill, C., Sher, S., Young, I. T., and Eden, M. (1978). Loss of division potential *in vitro:* Aging or differentiation? *Science* **202**, 1158-1163.

Bhat, S. P., Jones, R. E., Sullivan, M. A., and Piatigorsky, J. (1980). Chicken lens crystallin DNA sequences show at least two delta crystallin genes. *Nature (London)* **284**, 234-248.

Bito, L. Z. (1974). The effects of experimental uveitis on anterior uveal prostaglandin transport and aqueous humor composition. *Invest. Ophthalmol.* **13**, 959-966.

Bloemendal, H. (1977). The vertebrate eye lens. *Science* **197**, 127-138.

Bloom, G. D., and Balazs, E. A. (1965). An electron microscopic study of hyalocytes. *Exp. Eye Res.* **4**, 249-255.

Boeve, W. J. (1927). Het gedrag der buiten get oog gebrachte, verwonde lens in verschillende vloeistoffen. Dissertation, Groningen, Den Haag.

Bomboy, J. D., and Salmon, W. D. (1975). Somatomedin. *Clin. Orthop.* **108**, 228-240.

Bradley, R. F., Rees, S. B., and Fager, C. A. (1965). Pituitary ablation in the treatment of diabetic retinopathy. *Med. Clin. North Am.* **49**, 1105-1124.

Brem, S., Brem, H., Folkman, J., Finkelstein, D., and Patz, A. (1976). Prolonged tumor dormancy by prevention of neovascularization in the vitreous. *Cancer Res.* **36**, 2807-2812.

Brem, S., Preis, I., Langer, R., Brem, H., Folkman, J., and Patz, A. (1977). Inhibition of neovascularization by an extract derived from vitreous. *Am. J. Ophthalmol.* **84**, 323-328.

Briggs, R. C., Wainwright, N., and Rothstein, H. (1977). Biochemical events associated with healing of a chemical injury in the rabbit lens. *Exp. Eye Res.* **24**, 523-539.

Bryan, A., Leinfelder, P. J., and Meltzer, M. (1960). Effects of gases upon cell growth. *AMA Arch. Ophthalmol.* **63**, 132-139.

Buzney, S. M., and Massicotte, S. J. (1979). Retinal vessels: Proliferation of endothelium *in vitro. Invest. Ophthalmol. Visual Sci.* **18**, 1191-1195.

Buzney, S. M., Frank, R. N., and Robison, W. G. (1975). Retinal capillaries: Proliferation of mural cells *in vitro. Science* **190**, 985-986.

Caldwell, P. R. B., Seegal, B. C., Hsu, K. C., Das, M., and Soffer, R. L. (1976). Angiotensin-converting enzyme: Vascular endothelial localization. *Science* **191**, 1050-1051.

Cameron, I., and Jeter, J., Jr. (1971). Relationship between cell proliferation and cytodifferentiation in embryonic chick lens. *In* "Developmental Aspects of the Cell Cycle" (I. Cameron, G. Padilla, and A. Zimmerman, eds.), pp. 191-222. Academic Press, New York.

Campbell, F. W. (1957). The influence of a low atmospheric pressure on the development of the retinal vessels in the rat. *Trans. Ophthalmol. Soc. U.K.* **7**, 287-300.

Carpenter, G., and Cohen, S. (1978). Biological and molecular studies of the mitogenic effects of human epidermal growth factor. *In* "Molecular Control of Proliferation and Differentiation" (J. Papaconstantinou and W. J. Rutter, eds), pp. 13-31. Academic Press, New York.

Castellot, J. J., Karnovsky, M. J., and Spiegelman, B. M. (1980). Potent stimulation of vascular endothelial cell growth by differentiated 3T3 adipocytes. *Proc. Natl. Acad. Sci. U.S.A.* **77**, 6007-6011.

Cavallo, T., Sade, R. M., Folkman, J., and Cotran, R. S. (1972). Tumor angiogenesis: Rapid induction of endothelial cell mitoses demonstrated by autoradiography. *J. Cell Biol.* **54**, 408–420.

Cavallo, T., Sade, R. M., Folkman, J., and Cotran, R. S. (1973). Ultrastructural and autoradiographic studies of the early proliferative response in tumor angiogenesis. *Am. J. Pathol.* **70**, 345–362.

Chen, C. H., and Chen, S. C. (1980). Angiogenic activity of vitreous and retinal extract. *Invest. Ophthalmol. Visual Sci.* **19**, 596–602.

Chen, C. H., and Patz, A. (1976). Components of vitreous-soluble proteins: Effect of hyperoxia and age. *Invest. Ophthalmol.* **15**, 228–232.

Church, R. L. (1980). Cell hybrids in ocular tissues. *Curr. Top. Eye Res.* **13**, 41–89.

Clarkson, B., and Baserga, R., eds. (1974). "Control of Proliferation in Animal Cells." Cold Spring Harbor Lab., Cold Spring Harbor, New York.

Clarkson, J. G., Green, W. R., and Massof, D. (1977). A histopathologic review of 168 cases of preretinal membrane. *Am. J. Ophthalmol.* **84**, 1–17.

Clayton, R. M., Thomson, I., and de Pomerai, D. I. (1979). Relationship between crystallin mRNA expression in retina cells and their capacity to re-differentiate into lens cells. *Nature (London)* **282**, 628–629.

Cleary, P. E., and Ryan, S. J. (1979). Experimental posterior penetrating eye injury in the rabbit. I. Method of production and natural history. *Br. J. Ophthalmol.* **63**, 306–311.

Cogan, D. G., and Kuwabara, T. (1967). The mural cell in perspective. *Arch. Ophthalmol.* **78**, 133–139.

Cogan, D. G., Toussaint, D., and Kuwabara, T. (1961). Studies of retinal vascular patterns. IV. Diabetic retinopathy. *Arch. Ophthalmol.* **66**, 366–378.

Cohen, A. (1961). Electron microscopic observations of the developing mouse eye. I. Basement membranes during early development and formation. *Dev. Biol.* **3**, 297–316.

Cohen, M. P., Frank, R. A., and Khalifa, A. A. (1980). Collagen production by cultured retinal capillary pericytes. *Invest. Ophthalmol. Visual Sci.* **19**, 90–94.

Condon, P. I., and Sergeant, G. R. (1972). Ocular findings in homozygous sickle cell anemia in Jamaica. *Am. J. Ophthalmol.* **73**, 533–543.

Connelly, T. G. (1977). Pituitary enhancement of Wolffian lens regeneration *in vitro:* Spatial and temporal requirements. *J. Exp. Zool.* **200**, 359–364.

Connelly, T. G., Ortiz, J. R., and Yamada, T. (1973). Influence of the pituitary on Wolffian lens regeneration. *Dev. Biol.* **31**, 301–315.

Constant, M. A. (1958). *In vitro* lens studies: I. Technique and preliminary results of maintenance of lens culture for 24 hours. *Am. J. Ophthalmol.* **45**, 372–383.

Cotlier, E. (1962). Hypophysectomy effect on lens epithelium mitosis and galactose cataract development in rats. *Arch. Ophthalmol.* **67**, 176–182.

Coulombre, A. (1969). Regulation of ocular morphogenesis. *Invest. Ophthalmol.* **8**, 25–31.

Coulombre, J., and Coulombre, A. (1963). Lens development: Fiber elongation and lens orientation. *Science* **142**, 1489–1494.

Courtois, Y., Simmoneau, L., Tassin, J., Laurent, M. V., and Malaise, E. (1978). Spontanious transformation of bovine lens epithelial cells: Kinetic analysis and differentiation in monolayers and in nude mice. *Differentiation (Berlin)* **10**, 23–30.

Coulter, J. B., and Knebel, R. L. (1977). Entry of insulin into aqueous humor of rabbits after feeding. *Ophthalmic Res.* **9**, 269–275.

Coulter, J. B., Engelke, J. A., and Eaton, D. K. (1980). Insulin concentrations in aqueous humor after paracentesis and feeding of rabbits. *Invest. Ophthalmol. Visual Sci.* **19**, 1524-1526.

Crawford, B. (1979). Cloned pigmented retinal epithelium: The role of microfilaments in the differentiation of cell shape. *J. Cell Biol.* **81**, 301-315.

Creighton, M. O., Mousa, G. T., and Trevithick, J. R. (1976). Differentiation of rat lens epithelial cells in tissue culture. I. Effects of cell density, medium, and embryonic age of initial culture. *Differentiation (Berlin)* **6**, 155-167.

Cuny, R., and Zalik, S. E. (1981). Effect of bovine pituitary hormone preparations on newt lens regeneration *in vitro:* Stimulation by thyrotropin. *Dev. Biol.* **81**, 23-35.

Davis, M. D. (1974). Definition, classification, and course of diabetic retinopathy. *In* "Diabetic Retinopathy" (J. R. Lynn, W. B. Snyder, and A. Vaiser, eds), pp. 7-33. Grune and Stratton, New York.

Davis, M. D., and Engerman, R. L. (1980). Pathogenesis of diabetic retinopathy. *In* "Diabetic Renal-Retinal Syndrome" (E. A. Friedman and F. A. L'Esperance, eds.), pp. 69-76. Grune and Stratton, New York.

de Asua, L. H., Levi-Montalcini, R., Shields, R., and Iacobelli, S., eds. (1980). "Control Mechanisms in Animal Cells—Specific Growth Factors." Raven, New York.

Deckert, T., Simonsen, S., and Poulsen, J. E. (1967). Prognosis of proliferative retinopathy in juvenile diabetics. *Diabetes* **16**, 728-733.

Dilley, K. J., Bron, A. J., and Habgood, J. O. (1976). Anterior polar and posterior subcapsular cataract in a patient with retinitis pigmentosa: A light-microscopic and ultrastructural study. *Exp. Eye Res.* **22**, 155-167.

Dische, Z., and Borenfreund, E. (1954). Composition of the polysaccharide of the lens capsule and its topical distribution. *Am. J. Ophthalmol.* **38**, 165-173.

Dische, Z., and Zelmenis, G. (1965). The content and structural characteristics of the collagenous protein of rabbit lens capsules at different ages. *Invest. Ophthalmol.* **4**, 174-180.

Dobree, J. H. (1964). Proliferative diabetic retinopathy: Evolution of the retinal lesions. *Br. J. Ophthalmol.* **48**, 637-649.

Eakins, K. E. (1973). Ocular effects. *In* "The Prostaglandins" (I. Ramwell, ed.), Vol. 1, pp. 219-237. Plenum, New York.

Ehrmann, R. L., and Knoth, M. (1968). Choriocarcinoma: Transfilter stimulation of vasoproliferation in the hamster cheek pouch—Studied by light and electron microscopy. *J. Natl. Cancer Inst.* **41**, 1329-1341.

Engerman, R. L., and Bloodworth, J. M. B. (1965). Experimental diabetic retinopathy in dogs. *Arch. Ophthalmol.* **73**, 205-210.

Engerman, R. L., Pfaffenbach, D., and Davis, M. D. (1967). Cell turnover in capillaries. *Lab Invest.* **17**, 738-743.

Farnsworth, P. N., Burke, P., Dotto, M. E., and Cinotti, A. A. (1977). Ultrastructural abnormalities in a Marfan's syndrome lens. *AMA Arch. Ophthalmol.* **95**, 1601-1606.

Finkelstein, D., Brem, S., Patz, A., Folkman, J., Miller, S., and Chen, C. H. (1977). Experimental retinal neovascularization induced by intravitreal tumors. *Am. J. Ophthalmol.* **83**, 660-664.

Fischer, A. (1922). A three months old strain of epithelium. *J. Exp. Med.* **35**, 367-373.

Fischer, F. (1974). Retinopathia diabetica proliferans: Im Alter von 20 bis 30 Jahren. *Klin. Monatsbl. Augenheilkd.* **165**, 767-770.

Foidart, J. M., and Reddi, A. H. (1980). Immunofluorescent localization of type IV collagen and laminin during endochrondral bone differentiation and regulation by pituitary growth hormone. *Dev. Biol.* **75**, 130-136.

Folkman, J. (1974). Tumor angiogenesis. *Adv. Cancer Res.* **19**, 331-358.

Folkman, J., Merler, E., Abernathy, C., and Williams, G. (1971). Isolation of a tumor factor responsible for angiogenesis. *J. Exp. Med.* **133**, 275-288.

Folkman, J., Haudenschild, C. C., and Zetter, B. R. (1979). Long-term culture of capillary endothelial cells. *Proc. Natl. Acad. Sci. U.S.A.* **76**, 5217-5221.

Font, R. L., and Brownstein, S. (1974). A light and electron microscopic study of anterior subcapsular cataracts. *Am. J. Ophthalmol.* **78**, 972-984.

Frank, R. N. (1980). Tissue culture and biochemical studies of retinal microvessels. *In* "Diabetic Renal-Retinal Syndrome" (E. A. Friedman and F. A. L'Esperance, eds.), pp. 123-134. Grune and Stratton, New York.

Frank, R. N., Kinsey, V. E., Frank, K. Q., Mikus, K. P., and Randolph, A. (1979). *In vitro* proliferation of endothelial cells from kitten retinal capillaries. *Invest. Ophthalmol. Visual Sci.* **18**, 1195-1200.

Froesch, E. R., Zapf, J., Rinderknecht, E., Morel, B., Schoenle, E., and Humbel, R. E. (1979). Insulinlike growth factor (IGF-NSILA): Structure, function and physiology. *In* "Hormones and Cell Culture, Book A" (G. H. Sato and R. Ross, eds.), pp. 61-77. Cold Spring Harbor Lab., Cold Spring Harbor, New York.

Geeraets, W. J. (1972). Observations of lens epithelium in cell cultures. *Med. Col. Virginia Q.* **8**, 264-273.

Genis-Galvez, J., and Castro, J. (1971). Protein biosynthesis after lens rotation: An immunoelectrophoretic analysis in the chick embryo. *J. Exp. Zool.* **177**, 313-317.

Gimbrone, M. A. (1977). Culture of vascular endothelium and atherosclerosis. *In* "International Cell Biology 1976-1977" (B. R. Brinkley and K. R. Porter, eds.), pp. 649-658. Rockefeller Univ. Press, New York.

Gimbrone, M. A., Cotran, R. S., Leapman, S. P., and Folkman, J. (1974). Tumor growth and neovascularization: An experimental model using the rabbit cornea. *J. Natl. Cancer Inst.* **52**, 413-419.

Glaser, B. M., D'Amore, P. A., Michels, R. G., Brunson, S. K., Fenselau, A. H., Rice, T., and Patz, A. (1980a). The demonstration angiogenetic activity from ocular tissues: Preliminary report. *Ophthalmology* **87**, 440-446.

Glaser, B. M., D'Amore, P. A., Michels, R. G., Patz, A., and Fenselau, A. H. (1980b). Demonstration of vasoproliferative activity from mammalian retina. *J. Cell Biol.* **84**, 298-304.

Gloor, B. P. (1969). Mitotic activity in the cortical vitreous cells (hyalocytes) after photocoagulation. *Invest. Ophthalmol.* **8**, 633-646.

Golichenkov, V. A., Sakharova, N. Y., and Popov, V. V. (1966). Autonomy of the seasonal mitotic activity of the lenticular epithelium of the grass frog. *Dokl. Akad. Nauk USSR (Eng. transl.)* **166**, 43-44.

Gorthy, W. C. (1978). Cataracts in the aging rat lens: Morphology and acid phosphatase histochemistry of incipient forms. *Exp. Eye Res.* **27**, 301-322.

Gorthy, W. C. (1979). Developmental ocular abnormalities in rats with X-ray induced cataract mutation. *Invest. Ophthalmol.* **18**, 939-946.

Gorthy, W. C., and Abdelbaki, Y. Z. (1974). Morphology of a hereditary cataract in the rat. *Exp. Eye Res.* **19**, 147-156.

Gospodarowicz, D., and Ill, C. R. (1980a). Do plasma and serum have different abilities to promote cell growth? *Proc. Natl. Acad. Sci. U.S.A.* **77**, 2726-2730.

Gospodarowicz, D., and Ill, C. (1980b). The extracellular matrix and the control of proliferation of corneal endothelial and lens epithelial cells. *Exp. Eye Res.* **31**, 181-199.

Gospodarowicz, D., Mescher, A. L., Brown, K. D., and Birdwell, C. R. (1977). The role of fibroblast growth factor and epidermal growth factor in the proliferative response of the corneal and lens epithelium. *Exp. Eye Res.* **26**, 91-97.

Gospodarowicz, D., Brown, K. S., Birdwell, C. R., and Zetter, B. R. (1978a). The control of proliferation of vascular endothelial cells of human origin. I. Characterization of the response of human umbilical endothelial cells to the fibroblast growth factor, epidermal growth factor, and thrombin. *J. Cell Biol.* **77**, 774–788.

Gospodarowicz, D., Moran, J. S., and Mescher, A. L. (1978b). Cellular specificities of fibroblast growth factor and epidermal growth factor. *In* "Molecular Control of Proliferation and Differentiation" (J. Papaconstantinou and W. J. Rutter, eds.), pp. 33–79. Academic Press, New York.

Gospodarowicz, D., Vlodavsky, I., and Savion, N. (1981). The role of fibroblast growth factor and the extracellular matrix in the control of proliferation of corneal endothelial cells. *Vision Res.* **21**, 81–104.

Greenblatt, M., and Shubic, P. (1968). Tumor angiogenesis: Transfilter diffusion studies in the hamster by the transparent chamber technique. *J. Natl. Cancer Inst.* **41**, 111–124.

Grimes, P., and von Sallman, L. (1968). Lens epithelium proliferation in sugar cataracts. *Invest. Ophthalmol.* **7**, 535–543.

Grobstein, C. (1954). X. Tissue interaction in the morphogenesis of mouse embryonic rudiments *in vitro*. *In* "Aspect of Synthesis and Order in Growth, 13th Growth Symposia" (D. Rudnick, ed.), pp. 233–255. Princeton Univ. Press, Princeton, New Jersey.

Grobstein, C. (1962). Interactive processes in cytodifferentiation. *J. Cell Comp. Physiol.* **60**, 35–48.

Gross, J., Azizkhan, R. G., Biswas, C., Bruns, R. R., Hsieh, D. S. T., and Folkman, J. (1981). Inhibition of tumor growth, vascularization, and collagenolysis in the rabbit cornea by medroxyprogesterone. *Proc. Natl. Acad. Sci. U.S.A.* **78**, 1176–1180.

Gulati, A. K., and Reyer, W. (1980). Role of neural retina and vitreous body during lens regeneration: Transplantation and autoradiography. *J. Exp. Zool.* **214**, 109–118.

Gyllensten, L. J., and Hellstrom, B. E. (1954). Experimental approach to the pathogenesis of retrolental fibroplasia. I. Changes in the eye induced by exposure of newborn mice to concentrated oxygen. *Acta Pediatr. Suppl.* **43**, 7–17.

Ham, R. G., and McKeehan, W. L. (1978). Development of improved media and culture conditions for clonal growth of normal diploid cells. *In Vitro* **14**, 11–22.

Hamada, Y., and Okada, T. S. (1978). *In vitro* differentiation of cells of the lens epithelium of human fetus. *Exp. Eye Res.* **26**, 91–97.

Hamada, Y., Watanabe, K., Aoyama, H., and Okada, T. S. (1979). Differentiation and dedifferentiation of rat lens epithelial cells in short- and long-term cultures. *Dev. Growth Differ.* **21**, 205–220.

Hanna, C., and O'Brien, J. (1961). Cell production and migration in the epithelial layer of the lens. *Arch. Ophthalmol.* **66**, 96–99.

Harding, C. V., and Srinivasan, B. B. (1960). Stimulation of DNA synthesis and mitosis by injury. *Ann. N. Y. Acad. Sci.* **90**, 610–613.

Harding, C. V., and Srinivasan, B. B. (1961). A propagated stimulation of DNA synthesis and cell division. *Exp. Cell Res.* **25**, 326–340.

Harding, C. V., and Thayer, M. N. (1964). DNA synthesis and cell division in the cultured ocular lens. *Invest. Ophthalmol.* **3**, 302–313.

Harding, C. V., Donn, A., and Srinivasan, B. D. (1959). Incorporation of thymidine by injured lens epithelium. *Exp. Cell Res.* **18**, 582–585.

Harding, C. V., Rothstein, H., and Newman, M. B. (1962). The activation of DNA synthesis and cell division in rabbit lens *in vitro*. *Exp. Eye Res.* **1**, 457–465.

Harding, C. V., Reddan, J. R., Unakar, N. J., and Bagchi, M. (1971). The control of cell division in the ocular lens. *Int. Rev. Cytol.* **31**, 215–300.

Hasegawa, M. (1965). Restitution of the eye from the iris after removal of the retina and lens together with the eye-coats in the newt *Triturus pyrrhogaster*. *Embryologia* **8**, 362—386.

Hassell, J. R., Robey, P. G., Barrach, H. J., Wilczek, J., Rennard, S. I., and Martin, G. R. (1980). Isolation of a heparan sulfate-containing proteoglycan from basement membrane. *Proc. Natl. Acad. Sci. U.S.A.* **77**, 4494-4498.

Hay, E. D. (1978). Cell matrix interaction in embryonic induction. *In* "International Cell Biology" (B. R. Brinkley and K. R. Porter, eds.), pp. 50-57. Rockefeller Univ. Press, New York.

Hayden, J. H., and Rothstein, H. (1979). Complete elimination of mitosis and DNA synthesis in the lens of the hypophysectomized frog: Effects on cell migration and fiber growth. *Differentiation (Berlin)* **15**, 153-160.

Hayden, J. H., Rothstein, H., Worgul, B. V., and Merriam, G. R. (1980). Hypophysectomy exerts a radioprotective effect on frog lens. *Experientia* **36**, 116-118.

Hayflick, L. (1965). The limited *in vitro* lifetime of human diploid cell strains. *Exp. Cell Res.* **37**, 614-636.

Hendrix, R. W., and Zwaan, J. (1974). Cell shape regulation and cell cycle in embryonic lens cells. *Nature (London)* **247**, 145-147.

Hendrix, R. W., and Zwaan, J. (1975). The matrix of the optic vesicle-presumptive lens interface during induction of the lens in the chicken embryo. *J. Embryol. Exp. Morphol.* **33**, 1023-1049.

Henkind, P. (1978). Ocular neovascularization. *Am. J. Ophthalmol.* **85**, 287-301.

Hennings, H., Michael, D., Cheng, C., Steinert, P., Holbrook, K., and Yerspa, S. H. (1980). Calcium regulation of growth and differentiation of mouse epidermal cells in culture. *Cell* **19**, 245-254.

Hintz, R. L., Clemmons, D. R., Underwood, L. E., and Van Wyk, J. J. (1972). Competitive binding of somatomedin to the insulin receptors of adipocytes, chondrocytes, and liver membranes. *Proc. Natl. Acad. Sci. U.S.A.* **69**, 2351-2353.

Hollenberg, M. D. (1975). Receptors for insulin and epidermal growth factor: Relation to synthesis of DNA in cultured rabbit lens epithelium. *Arch. Biochem. Biophys.* **171**, 371-377.

Holley, R. W. (1980). Control of growth of kidney epithelial cells. *In* "Control Mechanisms in Animal Cells" (L. deAsua, R. Levi-Montalcini, R. Shields, and S. Iocabelli, eds.), pp. 15-25. Raven, New York.

Holley, R. W., Bohlen, P., Fava, R., Baldwin, J. G., Kleeman, G., and Armour, R. (1980). Purification of kidney epithelial cell growth inhibitors. *Proc. Natl. Acad. Sci. U.S.A.* **77**, 5989-5992.

Howard, A. (1952). Whole-mounts of rabbit lens epithelium for cytological study. *Stain Technol.* **27**, 313-315.

Jacobson, A. G. (1966). Inductive processes in embryonic development. *Science* **152**, 25-34.

Jaffe, E. A., Nachman, R. L., Becker, C. G., and Minick, C. R. (1973). Culture of human endothelial cells derived from umbilical veins. *J. Clin. Invest.* **52**, 2745-2756.

James, W. A. (1980). Historical aspects of diabetic retinopathy. *In* "Diabetic Renal-Retinal Syndrome" (E. A. Friedman and F. A. L'Esperance, eds.), pp. 37-42. Grune and Stratton, New York.

Kaighn, M. E. (1980). Human prostatic epithelial cell culture models. *Invest. Urol.* **17**, 382-385.

Kampik, A., Green, W. R., Michels, R. G., and Nase, P. K. (1980). Ultrastructural features of progressive idiopathic epiretinal membrane removed by vitreous surgery. *Am. J. Ophthalmol.* **90**, 810-816.

Kefalides, N. A., ed. (1978). "Biology and Chemistry of Basement Membranes." Academic Press, New York.

Kinsey, V. E., and Reddy, D. V. N. (1965). Studies on the crystalline lens. XI. The relative role or the epithelium and capsule in transport. *Invest. Ophthalmol.* **4,** 104-116.

Kinsey, V. E., and Zacharias, L. (1949). Retrolental fibroplasia. *J. Am. Med. Assoc.* **139,** 572-578.

Kinsey, V. E., Wachtl, C., Constant, M. A., and Camacho, E. (1955). Studies on the crystalline lens. VI. Mitotic activity in the epithelia of lenses cultured in various media. *Am. J. Ophthalmol.* **40,** 216-223.

Kinsey, V. E., Jacobus, J. T., and Hemphill, F. M. (1956). Retrolental fibroplasia: Cooperative study of retrolental fibroplasia and the use of oxygen. *AMA Arch. Ophthalmol.* **56,** 481-543.

Kinsey, V. E., Arnold, H. J., Kalina, R. E., Stern, L., Stahlman, M., Odell, G., Driscoll, J. M., Eliott, J. H., Payne, J., and Patz, A. (1977). PaO_2 levels and retrolental fibroplasia: A report of the cooperative study. *Pediatrics* **60,** 655-668.

Kirby, D. B. (1927). The cultivation of lens epithelium *in vitro. J. Exp. Med.* **45,** 1009-1016.

Kirby, D. B. Estey, K., and Tabor, F. (1929). Cultivation of lens epithelium *in vitro.* A further study. *Arch. Ophthalmol.* **1,** 358-365.

Kirby, D. B., Estey, K., and Wiener, R. von E. (1932). A study of the effect of changes in the nutrient medium on lens epithelium cultivated "*in vitro.*" *Trans. Am. Acad. Ophthalmol. Otolaryngol.* **37,** 196-212.

Kirkland, W. L., Yang, N., Jorgensen, T., Longley, C., and Furmanski, P. (1979). Growth of normal and malignant human mammary epithelial cells in culture. *J. Natl. Cancer Inst.* **63,** 29-41.

Kohner, E. M. (1974). A controlled trial of pituitary ablation. *In* "Diabetic Retinopathy" (J. R. Lynn, W. B. Snyder, and A. Vaiser, eds.), pp. 205-226. Grune and Stratton, New York.

Kohner, E. M., Joplin, G. F., Blach, R. K., Cheng, H., and Fraser, T. R. (1972). Pituitary ablation in the treatment of diabetic retinopathy. *Trans. Ophthalmol. Soc. U.K.* **92,** 79-90.

Kohner, E. M., Shilling, J. S., and Hamilton, A. M. (1976a). The role of avascular retina in new vessel formation. *Metab. Ophthalmol.* **1,** 15-23.

Kohner, E. M., Hamilton, A., Joplin, G. F., and Fraser, T. R. (1976b). Florid diabetic retinopathy and its response to treatment by photocoagulation or pituitary ablation. *Diabetes* **25,** 104-110.

Kroll, A. J., and Machemer, R. (1968). Experimental retinal detachment in the owl monkey. III. Electron microscopy of retina and pigment epithelium. *Am. J. Ophthalmol.* **66,** 410-427.

Kuck, J. F. R. (1970). Chemical constituents of the lens. *In* "Biochemistry of the Eye" (C. N. Graymore, ed.), pp. 183-260. Academic Press, New York.

Kuwabara, T., and Cogan, D. C. (1963). Retinal vascular patterns. VI. Mural cells of the retinal capillaries. *Arch. Ophthalmol.* **69,** 492-502.

Kuwabara, T., and Weidman, T. A. (1974). Development of the prenatal rat retina. *Invest. Ophthalmol.* **13,** 725-739.

Laatikainen, L., and Blach, R. K. (1977). Behavior of the iris vasculature in central retinal vein occlusion: A fluorescein angiographic study of the retina and the iris. *Br. J. Ophthalmol.* **61,** 272-277.

Lajtha, L. G. (1963). On the concept of the cell cycle. *J. Cell. Comp. Physiol.* **62,** 143-145.

Langer, R. L., Conn, H., Vacanti, J., Haudenschild, C., and Folkman, J. (1980). Control of

tumor growth in animals by infusion of an angiogenesis inhibitor. *Proc. Natl. Acad. Sci. U.S.A.* **77,** 4331-4335.
Laqua, H., and Machemer, R. (1975a). Glial cell proliferation in retinal detachment. (Massive periretinal proliferation.) *Am. J. Ophthalmol.* **80,** 602-618.
Laqua, H., and Machemer, R. (1975b). Clinical-pathological correlation in massive periretinal proliferation. *Am. J. Ophthalmol.* **80,** 913-929.
Lee, P. F., McMeel, J. W., Schepens, C. L., and Field, R. A. (1966). A new classification of diabetic retinopathy. *Am. J. Ophthalmol.* **62,** 207-219.
Leffert, H. L., and Koch, K. S. (1977). Control of animal cell proliferation. *In* "Growth, Nutrition and Metabolism of Cells in Culture" (G. H. Rothblat and V. J. Cristofalo, eds.), pp. 226-294. Academic Press, New York.
Leffert, H. L., and Koch, K. S. (1980). Ionic events at the membrane initiate rat liver regeneration. *Ann. N. Y. Acad. Sci.* **339,** 201-215.
L'Esperance, F. A. (1978). Diabetic retinopathy: Epidemiology and incidence. *Med. Clin. North Am.* **62,** 767-785.
L'Esperance, F. A. (1980). Natural history of diabetic retinopathy. In "Diabetic Renal-Retinal Syndrome" (E. A. Friedman and F. A. L'Esperance, eds.), pp. 43-68. Grune and Stratton, New York.
Luft, R., Olivecrona, H., Ikkos, D., Kornerup, T., and Ljunggren, H. (1955). Hypophysectomy in man: Further experience in severe diabetes mellitis. *Br. Med. J.* **2,** 752-756.
Lynn, J. R., Snyder, W. B., and Vaiser, A., eds. (1974). "Diabetic Retinopathy." Grune and Stratton, New York.
McAvoy, J. W. (1980). Induction of the eye lens. *Differentiation (Berlin)* **17,** 137-149.
Machemer, R. (1968). Experimental retinal detachment in the owl monkey. II. Histology of the retina and pigment epithelium. *Am. J. Ophthalmol.* **66,** 396-410.
Machemer, R., and Laqua, H. (1975). Pigment epithelium proliferation in retinal detachment. (Massive periretinal proliferation.) *Am. J. Ophthalmol.* **80,** 1-23.
Machemer, R., and Norton, E. W. D. (1968). Experimental retinal detachment in the owl monkey. I. Method of production and clinical picture. *Am. J. Ophthalmol.* **66,** 388-396.
Machemer, R., van Horn, D., and Aabert, T. (1978). Pigment epithelial proliferation in human retinal detachment with massive periretinal proliferation. *Am. J. Ophthalmol.* **85,** 181-191.
McDevitt, D., and Yamada, T. (1969). Acquisition of antigenic specificity by amphibian lens epithelial cells in culture. *Am. Zool.* **9,** 1130-1131.
McDevitt, D., Meza, I., and Yamada, T. (1969). Immunofluorescence localization of the crystallins in amphibian lens development, with special reference to the gamma crystallins. *Dev. Biol.* **19,** 581-607.
McMeel, J. W. (1971). Diabetic retinopathy: Fibrotic proliferation and retinal detachment. *Trans. Am. Ophthalmol. Soc.* **67,** 440-493.
Mamo, J. G., and Leinfelder, P. J. (1958). Growth of lens epithelium in culture. I. Characteristics of growth. *AMA Arch. Ophthalmol.* **59,** 417-419.
Mandelcorn, M. S., Machemer, R., Fineberg, E., and Hersch, S. B. (1975). Proliferation and metaplasia of intravitreal retinal pigment epithelium cell autotransplants. *Am. J. Ophthalmol.* **80,** 227-237.
Mann, I. (1948). Tissue culture of the mouse lens epithelium. *Br. J. Ophthalmol.* **32,** 591-596.
Mann, I. (1950). "The Development of the Human Eye." Grune and Stratton, New York.
Martin, G. M., Sprague, C. A., and Epstein, C. P. (1970). Replicative life-span of culti-

vated human cells. Effects of donor's age, tissue, and genotype. *Lab. Invest.* **23,** 86–92.

Masterson, E., and Russell, P., eds. (1981). Ocular tissue culture. *Vision Res.* **21,** 1–174.

Meezan, E., Brendel, K., and Carlson, E. C. (1974). Isolation of a purified preparation of metabolically active retinal blood vessels. *Nature (London)* **251,** 65–67.

Messier, B., and Leblond, C. P. (1960). Cell proliferation and migration as revealed by radioautography after injection of thymidine-H^3 into male rats and mice. *Am. J. Anat.* **106,** 247–265.

Michaelson, I. C. (1948). The mode of development of the vascular system of the retina with some observations on its significance for certain retinal diseases. *Trans. Ophthalmol. Soc. U.K.* **68,** 137–180.

Mikulicich, A., and Young, R. (1963). Cell proliferation and displacement in the lens epithelium of young rats injected with tritiated thymidine. *Invest. Ophthalmol.* **2,** 344–354.

Modak, S., Morris, G., and Yamada, T. (1968). DNA synthesis and mitotic activity during early development of the chick lens. *Dev. Biol.* **17,** 544–561.

Mueller-Jensen, K., Machemer, R., and Azarnia, R. (1974). Autotransplantation of retinal pigment epithelium in intravitreal diffusion chamber. *Am. J. Ophthalmol.* **80,** 530–537.

Newsome, D. A., and Kenyon, K. R. (1973). Collagen production *in vitro* by the retinal pigmented epithelium of the chick embryo. *Dev. Biol.* **32,** 387–400.

Numata, T., Constable, I. J., and Whiteny, D. E. (1975). Physical properties of experimental vitreous membranes. I. Tensile strength. *Invest. Ophthalmol.* **14,** 148–152.

Ogata, T., and Matsui, M. (1972). Electron microscopic studies on the capsule and epithelial cells of anterior polar cataract of human lens. *Acta. Soc. Ophthalmol. Jpn.* **76,** 1286–1297.

Okada, T. S. (1977). A demonstration of lens-forming cells in neural retina in clonal cell culture. *Dev. Growth Differ.* **19,** 47–55.

Okada, T. S., Eguchi, G., and Takeichi, M. (1973). The retention of differentiated properties of lens epithelial cells in clonal cell culture. *Dev. Biol.* **34,** 321–333.

Owens, R. B. (1976). Selective cultivation of mammalian epithelial cells. *Methods Cell Biol.* **15,** 341–344.

Pardee, A. B. (1974). A restriction point for control of normal animal cell proliferation. *Proc. Natl. Acad. Sci. U.S.A.* **71,** 1286–1290.

Patz, A. (1980). I. Studies on retinal neovascularization: Friedenwald lecture. *Invest. Ophthalmol. Visual Sci.* **19,** 1133–1138.

Patz, A., Eastham, A., Higginbotham, D. H., and Kleh, T. (1953). Oxygen studies in retrolental fibroplasia. II. The production of the microscopic changes of retrolental fibroplasia in experimental animals. *Am. J. Ophthalmol.* **36,** 1511–1522.

Pearce, T., and Zwaan, J. (1970). A light and electron microscopic study of cell behavior and microtubules in the embryonic chicken lens using Colcemid. *J. Embryol. Exp. Morphol.* **23,** 491–507.

Peehl, D. M., and Ham, R. G. (1980a). Growth and differentiation of human keratinocytes without a feeder layer or conditioned medium. *In Vitro* **16,** 516–525.

Peehl, D. M., and Ham, R. G. (1980b). Clonal growth of human keratinocytes with small amounts of dialyzed serum. *In Vitro* **16,** 526–540.

Phillips, L. S., and Vassilopoulou-Sellin, R. (1980a). Somatomedins. *N. Engl. J. Med.* **302,** 371–380.

Phillips, L. S., and Vassilopoulou-Sellin, R. (1980b). Somatomedins. *N. Engl. J. Med.* **302,** 438–445.

Philpott, G. W. (1970). Growth and cytodifferentiation of embryonic chick lens epithelial cells *in vitro*. *Exp. Cell Res.* **59**, 57-68.

Philpott, G. W., and Coulombre, A. (1965). Lens development. II. The differentiation of embryonic chick lens epithelial cells *in vitro* and *in vivo*. *Exp. Cell Res.* **38**, 635-644.

Piatigorsky, J. (1980). Intracellular ions, protein metabolism and cataract formation. *Curr. Top. Eye Res.* **3**, 1-39.

Piatigorsky, J., and Rothschild, S. (1972). Loss during development of the ability of chick embryonic lens cells to elongate in culture: Inverse relationship between cell division and elongation. *Dev. Biol.* **28**, 382-389.

Piatigorsky, J., Rothschild, S., and Wollberg, M. (1973). Stimulation by insulin of cell elongation and microtubule assembly in embryonic chick-lens epithelia. *Proc. Natl. Acad. Sci. U.S.A.* **70**, 1195-1198.

Piatigorsky, J., Shinohara, T., Bhat, S. P., Reszelbach, R., Jones, R. E., and Sullivan, M. A. (1980). Correlated changes in delta crystallin synthesis and ion concentration in the embryonic chick lens: Summary, current experiments, and speculations. *Ann. N. Y. Acad. Sci.* **339**, 265-279.

Prescott, D. M. (1976). "Reproduction of Eukaryotic Cells." Academic Press, New York.

Price, F. M., Comalier, R. F., Gantt, R., Taylor, W. G., Smith, G. H., and Sanford, K. K. (1980). A new culture medium for human skin epithelial cells. *In Vitro* **16**, 147-158.

Puck, T. T., and Markus, P. I. (1955). A rapid method for viable cell titration and clone production with HeLa cells in tissue culture: The use of X-irradiated cells to supply conditioning factors. *Proc. Natl. Acad. Sci.* **41**, 432-437.

Radtke, N. D., Tano, Y., Chandler, D., and Machemer, R. (1981). Stimulation of massive periretinal proliferation by intravitreal autotransplantation of retinal pigment epithelial cells in rabbits. *Am. J. Ophthalmol.* **91**, 76-87.

Rafferty, K. A., Jr. (1980). Methods in hand: Classic tissue culture. *In* "Birth Defects: Original Article Series. XVI. No. 2" (B. S. Danes, ed), pp. 5-34. A. R. Liss, Inc., New York.

Rafferty, N. S. (1963). Studies of an injury-induced growth in the frog lens. *Anat. Rec.* **146**, 299-312.

Rafferty, N. S. (1965). Propagation and prolongation of mitotic activity in the formation of injury-induced lentomas in *Rana pipiens*. *Anat. Rec.* **153**, 111-128.

Rafferty, N. S. (1967). Proliferative response in experimentally injured frog lens epithelium: Autoradiographic evidence for movement of DNA synthesis toward the injury. *J. Morphol.* **121**, 295-309.

Rafferty, N. S. (1973). Experimental cataract and wound healing in mouse lens. *Invest. Ophthalmol.* **12**, 156-160.

Rafferty, N. S., and Goossens, W. (1977). Ultrastructure of traumatic cataractogenesis in the frog: A comparison with mouse and human lens. *Am. J. Anat.* **148**, 385-408.

Rafferty, N. S., and Smith, R. (1976). Analysis of cell populations of normal and injured mouse lens epithelium. I. Cell cylce. *Anat. Rec.* **186**, 105-114.

Rafferty, N. S., Goossens, W., and Roth, A. (1979). Cytoplasmic filaments in aging lens of rodents. *Ophthalmic Res.* **11**, 276-282.

Reddan, J. R. (1974). Development and structure of the lens. *In* "Cataract and Abnormalities of the Lens" (J. Bellows, ed.), pp. 29-42. Grune and Stratton, New York.

Reddan, J. R., and Rothstein, H. (1965). Influence of temperature on wound healing in a poikilotherm. *Exp. Cell Res.* **40**, 442-445.

Reddan, J. R., and Rothstein, H. (1966). Growth dynamics of an amphibian tissue. *J. Cell. Physiol.* **67**, 307-318.

Reddan, J. R., and Unakar, N. J. (1976). Electron microscopy of cultured mammalian lenses. II. Changes preceding and accompanying insulin-induced mitosis. *Invest. Ophthalmol.* **15,** 411–417.

Reddan, J. R., and Wilson, D. C. (1978). Insulin growth factor triggers mitosis in the epithelium of mammalian lenses cultured in a serum-free medium. *J. Cell Biol.* **67a,** CU-307.

Reddan, J. R., and Wilson, D. C. (1979). Insulin growth factors I and II stimulate mitosis in the epithelia of mammalian lenses cultured in a serum-free medium. *J. Cell. Biol.* **83,** CU-469.

Reddan, J. R., and Wilson-Dziedzic (1982). Both insulin growth factor and epidermal growth factor trigger mitosis in lenses cultured in a serum-free medium. *Invest. Ophthalmol. Visual Sci.* (in press).

Reddan, J. R., Crotty, M. W., and Harding, C. V. (1970a). Characterization of macromolecular synthesis in the epithelium of cultured rabbit lenses. *Exp. Eye Res.* **9,** 165–174.

Reddan, J. R., Harding, C. V., Unakar, N. J., and Bell, R. M. (1970b). Electron microscopy of cultured mammalian lenses. I. Initial changes which precede and accompany the stimulation of DNA synthesis and mitosis. *Invest. Ophthalmol.* **9,** 496–515.

Reddan, J. R., Harding, C. V., Rothstein, H., Crotty, M. W., Lee, P., and Freeman, N. (1972). Stimulation of mitosis in the vertebrate lens in the presence of insulin. *Ophthalmic Res.* **3,** 65–82.

Reddan, J. R., Unakar, N. J., Harding, C. V., Bagchi, M., and Saldana, G. (1975). Induction of mitosis in the cultured rabbit lens initiated by the addition of insulin to medium KEI-4. *Exp. Eye Res.* **20,** 45–61.

Reddan, J. R., Unakar, N. J., Krasicky, E., and Wilson, D. C. (1976a). Triggering of mitosis in the cultured mammalian lens by proteolytic enzymes. *Doc. Ophthalmol. Proc. Ser.* **8,** 67–73.

Reddan, J. R., Harding, C., Harding, D., Weinsieder, A., Unakar, N., Shapiro, R., and Mathews, C. (1976b). Seasonal mitotic activity and wound healing in a teleost ocular lens. *Experientia* **31,** 1026–1027.

Reddan, J. R., Weinsieder, A., and Wilson, D. C. (1979). Aqueous humor from traumatized eyes triggers cell division in the epithelia of cultured lenses. *Exp. Eye Res.* **28,** 267–276.

Reddan, J. R. Friedman, T. B., Mostafapour, M. K., Sutherland, S. H., Bondy, R. L., McGee, S. J., and Goldenberg, E. M. (1980). Establishment of epithelial cell lines from individual rabbit lenses. *J. Tissue Culture Methods* **6,** 57–60.

Reddan, J. R., Friedman, T. B., Mostafapour, M. K., Bondy, R. L., Sutherland, S. H., McGee, S. J., and Goldenberg, E. M. (1981a). Donor age influences the growth of rabbit lens epithelial cells *in vitro*. *Vision Res.* **21,** 11–23.

Reddan, J. R., McGee, S., Goldenberg, E., Mostafapour, K., Dziedzic, D., and DeHart, D. (1981b). Laminin permits growth of lens cells in a serum-free medium. *Invest. Ophthalmol. Visual Sci.* (*Suppl.*) **20,** 90.

Reddan, J. R., Dziedzic, D. C., and McGee, S. J. (1982). Thrombin induces cell division in rabbit lenses cultured in a completely defined serum-free medium. *Invest. Ophthalmol. Visual Sci.* (in press). (8)

Reddan, J. R., DeHart, D. J. and Sackman, J. E. (1981). Influence of extracellular matrix on the growth of cultured mammalian epithelial cells. *J. Cell Biol.* **91,** 148a.

Reddy, D. (1968). Composition of normal aqueous humor. *In* "Biochemistry of the Eye" (M. Dardenne and J. O. Nordman, eds.), pp. 167–186. Karger, Basel.

Reyer, R. W. (1971). DNA synthesis and the incorporation of labelled iris cells into the lens during lens regeneration in adult newts. *Dev. Biol.* **24,** 535–558.

Reyer, R. W. (1977a). Morphological evidence for lens differentiation from intraocular implants of lens epithelium in *Ambystoma maculatum*. *Exp. Eye Res.* **24,** 511–522.

Reyer, R. W. (1977b). Repolarization of reversed regenerating lenses in adult newts, *Notophthalmus viridescens*. *Exp. Eye Res.* **24**, 501–519.

Reyer, R. W. (1977c). The amphibian eye: Development and regeneration. In "Handbook of Sensory Physiology. VII/5. The Visual System in Vertebrates." (F. Crescitelli, ed.), pp. 309–390. Springer-Verlag, Berlin and New York.

Rheinwald, J. G. (1979). The role of terminal differentiation in the finite culture lifetime of the human epidermal keratinocyte. *Int. Rev. Cytol. Suppl. 10,* 25–33.

Rheinwald, J. G. (1980). Serial cultivation of normal human epidermal keratinocytes. *Methods Cell Biol.* **21**, 230–255.

Rheinwald, J. G., and Green, H. (1975). Serial cultivation of strains of human epidermal keratinocytes: The formation of keratinizing colonies from single cells. *Cell* **6**, 317–330.

Rinderknecht, E., and Humbel, R. E. (1976). Polypeptides with nonsuppressible insulin-like and cell growth promoting activities in human serum: Isolation, chemical characterization, and some biological properties of forms I and II. *Proc. Natl. Acad. Sci. U.S.A.* **73**, 2365–2369.

Rinderknecht, E., and Humbel, R. E. (1978a). The amino acid sequence of human insulin-like growth factor I and its structural homology with proinsulin. *J. Biol. Chem.* **253**, 2769–2776.

Rinderknecht, E., and Humbel, R. E. (1978b). Primary structure of human insulin-like growth factor II. *FEBS Lett.* **89**, 283–286.

Rink, H., and Vornhagen, R. (1980). Rat lens epithelial cells *in vitro*. II. Changes of protein patterns during aging and transformation. *In Vitro* **16**, 277–280.

Robb, R. M., and Marchevsky, A. (1978). Pathology of the lens in Down's syndrome. *Arch. Ophthalmol.* **96**, 1039–1042.

Rosenbaum, D. M., and Rothstein, H. (1972). Mitotic variations in the lens epithelium of the frog. *Ophthalmic Res.* **3**, 95–107.

Rothstein, H., and Harding, C. V. (1962). Injury-induced synthesis of deoxyribonucleic acid in the lens of the sea bass. *Nature (London)* **194**, 294–295.

Rothstein, H., Reddan, J., and Weinsieder, A. (1965). Response to injury in the lens epithelium of the bullfrog (*Rana catesbiana*). II. Spatiotemporal patterns of DNA synthesis and mitosis. *Exp. Cell Res.* **37**, 440–451.

Rothstein, H., Fortin, J., and Youngerman, M. L. (1966). Synthesis of macromolecules in epithelial cells of the cultured amphibian lens (DNA and RNA). *Exp. Cell Res.* **44**, 303–311.

Rothstein, H., Fortin, J., and Bagchi, M. (1967). Influence of actinomycin-D upon repair of lenticular injuries. *Exp. Eye Res.* **6**, 292–296.

Rothstein, H., Van Buskirk, R., and Reddan, J. (1976). Hypophysectomy inhibits wound hyperplasia in the adult frog lens. *Ophthalmic Res.* **8**, 43–54.

Rothstein, H., Van Wyk, J. J., Hayden, J. H., Gordon, S. R., and Weinsieder, A. (1980). Somatomedin C: Restoration *in vivo* of cycle traverse in G_0/G_1 blocked cells of hypophysectomized animals. *Science* **208**, 410–412.

Rozengurt, E., and Mendoza, S. (1980). Monovalent ion fluxes and the control of cell proliferation in cultured fibroblasts. *Ann. N. Y. Acad. Sci.* **339**, 175–190.

Rubin, N. A., Tarsio, J. F., Borthwick, A. C., Gregerson, D. S., and Reid, T. W. (1981). Identification and characterization of a growth factor secreted by an established cell line of human retinoblastoma maintained in serum-free medium. *Vision Res.* **21**, 105–112.

Russell, P., Fukui, H. N., Tsunematsu, Y., Huang, F. L., and Kinoshita, J. H. (1977). Tissue culture of lens epithelial cells from normal and Nakano mice. *Invest. Ophthalmol. Visual Sci.* **16**, 243–246.

Salmon, W. D., Jr., and Daughaday, W. H. (1957). A hormonally controlled serum factor which stimulates sulfate incorporation by cartilage *in vitro*. *J. Lab. Clin. Med.* **49**, 825–836.

Sanger, J. W., and Sanger, J. M. (1976). Actin localization during cell division. *In* "Cell Motility" (R. Goldman, T. Pollard, and J. Rosenbaum, eds.), pp. 1295–1316. Cold Spring Harbor Lab., Cold Spring Harbor, New York.

Schaper, A. (1987). Die fruhesten differenzirungs vorgange im Centralnervensystem. *Arch. Entwicklungsmech. Org.* **5**, 81–132.

Schwartz, S. M., and Benditt, E. P. (1973). Cell replication in the aortic endothelium: A new method for study of the problem. *Lab. Invest.* **28**, 699–707.

Shabo, A. L., and Maxwell, D. S. (1976). Insulin-induced immunogenic retinopathy resembling the retinitis proliferans of diabetes. *Trans. Am. Acad. Ophthalmol. Otol.* **81**, 497–508.

Shabo, A. L., and Maxwell, D. S. (1977). Experimental immunogenic proliferative retinopathy in monkeys. *Am. J. Ophthalmol.* **83**, 471–480.

Shabo, A. L., Maxwell, D. S., and Shintako, P. I. (1980). Animal models of proliferative disease. *In* "Diabetic Renal-Retinal Syndrome" (E. A. Friedman and F. A. L'Esperance, eds.), pp. 113–122. Grune and Stratton, New York.

Shapiro, A. L., Siegel, I. M., Scharff, M. D., and Robbins, E. (1969). Characteristics of cultured lens epithelium. *Invest. Ophthalmol.* **8**, 393–400.

Shilling, J. S., and Kohner, E. M. (1976). New vessel formation in retinal branch vein occlusion. *Br. J. Ophthalmol.* **60**, 810–815.

Shinohara, T., and Piatigorsky, J. (1980). Anion and cation effects on delta crystallin synthesis in the cultured embryonic chick lens and in a reticulocyte lysate. *Exp. Eye Res.* **30**, 351–360.

Silver, J., and Hughes, A. (1973). The role of cell death during morphogenesis of the mammalian eye. *J. Morphol.* **140**, 159–170.

Speiser, P., Gittelsohn, A. M., and Patz, A. (1968). Studies on diabetic retinopathy. *Arch. Ophthalmol.* **80**, 332–337.

Spemann, H. (1905). Ueber Linsenbildung nach experimenteller Entfernung der Primaren Lensenbildungszellen. *Zool. Anz.* **28**, 419–432.

Sreevalsan, T., Rozengurt, J., Taylor-Papadimitriou, J., and Burchell, J. (1980). Differential effect of interferon on DNA synthesis, 2-deoxyglucose uptake and ornithine decarboxylase activity in 3T3 cells stimulated by polypeptide growth factors and tumor promoters. *J. Cell. Physiol.* **104**, 1–9.

Stone, L. S. (1957). Regeneration of iris and lens in hypophysectomized adult newts. *J. Exp. Zool.* **136**, 17–34.

Stoner, G. D. (1980). Explant culture of human peripheral lung. *Methods Cell Biol.* **21**, 65–77.

Striker, G. E., Harlan, J. M., and Schwartz, S. M. (1980). Human endothelial cells *in vitro*. *Methods Cell Biol.* **21**, 135–151.

Sugita, G., Tano, Y., Machemer, R., Abrams, G., Claflin, A., and Fiorentino, G. (1980). Intravitreal autotransplantation of fibroblasts. *Am. J. Ophthalmol.* **89**, 121–130.

Takano, K., Hizuka, N., Shizume, K., Hasumi, Y., and Tsushima, T. (1980). Effect of nutrition on growth and somatomedin A levels in the rat. *Acta Endocrinol. Logica* **94**, 321–326.

Tamura, S. (1965). Long-term cultures of epithelial cells of a rabbit lens. *Jpn. J. Ophthalmol.* **9**, 177–181.

Tano, Y., Chandler, D., and Machemer, R. (1980a). Treatment of intraocular proliferation with intravitreal injection of triamcinolone acetonide. *Am. J. Ophthalmol.* **90**, 810–816.

Tano, Y., Sugita, G., Abrams, G., and Machemer, R. (1980b). Inhibition of intraocular cellular proliferation with intravitreal corticosteroids. *Am. J. Ophthalmol.* **89,** 131-136.

Tassin, J., Simonneau, L., and Courtois, Y. (1976). Epithelial lens cells: A model for studying *in vitro* aging and differentiation. *In* "Biology of the Epithelial Lens Cells" (Y. Courtois and F. Regnault, eds.), pp. 146-162. INSERM, Paris.

Tassin, J., Malaise, E., and Courtois, Y. (1979). Human lens cells have an *in vitro* proliferative capacity inversely proportional to the donor age. *Exp. Cell Res.* **123,** 388-392.

Taylor-Papadimitriou, J., Shearer, M., and Watling, D. (1978). Growth requirements of calf lens epithelium in culture. *J. Cell Physiol.* **95,** 95-104.

Taylor-Papadimitriou, J., Purkis, P., and Fentiman, I. S. (1980). Cholera toxin and analogues of cyclic AMP stimulate the growth of cultured human mammary epithelial cells. *J. Cell. Physiol.* **102,** 317-321.

Terry, T. L. (1942). Fibroblastic overgrowth of persistent tunica vasculosa lentis in infants born prematurely. *Trans. Am. Ophthalmol. Soc.* **40,** 262-284.

Thesleff, I., Barrach, H. J., Foidart, J. M., Vaheri, A., Pratt, R. M., and Martin, G. R. (1981). Changes in the distribution of type IV collagen, laminin, proteoglycan, and fibronectin during mouse tooth development. *Dev. Biol.* **81,** 182-192.

Timpl, R., Rohde, H., Rennard, S. I., Foidart, J. M., and Martin, G. R. (1979). Laminin: A glycoprotein from basement membranes. *J. Biol. Chem.* **254,** 9933-9937.

Todaro, G. J., De Larco, J. E., Marquardt, H., Bryant, M. L., Sherwin, S. A., and Sliski, A. H. (1979). Polypeptide growth factors produced by tumor cells and virus-transformed cells: A possible growth advantage for the producer cells. *In* "Hormones and Cell Culture" (G. H. Sato and R. Ross, eds.), pp. 113-127. Cold Spring Harbor Lab., Cold Spring Harbor, New York.

Topping, T. M., Abrams, G. W., and Machemer, R. M. (1979). Experimental double-perforating injury of the posterior segment in rabbit eyes. *Arch. Ophthalmol.* **97,** 735-742.

Trese, M. R., Spitznas, M., Foos, R. Y., and Hall, M. O. (1980). Experimental tractional retinal detachment in rabbits: Clinical picture and histopathologic features. *A. Graefes Arch. Klin Ophthalmol.* **214,** 213-222.

Treton, J. A., and Courtois, Y. (1981). Evolution of the distribution, proliferation, and UV-repair capacity of rat lens epithelium cells as a function of maturation and aging. *Mech. Ageing Dev.* **15,** 251-267.

Tripathi, B., Ashton, N., and Knight, G. (1973). Effect of oxygen on the developing retinal vessels of the rabbit. III. Mode of growth of rabbit retinal vessels in tissue culture. *Exp. Eye Res.* **15,** 321-351.

Unakar, N., Harding, C., Reddan, J., and Shapiro, R. (1973). Characterization of wound healing in the rabbit lens. I. Light and electron microscopic observations. *J. Microsc. (Paris)* **16,** 309-320.

Unakar, N. J., Binder, L. I., Reddan, J. R., and Harding, C. V. (1975). Histochemical localization of acid phosphatase in the injured rabbit lens. *Ophthalmic Res.* **7,** 158-169.

Unakar, N. J., Smart, T., Reddan, J. R., and Devlin, I. (1979). Regression of cataracts in the offspring of galactose-fed rats. *Ophthalmic Res.* **11,** 52-64.

Unger, W. G., Cole, D. F., and Hammond, B. (1975). Disruption of the blood-aqueous barrier following paracentesis in the rabbit. *Exp. Eye Res.* **20,** 255-270.

Urrets-Zavalia. A. (1977). "Diabetic Retinopathy." Masson Publishing U.S.A., Inc., New York.

Valone, J. A., and McNeel, W. (1978). Severe adolescent-onset proliferative diabetic retinopathy. *Arch. Ophthalmol.* **96,** 1349-1353.

Van Buskirk, R., Worgul, B., Rothstein, H., and Wainwright, N. (1975). Mitotic variations in the lens epithelium of the frog. III. Somatotropin. *J. Gen. Comp. Endocrinol.* **25,** 52-59.

van Buul-Offers, S., Dumoleinj, L., Hackeng, W. *et al.* (1979). The Snell dwarf mouse: Interrelationship of growth in length and weight, serum somatomedin activity and sulfate incorporation in costal cartilage during growth hormones, thyroxine and somatomedin treatment. *In* "Somatomedins and Growth" (G. Giordano, J. J. Van Wyk, and F. Minuto, eds.), pp. 281-283. Academic Press, New York.

van Horn, D. L., Aaberg, T. M., Machemer, R., and Fenzl, R. (1977). Glial cell proliferation in human retinal detachment with massive periretinal proliferation. *Am. J. Ophthalmol.* **84,** 383-393.

Van Wyk, J., Underwood, L., Lister, R., and Marshall, R. (1973). The somatomedins. *Am. J. Dis. Child.* **126,** 705-711.

Varma, S., and Reddy, V. (1972). Phospholipid composition of aqueous humor, plasma and lens in normal and alloxan diabetic rabbits. *Exp. Eye Res.* **13,** 120-125.

Virchow, R. (1863). "Die Krankhaften Geschwulste." August Hirschwald, Berlin.

von Sallmann, L. (1952). Experimental studies on early lens changes after roentgen irradiation. III. Effect of X-irradiation on mitotic activity and nuclear fragmentation of lens epithelium in normal and cysteine-treated rabbit. *Arch. Ophthalmol.* **47,** 305-320.

von Sallmann, L., and Grimes, P. (1966). Effect of age on cell division, ^{3}H-thymidine incorporation, and diurnal rhythm in the lens epithelium of rats. *Invest. Ophthalmol.* **5,** 560-567.

von Sallmann, L., and Moore, D. (1948). Electrophoretic patterns of concentrated aqueous humor of rabbit, cattle, and horse. *Arch. Ophthalmol.* **40,** 279-284.

von Sallmann, L., Tobias, C., Anger, H., Welch, C., Kimura, S., Munoz, C., and Drungis, A. (1955). Effect of high energy pattern, x-rays and aging on lens epithelium. *AMA Arch. Ophthalmol.* **54,** 489-514.

von Sallmann, L., Caravaggio, L., Munoz, C. M., and Drungis, A. (1957). Species differences in the radiosensitivity of the lens. *Am. J. Ophthalmol.* **43,** 693-704.

von Sallmann, L., Grimes, P., and McElvain, N. (1962). Aspects of mitotic activity in relation to cell proliferation in the lens epithelium. *Exp. Eye Res.* **1,** 449-456.

Wachs, H. (1920). Restitution des Auges nach Extirpation von Retina und Lense bei Triton. *Arch. Entwicklungsmech. Org.* **46,** 328-390.

Wachtl, C., and Kinsey, V. E. (1958). Studies on the crystalline lens: VIII. A synthetic medium for lens culture and the effects of various constituents on cell division in the epithelium. *Am. J. Ophthalmol.* **46,** 288-292.

Wainwright, N., Rothstein, H., and Gordon, S. R. (1976). Mitotic variation in the lens epithelium of the frog. IV. Studies with isolated pituitary factors. *Growth* **40,** 317-328.

Wainwright, N., Hayden, J., and Rothstein, H. (1978). Total disappearance of cell proliferation in the lens of a hypophysectomized animal. *In vivo* and *in vitro* maintenance of inhibition with reversal by pituitary factors. *Cytobios* **23,** 79-92.

Walton, D. S., and Grant, W. M. (1968). Retinoblastoma and iris neovascularization. *Am. J. Ophthalmol.* **65,** 568-599.

Weibel, E. R., and Palade, G. E. (1964). New cytoplasmic components in arterial endothelia. *J. Cell Biol.* **23,** 101-112.

Weinsieder, A., and Roberts, L. (1979). Effects of thyroid hormones on cell proliferation in the cornea and lens of hypophysectomized adult frogs. *IRCS Med. Sci. Libr. Compend.* **7**, 578.

Weinsieder, A., and Roberts, L. (1980). Thyroidal hormones restore cell proliferation to the lenses of hypophysectomized adult frogs. *Gen. Comp. Endocrinol.* **40**, 268-274.

Weinsieder, A., Rothstein, H., and Drebert, D. (1973). Lenticular wound healing: Evidence for genomic activation. *Cytobiologie* **7**, 406-417.

Weinsieder, A., Briggs, R., Reddan, J., Rothstein, H., Wilson, D., and Harding, C. (1975). Induction of mitosis in ocular tissues by chemotoxic agents. *Exp. Eye Res.* **20**, 33-44.

Weinsieder, A., Reddan, J., and Wilson, D. (1976). Aqueous humor in lens repair and cell proliferation. *Exp. Eye Res.* **23**, 355-363.

Wise, G. N. (1956). Retinal neovascularization. *Trans. Am. Ophthalmol. Soc.* **54**, 729-826.

Wise, G. N., Dollery, C. T., and Henkind, P. (1971). "The Retinal Circulation." Harper, New York.

Wolbarsht, M. L., and Landers, M. B., III (1980). The rationale of photocoagulation therapy for proliferative diabetic retinopathy: A review and model. *Ophthal. Surg.* **11**, 235-245.

Wolfe, R. A., Sato, G. H., and McClure, B. D. (1980). Continuous culture of rat C6 glioma in serum-free medium. *J. Cell Biol.* **87**, 434-441.

Worgul, B., and Merriam, G. (1979). Effect of endotoxin-induced intraocular inflammation on the rat lens epithelium. *Invest. Ophthalmol. Visual Sci.* **18**, 401-408.

Worgul, B. V., and Rothstein, H. (1975). Radiation cataract and mitosis. *Ophthalmic Res.* **7**, 21-32.

Worgul, B., Bito, L., and Merriam, G., Jr. (1977). Intraocular inflammation produced by X-irradiation of the rabbit eye. *Exp. Eye Res.* **25**, 53-62.

Wrenn, J., and Wessels, N. (1968). An ultrastructural study of lens invagination in the mouse. *J. Exp. Zool.* **171**, 359-367.

Wright, A. D., Kohner, E. M., Oakley, N. W., Fraser, T. R., Joplin, G. F., and Hartog, M. (1965). Serum growth hormone levels and the response of diabetic retinopathy to pituitary ablation. *Br. Med. J.* **2**, 346-349.

Yamada, T. (1967). *Curr. Top. Dev. Biol.* **2**, 247-283.

Yamada, T. (1977). Control mechanisms in cell-type conversion in newt lens regeneration. *Monogr. Dev. Biol.* **13**, 1-126.

Yamamoto, Y. (1976). Growth of lens and ocular environment: Role of neural retina in the growth of mouse lens as revealed by an implantation experiment. *Dev. Growth Differ.* **18**, 273-278.

Young, R. W., and Ocumpaugh, D. E. (1966). Autoradiographic studies on the growth and development of the lens capsule in the rat. *Invest. Ophthalmol.* **5**, 583-593.

Zalokar, M. (1944). Contribution a l'etude de la regeneration du cristallin chez le triton. *Rev. Suisse Zool.* **51**, 444-521.

Zelenka, P., and Piatigorsky, J. (1974). Isolation and *in vitro* translation of delta-crystallin mRNA from embryonic chick lens fibers. *Proc. Natl. Acad. Sci. U.S.A.* **71**, 1896-1900.

Zelenka, P., and Piatigorsky, J. (1976). Reiteration frequency of delta crystallin DNA in lens and non-lens tissues of chick embryos. *J. Biol. Chem.* **251**, 4294-4298.

Zetter, B. R. (1980). Migration of capillary endothelial cells is stimulated by tumor-derived factors. *Nature (London)* **285**, 41-43.

Zwaan, J., and Kirkland, B. M. (1975). Malorientation of mitotic figures in the early lens rudiment in aphakic mouse embryos. *Anat. Rec.* **182**, 345-354.

Zwaan, J., and Pearce, T. (1970). Mitotic activity in the lens rudiment of the chicken embryo before and after the onset of the crystallin synthesis. *Wilhelm Roux' Arch. Entwicklungsmech. Org.* **164,** 313-320.

Zwaan, J., and Williams, R. M. (1968). Morphogenesis of the eye lens in a mouse strain with hereditary cataracts. *J. Exp. Zool.* **169,** 407-422.

Zwaan, J., Bryan, P., and Pearce, T. (1969). Interkinetic nuclear migration during the early stages of lens formation in the chick embryo. *J. Embryol. Exp. Morphol.* **21,** 71-83.

8

Retinoids in Ocular Tissues: Binding Proteins, Transport, and Mechanism of Action

GERALD J. CHADER

Cell Biology of the Eye

ISBN 0-12-483180-X

I. RETINOID STRUCTURE

Vitamins were first described as substances found in trace amounts in the diet that were needed for normal growth, function, and reproduction in an animal species. It soon became apparent that vitamins could be divided into two general classes: water-soluble and fat-soluble. About 1915, McCollum and his co-workers determined that a particular fat-soluble dietary component was important for normal growth of animals (McCollum and Davis, 1915) and coined the term "fat-soluble factor A" for this substance. Steenbock and others subsequently noted that the active factor A was particularly associated with yellow vegetables as well as with animal fats and fish liver oils. He made the suggestion that these yellow compounds or carotenoids could be converted to other compounds of high chemical potency and biological activity within the body. We now know that this is exactly the case. A range of carotenoids can act as precursors of retinoids in animal tissues. The most common and efficient precursor is all-*trans*-β-carotene which has the structure of two retinyl moieties attached head to head through a double bond as depicted in Fig. 1. Although they act as precursors of retinoids, α- and β-carotenes contain only one β-ionone ring; the second ring varies in structure and has no vitamin A activity. It is interesting to note that the carotene moiety, like all fat-soluble vitamins, is derived from isoprenoid building blocks. Similarly, steroids and steroid hormones are derived from the basic isoprene unit.

Retinoids occur in animal tissues in more than one chemical form (Fig. 1), the most important differences being at the C-15 position. The most commonly found type, vitamin A or retinol, has an alcoholic hydroxyl function at the R group. The aldehyde form, retinal, is used in the visual process, while retinoic acid is a normal further oxidative metabolite found in small quantities in tissues. Small differences in the β-ionone ring also occur (Fig. 1). Vitamin A_1 is the predominant form in mammals and most higher animals. Vitamin A_2 (dehydrovitamin A) is found more selectively, as in fish liver oils, and differs from vitamin A_1 by the presence of an additional double bond in the ring at

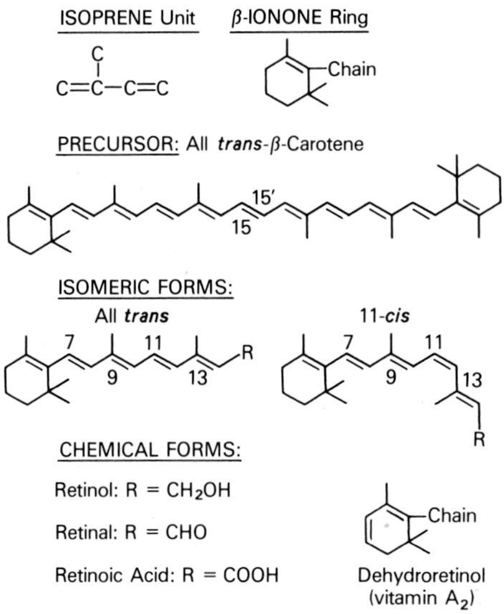

Fig. 1. Common retinoid structures.

the 3,4-position. The A_2 variant appears to be derived from the vitamin A_1 compound in most cases and to have about half the potency of vitamin A_1 in mammalian tissue studies.

Vitamin A also occurs in more than one steric configuration (Fig. 1). Cis and trans isomers of retinoids can differ in the side chain at the 7-, 9-, 11-, or 13-position. These isomers appear to serve specific physiological functions as does, for example, the 11-cis compound in the visual process.

II. RETINOID UPTAKE

A. Dietary Uptake and Serum Transport

Retinoids can be taken up directly from animal sources in the diet. Alternatively, carotenoids from yellow and green leafy vegetables can act as vitamin A precursors. In the gut, oxidative cleavage of the central double bond of the β-carotene or other carotenoid yields molecules of retinal by the action of β-carotene-15,15′-dioxygenase (Huang and Goodman, 1965). This is reduced in the intestinal mucosa to the more stable alcohol form, retinol. Dietary retinyl esters are also hydrolyzed in the intestine, and the free retinol absorbed into the mucosal cells.

Retinol is then esterified (mainly as the palmitate), complexed with other lipids and proteins in the form of chylomicra, and transported through the lymph and blood to the liver (Fig. 2). In the liver, it is stored predominantly in the ester form, apparently in a small number of specific fat-storing cells rather than in Kupffer or parenchymal cells (Hirosawa and Yamada, 1973). Upon demand, the ester is hydrolyzed to the free alcohol and complexed with the serum retinol-binding protein (RBP) (Kanai *et al.*, 1968) which is synthesized in the liver (Muto *et al.*, 1972). Release of retinol from the liver is always in concert with RBP in a 1:1 stoichiometric ratio. Little if any free retinol appears to be released or present in the circulating blood. The synthesis and release of RBP and its retinol ligand is largely dependent on the nutritional vitamin status of an animal. Vitamin A deficiency blocks RBP release from the liver, leading to increased stores of hepatic apo-RBP (i.e., free RBP). With vitamin A repletion, RBP can be rapidly mobilized into the circulation as a protein–vitamin (RBP-A) complex or holo-RBP. Smith *et al.* (1978) have demonstrated the influence of retinol on RBP synthesis and secretion using cultured liver cells. Thus, the control of RBP synthesis and secretion of the RBP-A moiety is dependent on a sophisticated system of extracellular nutritional and biochemical factors and intracellular control mechanisms.

In the serum of most species, RBP-A is transported as a ternary complex in association with plasma prealbumin (PA). Heller (1975b) was unable to find PA in bovine serum, however, and concluded that RBP-A circulated free and uncomplexed in this particular animal. PA is an important transport protein for thyroxine (T) in many species: Thus the final circulating complex is A-RBP-PA-T. Binding sites on the PA molecule for RBP-A and for T are independent and apparently noninteracting. It has been known that hyperthyroidism can lead to vitamin A deficiency and that plasma levels of RBP and PA are lowered in this condition (Smith and Goodman, 1971). More recently, Bhat and Cama (1978) have reported that thyroxine treatment leads to enhanced turnover of plasma RBP and PA and increased liver stores of vitamin A. An interesting and potentially important linkage between thyroid hormone transport and that of vitamin A thus is indicated.

B. Cellular Uptake

On a molecular level, the retinoid (retinol) first arrives at a target cell tightly bound to the serum RBP transport protein (Fig. 2). Studies from the laboratories of Heller (1975a) and of Maraini and co-workers (Maraini and Gozzoli, 1975) have helped our understanding of the ac-

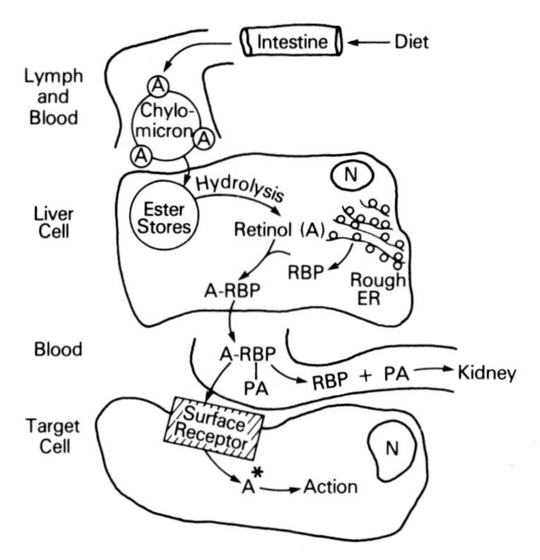

Fig. 2. Schema of vitamin A transport to a model target cell.

tual process of uptake of vitamin A into pigment epithelium (PE) cells. Similar results have also been obtained in a study utilizing intestinal mucosal cells (Rask and Peterson, 1976).

In studies by Heller (1975a), bovine or human RBP was iodinated to form [^{125}I]RBP, and the retinol-[^{125}I]RBP complex was presented to bovine PE cells *in vitro*. The retinol-[^{125}I]RBP was bound specifically to the PE cells in a temperature-dependent manner. Approximately 3.7–5.2 $\times$ 10^4 molecules of the complex bound to each PE cell; no binding was observed to isolated retinal photoreceptor outer segments. The bound retinol-[^{125}I]RBP was rapidly displaced by the addition of unlabeled retinol-RBP but not by apo-RBP, indicating specificity and reversibility. This study also showed that the binding was to surface membranes and was not due to endocytosis. Autoradiographic studies confirmed the presence of membrane receptors for the retinol-[^{125}I]RBP complex on PE cells (Bok and Heller, 1976). Receptors are confined to the plasma membrane surrounding the basal and lateral surface of the cell and are not found in the apical region. Bok and Heller also found little evidence of the presence of [^{125}I]RBP in PE cell cytoplasm. This strongly suggests that the RBP moiety itself is not translocated into the cell but rather, after losing its ligand, it remains in the serum and is processed elsewhere in the body.

In a parallel *in vitro* system, Maraini (Maraini and Gozzoli, 1975; Maraini, 1979) obtained results similar to those of Heller. He also found a very rapid uptake of [^{3}H]retinol into PE cells when it was bound to RBP, but little [^{3}H]retinol uptake when the vitamin was presented in the free form. The rapidity of uptake in this system suggests that an active transport process of retinol is present in PE cells when the retinol is bound to RBP.

From these data, it appears that specific membrane receptors (acceptors) are present on PE cells, which bind the retinol–RBP complex with high affinity but do not interact with either free retinol or RBP to any great extent. Retinol could then be translocated into the cell in a yet unknown but specific manner. This mechanism would allow for a quantitatively greater uptake of retinol as well as selectively of uptake, i.e., only into tissues which possess membrane acceptors for the retinol-RBP complex.

The situation in the cornea is of necessity quite different, since it is an avascular tissue. Vitamin A must therefore reach the corneal cells by diffusion through the tear film, through the aqueous humor, or from the scleral blood vessels in the limbus. Studies by Rask *et al.* (1980) point to delivery from the limbus region as the route of vitamin A entry. In their studies, Rask and co-workers found that bovine corneal epithelial cells readily accumulated [^{3}H]retinol complexed to serum RBP in a specific, saturable manner, as do PE cells and intestinal cells. Little or no RBP could be detected in either the tear fluid or aqueous humor, while both RBP and PA were found in the cornea in concentrations up to 2% of those in serum. Moreover the concentrations were greater in the peripheral than in the central region of the cornea, as would be expected if diffusion proceeded from the limbus vessels. Tanaka (1980) has recently showed that free [^{3}H]retinol injected into the aqueous humor can be quickly taken up into the various layers of the cornea. As assessed by autoradiography, silver grains were mainly localized in the cytoplasm of epithelial cells by 24 hr but also clustered around the nuclei of some cells. The silver grain distribution was more diffuse in the epithelium at this time. After experimental wounding of the cornea, silver grains were localized in the tip of the sliding epithelium, indicating that retinoids may play an important role in the wound-healing process.

III. RETINOID METABOLISM

Since all-*trans*-retinol is the retinoid form transported in the serum, it is also the form presented to target tissues for uptake and use in the cell. Once inside the cell, however, it must often undergo metabolic conversion before it can participate in a specific cellular process. This is best exemplified in the retina, where retinol must be converted to retinal before it can interact with opsin to form rhodopsin. The study of general vitamin A metabolism in epithelial tissues, however, has been mainly directed toward elucidation of a general active tissue metabolite of vitamin A and also toward understanding the mechanism of inactivation and removal of the vitamin from tissues.

Interest in the search for active forms of vitamin A was spurred in 1946 by the findings of Arens and von Dorp (1946) who reported that retinoic acid could support the growth of vitamin A-deficient rats, although it did not seem to be able to be converted to either retinal or retinol (Sharman, 1949; Dowling and Wald, 1960). Retinoic acid does not substitute for retinol in vision or reproduction, however, suggesting either that different retinoid forms are active in different cellular processes or that another metabolite derived from the alcohol or acid acts as an active form in most cell types. Attempts to identify such a tissue metabolite have been inconclusive. The situation is complicated by the fact that only a small amount of retinoic acid is formed from retinol in tissues and that this, in turn, is quickly converted to many other forms through isomerization, oxidation, decarboxylation, esterification, and conjugation (Zile *et al.*, 1980). One of these forms having high biological activity in some tissues is 13-*cis*-retinoic acid (Grubbs *et al.*, 1977). 5,8-Oxyretinoic acid has been identified in extracts from the small intestine of rats, although it is thought that this compound is formed during the extraction procedure and that the natural product *in vivo* is 5,6-epoxyretinoic acid (Napoli *et al.*, 1978).

Retinyl phosphate and retinyl phosphate sugar derivatives have also been isolated and characterized as authentic enzymatic tissue products (Peterson *et al.*, 1976; DeLuca, 1977) which may function in cellular glycosylation processes (Section V,B). Intriguing but scanty data have been published by Chen and Heller (1977) showing that retinoic acid can be reduced to a "retinol-like" compound in PE. A similar compound "X" has been proposed by DeLuca *et al.* (1979) in cultured 3T12 cells (Section V,B). Other metabolic products with some biological potency have been detected but, as with the retinol-like compound, it is not known if they lie on activation or inactivation pathways (Frolik *et al.*, 1978).

Since retinoic acid is not effective in all retinoid-requiring functions and since no other tissue metabolite has been identified which appears to act in this general manner, it could very well be that retinol, retinoic acid, and possibly yet unknown metabolities function independently in different metabolic processes in different tissues. Alternatively, it yet may be that a metabolite such as the glucosidic derivatives of retinyl phosphate or the retinol-like metabolite could be the key to understanding a unified theory of the mechanism of action of retinoids (other than in the visual process).

IV. PHYSIOLOGICAL ACTION OF RETINOIDS

In comparison to the information concerning its chemistry, biosynthesis, absorption, and transport, little is known about the precise cellular effects of vitamin A or its mechanism(s) of action. In general, retinoid action can be divided into two arbitrary categories: (1) functions for which only retinol will suffice and (2) functions in which retinoic acid can also be utilized (Table I). The two main functions associated with the obligatory need for retinol are reproduction (Thompson, 1969) and the visual process (Dowling and Wald, 1960). In the visual process in particular, vitamin A (retinol) is used in a manner analogous to that of a classical cofactor, with interconversion between it and the aldehyde form, retinal. It is tenaciously held onto and recycled by the retina-PE unit of the eye, as is the case with most true cofactors or coenzymes in tissues, and is minimally metabolized or degraded and removed. In other body functions (e.g., growth), either retinol or retinoic acid is effective. This is best seen in epithelial tissues where either of the retinoids promotes growth, induces differentiated characteristics, and maintains proper function. It is possible, however, that the retinol is converted to an active metabolite in these tissues, i.e., retinoic acid or other retinoid forms as discussed above.

Ocular function is perhaps the best body indicator of vitamin A sufficiency or deficiency and exemplifies (1) the need for retinol as a specific cofactor in some processes (e.g., the visual cycle) and (2) the use of either retinol or retinoic acid in other more general cellular processes, as in the cornea. International guidelines for vitamin A deficiency are based on eye signs, with night blindness the first and most sensitive

TABLE I

Retinoid Functions

Retinoid	Function
Retinol ↓	Specific: maintenance of specific tissue function as in vision or reproduction
Retinoic acid ↓	General: necessary for normal growth and differentiation of epithelial tissues and bone and their maintenance in a differentiated state

indicator. This condition is due to the lack of retinol as a cofactor in the visual cycle of the retina. The process is reversible in the early stages, with no overt changes in retina morphology. In somewhat more advanced stages of vitamin A deficiency, other ocular tissues become clinically involved. Overt cellular changes take place in the cornea and conjunctiva resulting in markedly altered epithelial morphology and function. Xerophthalmia or "dry eye" occurs when the goblet cells of the conjunctiva no longer produce the precorneal film that lubricates the exterior surface of the eye. Corneal cells then actually change in phenotype and take on a keratinized appearance. Either retinol or retinoic acid effectively promotes mucus production and counteracts this apparent keratinization. Thus the different tissues of the eye dramatically illustrate the pleiotropic effects of retinoids on cell function.

A. Visual Cycle

The visual process can be viewed as a series of reactions or steps involving retinoid transformations and complex interactions with protein moieties—a cycle designed to preserve the integrity of the retinoid.

On a molecular level, the retinoid (retinol) is first recognized and bound by the serum RBP transport protein and is translocated to the target cell (e.g., a PE cell). A membrane acceptor is next involved; the retinol alone is not recognized by the PE cell but rather only the RBP–retinol complex as discussed above. In the PE cell, retinol is esterified and stored. Upon demand, it is then hydrolyzed and further transported to the photoreceptor outer segment, probably bound to an intracellular transport protein. Along the way, oxidation and isomerization convert the retinoid to the 11-*cis*-retinal form which is suitable for binding to the opsin molecule. After opsin binding and participation in the light-dark visual cycle, it can then remain in the photoreceptor (probably as the aldehyde) or be reconverted to the alcohol and move back to the PE for storage and recycling (Section V,A).

Thus, in many ways, the use of retinoid in the visual cycle is similar to cofactor handling in tissues. Cofactors are specifically needed in cellular reactions, reversibly changed in form, and tenaciously conserved in the reaction cycle with little terminal metabolic transformation or inactivation. These conditions also hold true for retinoids in the photoreceptor–PE functional unit.

B. General Cell Effects

Growth and differentiation of many tissues are affected by retinoids. The processes of growth and of differentiation, although related, are certainly not identical and may be mutually exclusive under some circumstances. Growth is due to a quantitative increase in cell number or cell size or a combination of both factors. Differentiation reflects more the qualitative makeup of cellular constituents, although quantitative considerations are important as well. In ocular tissues, Friedenwald *et al.* (1945) first found a decrease in mitosis in the corneal epithelium of vitamin A-deficient rats. Similarly, Pirie and Overall (1972) showed that vitamin A deficiency in the lens epithelium was typified by nuclear abnormalities and a decreased number of dividing cells. These distinctions have also been examined by Zile, DeLuca, and co-workers who found that vitamin A deficiency interfered with DNA synthesis in intestinal mucosal cells (Zile *et al.,* 1977), as well as with expression of cellular differentiation. More recently, they found that vitamin A deficiency reduced the total number of cells in several tissues of the body, indicating a general role of vitamin A in the process of cellular proliferation (Zile *et al.,* 1979).

Retinoid control of differentiation has been best described in epithelial tissues, although other tissues such as bone appear to be equally affected (Fell, 1969). Many organs of the body contain or are lined with epithelial cells. The gastrointestinal tract, for example, has columnar epithelial cells (goblet cells) which are strongly influenced by vitamin A. Similarly, cells in the bladder, lung, and trachea are dependent on retinoid for the maintenance of columnar epithelial characteristics, with vitamin A deficiency rapidly leading to a stratified, squamous epithelium. In ocular tissues, smooth, normal corneal epithelial cells become thicker and roughened with vitamin A deficiency *in vivo* and *in vitro* (Aydelotte, 1963; Carter-Dawson *et al.,* 1980), while epithelial cells of the conjunctiva change from a secretory type to a keratinized type. Retinoids quickly reverse these effects except when they are in terminal stages. Pirie (1977), for example, found that topically applied retinoic acid could reverse the effects of xerosis and swelling of the cornea seen in xerophthalmia.

Conversely, cells which do not normally exhibit epithelial mucus-secreting characteristics can be induced to do so by vitamin A. Epidermal tissue of the chick embryo, for example, which normally produces keratin can be induced to change to columnar mucus-secreting cells by high amounts of retinoid in organ culture (Fell, 1957). The metaplastic change induced is virtually the reverse of that seen in mucus-secreting tissue with a vitamin A deficiency. In general, both retinoic acid and retinol are effective in promoting these changes.

Thus, both the processes of growth (cell number) and differentiation (functional characteristics) are affected by retinoids in many tissues.

C. Retinoids and Cancer

A very high percentage of cancer deaths result from metastasizing tumors of epithelial tissues. Even in the eye, retinoblastoma can be considered an epithelial-type cancer, since the retina is of neuroepithelial origin. Low serum retinol levels have been reported to be associated with an increased risk of cancer (Wald *et al.,* 1980).

Vitamin A has a profound effect on epithelial tissues, a deficiency leading to hyperkeratosis of the skin and metaplasia of the epithelia lining many internal organs. Such metaplastic changes are similar in many regards to neoplastic changes seen during the earlier stages of the cancer process. On the other hand, it has been known for many years that retinoid administration inhibits the development of several types of cancer *in vivo* (Saffiotti *et al.,* 1967), as well as *in vitro* (Crocker and Sanders, 1970). Jetten *et al.* (1979) have shown that retinoic acid induces long-term, stable differentiation of undifferentiated embryonal carcinoma cells in culture rather than only allowing selective growth of small numbers of already differentiated cells. These types of studies have led to the general postulate that retinoids may play a role in the prevention of chemical carcinogenesis and, in a broader sense, have prophylactic activity in oncology and dermatology (Sporn *et al.,* 1976; Mayer *et al.,* 1978).

The mechanism by which vitamin A exerts its protective effect is not known. It has been suggested that it induces the immune system *in vivo* but, although this may be true, it would not explain the striking results seen *in vitro.* Retinoid deficiency enhances the binding of a metabolite of the carcinogen benzo[*a*]pyrene to DNA in tracheal epithelial cells (Genta *et al.,* 1974), indicating possible action at the gene level. More recently it has been shown that retinoids block cell mitogenesis and transformation produced by a sarcoma growth factor (SGF) in murine sarcoma virus-transformed cells (Todaro *et al.,* 1978). In these cells, it is thought that damage (viral?) at the gene level could lead to the production of growth-promoting and transforming factors (e.g., SGF) or their membrane receptors, which are normally produced only during embryonic development. Natural suppressors of these growth and transforming factors may exist, which would be active at specific times of development and allow and maintain the normal differentiation of embryonal core cells. Retinoids may function in this capacity *in vivo* by directly antagonizing SGF effects at the gene level or elsewhere in the cell and promoting normal growth and differentiation.

Although it is not known whether retinoids act through normal cellular channels or by pharmacological mass action, they seem to be of future importance in both cancer prevention and treatment.

V. MOLECULAR ACTION OF RETINOIDS

The use of retinoids in molecular events of the visual process is quite well known. It is confined solely to the photoreceptor outer segment, however, with similar chemical transformations not observed in other tissues. Thus, the mechanism(s) of action of vitamin A in general cell types and tissues is poorly understood and still controversial. Similarly, there are no molecular markers of a general nature in cells (e.g., the induction of marker enzymes) which have been shown to be specifically dependent on retinoids. Other criteria must therefore be used in assessing retinoid action. Besides overt morphological changes, one of the most striking features induced by retinoids is the increase in cellular adhesivity observed with many cell types in culture (Dion *et al.*, 1978; Lotan, 1980). The mechanism for this increase is still obscure, but it does not appear to involve modulation of the cyclic nucleotide concentration (DeLuca *et al.*, 1979). To explain these specific effects, as well as the more general effects of retinoids, two theories have been proposed involving cytoplasmic and nuclear actions of retinoids. In the cytoplasm, it is likely that vitamin A is involved in glycosyltransferase reactions and the modification of cellular glycoconjugate biosynthesis (DeLuca, 1977; Wolf, 1977). In the nucleus, it has been proposed that vitamin A controls events in a manner analogous to the steroid hormones (Wolf and DeLuca, 1969; Chytil and Ong, 1978; Chader *et al.*, 1981). These mechanisms of action are certainly not mutually exclusive and may in fact function separately or together at different stages of development, to explain the highly variably and sometimes dramatically opposite results seen with retinoids in different tissues.

A. The Visual Process

The visual cycle, driven by alternating sequences of light and dark, results in the movement of retinoid between the retina and PE (Fig. 3). As mainly Wald has shown (Wald, 1968), it also results in chemical transformations and interconversions of retinoids that now constitute the best understood series of reactions involving retinoids (Table II).

In vision, the PE cell and the retinal photoreceptor outer segment function as a single unit. The PE cell stores retinoid in the ester form,

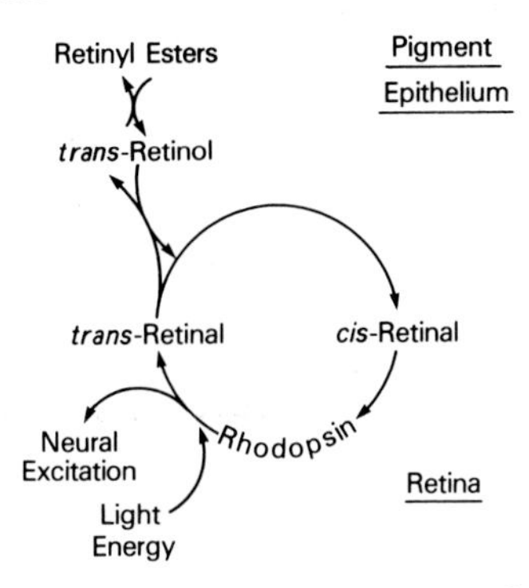

Fig. 3. Pathways of retinoid movement in the visual cycle.

generally as retinyl palmitate, until it is needed by the retinal photoreceptors. When called upon, the PE cell mobilizes the retinoid, with esterase action providing the free nonesterified retinoid. Isomerization and oxidation then occur, converting the retinoid to 11-*cis*-retinal, a form that can bind to opsin in the photoreceptor to form the visual protein, rhodopsin. The formation of rhodopsin takes place during dark adaptation and is the result of a Schiff base interaction (C=N) between the aldehyde function of the retinoid and a lysine residue of the opsin. With light, the Schiff base linkage is broken and a series of short-lived intermediates is formed (Table II). It is this series of reactions that forms the core of the visual process and results in neural impulses which the brain ultimately interprets as vision. The final products of this visual cascade are opsin and all-*trans*-retinal. The latter may remain in the photoreceptor outer segment and be recycled through further light–dark sequences or be reduced to retinol and move back to the PE cell for storage.

Since light triggers the cascade of retinoid intermediates, it is of interest to define the first change in the retinal moiety occurring after photon capture. It has been proposed by Warshel (1976) that the 11-*cis*-retinal in rhodopsin is initially converted to a "strained" all-*trans*-retinal upon bleaching. In a somewhat similar vein, Honig *et al.* (1979) have postulated that photon capture results in 11-cis isomerization to the all-trans configuration. Study of the initial events by the technique of Raman spectroscopy indicates that isomerization of the 11-*cis*-retinal chromophore to a distorted form of the all-trans isomer takes place within picoseconds of photic stimulation (Hayward *et al.*, 1981). The speed of this change is consistent with the hypothesis that it could indeed be the initial step in the visual process. Thus, a whole series of

TABLE II

Rhodopsin Intermediates

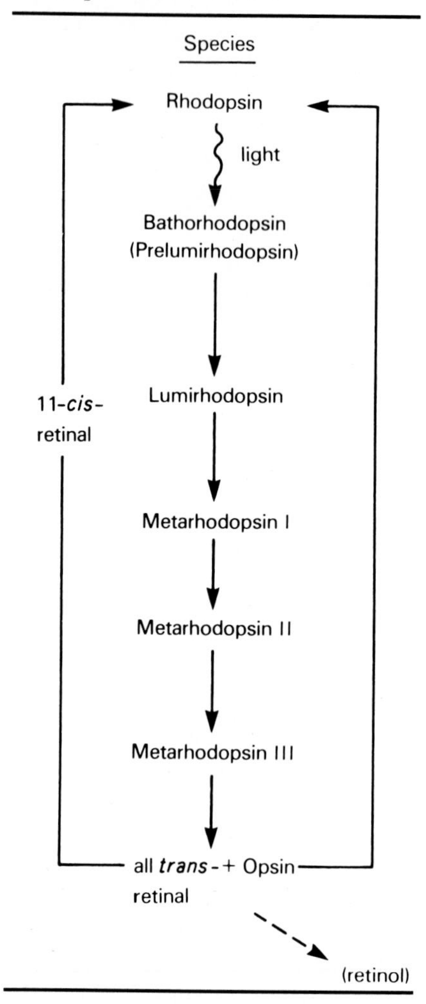

chemical and conformational changes take place in the retinoid ligand itself and in its interaction with the opsin protein moiety in the retina after bleaching, the primary event possibly being an isomeric conversion in retinoid conformation. No information is available as to whether similar changes occur in other tissues, although this seems unlikely because of the highly specialized, sensory nature of the retinal photoreceptor.

B. Cytoplasmic Effects: Glycoconjugate Biosynthesis

Early experiments with intestinal mucosal cells indicated that vitamin A could be involved in protein synthesis at the ribosomal level (Wolf and DeLuca, 1969). These studies showed that protein synthesis by membrane-bound ribosomes was reduced in vitamin A-deficient rats, probably as the result of a decrease in uncharged tRNA. More recently, attention has focused on the control by vitamin A of the biosynthesis of specific glycoconjugates in epithelial tissues.

Perhaps the most general and consistent histological observation in vitamin A deficiency is the decrease in Schiff–periodic acid staining in epithelial cells. On the other hand, vitamin A repletion increases such staining. Biochemical studies corroborate these findings and indicate an effect of vitamin A on glycoprotein synthesis (DeLuca, 1977). This effect is selective in that vitamin A does not uniformly stimulate incorporation of all monosaccharides into all glycoconjugates in epithelial tissues. For example, vitamin A increases incorporation of mannose but not galactose into rat liver conjugates (Hassell *et al.*, 1978). In the corneas of vitamin A-deficient rats, glucosamine incorporation into one specific glycopeptide is stimulated by vitamin A treatment (Kim and Wolff, 1974; Kiorpes *et al.*, 1979). The biosynthesis of specific high-molecular-weight glycoconjugates of the corneal epithelium has also been shown to be directly related to vitamin A levels administered to vitamin A-deficient rats (Hassell *et al.*, 1980). In cultured chondrocyte cells whose adhesivity to the substrate is increased by retinoids, Hassell *et al.* (1979) have shown that vitamin A enhances the cellular accumulation but not the synthesis of fibronectin. It probably does this by increasing the synthesis or availability of a fibronectin receptor (glycoprotein?) of the chondrocyte cell. Control of the biosynthesis of such glycoconjugates by retinoid may thus directly affect the mucus-secreting or keratinizing phenotypes of epithelial cells.

On a molecular basis, carbohydrate moieties must be activated before they can be incorporated into a glycoconjugate (e.g., glycoprotein). Dolichol, a long-chain isoprenol, is known to act as an intermediate in this process in some tissues (Waechter and Lennarz, 1976). As seen in Fig. 4, dolichol phosphate may react with an activated sugar species to form a sugar–lipid intermediate. The intermediate then allows insertion of the sugar into the core oligosaccharide of a glycoprotein. It is possible that retinol functions in a manner comparable to dolichol in some tissues (Fig. 4). Although enzymatic phosphorylation of retinol has not been observed as yet, DeLuca and his co-workers (Frot-Coutaz, 1976) have isolated the compound from hamster intestinal epithelium.

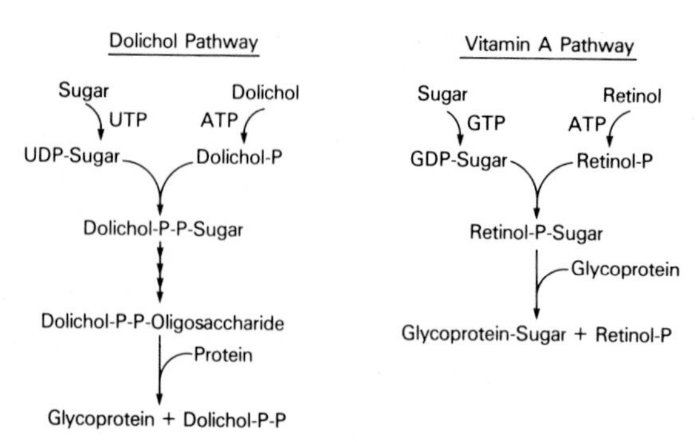

Fig. 4. Possible modes by which long-chain isoprenoids such as dolichol or retinoids could act as intermediates in glycoprotein biosynthesis.

The formation of a sugar–retinol phosphate has been identified and characterized in membrane fractions of several tissues (Peterson *et al.*, 1976; Masushige *et al.*, 1978). The transfer of mannose from mannosyl retinyl phosphate to protein in a membrane fraction of rat liver has also been reported (Rosso *et al.*, 1977).

Consideration of this pathway is complicated by the fact that retinoic acid substitutes for retinol in the maintenance of normal cell character and function in most epithelial tissues. It appears at first glance that formation of the sugar–retinoid phosphate moiety would necessitate the presence of an alcohol group in the retinoid molecule, thus obviating retinoic acid from consideration. Heller (Chen and Heller, 1977) reported, however, that retinoic acid was reduced to a retinol-like material after incubation with bovine PE cells. The material did not appear to be retinol itself, since it was not incorporated into rhodopsin. The possibility thus arises that the intracellular metabolism of retinoic acid in target tissues enables it to function on a common pathway with retinol. A similar retinol-like metabolite of retinoic acid (compound "X") has been proposed by DeLuca *et al.* (1979) to function in a cultured fibroblast system which responds to retinol or retinoic acid with increased adhesivity to the substratum. This reduced compound could then be phosphorylated and glycosylated, as is retinol in other systems.

A series of molecular steps is thus outlined by which vitamin A could control the synthesis of potentially important glycoproteins in target tissues and thus influence aspects of growth and differentiation. Whether these processes are actually operative to any extent in target tissues has yet to be determined.

C. Nuclear Effects: DNA and RNA Synthesis

It has long been known that vitamin A promotes tissue growth as discussed above, although direct evidence of its influence on actual nuclear events has been limited. Combined data from several sources, however, indicate that vitamin A may in fact directly influence DNA and RNA synthesis.

1. *DNA Synthesis*

In the cornea as well as in other epithelial tissues, a reduction in mitotic activity and DNA synthesis was found to accompany vitamin A deficiency (Sherman, 1961). Mitotic activity returned to a normal level with retinoid repletion. In vitamin A deficiency, a defect in DNA metabolism has also been reported in placental development (Takahashi *et al.*, 1975) and liver regeneration (Jayaram *et al.*, 1975). Zile *et al.* (1977) found that DNA synthesis in intestinal mucosal cells was impaired in vitamin A deficiency. More recently they showed that vitamin A deficiency decreased the number of cells (relative DNA content) of several other, nonepithelial, tissues of the rat (Zile *et al.*, 1979). This indicates that vitamin A deficiency not only affects cell proliferation and differentiation in renewing tissues such as intestinal epithelium, duct linings, and skin but also may play a much more general role in controlling DNA synthesis and other nuclear events in the body.

2. *RNA Synthesis*

Total cellular RNA and RNA of the nuclear compartment are increased by vitamin A repletion in several tissues of the vitamin A-deficient rat (Johnson *et al.*, 1969; Zile and DeLuca, 1970). In tracheal epithelium, there is an altered electrophoretic pattern of RNA in vitamin A deficiency, with a smaller proportion of RNA species with low electrophoretic mobility (possibly rRNA precursors) than in control animals (Kaufman *et al.*, 1972). In cell cultures of epidermis established in chemically defined medium, retinyl acetate addition causes an increase in both the total amount and relative concentration of RNA, an effect potentiated by insulin and cortisol (Sporn *et al.*, 1973). In isolated hepatic nuclei, Tsai and Chytil (1978) found that incorporation of [^{3}H]UTP into RNA was lowered in vitamin A deficiency. The RNA synthesized in nuclei of vitamin A-deficient rats also had a smaller proportion of higher-molecular-weight species than in controls. This was not due to increased RNase activity but rather to a smaller number of mRNA chains being elongated at a slower rate. In a somewhat similar vein, Jayaram and Ganguly (1977) reported that vitamin A deficiency led to a decrease in high-molecular-weight RNA in intestinal mucosa and testes but not in liver. In an ingenious series of studies, Ganguly and his co-workers (Ganguly *et al.*, 1978) also used the immature chick oviduct treated with estrogen as a model for studying retinoid effects on rapidly dividing and differentiating cells (Joshi *et al.*, 1976). They found that general growth of the tissue (e.g., wet weight, protein), as well as specific concentrations of DNA and RNA, was markedly influenced by vitamin A status. Retinoids may also be involved in negative control mechanisms at the nuclear level. Blalock and Gifford (1977) have reported that retinoic acid induces a putative repressor of interferon production in a transformed cell line that blocks interferon mRNA transcription. In spite of different effects in different tissues and cell types, therefore, the sum total of data strongly indicates that retinoids have a marked effect on transcriptional events in the nuclei of many tissues.

Interestingly, Tsai and Chytil found that hepatic nuclei of retinol-deficient animals given a supplementation of retinoic acid, which maintains general good health of the animal, demonstrated lower RNA synthesis than in retinol-sufficient animals. Similarly, Jayaram and Ganguly (1977) found that retinoic acid was ineffective in reversing the effects of retinol deficiency on RNA of the testes, although it was effective in intestinal mucosa. Retinoic acid was also only partially effective in reversing the effects of vitamin A deficiency on DNA and RNA content in the estrogen-primed chick oviduct (Joshi *et al.*, 1976). Thus, retinol and retinoic acid have markedly different effects in different tissues. Retinol appears to be specifically required for differentiation in liver and germinal epithelia, with retinoic acid able to substitute for the alcohol only in other (epithelial) tissue types.

VI. 2 S VITAMIN A-BINDING PROTEINS (RECEPTORS)

A. General Receptor Considerations: Steroid Hormone Correlates

Specific intracellular receptor proteins which bind hormones with high affinity and specificity have been known for many years. Toft and Gorski (1966) were the first to demonstrate such soluble binding proteins in the uterus for [^{3}H]estradiol. It was quickly shown that other steroid hormones had their own specific receptor proteins in target tissues, e.g., testosterone (dihydrotestosterone) in the prostate and cortisol in the liver. A unified concept of steroid hormone action soon emerged, exemplified by the two-step mechanism proposed by Jensen *et al.* (1968) for the interaction of estrogen with the uterus. This theory assigned a crucial role to the intracellular soluble binding protein or receptor mediating the hormone's action in the cell. In this scheme, the receptor binds the hormone as it enters the cytoplasm (step I) and translocates it to the nucleus. In the nucleus, it is bound to another nuclear receptor or acceptor (step II), which further facilitates the action of the hormone.

By definition then, these receptors are soluble proteins present in target tissues, which bind a hormone with high specificity and affinity. No specific function or enzymatic role has yet been assigned to the receptors as for, say, the membrane-bound neurotransmitter receptors, but it is assumed that they mediate or facilitate the mechanism of action of the particular hormone in the target cell. Several other binding species are present in most cells and in serum, which bind the hormone ligand with lower affinity and specificity and are thus not classified as receptors. Albumin is a good example of such a protein which binds hormone with little specificity and whose binding affinity is several log units lower than that seen in the hormone–receptor interaction.

B. Cellular Vitamin A-Binding Proteins

Like the discovery of specific intracellular binding proteins in 1966, which cast a new light on the study of the mechanism of action of steroid hormones, the finding by Bashor *et al.* (1973) of a specific, soluble cellular retinol-binding protein (CRBP) in several tissues of the rat was a milestone in the vitamin A field. A separate and distinct cellular retinoic acid-binding protein (CRABP) was soon described in chick embryo skin (Sani and Hill, 1974), punctuating the physiological significance of the acid *in vivo* and further indicating its possible role as a entity separate from retinol in cellular processes. Subsequent work, mainly from the laboratory of Chytil and co-workers, has further elucidated the tissue distribution of the binding proteins (Bashor and Chytil, 1975), their changing levels during development (Ong and Chytil, 1976), and differences in binding between normal and cancerous tissues (Ong *et al.*, 1975). The separate nature of the retinol receptor and the retinoic acid receptor has been firmly established by their isolation and partial characterization (Ong and Chytil, 1975; Ong and Chytil, 1978a,b; Ross *et al.*, 1978). Similar separation of retinol- and retinoic acid-binding proteins in retina was shown by Saari *et al.* (1978a) and will be further discussed below.

Retinoic acid receptors are not as widely distributed as retinol receptors in tissues. Although no firm rule can be established, CRABP is more commonly observed in embryonic or tumorous tissues than CRBP (Ong *et al.*, 1975; Ong and Chytil, 1976). CRBP is most often found in adult, nontransformed tissues. Binding proteins for both retinol (Maraini *et al.*, 1977) and retinoic acid (Sani, 1979) have been reported to be present as integral components of the plasma membranes of PE cells and chick embryo skin, respectively. These binding proteins are very similar to their respective soluble binding proteins in size and properties. It is thus probable that soluble binding proteins may be at least partially membrane-associated under normal conditions *in vivo* and that only a portion of receptor binding is actually measured in the soluble phase. No systematic examination of all the cellular components required for [^{3}H]retinoid-binding activity has yet been reported.

C. 2 S Retinol Receptor of Retina and PE

In the eye, specific retinol receptors were first described in 1975 for the retina and the PE–choroid unit of the chick embryo (Wiggert and Chader, 1975). The chick embryo retina is particularly suited for these studies, since it is avascular, easy to dissect, and would not be expected to contain serum RBP to interfere with the determination of intracellular receptor binding. Moreover, use of the embryonic retina offers a good opportunity to study the involvement of retinoids in events during the crucial early phases of tissue development.

On 5–20% sucrose gradients, a single peak of [^{3}H]retinol binding is observed after incubation of the supernatant fraction of chick retina with 100 n*M* [^{3}H]retinol for 2 hr at 4°C (Fig. 5). Binding is reversible and saturable, since the addition of nonradiolabeled retinol before or after addition of the [^{3}H]retinol eliminates the gradient peak. This is most probably due to competition for receptor (CRBP) binding site(s) and dilution of the radiolabel in the peak. The CRBP of the retina sediments at about 2 S on sucrose gradients, indicating an apparent molecular weight of about 14,000–20,000 (Wiggert and Chader, 1975) in agreement with sedimentation values in various tissues of the rat (Bashor *et al.*, 1973) and of the chick (Fig. 6). A second binding species is seen in liver, brain, and the PE–choroid complex of the chick, which sediments at about 5 S. This corresponds to the sedimentation value of albumin and has relatively nonspecific low-affinity but high-capacity binding characteristics. As exemplified by binding in a chick PE–

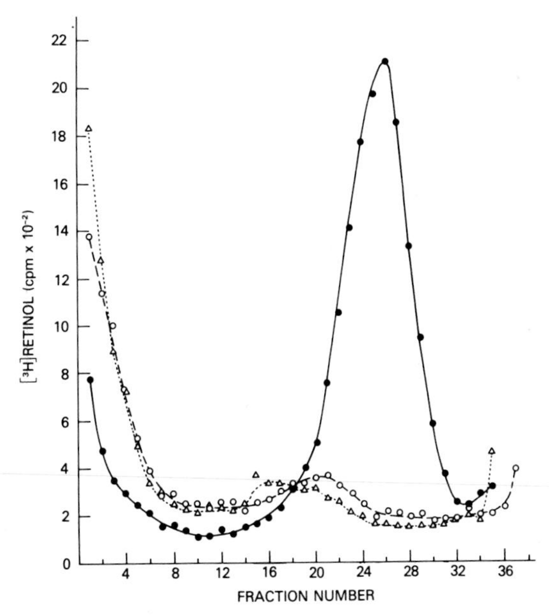

Fig. 5. Sucrose density gradient (5–20%) pattern of the 100,000 *g* supernatant fraction (cytosol) of chick embryo retina. ●, Incubation with 100 n*M* [^{3}H]retinol for 120 min at 4°C; ○, incubation with [^{3}H]retinol for 30 min at 4°C and then 8 μM nonradiolabeled retinol added and incubation continued for 90 min at 4°C; △, incubation with [^{3}H]retinol for 30 min at 4°C and then 8 μM nonradiolabled retinol added and incubation continued for 90 min at 25°C. (From Wiggert *et al.*, 1976.)

choroid preparation (Fig. 7), incubation with low (50 n*M*) concentrations of [^{3}H]retinol resulted in the binding of virtually all the radiolabel at the 2 S position. Increasing concentrations of [^{3}H]retinol led to some binding in the 5 S region as the specific 2 S receptor binding sites became saturated (Wiggert *et al.*, 1976). Heller (1976) has reported the presence of high-molecular-weight lipoglycoproteins (MW 1.5×10^6) in PE cell and photoreceptor cell fractions that bind [^{3}H]retinol. The relation of these large molecules to 2 S CRBP is not clear at present.

An important question that had to be considered early in the study of the retinol receptor was its relationship to serum RBP. Both of these are small proteins, sedimenting at 2 S and having a high affinity for retinol. It was thus possible that the intracellular receptor was in fact RBP or a close derivative thereof. The presence of the RBP could be accounted for by the endogenous presence of serum in most tissues or possibly by specific uptake mechanisms operative *in vivo* for the serum

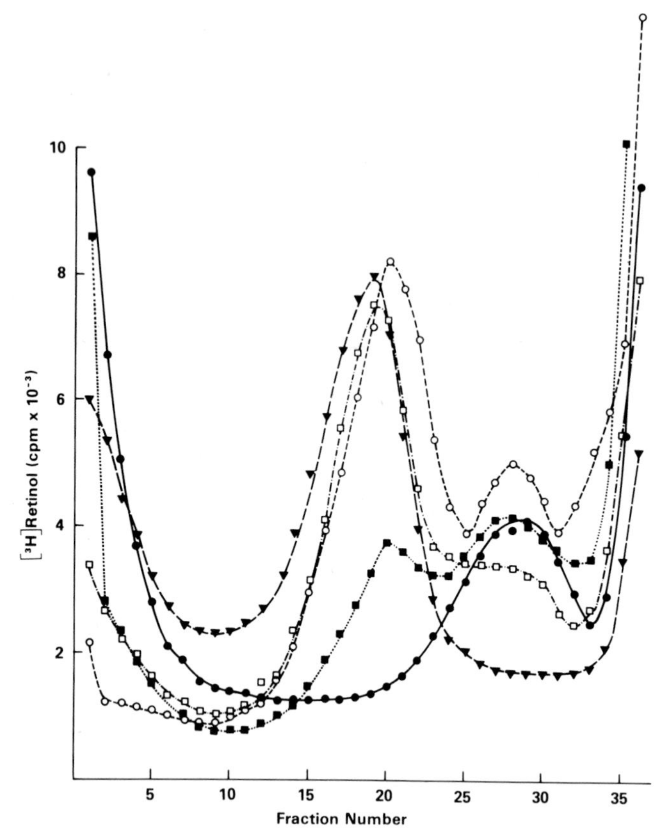

Fig. 6. Sucrose density gradient pattern of the cytosol fractions of chick embryo incubated with 1 μM [^{3}H]retinol. ●, Retina; ○, pigment epithelium; ■, liver; □, brain; ▼, serum. (From Wiggert and Chader, 1975.)

protein. Serum contamination in chick embryo retinas is obviated, since the retina is avascular, as pointed out above. The possibility of specific RBP uptake is in itself interesting, since transport from choroid to retina would indicate selective transport of the RBP moiety through the PE cell into the retina. Some previous evidence, on the surface, tends to support this notion. Abe *et al.* (1975) found high amounts of immunoreactivity to serum RBP in the chick embryo eye. The peak concentration was at 6 days of embryonation, however, and was virtually absent by day 12, the age of the embryonic tissues used in the chick embryo studies. PE cytosol on the other hand is usually heavily contaminated with blood (see Fig. 7) because of the tenacious

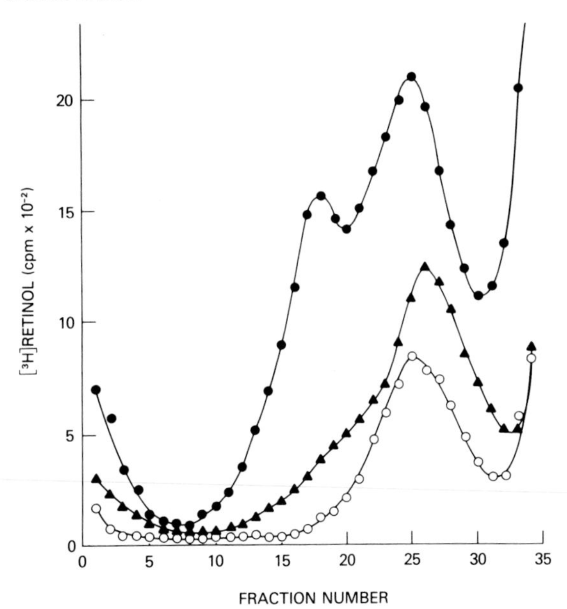

Fig. 7. Sucrose density gradient pattern of chick embryo pigment epithelium cytosol incubated with [^{3}H]retinol at 2×10^{-7} *M* (●), 1×10^{-7} *M* (▲), or 5×10^{-8} *M* (○). (From Wiggert *et al.*, 1976.)

binding of choroidal vessels to the PE cell layer. This problem can be overcome by painstaking dissection under a microscope or the use of PE cell cultures, as will be described in Section VI,E.

Several lines of evidence, however, indicate the separate natures of CRBP of the retina and serum RBP:

1. The fluorescence spectra for RBP and tissue CRBP are quite different (Abe *et al.*, 1977). Chick serum and purified chicken RBP exhibit a fluorescence peak with an excitation maximum at 330 nm and an emission maximum at 470 nm, while the maxima in retina cytosol are at 340 and 430 nm, respectively.
2. Chick serum demonstrates an endogenous fluorescence peak at 5 S, when subjected to sucrose gradient analysis (Abe *et al.*, 1977), that is consistent with the known molecular weight of 76,000 for the RBP-PA complex (Fig. 8A). Purified RBP (MW 20,000) exhibits a fluorescence peak at 2 S (Fig. 8C), but this peak moves to 5 S upon the addition of equimolar amounts of purified serum PA as a result of the formation of a RBP-PA complex. In contrast to this, the endogenous fluorescence peak seen with chick retina cytosol is at 2 S and not 5 S and coincides with the peak of added [^{3}H]retinol (Fig. 8B).

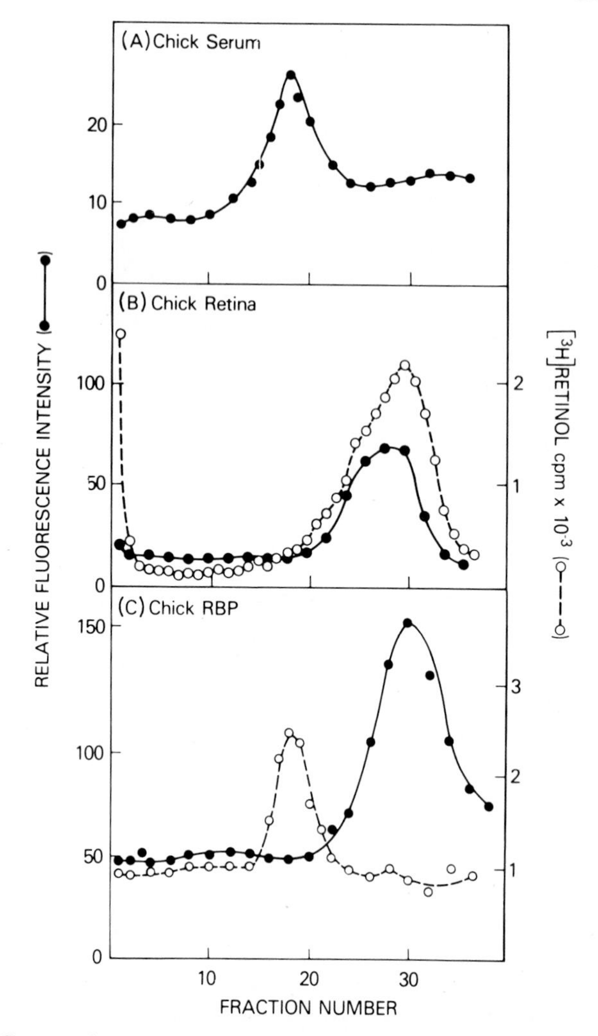

Fig. 8. Sucrose density gradient patterns. (A) Chick serum fluorescence pattern, excitation maximum 330 nm, emission maximum 470 nm. (B) Retinal cytosol fluorescence pattern (●) or binding pattern after incubation with 1×10^{-7} *M* [³H]retinal (○). (C) Purified chicken RBP fluorescence pattern alone (●) or in the presence of purified PA (○). (From Abe *et al.*, 1977.)

3. Similar data are obtained in [³H]retinol-binding studies (Fig. 9). Purified RBP binds a small amount of [³H]retinol, which results in a 2 S peak of radioactivity with sucrose gradient analysis (Abe *et al.*, 1977). The peak moves to 5 S with the addition of purified PA. The 2 S

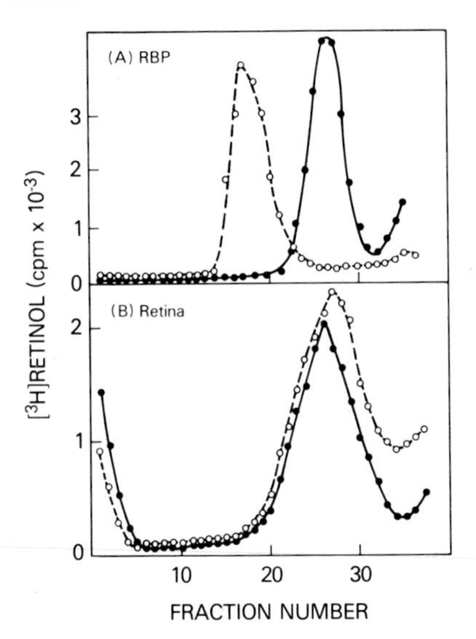

Fig. 9. Effects of prealbumin. Sucrose density gradient patterns after incubation with 1×10^{-7} *M* [^{3}H]retinol. (A) Purified chicken RBP alone (●) or in the presence of PA (○). (B) Chick retinal cytosol alone (○) or in the presence of PA (○). (From Abe *et al.*, 1977.)

peak in retinal supernatant does not shift to 5 S with the addition of PA.

4. Immunologically, no interaction of retinal CRBP is observed with either anti-chicken serum albumin or anti-chicken RBP by immunodiffusion studies, immunoprecipitation, or gradient analysis (Abe *et al.*, 1977).

Taken together, these data indicated the separate nature of the tissue soluble retinol receptor and serum RBP. The data did not preclude the possibility, however, that the receptor could be a modified form of RBP, having lost its antibody binding site and its PA binding site. This was an interesting possibility, since serum RBP has a molecular weight of about 21,000, while that of purified CRBP is only about 14,000 (Ong and Chytil, 1975; Ross *et al.*, 1978). In this regard, it has been found that, although purified CRBP has an amino acid composition different from that of serum RBP (Ross *et al.*, 1978), the differences observed are relatively small and not particularly striking. Cellular uptake and enzymatic processing of the corticosteroid-binding globulin (CBG) of serum had previously been suggested (Werthamer *et al.*, 1973). Saari *et al.* (1978) have pointed out, however, that the higher aspartic acid and lysine content of purified CRBP and threonine content of CRABP compared to that found in RBP preclude them from being proteolysis products of the serum protein.

D. 2 S Retinol Receptor Characteristics

1. *Protein Nature*

In early studies, it was found that incubation of chick retina cytosol with pronase along with [^{3}H]retinol greatly reduced 2 S binding on sucrose gradients (Wiggert and Chader, 1975). RNase, DNase, and *N*-ethylmaleimide were found to have little effect on binding, indicating the protein nature of the receptor and the lack of involvement of sulfhydryl groups in receptor binding. Actual separation of binding proteins in the retina (Saari and Futterman, 1976a) and other tissues (Ong and Chytil, 1975) have confirmed the protein nature of these soluble receptors.

2. *Specificity*

Citral, an octadiene similar in structure to vitamin A, is a well-known inhibitor of vitamin A action (Aydelotte, 1963). At concentrations used to block retinoid effects on cultured chick cornea and esophagus (Aydelotte, 1963), citral effectively abolishes 2 S [^{3}H]retinol binding in sucrose gradients of chick retina supernatant.

Addition of nonradiolabeled retinoic acid to cytosol preparations of chick retina or PE at concentrations 100-fold higher than that of the added [^{3}H]retinol had little if any effect on the 2 S [^{3}H]retinol receptor peak (CRBP) of sucrose gradients, indicating the specificity of [^{3}H]retinol binding (Wiggert and Chader, 1975) for the alcohol form of the retinoid molecule. Similarly, retinal competes very poorly for binding to the 2 S retinol receptor. Retinyl esters have a mixed effect on 2 S binding in chick retina and PE, however. Incubating retina cytosol or PE cytosol with 2×10^{-7} *M* [^{3}H]retinol and a 1000-fold higher concentration of nonradiolabeled retinyl acetate (2×10^{-4}*M*) abolished the 2 S receptor peaks, most probably as a result of competition for the CRBP-binding site(s). Alternatively, esterase activity may have released some free retinol that could compete with and dilute out the [^{3}H]retinol, even though the incubations were conducted at 4°C. Such studies probably indicate that chick PE and retina differ considerably in esterase activity; the differential binding observed may help to characterize these differences.

3. *Affinity*

Differences between specific (2 S) and nonspecific (5 S) binding are graphically demonstrated in Fig. 10. Tissue 2 S CRBP binding is high

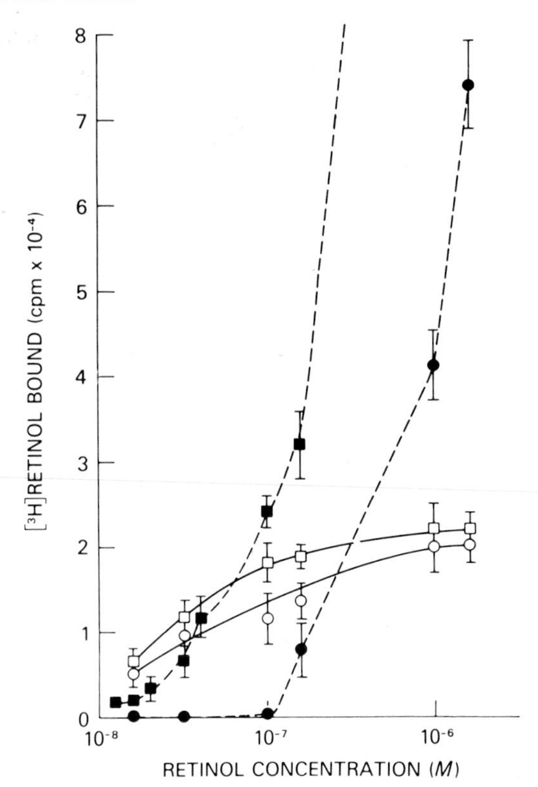

Fig. 10. Concentration-dependent pattern of [^{3}H]retinol binding: 2S receptor of chick retina (□), 2 S receptor of chick PE (○), 5 S peak of chick PE (●); 5 S peak of chick serum (■). (From Wiggert *et al.*, 1977a.)

even at a low [^{3}H]retinol concentration, with binding site saturation seen in the retina and PE above 5×10^{-7} *M* ligand. Binding in serum is low at a low ligand concentration but appears to be a nonsaturable process at least at the probable physiological range of 10^{-8}–10^{-6} *M*. The 5 S peak of the PE does not appear until most of the 2 S binding sites are occupied, and then binding is nonsaturable, as seen in serum (Wiggert *et al.,* 1977a). Scatchard analysis and Bjerrum formation analysis indicates two classes of binding sites in the 2 S region of retina cytosol gradients with K_d values of 1×10^{-9} *M* and 4×10^{-8} *M*. Thus, although binding may be complex, it is of quite high affinity.

There also appears to be a distinct quantitative difference in receptor concentration in different tissues of the chick embryo. Binding in retina and PE supernatant preparations is 2.4 and 10.1 pmol/mg of total protein as compared to brain and liver with 1.8 and 1.7 pmol/mg total protein. PE thus has the highest concentration of retinol receptors, as one would expect if the receptors were actually involved in vitamin A uptake, translocation, and storage. It should be pointed out, however, that these results were obtained *in vitro* using a relatively high amount of [^{3}H]retinol (1 μM) and have yet to be confirmed *in vivo* with other more physiological concentrations of retinol and at different stages of development.

E. 2 S Retinoic Acid Receptors of Retina and PE

A retinoic acid receptor was first described in the retina by Saari and Futterman (Futterman *et al.*, 1976; Saari and Futterman, 1976a) and separated from the retinol receptor by DEAE-cellulose chromatography . It was found to be specific for retinoic acid and in an approximately fivefold higher concentration in bovine retina than the retinol receptor. By sucrose gradient analysis (Fig. 11), a 2 S peak of bound

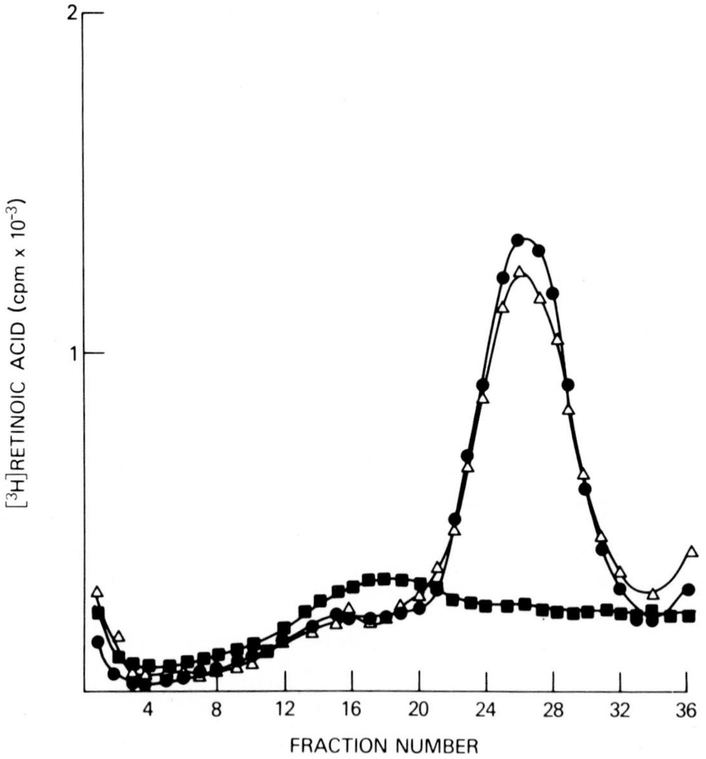

Fig. 11. Sucrose gradient analysis of [^{3}H]retinoic acid receptors of bovine retina cytosol. 1×10^{-7} M ^{3}H-retinoic acid alone (●) or in the presence of 1.4×10^{-4} M non-radiolabeled retinoic acid (■) or retinol (△).

[³H]retinoic acid (CRABP) is detected in bovine retina cytosol (Wiggert *et al.,* 1978a). Competition studies clearly demonstrate that, although it sediments at 2 S, like CRBP, the two proteins are separate and distinct. Embryonic retina (chick and bovine) also exhibit significant [³H]retinoic acid binding.

High [³H]retinoic acid-binding activity is observed in the soluble fraction of PE–choroid preparations (Wiggert *et al.,* 1978b). In the main, this is probably due to the choroidal tissue rather than the PE. Saari *et al.* (1977) found only low amounts of CRABP in PE cell extracts in a carefully controlled study. Our laboratory further examined this point using cultured chick PE cells and PE tissue cleanly dissected from the chick embryo. In culture, chick PE cells differentiate very well, maintaining most of the differentiated morphological characteristics seen *in vivo* (Redfern *et al.,* 1976). These cells exhibit striking [³H]retinol binding (Fig. 12) but are totally lacking in [³H]retinoic acid

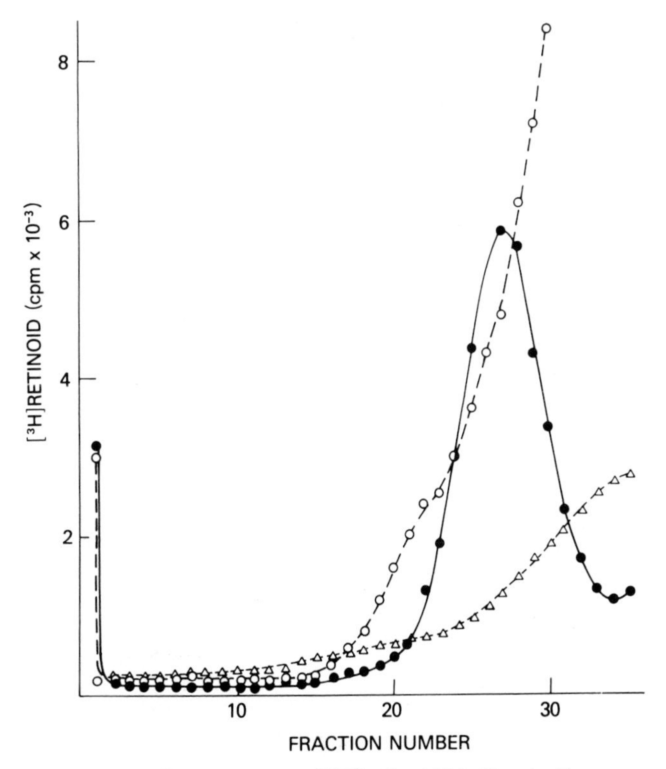

Fig. 12. Sucrose gradient patterns of [³H]retinoid binding in the supernatant fraction of chick embryo PE cells grown in culture. Incubation was with 2.8×10^{-7} *M* [³H]retinol alone (●) or in the presence of 5.6×10^{-5} *M* nonradiolabeled retinol (○). Incubation with 2.3×10^{-8} *M* [³H]retinoic acid (△). (From Wiggert *et al.,* 1979b.)

binding (Wiggert *et al.*, 1979a). When sheets of PE cells were carefully dissected from the underlying choroid of 12-day-old chick embryos and analyzed for [^{3}H]retinoid binding, only CRBP was found. Conversely, the carefully dissected choroidal tissue exhibited high CRABP but low CRBP activity. The absence of a 2 S receptor for retinoic acid in PE is compatible with the lack of such a receptor in many other adult differentiated tissues (Ong and Chytil, 1975). Its presence in the retina and choroid, however, indicate that retinoic acid and its binding proteins are important in cellular metabolism in ocular function.

F. Retinoid Receptor Isolation and Characterization

Both the retinol receptor (CRBP) and the retinoic acid receptor (CRABP) have been isolated and studied in several laboratories. Ong and Chytil (1975) partially purified the soluble retinol receptor from rat testes and confirmed the earlier binding specificity studies performed on whole cytosol preparations. More recently, they purified the rat liver protein to homogeneity and found it to be a single polypeptide chain of molecular weight 14,600 (Ong and Chytil, 1978a). The K_d of the purified protein was determined to be 1.6×10^{-8} M, a value reasonably close to that determined for the receptor of the chick embryo retina (Wiggert *et al.*, 1977a). Specific antibodies to the isolated CRBP were found to cross-react in several tissues, suggesting that CRBP was not tissue-specific (Ong and Chytil, 1979). Goodman and his colleagues (Ross *et al.*, 1978) have fully purified the protein from rat testes and find its properties to be very close to those of the liver protein. The purified CRABP of rat testes also has characteristics quite similar to those of the retinol receptor (Ong and Chytil, 1978b). A molecular weight of 14,600, high percentages of acidic amino acids, and strikingly similar absorption spectra characterize both proteins.

In the retina, CRBP and CRABP were first separated by Saari and Futterman (1976a). Two closely related polypeptides of about molecular weight 17,500 were found by gel electrophoresis for CRBP (Saari and Futterman, 1976b), although the fully purified protein appears to have a molecular weight of about 16,600 (Saari *et al.*, 1978a). CRABP is of comparable size (MW 16,300) and of similar but not identical amino acid composition (Table III).

If one compares the amino acid compositions of the retinol receptors of various tissues (Table III), one sees that they are quite similar to each other but different from serum RBP. The retinoic acid receptors of bovine retina and testes are also virtually identical. This, and the similar antigenicity of RBPs in various tissues and from species to

TABLE III

Amino Acid Compositions of Retinoid-Binding Proteins[a]

Amino acid	Retinol receptor			Retinoic acid receptor		Serum RBP[d]	11-*cis*-Retinal receptor[f]	Opsin[g]
	Retina[b]	Liver[c]	Testes[d]	Retina[b]	Testes[e]			
Lysine	9	9	6	6	7	6	5	4
Histidine	2	2	2	1	2	1	2	2
Arginine	5	4	4	6	5	7	6	2
Tryptophan	2	2	1	2	2	2	1	3
Aspartic acid	12	13	11	12	12	15	9	7
Threonine	6	5	5	10	9	5	5	7
Serine	5	5	7	5	5	6	5	4
Glutamic acid	16	15	14	13	13	12	16	10
Proline	3	3	4	3	4	3	5	6
Glycine	8	9	13	8	7	7	7	7
Alanine	5	5	5	7	7	8	7	9
Cysteine	2	2	2	1	2	3	3	1
Valine	9	6	6	8	6	7	6	9
Methionine	3	3	1	1	2	2	2	3
Isoleucine	4	3	4	4	3	2	3	6
Leucine	8	8	8	7	8	7	10	9
Tyrosine	2	2	2	2	2	4	4	5
Phenylalanine	3	4	3	4	3	6	6	10

[a] Recalculated from literature sources as moles of a particular amino acid per 100 mol of total amino acid composition.
[b] Saari *et al.* (1978a).
[c] Ong and Chytil (1978b).
[d] Ross *et al.* (1978).
[e] Ong and Chytil (1978a).
[f] Stubbs *et al.* (1979).
[g] Plantner and Kean (1976).

species, indicates that the composition and structure of this protein is highly conserved and must be of widespread importance to cellular function. Small but discrete differences occur between the two classes of proteins (i.e., CRBP versus CRABP), however, as exemplified by the higher threonine content of CRABP. These differences could lead to or account in some measure for functional diversity *in vivo* through differences in hydrophobic binding, phosphorylation, etc.

An 11-*cis*-retinal-binding protein is also present in retina (Futterman *et al.*, 1977) and has been isolated and characterized (Stubbs *et al.*, 1979). This protein is quite different from either CRBP or CRABP in that its preferred ligands are 11-*cis*-retinal and 11-*cis*-retinol. It is also over double the size of the retinol- and retinoic acid-binding proteins, having an estimated molecular weight of about 50,000. The protein is specific to the retina (Futterman and Saari, 1977) and is found in quite high concentration—about 1 mol of binding protein per mole of rhodopsin. As seen in Table III, its amino acid composition differs from that of CRBP and CRABP, as well as from that of rhodopsin and serum RBP. The amino acid composition of the high-molecular-weight lipoglycoprotein of the PE cell and photoreceptor described by Heller (1976) is also different from the composition of 2 S CRBP and CRABP.

G. Retinoid Receptors of the Cornea

A specific receptor for retinol is present in the cytosol fraction of bovine cornea epithelium (Fig. 13) that is qualitatively indistinguishable from that in the retina and PE (Wiggert *et al.*, 1977b). It is protein in nature, sediments at 2 S, and is quite specific for retinol. Computer analysis of binding data indicates a molecular weight of 14,950 for the receptor; $K_a = 5.3 \times 10^7$, and $G^0 = -8.53$ kcal/mol. The high association constant and change in standard free energy indicates strong binding between receptor and ligand. 2 S receptor binding is also observed in isolated corneal stroma and endothelium. The binding situation is complicated, however, by the presence of other lower-affinity, high-capacity binding species in these layers of the cornea.

Oddly, no 2 S [^{3}H]retinoic acid binding peak is observed in the soluble fraction of bovine corneal epithelium (Wiggert *et al.*, 1978a). One might expect it to be present, since topically applied retinoic acid is effective against the corneal signs of xerophthalmia (Pirie, 1977), although, as pointed out above, CRABP is of much more limited distribution than CRBP. Bovine corneal endothelium contains a limited number of specific, 2 S [^{3}H]retinoic acid-binding sites. Although the

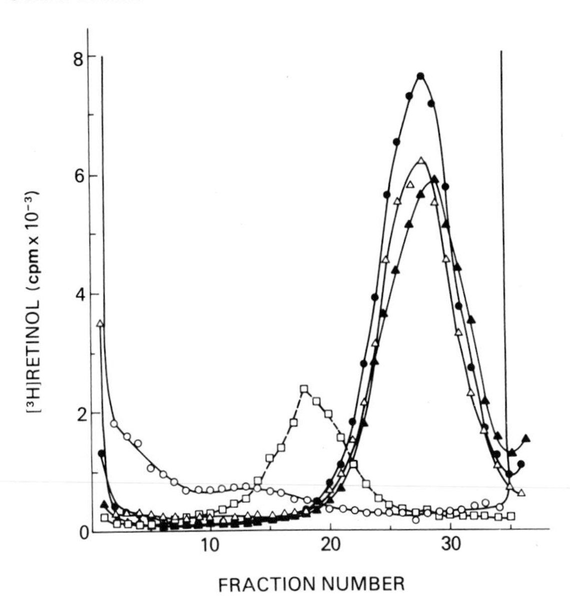

Fig. 13. Sucrose gradient density patterns of [^{3}H]retinol receptors of bovine corneal epithelium. Incubation with 2×10^{-7} *M* [^{3}H]retinol alone (●) or in the presence of 2.8×10^{-4} *M* nonradiolabeled retinol (○), retinyl palmitate (▲), or retinoic acid (△). Bovine serum incubated with 2×10^{-7} *M* [^{3}H]retinol (□). (From Wiggert *et al.*, 1977b.)

cornea is usually considered a single tissue, differences in receptor distribution in the three layers of the cornea may indicate significant differences in retinoid metabolism and effects in the different strata.

Also in contrast to corneal epithelium, conjunctival epithelium exhibits specific 2 S [^{3}H]retinoic acid binding as well as [^{3}H]retinol binding (Fig. 14). It will thus be interesting as well, in the future, to compare retinol and retinoic acid metabolism and action in corneal versus conjunctival epithelium.

VII. 7 S BINDING PROTEIN OF RETINA

If one subjects the supernatant fraction of bovine retina to sucrose gradient analysis as described above for chick retina, one not only sees the 2 S [^{3}H]retinol receptor but also a binding species sedimenting at approximately 7 S as shown in Fig. 15 (Wiggert *et al.*, 1977c). These coincide with fluorescent peaks (excitation, 340 nm, emission 430 nm) in the same regions. Similar large 7 S peaks are seen in retinal supernatant fractions from rhesus monkey, pig, and human, although, as

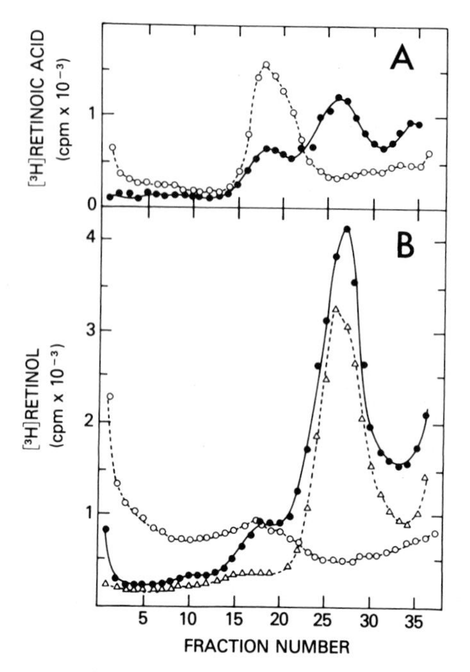

Fig. 14. Conjunctival [^{3}H]retinoid receptors as assessed by sucrose gradient analysis. (A) Incubation with 20 nM [^{3}H]retinoic acid alone (●) or with 20 μM nonradiolabeled retinoic acid (○). (B) Incubation with 20 nM [^{3}H]retinol alone (●) or in the presence of 20 μM nonradiolabeled retinol (○) or retinoic acid (△). (From Wiggert *et al.*, 1978a.)

with chick embryo, no 7 S binding was observed in adult chicken, guinea pig, or rat retina. As yet, it is not known why retinas of some species exhibit 7 S binding while others do not. Only small and variable amounts of 7 S binding are observed in the PE of several species tested, probably as a result of rod outer segment (ROS) contamination.

A. Characteristics

The 7 S binding species is protein in nature, since it is affected by pronase and does not appear to be a simple multimer of the 2 S receptor as assessed by the use of chaotropic agents (Wiggert *et al.*, 1977c). On sucrose gradients, much higher binding is usually observed at 7 S than at 2 S, although this is somewhat variable. The 7 S species is observed only on sucrose density gradient analysis and not by other techniques such as gel filtration, indicating that its affinity for retinol is considerably lower than that of the 2 S receptor. The specificity of 2 and 7 S binding also differs significantly. Nonradiolabeled retinol effectively

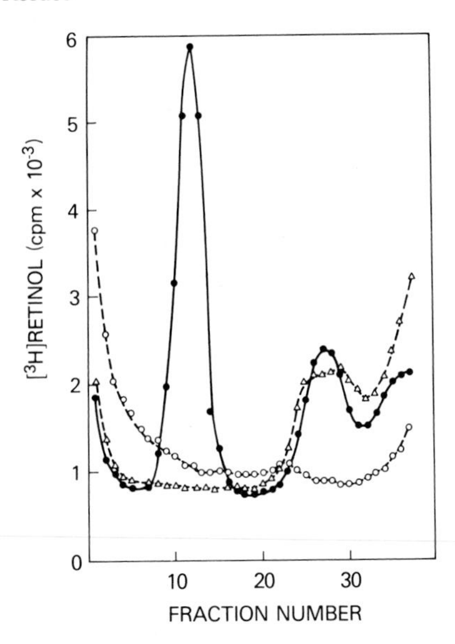

Fig. 15. Sucrose density gradient patterns of [^{3}H]retinol receptors of bovine retina. Incubation with 2.8×10^{-7} *M* [^{3}H]retinol alone (●) or in the presence of 2.8×10^{-4} *M* nonlabeled retinol (○) or retinoic acid (△). (From Wiggert *et al.*, 1977c.)

dilutes out [^{3}H]retinol binding for both the 2 and 7 S species, indicating a limited number of such binding sites for each species. In contrast, excess nonradiolabeled retinoic acid does not affect 2 S binding but does abolish 7 S binding. Thus, significant differences in 2 and 7 S binding are observed, demonstrating the separate natures of the binding proteins in retina.

The 7 S species is not unique to the retina. Bovine brain exhibits significant 7 S binding as well as 2 S binding (Wiggert *et al.*, 1978b). Other tissues such as liver do not exhibit 7 S binding protein. The significance of the possible restriction of the 7 S retinol binding protein to only neural tissue has yet to be determined.

B. Distribution

In the retina, several lines of evidence point to the compartmentalization of the 7 S species in photoreceptor cell ROS:

1. *Fractionation Study*

Bovine ROS purified by the commonly used sucrose floatation gradient method exhibit neither 2 nor 7 S receptor binding (Wiggert *et al.*, 1978b). Most of the membranes are ruptured by this procedure, how-

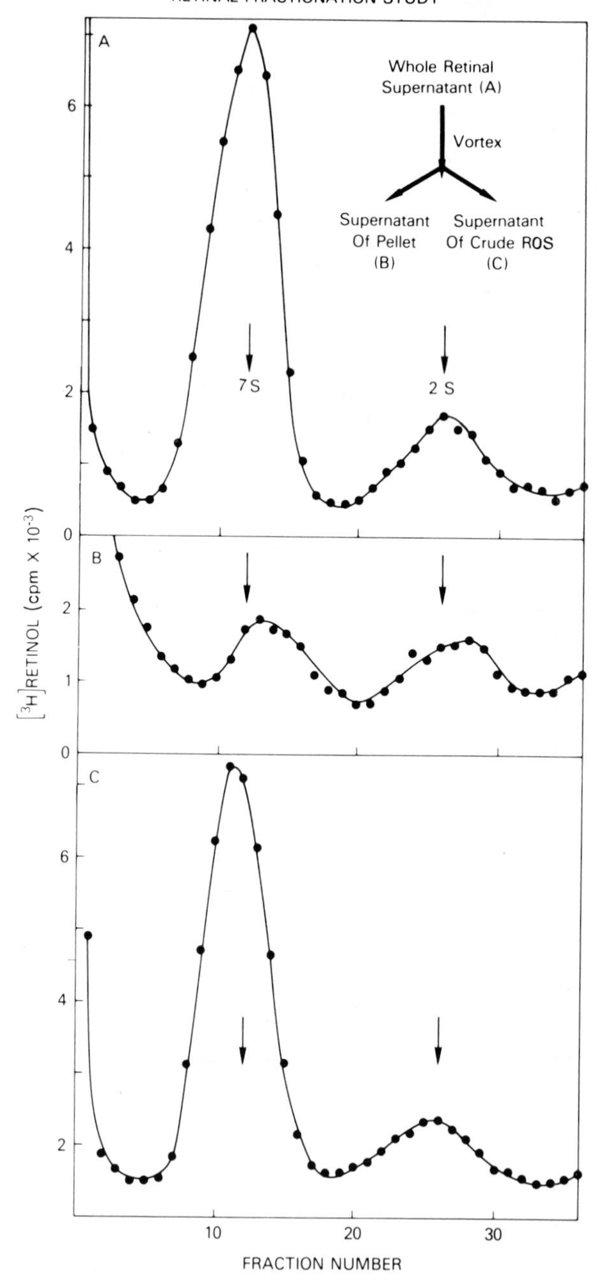

Fig. 16. Differential distribution of [^{3}H]retinol receptors in fractions of bovine retina as assessed by sucrose gradient analysis. (A) Pattern of retinal supernatant. (B) Pattern of outer-segment-depleted fraction of retina. (C) Pattern of outer-segment enriched fraction of retina. (From Wiggert *et al.*, 1978b.)

ever, and one might expect to lose proteins that are soluble or loosely bound to membranes. A different picture is observed if one does not purify the photoreceptor units but rather only crudely separates the retina into an outer segment-enriched fraction and an outer segment-depleted fraction (Wiggert *et al.,* 1978b). This is easily done by gentle vortexing, slow centrifugation, and decantation of the many outer segments which have broken off and remain suspended in the supernatant fluid. As seen in Fig. 16, the outer segment-depleted fraction (B) shows a marked decrease in 7 S binding, whereas the outer segment-enriched fraction (C) shows strong 7 S binding.

2. *Ontogeny Study*

In development, most of the inner layers of the retina differentiate first, making the ROS the last major element to appear. Thus, early in fetal development, retinas are naturally rodless. As seen in Fig. 17, retinas of newborn animals have much the same binding pattern as that seen in the adult but, in the rodless fetal retina, only 2 S binding is observed (Wiggert *et al.,* 1978b).

3. *Retinal Degeneration Study*

In human retinitis pigmentosa, ROS selectively degenerate during the course of the disease, leaving much of the inner, neural part of the retina virtually intact. Within the confines of the pathology of the disease then, this affords another example of a naturally occurring rodless retina. The retinas of two such cases were examined and were found to exhibit virtually no 7 S binding (Bergsma *et al.,* 1977). Binding in the 2 S region was indistinguishable from that seen in normal human controls.

Taken separately, none of the three lines of evidence cited above conclusively prove the selective compartmentalization of the 7 S binding species in ROS. They are consistent with such compartmentalization, however, and taken together support such a postulate. If this assumption is correct, then the 7 S species serves as the first known example of a receptor protein mainly restricted to a specific subcellular organelle (the outer segment) of a cell probably being attached to the membrane surface within the subretinal space (SRS). Sequestering the 7 S species in an area as specialized as the ROS-PE unit strongly implicates the receptor in the visual cycle in some capacity.

C. Light Effects

The SRS compartmentalization postulate is strengthened by the finding of differences in 7 S binding in light and dark (Wiggert *et al.,* 1979a). Figure 18 shows typical [^{3}H]retinol-binding patterns in soluble retinal

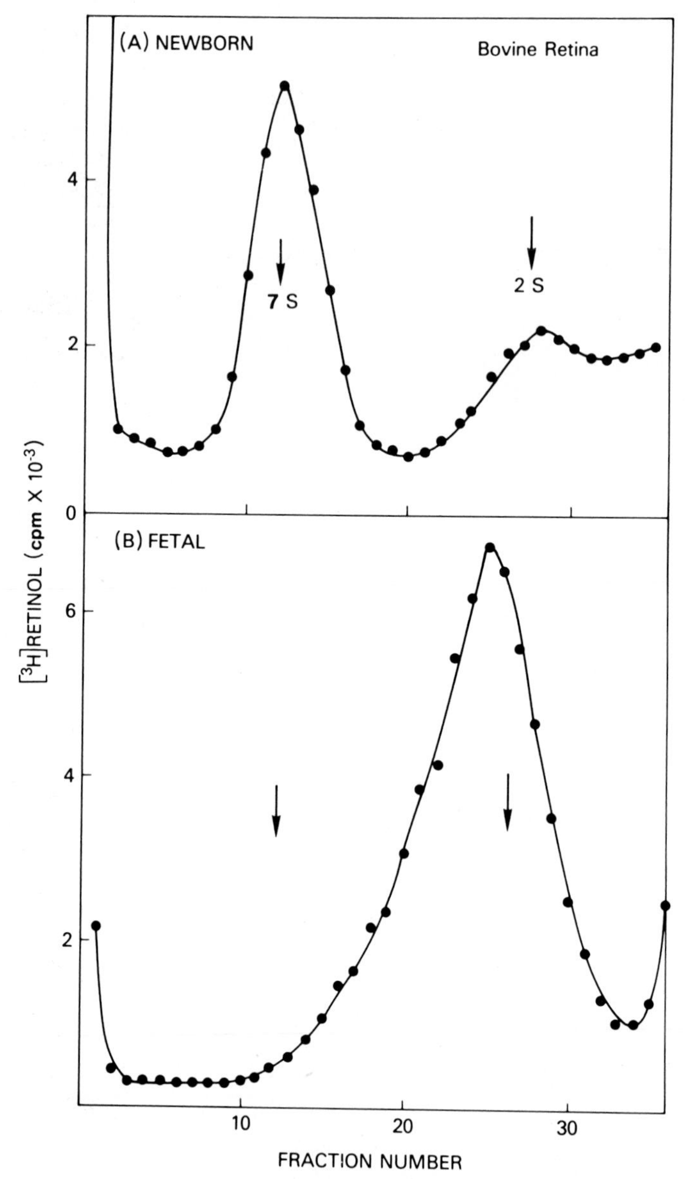

Fig. 17. [^{3}H]Retinol receptors of newborn (A) and fetal (B) bovine retina as assessed by sucrose gradient analysis. Incubation was with 2×10^{-7} *M* [^{3}H]retinol. (From Wiggert *et al.*, 1978b.)

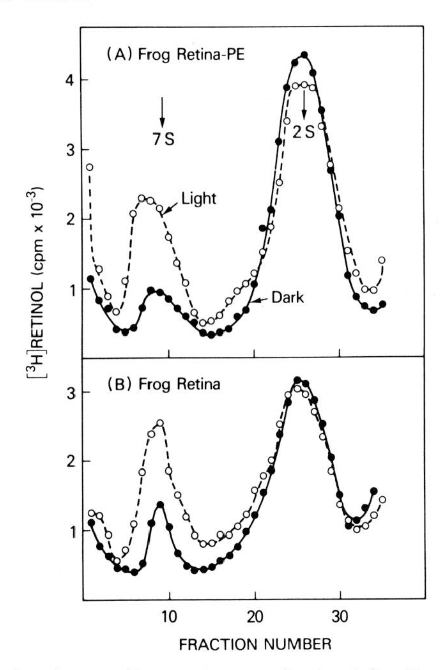

Fig. 18. Sucrose density gradient patterns of cytosol fractions of frog tissues incubated with 280 n*M* [³H]retinol. (A) Retina-PE-choroid unit. Frogs were dark-adapted (●) or light-adapted (○) prior to tissue dissection. (B) Isolated retina. Frogs were dark-adapted, retinas dissected, and one-half bleached *in vitro* (○), while the other half remained dark-adapted (●). (From Wiggert *et al.*, 1979a.)

preparations of light- and dark-adapted frogs. In the whole retinal-PE unit, little light-dark difference in binding is observed in the 2 S region (Fig. 18A). A much larger 7 S binding peak (three- to fourfold) is observed in the light-adapted state, however. Similar results are obtained in isolated retinal preparations, light- or dark-adapted *in vivo* or *in vitro* (Fig. 18B). In agreement with data on fractionation of the bovine retina given above, frog ROS exhibit strong 7 S binding; this is greatly enhanced in bleached ROS. The effect is specific for [³H]retinol, since no light-dark differences are observed in [³H]retinoic acid binding in either the 2 or 7 S gradient region. The effect is also specific to the retina, since no [³H]retinol-binding differences are observed with soluble brain fractions of light- or dark-adapted frogs.

These data indicate the possibility that the 7 S receptor protein is more tightly membrane bound in the dark-adapted retina. Bleaching may change the conformational state of the photoreceptor membranes, weakening the binding and allowing more receptor to be extracted into the homogenizing buffer. Other explanations may of course be possible but would invoke more complex processes of light activation or masking of binding in the dark and seem less likely at the present time.

VIII. MODEL FOR RETINOID MOVEMENT IN PE AND RETINA

Implicit in the visual cycle depicted in Fig. 3 is the uptake of retinol into the PE and its translocation to the retina. As a first step in this process, vitamin A uptake into the PE cell appears to be controlled by recognition sites (acceptors) on the plasma membrane (Fig. 19). CRBP could very well be associated with acceptor areas on the internal surface of the cell. After interaction of the acceptors with the serum RBP–retinol complex, vitamin A could be released from the RBP, cross the membrane, and be bound to CRBP. Interaction of retinol with CRBP under these circumstances could make it a soluble component of the cytosol, allowing for movement of the vitamin to PE cell storage areas or to the retina. Berman *et al.*, (1980) have reported that retinol bound to CRBP is esterified by PE cell microsomes about threefold more effectively than retinol not bound to CRBP. This might indicate an enzymatic as well as vectorial role for the receptor protein in the PE cell. This is not a simple situation, however, since it has been recently shown that retinyl palmitate, the major storage form of retinol in the

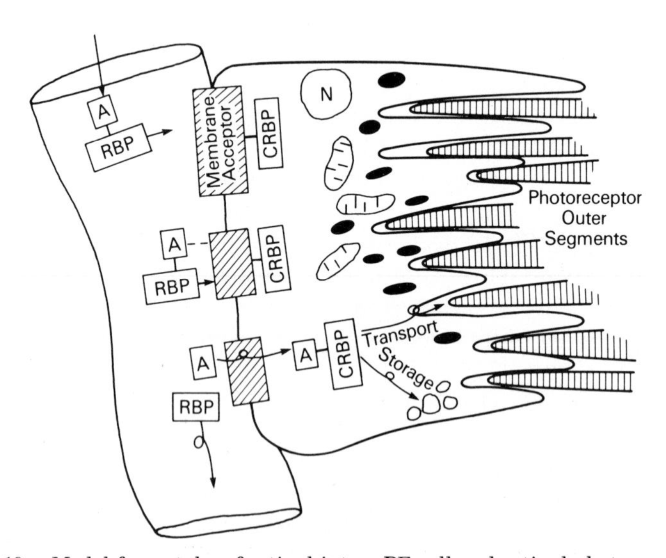

Fig. 19. Model for uptake of retinol into a PE cell and retinol photoreceptors.

PE, is preferentially bound to a soluble PE cell protein which sediments at 6 S (Wiggert *et al.*, 1981). The protein is not found in retina cytosol.

Photoreceptor membranes are fragile and easily damaged, and retinol is a highly toxic as well as labile substance. The binding of retinol to CRBP could afford protection for both the outer segment membranes and the retinoid moiety during transport between PE and photoreceptor in the visual cycle (Fig. 20). In this scheme, the 7 S receptor in particular may function in this capacity. It is predominantly associated with ROS in the retina and may be an integral part of the plasma membrane probably in the subretinal space (SRS). It is known that strong bleaching drives vitamin A from ROS to the PE cell; the weaker 7 S binding to bleached ROS membranes is consistent with the theory that the binding protein could function as an intermediate in the process. The actual routes of vitamin A transport and the extent of this transport are unknown. As depicted on the left-hand side of Fig.

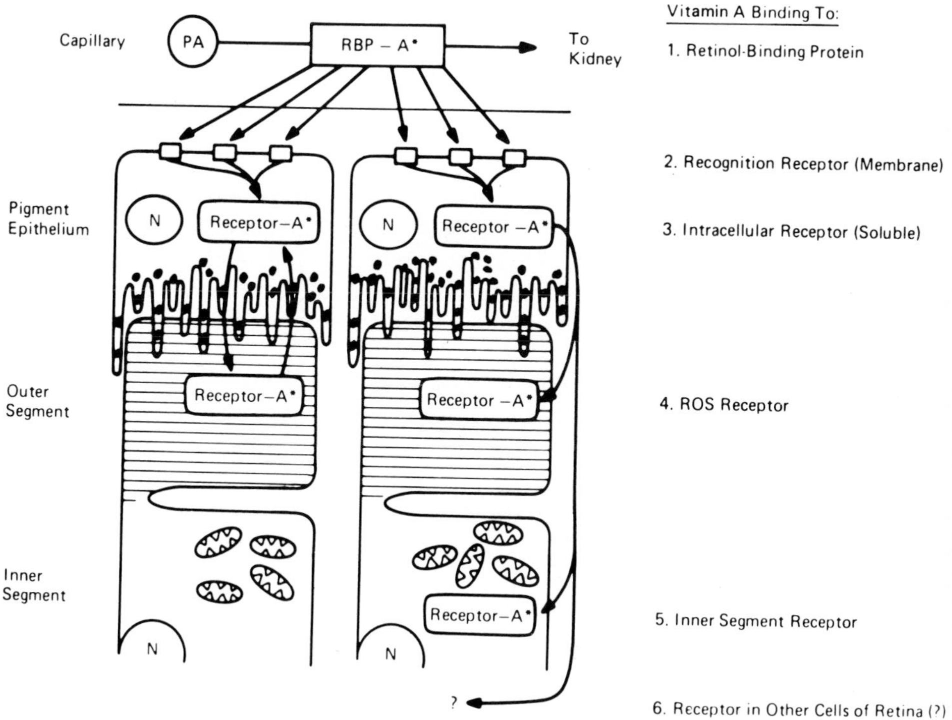

Fig. 20. Model for vitamin A uptake and transport in PE and retina.

20, vitamin A (alcohol, ester, etc.) cycles between the PE cell and photoreceptor. Movement of retinoid to other retinal elements (right-hand side) is still theoretical and will be a fertile area for future research.

IX. NUCLEAR INTERACTIONS OF RETINOIDS AND RECEPTOR PROTEINS

The study of retinoid effects and binding to intracellular receptor proteins in several cultured cell lines has given a good deal of insight as to the possible mode of vitamin A action in normal and transformed cells (Lotan, 1980). A well-characterized retinoblastoma cell line has also been available for such study (Reid *et al.*, 1974), which offers the dual opportunity of examining such effects on (1) a human cancer cell line under well-controlled conditions and on (2) cells that are of retinal, neuroepithelial origin.

A. Nuclear Interactions of Retinoblastoma Cells

Most tissues and cultured cells contain either CRBP or CRABP (Ong and Chytil, 1975). The WERI-Rbl retinoblastoma cell line, for example, contains only the CRABP (Saari *et al.*, 1978b). Y-79 retinoblastoma cells, however, demonstrate both [^{3}H]retinol and [^{3}H]retinoic acid receptors (Wiggert *et al.*, 1977d; Saari *et al.*, 1978b). Receptor binding is severalfold higher in these cells (Wiggert *et al.*, 1977d) than in bovine retina (Bergsma *et al.*, 1977) or other tissues and organs (Ong and Chytil, 1975). [^{3}H]Retinol binding is specific for the alcohol function, with little or no competition observed with retinyl palmitate or retinoic acid (Fig. 21A).

If the intact cells are first incubated with 1 μM [^{3}H]retinol, substantial uptake and specific binding is observed (Fig. 21B). Likewise, specific [^{3}H]retinoic acid receptors are observed at 2 S when the cytosol is incubated with [^{3}H]retinoic acid and analyzed (Fig. 21C) or when the intact cells are incubated with 1 μM [^{3}H]retinoic acid and then homogenized and analyzed for receptor binding (Fig. 21D). CRBP and CRABP can be separated by isoelectric focusing and polyacrylamide slab gel electrophoresis (Russell *et al.*, 1978). This physical separation coupled with the gradient specificity studies firmly establishes the separate nature of the two proteins in retinoblastoma cells. Significant differences in receptor specificity are seen between the retinoblastoma cells and binding proteins in other cells. The dimethylacetylcyclopen-

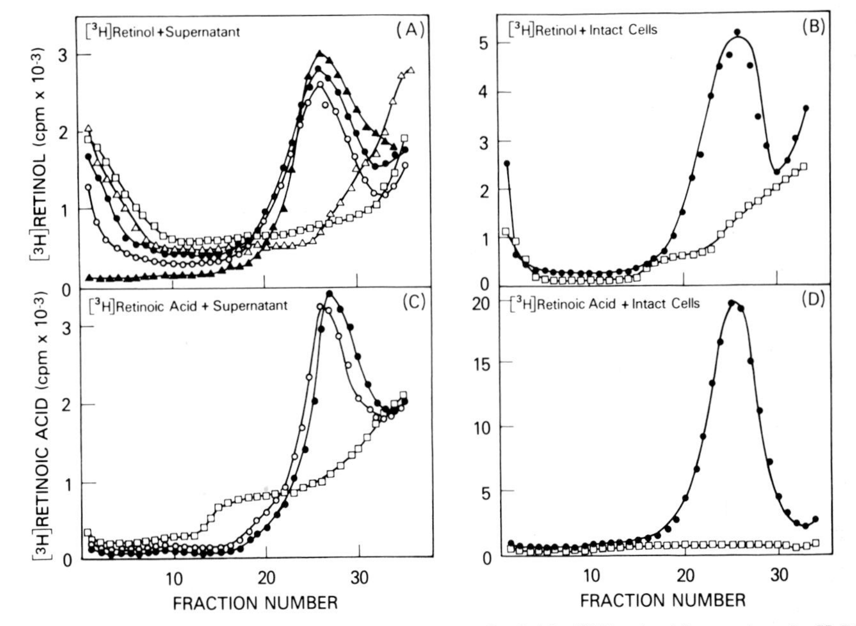

Fig. 21. Sucrose density gradient patterns of soluble [^{3}H]retinoid receptors in Y-79 retinoblastoma cells. (A) Incubation of 100,000 *g* supernatant fraction with 2.8×10^{-7} *M* [^{3}H]retinol alone (●) or in the presence of 5.6×10^{-5} *M* nonradiolabeled retinol (△), retinyl acetate (□), retinyl palmitate (▲), or retinoic acid (○). (B) Incubation of intact cells with 1×10^{-6} *M* [^{3}H]retinol alone (●) or in the presence of 2×10^{-4} *M* nonradiolabeled retinol (□) for 30 min at 37°C prior to homogenization. (C) Incubation of 100,000 *g* supernatant with 2.3×10^{-8} *M* [^{3}H]retinoic alone (●) or in the presence of 4.6×10^{-6} *M* nonradiolabeled retinoic acid (□) or retinol (○). (D) Incubation of intact cells with 1×10^{-6} *M* [^{3}H]retinoic acid alone (●) or in the presence of 2×10^{-4} *M* nonradiolabeled retinoic acid (□) prior to homogenization. (From Wiggert *et al.*, 1977d.)

tenyl (DACP) analog of retinoic acid competes well for [^{3}H]retinoic acid receptor binding in retinoblastoma supernatant, while the trimethylmethoxyphenyl (TMMP) does not (Fig. 22). Both of these analogs, however, compete for CRABP-binding sites in rat testes supernatant (Chytil and Ong, 1976).

^{3}H-Labeled receptor binding can thus be studied either in the isolated soluble fraction after homogenization or after cells have taken up retinoid through a somewhat more physiological delivery route, i.e., through the membrane of the intact cell. In this regard, Saari *et al.* (1980) have reported that the major component of retinoid uptake by this method is not receptor-mediated, in agreement with the picomole quantity of specifically bound retinoid assessed by the gradient method (see below). Total uptake was found to be less when retinol was com-

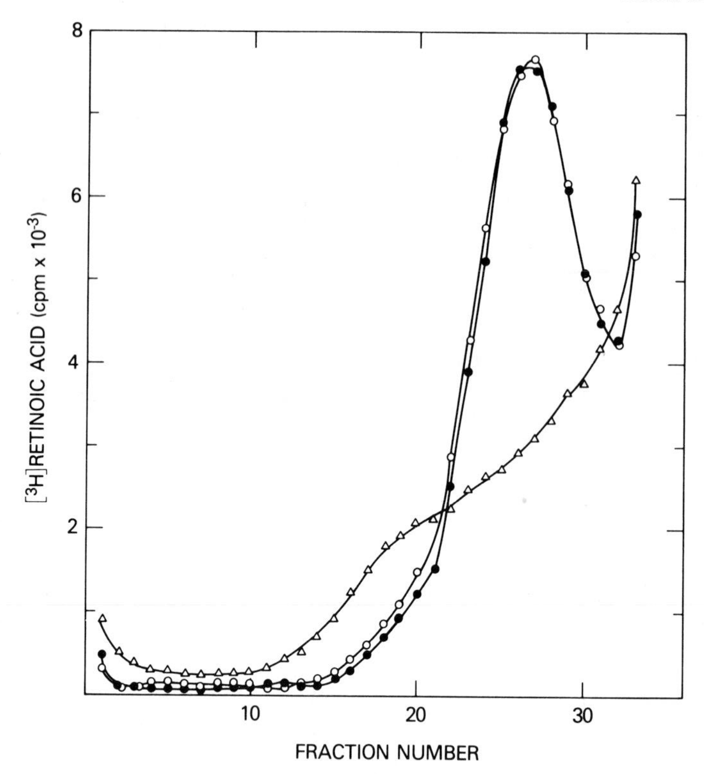

Fig. 22. Sucrose gradient density patterns of [^{3}H]retinoic acid binding in Y-79 retinoblastoma cells. Incubation with 2.3×10^{-6} *M* [^{3}H]retinoic acid alone (●) or in the presence of 4.6×10^{-6} *M* nonradiolabeled DACP analog (△) or TMMP analog (○). (B. Wiggert and G. Chader, unpublished observations.)

plexed to serum RBP prior to delivery to the cells, but one would expect that specific intracellular binding sites are still saturated under these conditions. As seen in Fig. 23, retinoids have a profound acute effect on retinoblastoma cell viability (Chader *et al.,* 1981). The most effective retinoids in killing the cells are those with a free hydroxyl function (e.g., all-*trans*-retinol, 9-*cis*-retinol). Blocking this function (e.g., the acetate derivative) exerts a partial protective effect on cell viability. Full oxidation to the carboxyl function further protects the cells, with cell death observed only at very high retinoid concentrations. The effect on viability is mirrored to some extent in cellular uptake and binding of retinoid from the culture medium. After incubation of intact cells with 1 μM radiolabeled retinoid, threefold more [^{3}H]retinol than [^{3}H]retinoic acid is bound to the intracellular receptor (45 versus 14

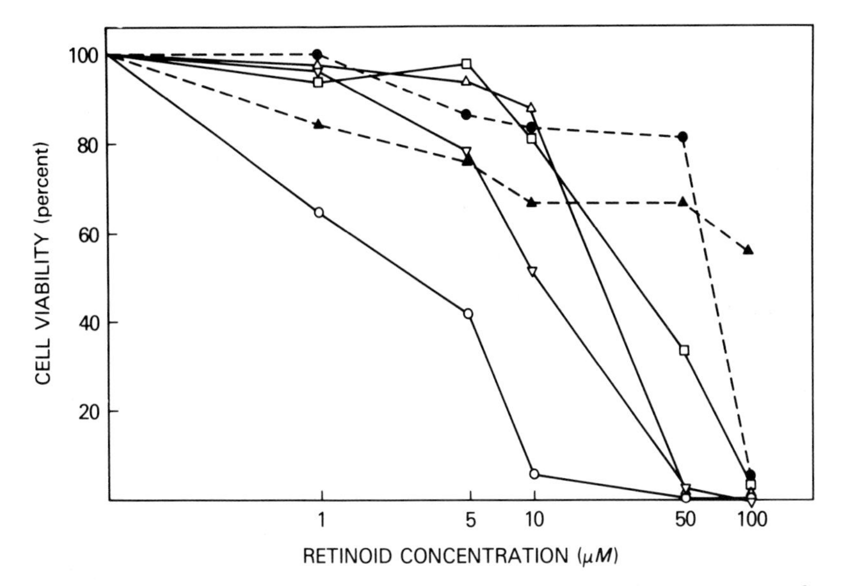

Fig. 23. Effect of retinoids of Y-79 retinoblastoma cell viability over a 24-hour period. ●, Retinoic acid; ▲, 13-*cis*-retinoic acid; ○, retinol; ▽, 9-*cis*-retinol; △, retinyl acetate; □, 13-*cis*-retinyl acetate. (From Chader *et al.*, 1981.)

pmol/ng soluble protein, respectively). The preferential action of retinol on cell death could be effected in several ways. Among them, one could consider direct membrane solubilization through detergent-like action, cytoplasmic effects (glycosylation, etc.), or control of nuclear events (mRNA, etc.). In analogy to the steroid hormone system, it was logical to examine the latter possibility to see if retinoid (either retinol or retinoic acid) was taken up into the nuclear compartment.

Significant differences are seen in uptake and binding of [^{3}H]retinoic acid in retinoblastoma cell nuclei (Wiggert *et al.*, 1977d). When nuclei are isolated, the nuclear supernatant fraction prepared and then incubated with [^{3}H]retinol or [^{3}H]retinoic acid, no specific binding of either retinoid is noted (Fig. 24). When intact cells are first incubated with 1 μM [^{3}H]retinoid at 37°C for 1 hr and the nuclei subsequently isolated, no nuclear [^{3}H]retinol binding peak is observed, but a distinct 2 S peak of bound [^{3}H]retinoic acid is found. Binding to the soluble nuclear retinoic acid-binding protein (NRABP) is specific and saturable (Wiggert *et al.*, 1977d). It is particularly interesting that NRABP is observed only in the soluble nuclear fraction when intact cells are incubated with [^{3}H]retinoic acid prior to isolation of the nuclei. This is a situation very similar to that seen with the steroid hormones. Nu-

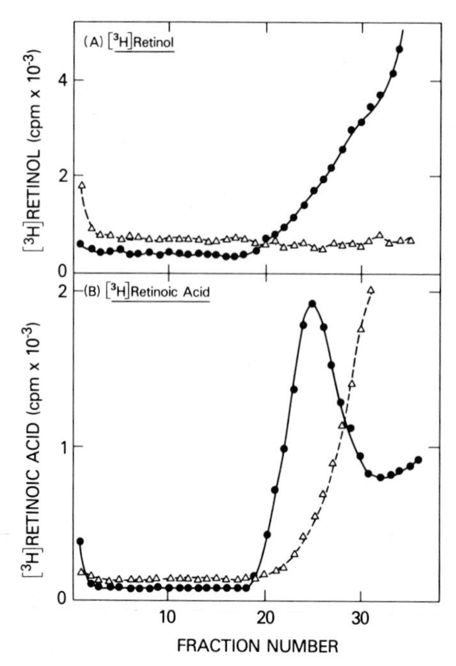

Fig. 24. Sucrose gradient patterns of retinoblastoma cell nuclear fractions. (A) Nuclear supernatant prepared and incubated with 2.8×10^{-7} M [^{3}H]retinol. (△). Nuclear supernatant prepared after intact cells were incubated with 1 μM [^{3}H]retinol for 1 hr at 37°C (●). (B) Nuclear supernatant prepared and incubated with 2.3×10^{-6} M [^{3}H]retinoic acid (△). Nuclear supernatant prepared after intact cells were incubated with 1 M [^{3}H]retinoic acid for 1 hr at 37°C (●). (From Wiggert *et al.*, 1977d.)

clear estrogen receptor, for example, is seen only after incubation of intact target cells with estrogen and does not appear to be in the nuclear compartment prior to estrogen treatment. This implies that both the steroid hormones and retinoic acid may have to interact with a cytoplasmic component (e.g., receptor) prior to nuclear translocation.

Nuclear uptake of [^{3}H]retinoic acid can also be demonstrated by the technique of autoradiography (Russell *et al.*, 1980; Chader *et al.*, 1981). Incubation of cells with [^{3}H]retinol for 1 hr resulted in a generally denser labeling pattern over cytoplasmic areas than over nuclear areas, leading to a halo or doughnut effect (Fig. 25A). With [^{3}H]retinoic acid, the effect is just the opposite, the heaviest labeling being observed over nuclei (Fig. 25B). All cells do not show this effect, indicating the possibility of a somewhat heterogeneous cell population or time course of uptake.

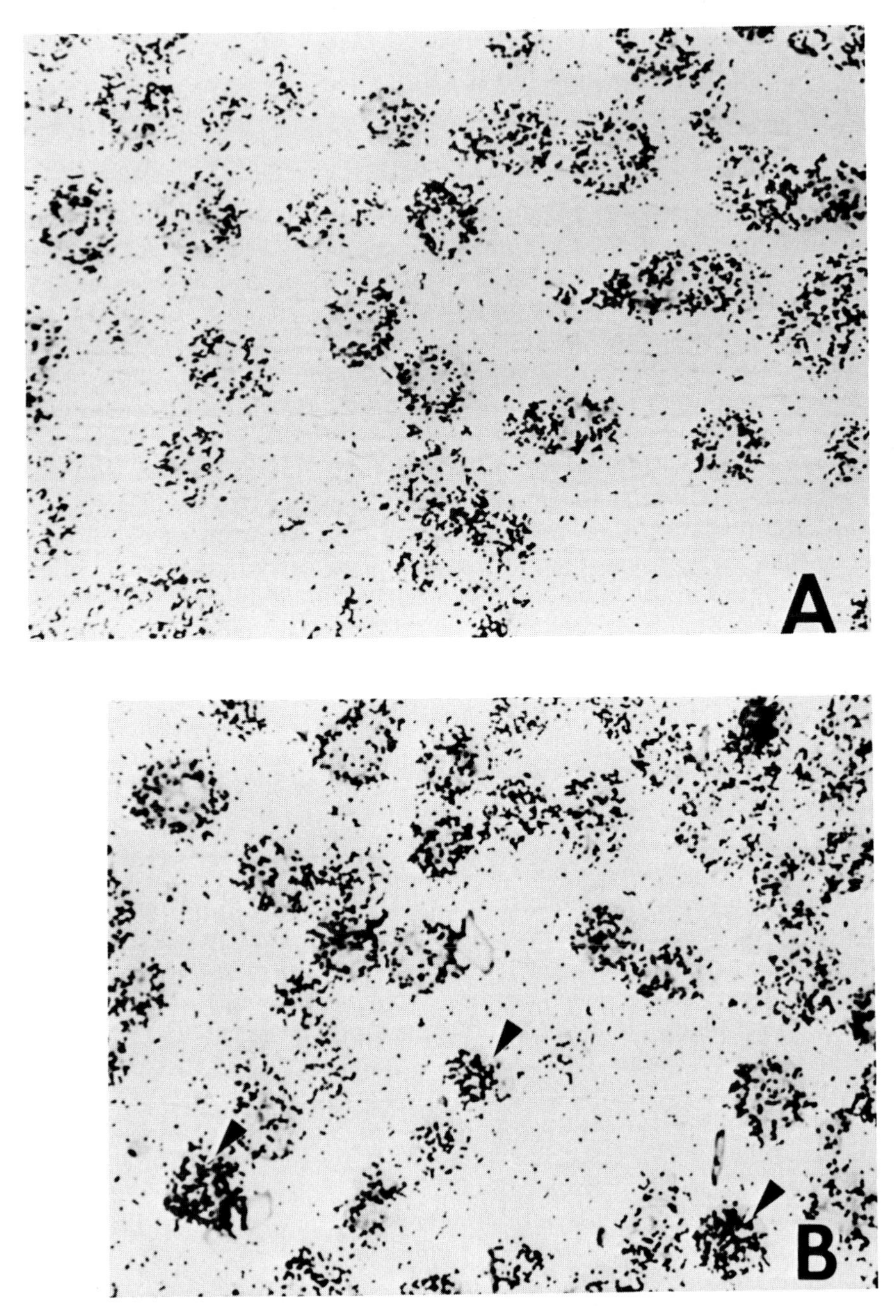

Fig. 25. Autographs of retinoblastoma cells incubated for 1 hr with 1 μM [^{3}H]retinol (A) or 1 μM [^{3}H]retinoic acid (B). ×960. (From Chader *et al.*, 1981.)

B. Nuclear Interactions in Other Tissues

1. Retinoic Acid

NRABPs have been detected in other cells and tissues. In an autoradiography study of keratoacanthoma tumors, Prutkin and Bogart (1970) found that topically applied [^{14}C]retinoic acid was readily taken up into the cells and that at 24 hr most of the silver grains were located over the cytoplasm. By 48 hr though most of the silver grains were over the nuclei. Prior application of actinomycin D to the tissue abolished the preferential nuclear accumulation, further suggesting involvement of retinoid in nuclear function (Prutkin, 1971). Jetten and Jetten (1979) have reported the presence of 2 S NRABP in embryonal carcinoma cells stimulated to differentiate by retinoic acid and its analogs. As with retinoblastoma cells, binding of [^{3}H]retinoic acid was observed in the soluble nuclear fraction only when intact cells were preincubated with the radiolabeled retinoid. A 2 S NRABP has also been observed in the nuclei of chick embryo skin and Lewis lung tumor (Sani, 1977; Sani and Donovan, 1979). Unlike retinoblastoma and the embryonal carcinoma cells, NRABP appears to be present in the nucleus of these cells prior to incubation of the intact cells or tissue with retinoid, indicating distinct differences in compartmentalization and nuclear uptake in the limited numbers of tissue analyzed to date.

2. Retinol

Binding proteins for [^{3}H]retinol are also present in nuclei of some tissues. Rao *et al.* (1979) have isolated such proteins from the nuclear soluble fraction (nucleosol) of the oviduct magnum of laying hens. Partial characterization by gel filtration and gel electrophoresis suggests a molecular weight of about 14,500, a value similar to that for the cytosolic CRBP. After treatment of oviduct minces with [^{3}H]retinyl acetate, much of the radioactivity in the nuclear fraction was associated with the chromatin fraction. In parallel experiments, Takase *et al.* (1979) found that purified CRBP mediated the binding of [^{3}H]retinol to isolated nuclei from livers of vitamin A-deficient rats. Purified CRABP complexed with retinoic acid did not diminish CRBP–retinol binding. Nuclear binding sites for free retinol were not found, reminiscent of the effect of serum RBP on retinol uptake by PE cell membrane acceptors. It is yet not known if the CRBP actually enters the nuclei or possibly remains at the membrane and only facilitates the binding of [^{3}H]retinol to other intranuclear acceptors.

X. MODEL FOR NUCLEAR INTERACTION WITH RETINOIDS

Studies on retinoblastoma cells indicate that:

1. [^{3}H]Retinoic acid but not [^{3}H]retinol is preferentially taken up or bound to the cell nucleus.
2. Specific soluble binding proteins (NRABP) for [^{3}H]retinoic acid can be detected in the nucleosol fraction, but that these are not present (at least in soluble form) prior to retinoid interaction in the cytoplasm.

At least in this cell line therefore, one could propose a working model for future studies, which has striking similarities to the system worked out for the steroid hormones. With the hormones, the steroid molecule enters a target cell and interacts with its cytoplasmic receptor. It then can be translocated to the nucleus and interact with nuclear acceptors, soluble and/or chromatin-bound. With retinoblastoma cells in culture (Fig. 26A), [^{3}H]retinoic acid enters the cell by a yet unknown mechanism, as well as probably being enzymatically formed in the cytoplasm from [^{3}H]retinol. It is bound to CRABP, translocated to the nucleus where it is bound to soluble acceptors (NRABP) that may or may not be similar to the cytoplasmic receptors. It is not yet known if

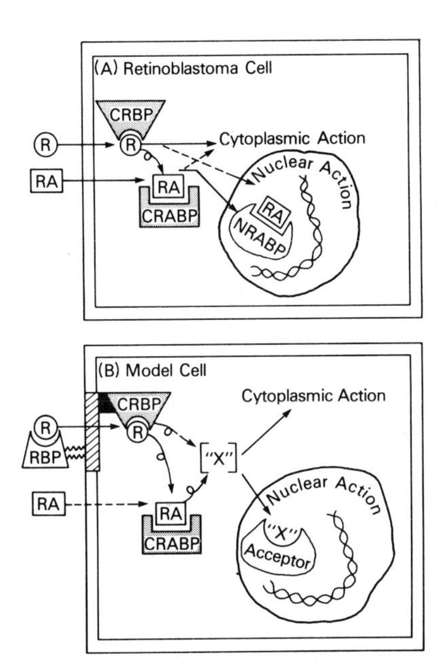

Fig. 26. Possible mechanisms for translocation and action of retinoids in a retinoblastoma cell (A) and a model cell (B).

retinoic acid or retinol can influence events in the cytoplasm (e.g., glycosylation) or if retinoic acid has a direct nuclear effect (e.g., mRNA synthesis).

A somewhat more complex scheme must be envisioned for a model target cell (Fig. 26B). Hormones are synthesized in the body and, upon demand, are transported to target tissues bound to specific serum proteins. Vitamin A is stored in the liver and, likewise upon demand, is transported to its target bound to serum RBP. Retinoid most probably enters the cell as retinol through interaction at the plasma membrane acceptor site, as in PE cells, and is bound to CRBP. Little if any retinoic acid would normally enter the cell from the blood supply but rather would be formed enzymatically within the cell and bound to its receptor, CRABP. The general hypothetical metabolite "X" may then be formed from retinol, retinoic acid, or both. This metabolite could then effect cytoplasmic changes or affect nuclear processes. Alternatively, retinol or retinoic acid itself could function as the hypothetical metabolite "X", affecting cellular processes separately or in concert. There is no compelling evidence that retinol and retinoic acid have identical functions in cells. The pleiotropic action of retinoids, the large number and variety of cells affected by them, and the appearance or disappearance of intracellular receptors during ontogenesis and oncogenesis more probably indicate different modes of action in the different cell types. It would also be interesting to consider that different retinoids could have separate but complementary actions even in the same cell type.

In a broad sense, a hormone can be defined as a chemical substance formed in one organ and carried in the blood to another part of the body where it affects cellular activity.

A vitamin can be also defined as an organic substance present in natural foodstuffs or made in the body in minute quantity that is essential to normal cellular metabolism.

Whether one considers vitamin A a vitamin, hormone, cofactor, or chromophore as in the visual cycle thus depends on the specific definition one chooses and the cellular processes involved. In any event, retinoids affect many body functions, as perhaps best exemplified by their varied and singularly important effects on ocular tissues.

REFERENCES

Abe, T., Muto, Y., and Hosoya, N. (1975). Demonstration of immunoreactive retinol-binding protein and prealbumin in developing chicken embryo. *Life Sci.* **17**, 1579–1588.

Abe, T., Wiggert, B., Bergsma, D., and Chader, G. (1977). Vitamin A receptors. I. Comparison of retinol binding to serum retinol-binding protein and to tissue receptors in chick retina and pigment epithelium. *Biochim. Biophys. Acta* **498,** 355-365.

Arens, J., and van Dorp, D. (1946). Synthesis of some compounds possessing vitamin A activity. *Nature (London)* **157,** 190-192.

Aydelotte, M. (1963). The effects of vitamin A and citral on epithelial differentiation *in vitro.* II. The chick oesophageal and corneal epithelia and epidermis. *J. Embryol. Exp. Morphol.* **2,** 621-635.

Bashor, M., and Chytil, F. (1975). Cellular retinol-binding protein. *Biochim. Biophys. Acta* **411,** 87-96.

Bashor, M., Toft, D., and Chytil, F. (1973). *In vitro* binding of retinol to rat-tissue components. *Proc. Natl. Acad. Sci. U.S.A.* **70,** 3483-3487.

Bergsma, D., Wiggert, B., Funahashi, M., Kuwabara, T., and Chader, G. (1977). Vitamin A receptors in normal and dystrophic human retina. *Nature (London)* **265,** 66-67.

Berman, E. *et al.* (1980). Enzymatic esterification of vitamin A in the pigment epithelium of bovine retina. *Biochim. Biophys. Acta* **630,** 36-46.

Bhat, M., and Cama, H. (1978). Thyroidal control of hepatic release and metabolism of vitamin A. *Biochim. Biophys. Acta* **541,** 211-222.

Blalock, J., and Gifford, G. (1977). Retinoic acid (vitamin A acid) induced transcriptional control of interferon production. *Proc. Natl. Acad. Sci. U.S.A.* **74,** 5382-5386.

Bok, D., and Heller, J. (1976). Transport of retinol from the blood to the retina: An autoradiographic study of the pigment epithelial cell surface receptor for plasma retinol-binding protein. *Exp. Eye Res.* **22,** 395-402.

Carter-Dawson, L., Tanaka, T., Kuwabara, T., and Bieri, J. (1980). Early corneal changes in vitamin A deficiency. *Exp. Eye Res.* **30,** 261-268.

Chader, G., Wiggert, B., Russell, P., and Tanaka, M. (1981). Retinoid-binding proteins of retina and retinoblastoma cells in culture. *Ann. N.Y. Acad. Sci.* **359,** 115-133.

Chen, C.-C., and Heller, J. (1977). Uptake of retinol and retinoic acid from serum retinol-binding protein by retinal pigment epithelial cells. *J. Biol. Chem.* **252,** 5216-5221.

Chytil, F., and Ong, D. (1976). Mediation of retinoic acid-induced growth and anti-tumor activity. Nature *(London)* **260,** 49-51.

Chytil, F., and Ong, D. (1978). Cellular vitamin A binding proteins. *Vitam. Horm. (N.Y.)* **36,** 1-32.

Crocker, T., and Sanders, L. (1970). Influence of vitamin A and 3,7-dimethyl-2,6-octadienal (citral) on the effect of benzo[*a*]pyrene on hamster trachea in organ culture. *Cancer Res.* **30,** 1312-1318.

DeLuca, L. (1977). The direct involvement of vitamin A in glycosyl transfer reactions of mammalian membranes. *Vitam. Horm. (N.Y.)* **35,** 1-57.

DeLuca, L., Adamo, S., Bhat, P., Sasak, W., Silverman-Jones, C., Akalovsky, I., Frot-Coutaz, J., Fletcher, R., and Chader, G. (1979). Recent developments in studies on biological functions of vitamin A in normal and transformed tissues. *Pure Applied Chem.* **51,** 581-591.

Dion, L., Blalock, J., and Gifford, G. (1978). Retinoic acid and the restoration of anchorage dependent growth to transformed mammalian cells. *Exp. Cell Res.* **117,** 15-22.

Dowling, J., and Wald, G. (1960). The biological function of vitamin A acid. *Proc. Natl. Acad. Sci. U.S.A.* **46,** 587-592.

Fell, H. (1957). The effect of excess vitamin A on cultures of embryonic chicken skin explanted at different stages of differentiation. *Proc. R. Soc. London Ser.* B **146,** 242.

Fell, H. (1969). The direct action of vitamin A on skeletal tissue *in vitro. In* "The Fat-Soluble Vitamins" (H. DeLuca and J. Suttie, eds.), pp. 187-202. Univ. of Wisconsin Press, Madison.

Friedenwald, J., Buschke, W., and Morris, M. (1945). Mitotic activity and wound healing in the corneal epithelium of vitamin A deficient rats. *J. Nutr.* **29,** 299–307.

Frolik, C., Tavela, T., Newton, D., and Sporn, M. (1978). *In vitro* metabolism and biological activity of all-*trans*-retinoic acid and its metabolites in hamster trachea. *J. Biol. Chem.* **253,** 7319–7324.

Frot-Coutaz, J., Silverman-Jones, C. and DeLuca, L. (1976). Isolation, characterization and biological activity of retinyl phosphate from hamster intestinal epithelium. *J. Lipid Res.* **17,** 220–230.

Futterman, S., and Saari, J. (1977). Occurrence of 11-*cis*-retinal-binding protein restricted to the retina. *Invest. Ophthalmol. Visual Sci.* **16,** 768–771.

Futterman, S., Saari, J., and Swanson, D. (1976). Retinol and retinoic acid-binding proteins in bovine retina: Aspects of binding specificity. *Exp. Eye Res.* **22,** 419–424.

Futterman, S., Saari, J., and Blair, S. (1977). Occurrence of a binding protein for 11-*cis*-retinal in the retina. *J. Biol. Chem.* **252,** 3267–3271.

Ganguly, J., Sarada, K., Murthy, S., and Kumar, T. (1978). Some recent observations on the systemic mode of action of vitamin A. *Curr. Sci.* **47,** 292–295.

Genta, V., Kaufman, C., Harris, J., Smith, J., Sporn, M., and Saffiotti, U. (1974). Vitamin A deficiency enhances the binding of benzo[*a*]pyrene to tracheal epithelial DNA. *Nature (London)* **247,** 48–49.

Grubbs, C., Moon, R., Squire, R., Farrow, G., Stinson, S., Goodman, D., Braren, C., and Sporn, M. (1977). 13-*Cis* retinoic acid: Inhibition of bladder carcinogenesis induced in rats by *N*-butyl-*N*-(4-hydroxybutyl) nitrosamine. *Science* **198,** 743–744.

Hassell, J., Silverman-Jones, C., and DeLuca, L. (1978). The *in vivo* stimulation of mannose incorporation into mannosylretinyl phosphate, dolichylmannosyl phosphate and specific glycopeptides of rat liver by high doses of retinyl palmitate. *J. Biol. Chem.* **253,** 1627–1631.

Hassell, J., Pennypacker, J., Kleinman, H., Prat, R., and Yamada, K. (1979). Enhanced cellular fibronectin accumulation in chondrocytes treated with vitamin A. *Cell* **17,** 821–826.

Hassell, J., Newsome, D., and DeLuca, L. (1980). Increased biosynthesis of specific glycoconjugates in rat corneal epithelium following treatment with vitamin A. *Invest. Ophthalmol Visual Sci.* **19,** 642–647.

Hayward, G., Carlsen, W., Siegman, A., and Stryer, L. (1981). Retinal chromophore of rhodopsin photoisomerizes within picoseconds. *Science* **211,** 942–944.

Heller, J. (1975a). Interactions of plasma retinol-binding protein with its receptor: Specific binding of bovine and human retinol-binding protein to pigment epithelium cells from bovine eyes. *J. Biol. Chem.* **250,** 3613–3619.

Heller, J. (1975b). Characterization of bovine plasma retinol-binding protein and evidence for lack of binding between it and other bovine plasma proteins. *J. Biol. Chem.* **250,** 6549–6554.

Heller, J. (1976). Intracellular retinol-binding proteins from bovine pigment epithelial and photoreceptor cell fractions. *J. Biol. Chem.* **251,** 2952–2957.

Hirosawa, K., and Yamada, K. (1973). The localization of the vitamin A in the mouse liver as revealed by electron microscope radioautography. *J. Electron Microsc.* **22,** 337–346.

Honig, B., Ebrey, T., Callender, R., Dinur, U., and Ottolenghi, M. (1979). Photoisomerization, energy storage and charge separation: A model for light energy transduction in visual pigments and bacteriorhodopsin. *Proc. Natl. Acad. Sci. U.S.A.* **76,** 2503–2506.

Huang, H., and Goodman, D. (1965). Vitamin A and carotenoids. I. Intestinal absorption and metabolism of ^{14}C-labeled vitamin A alcohol and β-carotene in the rat. *J. Biol. Chem.* **240,** 2839-2844.

Jayaram, M., and Ganguly, J. (1977). Effect of vitamin A nutritional status on the ribonucleic acids of liver, intestinal mucosa and testes of rats. *Biochem. J.* **166,** 339-346.

Jayaram, M., Sarada, K., and Ganguly, J. (1975). Effect of depletion of vitamin A, followed by supplementation with retinyl acetate or retinoic acid on regeneration of rat liver. *Biochem. J.* **146,** 501-504.

Jensen, E., Suzuki, T., Kawashima, T., Stumpf, W., Jungblut, P., and DeSombre, E. (1968). A two step mechanism for the interaction of estradiol with rat uterus. *Proc. Natl. Acad. Sci. U.S.A.* **59,** 632-638.

Jetten, A., and Jetten, M. (1979). Possible role of retinoic acid binding protein in retinoid stimulation of embryonal carcinoma cell differentiation. Nature (*London*) **278,** 180-182.

Jetten, A., Jetten, M., and Sherman, M. (1979). Stimulation of differentiation of several murine embryonal carcinoma cell lines by retinoic acid. *Exp. Cell Res.* **124,** 381-391.

Johnson, B., Kennedy, M., and Chiba, N. (1969). Vitamin A and nuclear RNA synthesis. *Am. J. Clin. Nutr.* **22,** 1048-1058.

Joshi, P., Murthy, S., and Ganguly, J. (1976). Effect of vitamin A nutritional status on the growth of estrogen primed chick oviduct. *Biochem. J.* **154,** 249-251.

Kanai, M., Raz, A., and Goodman, DeW. (1968). Retinol-binding protein: The transport protein for vitamin A in human plasma. *J. Clin. Invest.* **47,** 2025-2044.

Kaufman, D., Baker, M., Smith, J., Henderson, W., Harris, C., Sporn, M., and Saffiotti, U. (1972). RNA metabolism in tracheal epithelium: Alteration in hamsters deficient in vitamin A. Science **177,** 1105-1108.

Kim, Y.-C., and Wolf, G. (1974). Vitamin A deficiency and the glycoproteins of rat corneal epithelium. *J. Nutr.* **104,** 710-718.

Kiorpes, T., Kim, Y.-C., and Wolf, G. (1979). Stimulation of the synthesis of specific glycoproteins in corneal epithelium by vitamin A. *Exp. Eye Res.* **28,** 23-35.

Lotan, R. (1980). Effects of vitamin A and its analogs (retinoids) on normal and neoplastic cells. *Biochim. Biophys. Acta* **605,** 33-91.

McCollum, E., and Davis, M. (1915). The nature of the dietary deficiencies of rice. *J. Biol. Chem.* **23,** 181-190.

Maraini, G. (1979). Binding proteins for retinol in retina and pigment epithelium. *Curr. Top. Eye Res.* **1,** 143-174.

Maraini, G., and Gozzoli, F. (1975). Binding of retinol to isolated retinal pigment epithelium in the presence and absence of retinol-binding protein. *Invest. Ophtalmol.* **14,** 785-787.

Maraini, G. *et al.* (1977). Identification of a membrane protein binding the retinol in retinal pigment epithelium. *Nature* **265,** 68-69.

Masushige, S., Schreiber, J., and Wolf, G. (1978). Identification and characterization of mannosyl retinyl phosphate occurring in rat liver and intestine *in vivo*. *J. Lipid Res.* **19,** 619.

Mayer, H. Bollag, W. Hanni, R., and Ruegg, R. (1978). Retinoids: A new class of compounds with prophylactic and therapeutic activities in oncology and dermatology. Experientia **34,** 1105-1119.

Muto, Y., Smith, J., Milch, P., and Goodman, DeW. (1972). Regulation of retinol binding protein metabolism by vitamin A status in the rat. *J. Biol. Chem.* **247,** 2542-2550.

Napoli, J., McCormick, A., Schnoes, H., and DeLuca, H. (1978). Identification of 5,8-

oxyretinoic acid isolated from small intestine of vitamin A-deficient rats dosed with retinoic acid. *Proc. Natl. Acad. Sci. U.S.A.* **75**, 2603-2605.

Ong, D., and Chytil, F. (1975). Retinoic acid-binding protein in rat tissue: Partial purification and comparison to rat tissue retinol-binding protein. *J. Biol. Chem.* **250**, 6113-6117.

Ong, D., and Chytil, F. (1976). Changes in levels of cellular retinol- and retinoic acid-binding proteins of liver and lung during perinatal development of rat. *Proc. Natl. Acad. Sci. U.S.A.* **73**, 3976-3978.

Ong, D., and Chytil, F. (1978a). Cellular retinoic acid-binding protein from rat testes: Purification and characterization. *J. Biol. Chem.* **253**, 4551-4554.

Ong, D., and Chytil, F. (1978b). Cellular retinol-binding protein from rat liver: Purification and characterization. *J. Biol. Chem.* **253**, 828-832.

Ong, D., and Chytil, F. (1979). Immunochemical comparison of vitamin A binding proteins of rat. *J. Biol. Chem.* **254**, 8733-8735.

Ong, D., Page, D., and Chytil, F. (1975). Retinoic acid binding protein: Occurrence in human tumors. *Science* **190**, 60-61.

Peterson, P., Rask, L., Helting, T., Ostberg, L., and Fernstend, Y. (1976). Formation and properties of retinylphosphate galactose. *J. Biol. Chem.* **251**, 4986-4995.

Pirie, A.(1977). Effects of locally applied retinoic acid on corneal xerophthalmia in the rat. *Exp. Eye Res.* **25**, 297-302.

Pirie, A., and Overall, M. (1972). Effect of vitamin A deficiency on the lens epithelium of the rat. *Exp. Eye Res.* **13**, 105-109.

Plantner, J., and Kean, E. (1976). Carbohydrate composition of bovine rhodopsin. *J. Biol. Chem.* **251**, 1548-1552.

Prutkin, L. (1971). The effect of actinomycin D on the incorporation of labeled vitamin A acid in normal and tumor epithelium. *J. Invest. Dematol.* **57**, 323-329.

Prutkin, L., and Bogart, L. (1970). The uptake of labeled vitamin A acid in keratoacanthoma. *J. Invest. Dermatol.* **55**, 249-255.

Rao, M., Prasad, V., Padmanaban, G., and Ganguly, J. (1979). Isolation and characterization of binding proteins for retinol from the cytosol, nucleosol and chromatin of the oviduct magnum of laying hens. *Biochem. J.* **183**, 501-506.

Rask, L., and Peterson, P. (1976). *In vitro* uptake of vitamin A from the plasma retinol-binding protein to mucosal epithelial cells from the monkey's small intestine. *J. Biol. Chem.* **251**, 6360-6366.

Rask, L., Geijor, C., Bill, A., and Peterson, P. (1980). Vitamin A supply of the cornea. *Exp. Eye Res.* **31**, 201-211.

Redfern, N., Israel, P., Bergsma, D., Robison, W., Whikehart, D., and Chader, G. (1976). Neural retinal and pigment epithelial cells in culture: Patterns of differentiation and effects of prostaglandins and cyclic AMP on differentiation. *Exp. Eye Res.* **22**, 559-568.

Reid, T., Albert, D., Rabson, A., Russell, P., Craft, J., Chu, E., Tralka, T., and Wilcox, J. (1974). Characterization of an established cell line of retinoblastoma. *J. Natl. Cancer Inst.* **53**, 347-360.

Ross, A., Takahashi, Y., and Goodman, DeW. (1978). The binding protein for retinol from rat testes cytosol. *J. Biol. Chem.* **253**, 6591-6598.

Rosso, G., Masushige, S., Quill, H., and Wolf, G. (1977). Transfer of mannose from mannosyl retinyl phosphate to protein. *Proc. Natl. Acad. Sci. U.S.A.* **74**, 3762-3766.

Russell, P., Wiggert, B., and Chader, G. (1978). Separation of retinoid receptors from cultured retinoblastoma cells. *Biochim. Biophys. Acta* **543**, 586-589.

Russell, P., Wiggert, B., Derr, J., Albert, D., Craft, J., and Chader, G., (1980). Nuclear uptake of retinoids: Autoradiographic evidence in retinoblastoma cells *in vitro*. *J. Neurochem.* **34,** 1557-1560.

Saari, J., and Futterman, S. (1976a). Separable binding proteins for retinoic acid and retinol in bovine retina. *Biochim. Biophys. Acta* **444,** 789-793.

Saari, J., and Futterman, S. (1976b). Retinol-binding protein in bovine retina: Isolation and partial characterization. *Exp. Eye Res.* **22,** 425-433.

Saari, J., Bunt, A., Futterman, S., and Berman, E. (1977). Localization of cellular retinol-binding protein in bovine retina and retinal pigment epithelium with a consideration of the pigment epithelium isolation technique. *Inve. Ophthalmol. Visual Sci.* **16,** 797-806.

Saari, J. Futterman, S., and Bredberg, L. (1978a). Cellular retinol- and retinoic acid-binding proteins of bovine retina: Purification and properties. *J. Biol. Chem.* **253,** 6432-6436.

Saari, J., Futterman, S., Stubbs, G., Hefferman, J. Bredberg, L., Chan, K., and Albert, D. (1978b). Cellular retinol- and retinoic acid-binding proteins in transformed mammalian cells. *Invest. Ophthalmol. Visual Sci.* **17,** 988-992.

Saari, J. Bredberg, L., and Futterman, S. (1980). Esterification of retinol by retinoblastoma cell lines WER-Rbl and Y-79. *Invest. Ophthalmol. Visual Sci.* **19,** 1301-1308.

Saffiotti, V. Montesano, R., Sellakumar, A., and Borg, S. (1967). Experimental cancer of the lung: Inhibition by vitamin A of the induction of tracheobronchial squamous metaplasia and squamous cell tumors. Cancer (*Philadelphia*) **20,** 857-864.

Sani, B. (1977). Localization of retinoic acid-binding protein in nuclei. *Biochem. Biophys. Res. Commun.* **75,** 7-12.

Sani, B. (1979). Retinoic acid-binding protein: A plasma membrane component. *Biochem. Biophys. Res. Commun.* **91,** 502-507.

Sani, B., and Donovan, M. (1979). Localization of retinoic acid-binding protein in nuclei and the nuclear uptake of retinoic acid. *Cancer Res.* 2492-2496.

Sani, B., and Hill, D. (1974). Retinoic acid: A binding protein in chick embryo metatarsal skin. *Biochem. Biophys. Res. Commun.* **61,** 1276-1282.

Sharman, I. (1949). The biological activity and metabolism of vitamin A acid. *Br. J. Nutr.* **3,** viii.

Sherman, B. (1961). The effect of vitamin A on epithelial mitosis *in vitro* and *in vivo*. *J. Invest. Dermatol.* **37,** 469-480.

Smith, F., and Goodman, DeW. (1971). The effects of diseases of the liver, thyroid and kidneys on the transport of vitamin A in human plasma. *J. Clin. Invest.* **50,** 2426-2432.

Smith, J., Borek, C., and Goodman, DeW. (1978). Regulation of retinol-binding protein metabolism in c ltured rat liver cell lines. Cell **15,** 866-873.

Sporn, M., Dunlop, N., and Yuspa, S. (1973). Retinyl acetate: Effect on cellular content of RNA in epidermis in cell culture in chemically defined medium. *Science* **182,** 722-723.

Sporn, M., Dunlop, N., Newton, D., and Smith, J. (1976). Prevention of chemical carcinogenesis by vitamin A and its synthetic analogs (retinoids). *Fed. Proc. Fed. Am. Soc. Exp. Biol.* **35,** 1332-1338.

Stubbs, G., Saari, J., and Futterman, S. (1979). 11-*cis*-Retinal-binding protein from bovine retina: Isolation and partial characterization. *J. Biol. Chem.* **254,** 8529-8533.

Takahashi, Y., Smith, J., Winick, M., and Goodman, DeW. (1975). Vitamin A deficiency and fetal growth and development in the rat. *J. Nutr.* **105,** 1299-1310.

Takase, S., Ong, D., and Chytil, F. (1979). Cellular retinol-binding protein allows specific interaction of retinol with nucleus *in vitro*. *Proc. Natl. Acad. Sci. U.S.A.* **76**, 2204–2208.

Tanaka, M. (1980). Localization of ^{3}H-retinol in normal and sliding corneal epithelium and endothelium. *Jpn. J. Ophthalmol.* **24**, 60–66.

Thompson, J. (1969). The role of vitamin A in reproduction. *In* "The Fat-Soluble Vitamins" (H. DeLuca and J. Suttie, eds.), pp. 267–281. Univ. of Wisconsin Press, Madison.

Todaro, G., DeLarco, J., and Sporn, M. (1978). Retinoids block phenotypic cell transformation produced by sarcoma growth factor. *Nature (London)* **276**, 272–274.

Toft, D., and Gorski, J. (1966). A receptor molecule for estrogens: Isolation from the rat uterus and preliminary characterization. *Proc. Natl. Acad. Sci. U.S.A.* **55**, 1574–1581.

Tsai, C., and Chytil, F. (1978). Effect of vitamin A deficiency on RNA synthesis in isolated rat liver nuclei. *Life Sci.* **23**, 1461–1472.

Waechter, C., and Lennarz, W. (1976). The role of polyprenol-linked sugars in glycoprotein synthesis. *Annu. Rev. Biochem.* **45**, 95–112.

Wald, G. (1968). The molecular basis of visual excitation. Nature (*London*) **219**, 800–807.

Wald, N., Idle, M., Boreham, J., and Bailey, A. (1980). Low serum vitamin A and subsequent risk of cancer. Lancet **190**, 813–815.

Warshel, A. (1976). Bicycle-pedal model for the first step in the vision process. Nature (*London*) **260**, 678–683.

Werthamer, S., Samuels, A., and Amaral, L. (1973). Identification and partial purification of "transcortin"-like protein with human lymphocytes. *J. Biol. Chem.* **248**, 6398–6407.

Wiggert, B., and Chader, G. (1975). A receptor for retinol in the developing retina and pigment epithelium. *Exp. Eye Res.* **21**, 143–151.

Wiggert, B., Bergsma, D., and Chader, G. (1976). Retinol receptors of the retina and pigment epithelium: Further characterization and species variation. *Exp. Eye Res.* **22**, 411–418.

Wiggert, B., Bergsma, D., Lewis, M., Abe, T., and Chader, G. (1977a). Vitamin A receptors. II. Characteristics of retinol binding in chick retina and pigment epithelium. *Biochim. Biophys. Acta* **498**, 366–374.

Wiggert, B., Bergsma, D., Helmsen, R., Alligood, J., Lewis, M., and Chader, G. (1977b). Retinol receptors in corneal epithelium, stroma, and endothelium. *Biochim. Biophys. Acta* **491**, 104–113.

Wiggert, B., Bergsma, D., Lewis, M., and Chader, G. (1977c). Vitamin A receptors: Retinol binding in neural retina and pigment epithelium. *J. Neurochem.* **29**, 947–954.

Wiggert, B., Russell, P., Lewis, M., and Chader, G. (1977d). Differential binding to soluble nuclear receptors and effects on cell viability of retinol and retinoic acid in cultured retinoblastoma cells. *Biochem. Biophys. Res. Commun.* **79**, 218–225.

Wiggert, B., Bergsma, D., Helmsen, R., and Chader, G. (1978a). Vitamin A receptors: Retinoic acid binding in ocular tissues. *Biochem. J.* **169**, 87–94.

Wiggert, B., Mizukawa, A., Kuwabara, T., and Chader, G. (1978b). Vitamin A receptors: Multiple species in retina and brain and possible compartmentalization in retinal photoreceptors. *J. Neurochem.* **30**, 653–659.

Wiggert, B., Derr, J., Fitzpatrick M., and Chader, G. (1979a). Vitamin A receptors of the retina: Differential binding in light and dark. *Biochim. Biophys. Acta* **582**, 115–121.

Wiggert, B., Masterson, E., Israel, P., and Chader, G. (1979b). Differential retinoid binding in chick pigment epithelium and choroid. *Invest. Ophthalmol. Visual Sci.* **18,** 306-310.

Wiggert, B., Derr, J. Israel, P., and Chader, G. (1981). Cytosol binding of retinyl palmitate and palmitic acid in pigment epithelium and retina. *Exp. Eye Res.* **32,** 187-196.

Wolf, G. (1977). Retinol-linked sugars in glycoprotein synthesis. *Nutr. Rev.* **35,** 97-99.

Wolf, G., and DeLuca, L. (1969). Recent studies on some metabolic functions of vitamin A. *In* "The Fat-Soluble Vitamins" (H. DeLuca and J. Suttie, eds.), pp. 257-265. Univ. Wisconsin Press, Madison.

Zile, M., and DeLuca, H. (1970). Vitamin A and RNA synthesis in rat intestine. *Arch. Biochem. Biophys.* **140,** 210-214.

Zile, M., Bunge, E., and DeLuca, H. (1977). Effect of vitamin A deficiency on intestinal cell proliferation in the rat. *J. Nutr.* **107,** 552-560.

Zile, M., Bunge, E., and DeLuca, H. (1979). On the physiological basis of vitamin A-stimulated growth. *J. Nutr.* **109,** 1787-1796.

Zile, M., Schnoes, H., and DeLuca, H. (1980). Characterization of retinol β-glucuronide as a minor metabolite of retinoic acid in bile. *Proc. Natl. Acad. Sci. U.S.A.* **77,** 3230-3233.

9

Chromatic Organization of the Retina

ROBERT E. MARC

I. INTRODUCTION

The photoreceptor layer of the vertebrate eye is, in effect, a two-dimensional light-capture surface composed of discrete sensor ele-

Cell Biology of the Eye

ISBN 0-12-483180-X

ments (the rods and cones) each possessing a unique position on the surface. In addition to the rods, there is usually more than one class of cone, leading to the constraint that every receptor type cannot be represented at all points on the sensor surface. What, then, are the *optimal* placements of the various receptor types? Unfortunately, we cannot ask such a question, since it presumes that we have ways to evaluate the selective advantage of the many kinds of visual processing. At present these evaluations are not possible in the quantitative sense. The question we *may* ask is: What are the placements of the various receptor types? From such information we can begin to build a body of knowledge that may allow us to understand the structure and function of the eye.

Many areas of investigation suggest that knowledge of the organization of the receptor array would be of interest, of which I shall note only a few. First, the retinas of many vertebrates possess marvelous repeating mosaics of cones. It is difficult to believe that such order is not related to visual performance. Second, photopic vision is mediated by cones and typically involves large amounts of color processing. The encoding of color is achieved by differential combinations of signals from different spectral classes of cones. Since this system must be responsible for precise spatial and temporal processing, as well as color encoding, the kind, number, and placement of different cone types should have a profound influence on the level of performance of various visual tasks. Finally, Campbell and Green (1965) provided evidence that attenuation of human contrast sensitivity to high-spatial-frequencies due to limits of optical resolution were reasonably matched by similar limits in neural performance. What are the determinants of the neural limitations? For this we need to know many things about the structure of the retina, not the least of which concerns the placement of the various receptor types.

This question must be reduced to more basic ones. What spectral classes of cones exist in a given retina? How are the cones distributed, both locally and globally? This chapter will deal primarily with examples of vertebrate retinas where we have begun to answer some of these questions.

II. VERTEBRATE CONES

I here introduce a classificatory scheme that groups vertebrate retinas into three types. I will discuss only photoreceptors, though the classifications actually depend on many additional retinal features.

A. Type 1: Chondrichthyan Retinas

This group is composed of the elasmobranchs (sharks, skates, rays) and holocephalans (ratfishes, chimeras). Type-1 retinas are generally rod-rich. Ratfishes (Stell, Marc, unpublished data) and skates (Dowling and Ripps, 1970) are thought to lack cones. Sharks of the genera *Squalus* and *Mustelus* have few cones, with a probable rod/cone ratio of 50–100 (Stell, 1972; Stell and Witkovsky, 1973); the lemon shark, *Negaprion brevirostrus,* shows a more robust rod/cone ratio of 12 in its central retina (Gruber *et al.,* 1963). Finally, the diurnal stingrays, genus *Dasyatis,* show an abundance of cones, with a rod/cone ratio of 5 (Toyoda *et al.,* 1978). All cones from type-1 retinas are single and identical in shape. Nothing is known of the cone pigments. The existence of color-opponent horizontal cells in stingrays does suggest there are at least two spectral classes of cones in rays (Niwa and Tamura, 1975; Toyoda *et al.,* 1978). No further mention will be made of type-1 retinas in this chapter, as no chromatic mapping has been done.

B. Type 2: Pleomorphic Retinas

Animals with pleomorphic retinas include chondrostean, holostean, and teleostean fishes, amphibians, reptiles, and avians. In general type-2 retinas are cone-rich. The primary feature of this group is that there are different morphological classes of cones within a retina, and this morphological specialization is an important asset in analyzing the chromatic organization of the cone arrays. The major specialization is the division of cones into two categories: single and double. Double cones are pairs of functionally individual cones, as far as is known, that are held in close apposition by intercellular adhesion. Typical double cones are composed of members of unequal length, which shall be referred to as long members (LD) and short members (SD) of the double cones (Fineran and Nicol, 1974; Stell and Hárosi, 1976) in preference to the older "principal/accessory" terminology. Some fishes possess double cones in which both members are apparently structurally identical, and I will refer to these as twin cones. The convention of calling unequal pairs double cones and equal pairs twin cones is an admirable simplification and has been recently espoused by Munz and McFarland (1977). All other type-2 retinas possess double, rather than twin, cones. Single cones may also come in subtypes, and in some cases the appellations "long single (LS)" and "short single (SS)" cones are appropriate.

Another feature of type-2 retinas is the tendency in many fishes and some reptiles and avians to possess arrangements of small groups of cones into repeating units called mosaic units. This will be useful in analyzing chromatic organization. Depending on the species, there are also variations in receptor frequency over the retina.

C. Type 3: Mammalian Retinas

This category, in its strictest sense, contains only placental mammals. We have little information on which to evaluate the appropriate classifications of monotreme and marsupial mammals (see Walls, 1942; Crescitelli, 1972). Type-3 retinas contain only single cones, and all are morphologically alike; i.e., they are monomorphic rather than pleomorphic. Even though we know that many mammals (especially primates) have color vision, there are no morphological differences among spectral classes of cones that permit their identification. There are large changes in cone frequency over the retina in many type-3 retinas.

III. CONE DISTRIBUTION

Even without information about the spatial organization of spectral classes of cones, it is evident that there are two types of distributions that must be evaluated: *local* and *global*. The chromatic composition of small retinal regions may be of major importance for spatial and chromatic signal processing in response to small stimuli. The development and adult morphology of higher-order neural elements will be subject to constraints imposed by local receptor organization. Local organization can be random or nonrandom. If local organization appears to be nonrandom, it may then take the form of highly ordered structures such as repeating mosaic units common to type-2 retinas or critical spacing of certain cones, as will be shown for some type-2 and -3 retinas. There always exists the possibility that random groups of cones may be part of a larger form of local order, such as repeating clusters. The existence of such units has not been established.

As one examines different loci of some retinas, global changes in receptor frequency occur; i.e., the absolute numbers of receptors per unit area change with retinal eccentricity. The visual performance of an organism may thus depend on where a signal impinges upon the retina if receptor populations are not homogeneous with eccentricity. There may be alterations in the frequency of one spectral class of cones with respect to other classes, so that the chromatic performance of the system may vary with eccentricity. Both type-2 and -3 retinas can be found to show such changes in global organization.

IV. SPECTRAL IDENTIFICATION OF CONES

The three principal methods for characterizing the spectral identities of cones are microspectrophotometry (MSP), film densitometry, and cytochemistry. Each possesses unique advantages and disadvantages; none is an adequate substitute for any other.

A. Microspectrophotometry

MSP remains the primary method for spectral analysis of cones and is essential to a complete description of the spectral properties of cones. The history of MSP in visual cells has been reviewed by Liebman (1972). In principle, MSP is simple: Monochromatic beams of various wavelengths are passed through the outer segments of photoreceptors imaged in a microscope, and the fraction of light absorbed at each wavelength is measured with a photomultiplier. In practice the measurement of spectral absorbance in small structures containing a dilute solution of a photosensitive chromophore that changes its spectral absorbance upon photon capture is very difficult. For purposes of characterizing the spatial organization of spectral classes of cones, MSP has additional limitations.

In order to measure cone absorbance spectra one has to pass light through and measure from one outer segment. Conventional MSP involves (1) macerating isolated retinas into small pieces, so that cones become detached and absorbance can be measured by passing the beam through the side of the outer segment, or (2) using a flat piece of retina and focusing a beam of light through the neural side of the retina onto an individual cone, having the beam traverse the length of the outer segment in the natural path. The former is a transverse and the latter the axial approach. The transverse technique is particularly useful with large cones from type-2 retinas, and those of fishes in particular. Since cones from type-2 retinas are pleomorphic, it is possible to correlate the shape and size of a cone with its spectral absorbance. The most successful example of this approach is the study of cone spectral identities in the goldfish retina by Stell and Hárosi (1976). They identified LD cones as red-sensitive, SD cones as green-sensitive, SS cones as blue-sensitive, and some LS cones as either red- or green-sensitive. This effort permitted the assignment of color sensitivity to specific mosaic positions because the various cone types (LD, SD, SS, LS) are easily recognizable within the mosaic. This example also shows the limitations of MSP: The reliability of an assignment depends on adequate morphological preservation during MSP, the ease of correlation with cells preserved by conventional histological techniques, and the frequency with which cell types are encountered. There are cone types in the goldfish retina that were not encountered by MSP in the Stell and Hárosi study. Further, determining whether all LS cones were actually valid single cones or dissociated double cones was difficult. While this study exemplifies the most complete effort in spectroscopic classifications of cones, with photomicrographs of the actual cones measured, it strongly emphasizes the limits of the technique, even in type-2 retinas. By comparison, MSP studies lacking photomicrographs of the measured cells are of questionable value in analyzing chromatic organization. The transverse approach has been successfully used in primate (type-3) retinas (Bowmaker *et al.*, 1978), but since the cones are monomorphic the only information thus acquired will be (1) identification of the spectral classes of cones (even this may be incomplete) and (2) estimation of the frequencies of cone classes so measured.

Axial measures were introduced primarily to increase the path length of absorbing pigment in the small outer segments of mammalian cones (Marks *et al.*, 1964; Brown and Wald, 1964), but even though the morphological positions of the cones in the retina were preserved, no workers have been able to classify all the cones in a small patch of retina, such as a group of 10 cones. The practical limits of the axial approach have been discussed by Liebman (1972) and Bowmaker *et al.* (1978) from the view of the pigment spectroscopist, but it is disappointing that the preservation of local order appears to be of little use in MSP.

B. Film Densitometry

Film densitometry is an optical method that can be used to make spectroscopic measurements with single-cell resolution while preserving local order. The technique, introduced by Denton and Wyllie (1955) for measuring spectral absorbance in amphibian rods, uses a photographic emulsion as a sensor. Spectral light passing through the receptor layer is differentially absorbed by various classes of receptors, and the outer segments are photographed in this spectral light before and after selective bleaching. The point we are concerned with here is that local order is preserved. As promising as film densitometry seems to be, it suffers from many of the optical problems encountered in axial MSP: light scattering, photoproduct interference, the need for good alignment of outer segments, and the fact that outer segments of different cone classes in type-2 retinas lie in different focal planes. The major attempt to use film densitometry on cones (Scholes, 1975) will be discussed later, but certainly more attempts at film densitometry should be encouraged.

C. NBT Cytochemistry

There are a number of promising cytochemical methods that could be used for photoreceptor analysis in the future, but I shall describe only the use of the redox dye nitro blue tetrazolium (NBT) chloride. This dye was first synthesized by Tsou *et al.* (1965) as an indicator for dehydrogenase activity. NBT is a ditetrazole possessing methoxy- and nitro-substituted phenyl rings. One of many kinds of tetrazoles, it possesses a satisfactory combination of weak visible absorbance and fair solubility in the oxidized form, which converts to a high-extinction, insoluble blue product upon complete reduction of the molecule. Enoch (1963, 1964a,b, 1966) demonstrated, in an important but largely unheralded series of studies, that the light-dependent reduction of NBT to a blue product in rat rods was mediated by rhodopsin bleaching. The technique lay dormant for nearly a decade, and then Harry Sperling and I developed protocols making the approach feasible in goldfish and monkey cones. It appears that extensive bleaching in photoreceptors causes increases in the rate of oxidative metabolism, which can be measured using the rate of reduction of NBT. Upon reduction of a tetrazole ring, the ring opens and extends the molecule, causing a bathochromic shift in its spectral absorption. Four atoms of hydrogen are required for full reduction, yielding a product known generally as a diformazan, here abbreviated NBT-DF. Partial reduction is also possible, and the practical significance of this is discussed in Marc and Sperling (1976b).

NBT techniques are used primarily on isolated retinas exposed to spectral bleaching lights (5–10 min) while being physiologically maintained. Afterward the retinas are incubated in saline–substrate–buffer mixtures with 6 m*M* NBT for 5–10 min, fixed, and viewed as whole mounts (see Marc and Sperling, 1976a,b, 1977, for details). By selecting appropriate wavelengths and intensities of light, different spectral classes of cones can be stimulated to reduce NBT.

The advantages of NBT cytochemistry are that all cell types are easily sampled, their morphological relationships to each other are preserved, organization of spectral classes of monomorphic cones can be discerned, and the ability to classify cones spectrally is largely independent of the size of their outer segments or frequency of occurrence. The disadvantages are equally distinct: The identification of pigment types is only qualitative, the technique requires different incubation protocols for various species and may be inapplicable to some, and a complete spectral analysis requires the use of many wavelengths and intensities, thereby requiring many preparations. These shortcomings are minor compared to the usefulness of accurate spectral assignments as a function of position in the receptor array.

D. Miscellaneous Indicators

Incidental features of some photoreceptors may also serve as qualitative indicators of spectral identity, such as the existence of colored oil droplets within cones of some type-2 retinas. There have been many reviews and discussions regarding the functions of oil droplets (Muntz, 1972; Liebman, 1972; Bowmaker, 1977; Bowmaker and Knowles, 1977). In brief, each cone type may possess an oil droplet, the color of which is uniquely associated with that class of cone. In avians, for example, there are several classes of single cones, varying in length and the kind of oil droplet contained within the ellipsoid near the base of the outer segment, e.g., red, yellow, or clear. These colored structures serve as passive filters which will shape the spectral responses of the cones regardless of the pigments they contain. The oil droplets are, therefore, very useful for classifying the spatial distributions of different spectral classes of cones.

V. CHROMATIC ORGANIZATION OF VERTEBRATE CONES

MSP, film densitometry, NBT cytochemistry, and other structural indicators of spectral identity will be considered in the following review of the chromatic organization of vertebrate cones. Not all techniques are applicable to or have been used in these studies. In reality one should consider the results merely an initial effort in the study of photoreceptor distribution, since only in the past 15 years have there been serious attempts to determine the chromatic patterning of cones, and it certainly has been less than a decade since any adequate results were obtained.

A. Teleostean Fishes

The best studied retinal mosaics are those of teleostean fishes. Over a century ago, the crystalline order of photoreceptor organization was noted in teleosts (e.g., Hannover, 1844), and current anatomists continue to describe these arrays. One of the most complete summaries of cone patterning in fishes was compiled by Kjell Engstrom (1960, 1963a,b), and the discussion of mosaics by Crescitelli (1972) is highly recommended. Few scientists can fail to be impressed by the order present in the retinas of fishes. We have used the order and morphological differentiation of teleostean cones to uncover the spectral organization of the photoreceptor layer.

MSP of teleostean cones has concentrated primarily on the receptors of the goldfish *Carassius auratus*. I shall describe the chromatic organization of goldfish cones as a reference point and then compare the findings with those made in other teleostean fishes. MSP of isolated goldfish cones in many laboratories has shown the presence of three spectral classes: red-absorbing, green-absorbing, and blue-absorbing (Liebman and Entine, 1964; Marks, 1965; Svaetichin *et al.*, 1965; Hárosi and MacNichol, 1974; Hárosi, 1976; Stell and Hárosi, 1976). Hárosi (1976) has placed the absorption maxima at 453 nm (blue), 533 nm (green), and 620 nm (red). In early work, the correlation of pigment content with cone type was largely anecdotal and lacked morphological documentation. Double cones were described as red-green-absorbing pairs, but the relation between the members was largely unspecified. Some workers described the unequal double cones as twin cones, leading to confusion about the organization of the mosaic. The definitive findings were those of (1) Stell and Hárosi (1976) using MSP combined with rigorous morphological characterization, and (2) Marc and Sperling (1976a,b) using NBT, which allowed direct visualization of the cones selectively activated by red, green, or blue light. The LD cones were red-absorbing and the SD cones were green-absorbing in all cases. The same workers also showed that all SS cones were blue-absorbing. Stell and Hárosi (1976) provided convincing evidence that morphologically similar LS cones of both red- and green-absorbing varieties existed, a finding confirmed directly in the cone mosaic with NBT cytochemistry (Marc and Sperling, 1976b). Two other classes of cones, miniature short single (MSS) cones and miniature long single (MLS) cones were not successfully identified by MSP. NBT cytochemistry established that MSS cones were blue-absorbing (Marc and Sperling, 1976a,b) and that some MLS cones, at least, were red-absorbing (Marc, 1977).

The goldfish retina may be considered the standard with which to compare spectral classifications in other teleosts. The photoreceptors and their spectral classifications are shown in Fig. 1, and their idealized mosaic arrangement in Fig. 2. Typical NBT experiments on which these results are based are shown in Figs. 3–5. Upon exposure to spectral lights, different morphological classes of goldfish cones re-

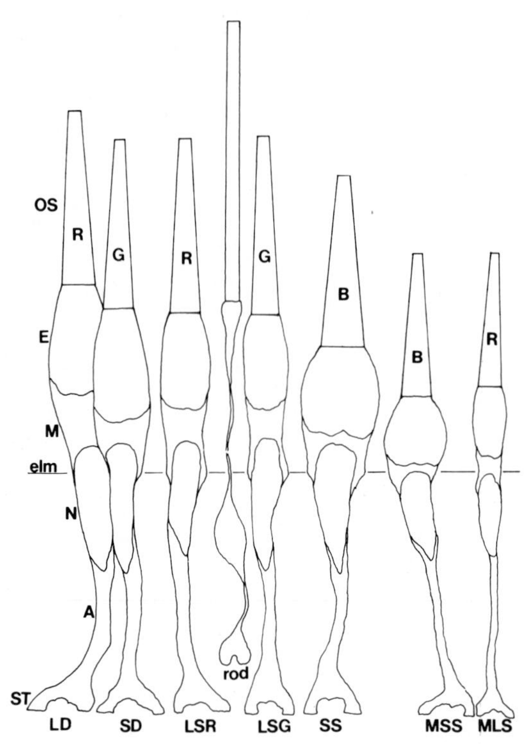

Fig. 1. Morphology of the photoreceptor types in the light-adapted goldfish retina. OS, Outer segment; E, ellipsoid; M, myoid; elm, external limiting membrane; N, nucleus; A, axon; ST, synaptic terminal. Structural classifications (LD, SD, etc.) are explained in the text. Color identities (R, red; G, green; B, blue) are indicated on the outer segments. A rod is shown for comparison. From Marc (1977).

spond in reliable ways: (1) Exposure to 1.2×10^4 quanta sec^{-1} μm^{-2} of narrow-band 650-nm light (Fig. 3a) evoked reduction of NBT to NBT-DF in LD cones and some LS cones (Fig. 4); (2) exposure to 3.0×10^4 quanta sec^{-1} μm^{-2} narrow-band 535-nm light evoked strong reduction in SD cones, weak reduction in LD cones (Fig. 3b), and no responses in SS or MSS cones (Fig. 5a); (3) exposure to 5.0×10^3 quanta sec^{-1} μm^{-2} 470-nm light evoked strong responses from SS and MSS cones (Fig. 5b).

The cones in the central retina of light-adapted goldfishes form repeating patterns where the double cones define the sides of a rhombus or square; the red- and green-absorbing members exhibit alternating symmetry across the square units (Fig. 2). Centered in each square is a blue-absorbing SS cone. MSS, LS, and MLS cones tend to be found at

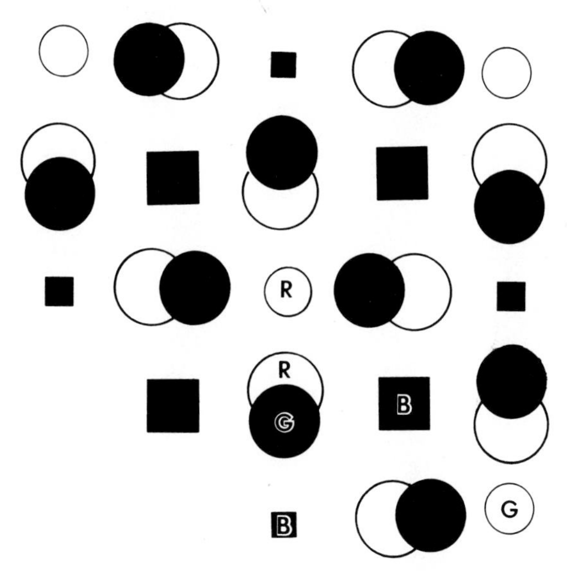

Fig. 2. Idealized mosaic organization showing cone symmetries, interrelating positions, and relative frequencies. Double cones are shown as paired circles (LD, open; SD, solid), LS cones as open circles, SS cones as large squares, and MSS cones as small squares. A unit of four double cones surrounding each SS cone occupies half the area of a square unit described by four adjacent SS cones, and one-quarter the area occupied by four MSS cones. Color assignments are indicated as R, G, and B. LS cones can be either R or G. This basic unit, modified by cell size, random intrusions from extra LS or MLS cones, and lattice defects is the basic organization found in most adult light-adapted goldfishes. Modified from Marc and Sperling (1976b), *Vision Res.* **16** 1211–1224. Copyright 1976, Pergamon Press, Ltd.

the corners of the square arrays, with less overall precision. Present calculations indicate the approximate frequencies of the cones to be 45% red-absorbing, 35% green-absorbing, and 20% blue-absorbing. While these data allow us to view the mosaic as a color-coded array, the frequencies are of less interest than the specificities of connections with retinal neurons. Stell and his colleagues have made extensive use of the specific chromatic identities of goldfish cones to analyze some of the color-specific contacts of horizontal and bipolar cells with cones (Stell and Lightfoot, 1975; Stell *et al.*, 1975; Stell *et al.*, 1977; Stell, 1978; Stell, 1980; Ishida *et al.*, 1980). These data have been seminal in understanding the partitioning of spectral signals among some of the neurons in the goldfish retina. I will not attempt to review them here but refer the reader to the above-mentioned works.

In the goldfish retina one finds small but measureable changes in

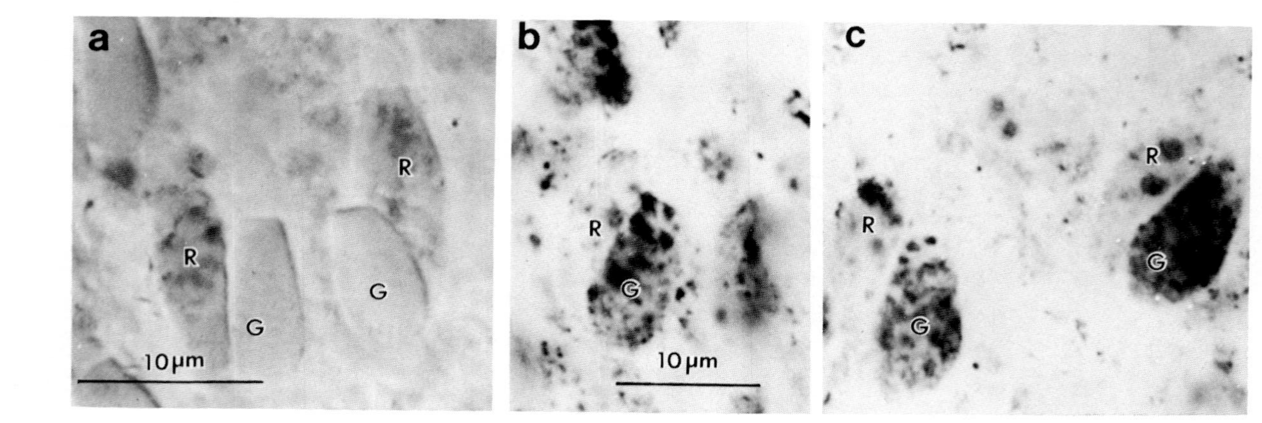

Fig. 3. Responses of goldfish double cones to chromatic stimulation. Color assignments are placed on cone ellipsoids, typically the only cell part visible in these whole-mount micrographs. Stimuli are described in the text. (a) Double cones in lateral view after exposure to red light. Only LD cones contain diformazan. From Marc and Sperling (1976a) *Science* **191,** 487–489. Copyright 1976 by the American Association for the Advancement of Science. (b) Double cone in lateral view after exposure to green light. SD cones contain more diformazan than LD cones. (c) Pair of double cones from a different field in the same preparation as in (b). Both (b) and (c) from Marc and Sperling (1976b), *Vision Res.* **16,** 1211–1224. Copyright 1976, Pergamon Press, Ltd.

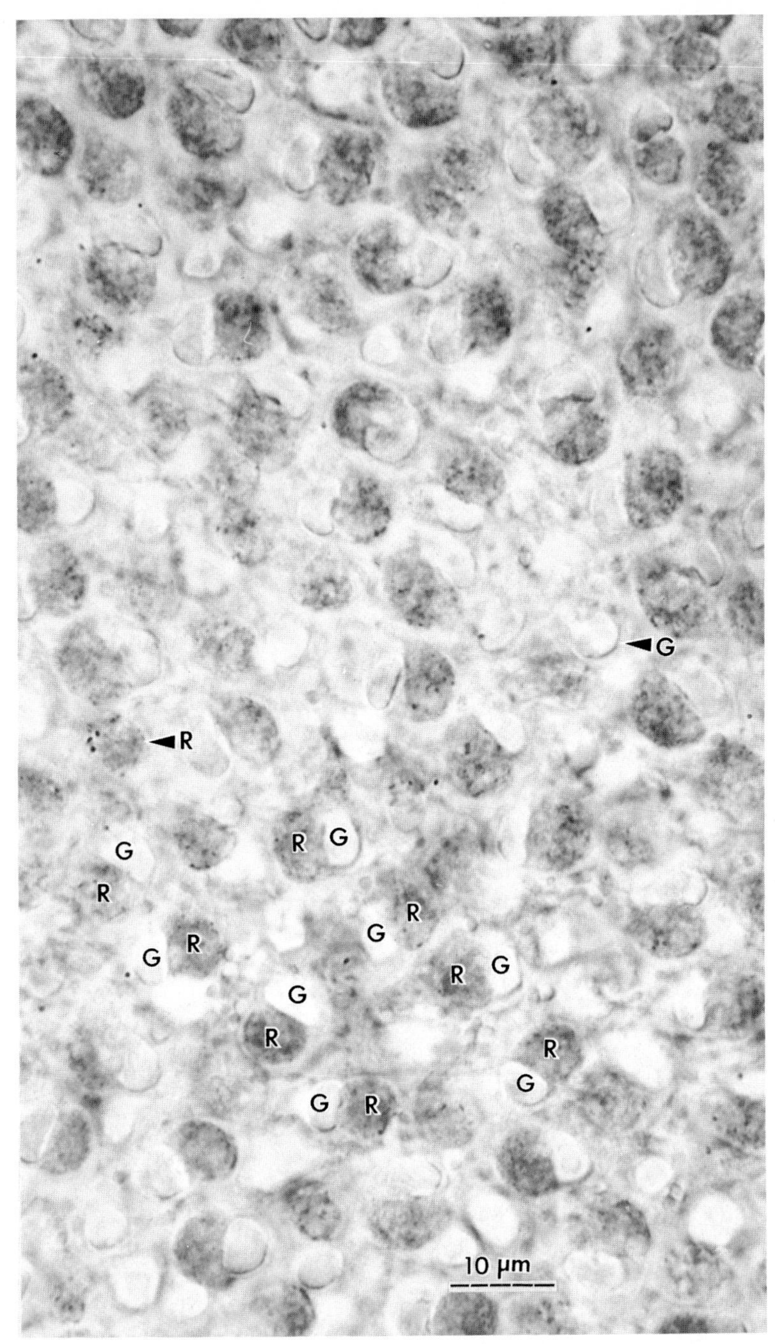

Fig. 4. Mosaic view of LD, SD, and LS cones after exposure to 650-nm light (8×10^4 quanta sec^{-1} μm^{-2}). Epoxy-embedded whole mount. Color assignments are indicated as R and G.

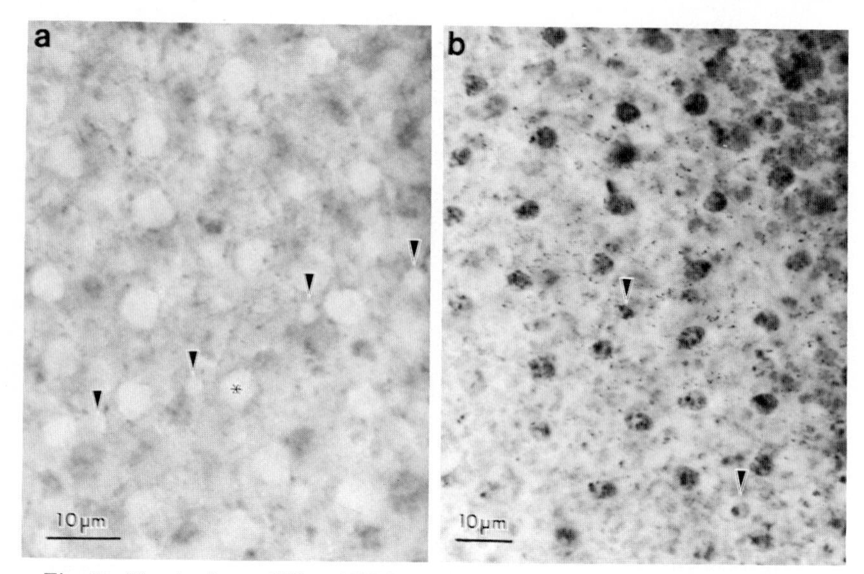

Fig. 5. Mosaic views of SS and MSS cones (focus set below the plane of the LD cones) after exposure to colored lights. Stimuli are described in the text. (a) Mosaic view of SS (asterisk) and MSS cones (arrowheads) after exposure to green light. No responses are found at this wavelength or at any longer wavelengths of equivalent flux density. From Marc and Sperling, *Vision Res.* **16,** 1211-1224. Copyright 1976, Pergamon Press, Ltd. (b) Mosaic view of SS and MSS cones (arrowheads) after exposure to blue light. Intense responses occur in both types of cones. From Marc and Sperling (1976a) *Science* **191,** 487-489. Copyright 1976 by the American Association for the Advancement of Science.

the global frequencies of cones: Double and SS cones are most frequent in the ventrocaudal to equatoriocaudal regions and fall to almost half this frequency in most of the dorsal retina (Marc and Sperling, 1976b). This is qualitatively similar to the results of Hester (1968) for double cones in the goldfish and the variations described by O'Connell (1963) for some marine teleosts. The gradient, however weak, is not without its interesting features, such as the fact that ventral cones in general are much smaller than dorsal cones, ventral retina lacks an occlusible tapetum, and the photopic sensitivity of ventral retina measured electroretinographically is greater than that of dorsal retina (Wheeler, 1978). Another global reorganization that is poorly understood is the variability of encountering certain single cones in different parts of the retina. For example, blue-sensitive MSS cones have been found to be totally absent from large areas of dorsal retina (Marc, unpublished data). What these intraretinal variations mean or even if they are reliable features of a goldfish retina are unknown at present but, as will be shown, bona fide intraretinal chromatic specializations do occur in some type-2 retinas. We do know that some freshwater fishes live in a photic environment that consists of bright light above (impinging on ventral retina) and dimmer, but often red-shifted, light from below impinging on dorsal retina. The global variations in cone organization in cyprinids may represent adaptations to the different intensities and spectral compositions of natural lighting.

The rudd, *Scardinius erythophthalmus,* is a cyprinid fish rather similar to the goldfish. Scholes (1975) described the possible spectral assignments of rudd cones based upon film densitometry. The rudd contains double cones and an assortment of single cones qualitatively similar to goldfish cones. Scholes refers to the LD cones as "principal" cones, SD cones as "accessory" cones, LS cones as "free principal" cones, SS cones as "single" cones, and what are likely homologs of MSS cones as "oblique" cones. Again, the terminology is a matter of preference, but Stell and Hárosi (1976) have made a convincing case for the use of more phenomenological labels. Film densitometry is technically very difficult, rendering the successful characterization of some cones all the more impressive. Scholes identified SD cones as green-absorbing and SS cones as blue-absorbing, but the presence of a red-absorbing pigment could not be detected in any cone type by film densitometry; he therefore assumed that LD and LS cones contained such a long-wave pigment. Leow and Dartnall (1976) confirmed by MSP that rudd double cones are red–green-absorbing pairs; they also claimed to have measured oblique cones and to have identified them as blue-absorbing. In view of the relative infrequency of oblique cones and notwithstanding the likelihood of the assignment, such a claim is difficult to accept in the absence of morphological evidence.

These characterizations are, in a qualitative sense, identical to those made for goldfish cones. There are sufficient similarities in rudd and goldfish mosaic organization to expect, as has been shown (Scholes, 1975; Ishida *et al.,* 1980), that the circuitries of some second-order neurons are similar in both species.

Following these initial efforts have come a series of studies on various teleosts showing by MSP that many of the identifications made by Scholes (1975), Stell and Hárosi (1976), and Marc and Sperling (1976a,b) are typical of most teleosts. In both freshwater and marine species, regardless of whether the chromophores of the pigments are retinal, dehydroretinal, or mixtures of both, the presence of double cones as I have defined them indicates that the longer member contains a visual pigment with an absorption maximum at wavelengths longer than the pigment contained by the shorter member (e.g., Leow and Lythgoe, 1978; Levine and MacNichol, 1979). Twin cones present more of a problem, since there is now good evidence that some pairs of morphologically indistinguishable cones contain different pigments or the same pigments, usually of the long- or middle-wave variety, sometimes both within a single retina (Leow and Lythgoe, 1978; Ali *et al.*, 1978; Levine and MacNichol, 1979; Fernald and Liebman, 1980). Levine and MacNichol (1979) chose to define cone pairs as twins only when they had spectroscopically shown that both members contained the same pigment, and reserved the term "double" for all cone pairs containing dissimilar pigments. Despite the attractiveness of this suggestion for photochemical classification, the term "twin" is, after all, a morphological term, and redefining for another purpose only adds to the confusion. What is needed is careful morphology to determine whether twin cones containing different pigments are truly monomorphic and, if so, strict attention must then be paid to higher-order measures, such as the neuronal connectivity of twin cones containing the same or different visual pigments. If a difference in the synaptic connections of twin cones containing the same or different visual pigments cannot be resolved, then the chance of uncovering the significance of such pairings for visual function is likely to be small. Less conceptual difficulty is encountered with the very common occurrence of single cones containing blue-absorbing visual pigments (Ali *et al.*, 1978; Levine and MacNichol, 1979; Fernald and Liebman, 1980). The SS cones of the goldfish are blue-absorbing, are unique in shape, and occupy the centers of the square mosaic units. Though many authors have described blue-absorbing cones, less attention has been paid to their shape or location. I suspect that the blue-sensitive cones of most fishes occupy the centers of a regular square array. That this is likely to be so is supported by further NBT studies on some marine teleosts (R. E. Marc, unpublished data) where the cones showing the highest blue-responsivity were small, single, and arrayed in a regular fashion (Fig. 6a and b). Fernald and Liebman (1980) have shown by MSP that blue-absorbing SS cones of *Haplochromis* form the centers of square arrays.

Despite this growing body of information on the pigment content of fish cones, most of our detailed information on the local and global distribution of these cone types is still restricted to the goldfish and the rudd. The literature in this area is rather new, however, and I am confident that many equivocations about organization necessary at this time will be resolved in a few years.

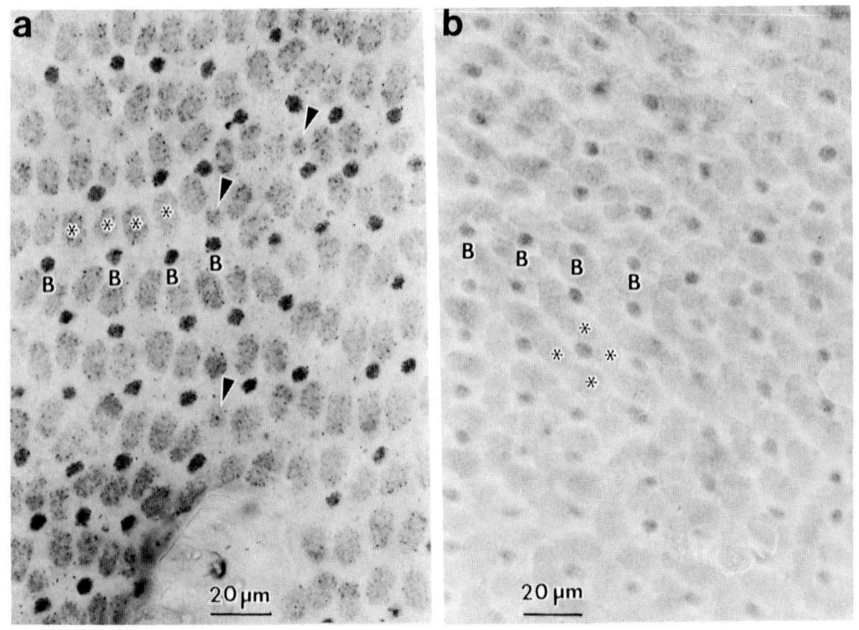

Fig. 6. Mosaic views of marine teleost retinas after exposure to 440-nm light (4×10^4 quanta sec^{-1} μm^{-2}). Blue-absorbing single cones are indicated by a "B" beneath their stained ellipsoids. Double cones are indicated by an asterisk directly on the ellipsoid region. Infrequent single cones are indicated by arrowheads. (a) Sculpin (*Enophrys bison*). Peripheral retina shows strongly blue-responsive single cones in a regular, albeit imprecise, array. (b) Shiner perch (*Cymatogaster aggregata*). Peripheral retina shows blue-responsive single cones in a very precise pattern. The primary difference between these fishes is that the perch dwells much more in open water and is a much more active animal than the sculpin.

B. Amphibians

Only anuran retinas have been examined in enough detail to warrant review. An extensive description of the visual pigments of frog (*Rana pipiens*) rods and cones was made by Liebman and Entine (1968). Anuran retinas in general contain two classes of rods, red rods and green rods, based *not* on their visual pigment content but rather on their appearance in the light microscope. In freshly mounted dark-adapted retinas, the presence of rhodopsin lends a briefly visible but strong reddish tint to the image of red rods. Viewed axially, green rods have a photostable green tint which may be optical in origin (Miller and Snyder, 1972). Liebman and Entine (1968) showed that red rods contained a green-absorbing visual pigment with an absorbance maximum (502 nm) and shape similar to that of extracted rhodopsin (Crescitelli, 1958); green rods contained a blue-absorbing pigment (absorbance maximum 432 nm) similar to the 433-nm pigment extracted from *Rana cancrivora* by Dartnall (1967). Denton and Wyllie (1955) used film densitometry of frog retinas to show that there were two classes of rods: green- and blue-absorbing. It is a departure from our emphasis on cones to discuss rods in such detail, but it is not at all clear that the green rod behaves as a typical rod (Goldstein and Wolf, 1973). The spatial distribution of green rods has been estimated three ways. Nilsson (1964), using electron microscopy, found that 15% of the photoreceptors in the posterior pole of the frog's eye were green rods. By observing the photostable green coloration of green rods in whole mounts, I calculate green rods to be 8–10% in midperiphery (Fig. 7a), consistent with the results of Fite and Carey (1973) using conventional morphology. Liebman and Leigh (1969) described a bright autofluorescence of green rods that was activated by intense blue light exposure (405- or 436-nm line of a mercury arc lamp used for fluoromicroscopy); dark-adapted, unexposed retinas showed no such fluorescence. I have reproduced that remarkable experiment (Fig. 7b), and it yields estimates of green rod density of 8–14%. The point of this diversion is to emphasize that, even when the short-wave signal is transduced by rodlike cells, blue-absorbing photoreceptors are always the least frequent color type.

An intriguing feature of some kinds of blue-absorbing cells in type-2 retinas is the presence of an oblique axon in some of them. This was mentioned by Leeper (1978) who noted that the blue-absorbing cones of turtles, the green rods (blue-absorbing) of frogs, the blue-absorbing MSS cones of goldfish, and the aptly named oblique cones of the rudd (which are likely to be blue-absorbing) all possessed axons which deviated from the paths taken by their fellow receptors. The oblique

---

Fig. 7. Whole mounts of frog (*R. pipiens*) retina, showing end-on views of both red rods (green-absorbing, rhodopsin-containing) and green rods (blue-absorbing, containing a 432-nm pigment). (a) Bright-field view. Light rods are bleached, green-absorbing red rods, and dark rods are bleached, blue-absorbing green rods showing the photostable dark coloration that gave rise to the name "green" rods. (b) End-on fluorescence micrograph of rods. Exposure to the violet (405-nm) line of the microscope mercury arc for 5 sec elicited this pattern of fluorescent photoreceptors. These are the blue-absorbing green rods. Liebman (1972) has shown that the fluorescence has the emission spectrum of vitamin A. The exciting filter allows a certain amount of visible blue light to pass through the specimen, causing a uniform gray background between the rods. The green-absorbing red rods, being largely unbleached, absorb much of this background light and appear as dark spots. This micrograph actually represents a combination of photically activated autofluorescence and film densitometry. Ultraviolet excitation at 365 nm.

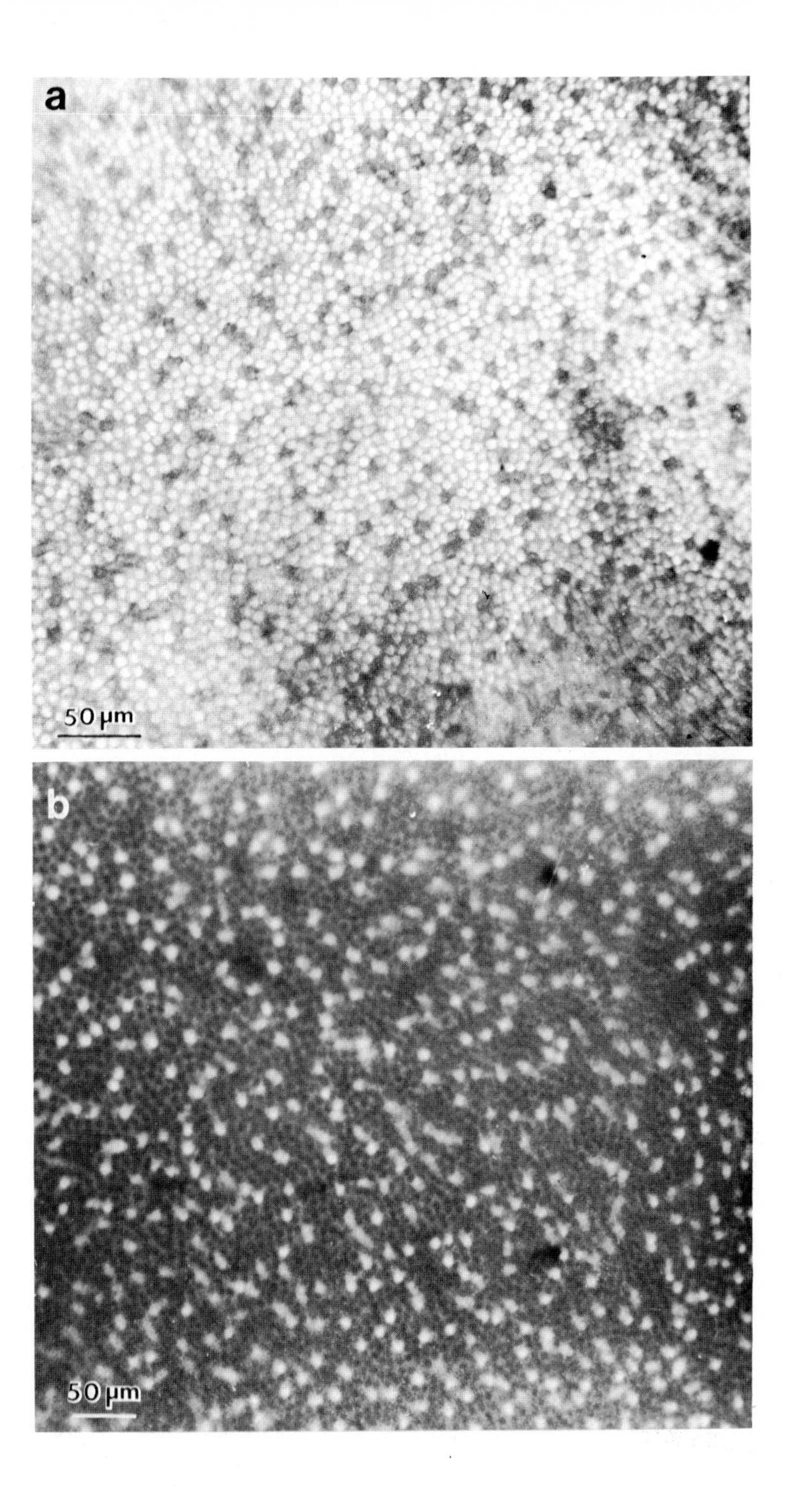
a
50 µm
b
50 µm

axons cut across the trajectories of other cone axons to reach a point in the outer plexiform layer several cone widths displaced from their expected terminations. Although the significance of this observation is not known, it is one of the consistencies that suggests a process of fundamental importance is occurring, if only we had the insight to interpret it.

Frog cones are either double or single. I might add that, from this point on, the morphological problem of twin cones will not appear again. Double cones are unequal, being classified as principal and accessory; I will continue to use the labels LD and SD, inasmuch as they are as applicable here as anywhere. The LD cone contains a clear oil droplet and a long-wave pigment in its outer segment, absorbing maximally at 575 nm; the SD cone lacks an oil droplet, possessing instead a glycogen-rich paraboloid, and has a rodlike pigment absorbing at 502 nm (Leibman and Entine, 1968). Single cones with oil droplets contain the long-wave pigment. Carey (1976) described single cones lacking oil droplets, and now Hárosi (1981) has reported that some oil droplet-free cones in the frog contain a blue-absorbing pigment. Clearly, the anuran retina deserves a closer examination. Nilsson (1964) could detect no obvious mosaic order in the frog retina, but no spectral information was available then. The absence of order cannot be accepted on the basis of visual appearance (see Section V,D) since many types of order are not readily visible.

On a global scale, there appears to be a linear zone at the horizontal meridion where cone density increases somewhat. In *R. pipiens* I have found that the long-wave cone frequency doubles from the far periphery in the dorsotemporal retina to the central pole. Fite and Cary (1973) described marked increases in cone frequency in a linear horizontal zone in both *Rana* and *Bufo*. An interesting intraretinal specialization in anurans was described by Reuter *et al.* (1971), where the dorsal retina of the bullfrog (*Rana catesbeiana*) was porphyopsin-enriched and the ventral retina rhodopsin-enriched; this constitutes a bathochromic shift in the sensitivity of the dorsal retina by up to 20 nm. It is not known whether this change applies to cones as well, but if there is a selective advantage to possessing this intraretinal difference, one might expect it to be expressed photopically as well.

C. Reptiles

There are a great number of species of reptiles that a visual scientist might be interested in. Walls (1942) discussed his view of the evolutionary importance of reptilian eyes, especially those of the Gekkonidae. Since we are interested in color, however, this discussion shall be limited to the chelonians (turtles), with a brief consideration of organizational features in geckos.

The cones of turtles are generally quite large, possessing 5- to 8-μm diameter ellipsoids, and turtle retinas have been extensively used in intracellular electrophysiology. The turtles most often used are *Pseudemys scripta* (the red-eared turtle) and *Chelydra serpentina* (the snapping turtle). Each possesses an abundance of cones, although *Chelydra* has about four times as many rods as *Pseudemys* (Underwood, 1970; Lasansky, 1971; Leeper, 1978), *Pseudemys* having roughly 10% rods. In both species there are double cones and single cones. In *Pseudemys* all single cones contain an oil droplet: either red, yellow, or colorless (Liebman and Granda, 1971; Leeper, 1978). The LD cones contain an orange oil droplet, and the accessory cones lack an oil droplet (Liebman and Granda, 1971; Baylor and Fettiplace, 1975), but careful observation shows that the SD cones are endowed with many yellow granules in their inner segments. The assignments of pigment content to cell type for *Pseudemys* are discussed thoroughly by Leeper (1978), and I will only summarize those results here: LD cones are red-absorbing, SD cones are green-absorbing, single cones with red oil droplets are red-absorbing, those with yellow oil droplets are green-absorbing, and those with colorless oil droplets are blue-absorbing. Red-absorbing cones are in the majority, and blue-absorbing cones in the minority, consistent with the organization found in fishes and amphibians. No critical analysis of the spatial organization of turtle cones has been made, but even a cursory examination in peripheral retina gives the impression that the oil droplets are critically rather than randomly spaced.

Brown (1969) described a striking linear area centralis in the retina of *Pseudemys:* The cone frequency rises from about 20,000 cones/mm^2 in the periphery to almost 60,000 cones/mm^2 in the area centralis. This is clearly an adaptation associated with improved photopic acuity and may be related to selective advantages in panoramic acuity at the air–water interface (Brown, 1969).

If, as seems probable, all cones with clear oil droplets are blue-absorbing, the data of Brown (1969) indicate that (1) blue-absorbing cones are the least frequent type and (2) the relative proportion of blue-absorbing cones has a different global distribution than the other spectral classes. Red- and green-absorbing cones dramatically increase in frequency in the area centralis (two to four times), but blue-absorbing cones remain roughly constant in frequency. The blue-absorbing cones constitute between 10 and 19% of all cones in the periphery but drop to no more than 6% in the area centralis. As will be discussed later, this indicates a negligible role for blue signals in acuity tasks and fixation.

Other than for the turtles, we have little information about the spectral organization of other reptilian retinas. Nocturnal geckos, however, are on the threshold of being characterized, primarily as a result of the persistence of Frederick Crescitelli at the University of California at Los Angeles. Nocturnal geckos (e.g., the genera *Gekko* and *Coleonyx*) possess a precise mosaic of rodlike photoreceptors: type-A single rods, type-B double rods, and type-C double rods. The problem of whether these are truly rods and how they may have evolved has been discussed by Walls (1942) and Crescitelli (1972, 1977). The mosaics are rows of type-B double rods whose unequal members alternate from side to side; interspersed between each pair of rows of type-B doubles is a row composed of both type-A single rods and type-C double rods, with no thoroughly defined sequence for the type-A and -C members (Dunn, 1966). The doubles are pairs of morphologically distinguishable rodlike cells, so that gecko retinas contain two kinds of double photoreceptors. Efforts have been made to characterize the photopigments by MSP (Crescitelli, 1977; Crescitelli *et al.,* 1977). Two pigments were found: a green-absorbing pigment (absorbance maximum 521 nm) and a blue-absorbing pigment (absorbance maximum 467 nm). Only isolated outer segments were measured, so the actual identities of the cell types are still in question. It is fairly certain that type-A single rods contain a green-absorbing pigment, although a few instances of isolated outer segments containing blue-absorbing pigment were found. The double rods (type unspecified) were, unfortunately, of all mixes: green plus green, green plus blue, and blue plus blue. As pointed out by Crescitelli (1977), no claims for specific identification can be made until better morphological methods are employed, but the likelihood of unequal pairs containing the same pigment is a serious deviation from the LD–long wave, SD–middle wave dichotomy seen in most type-2 retinas. Crescitelli (1977) also mentions that the neural connectivity of such cells may be a clue to their function, as I have suggested for teleostean twin cones.

D. Avians

On ethological grounds alone, how avians code spectral information is a major issue, and there has been a fair amount of anatomical, spectroscopic, and behavioral investigation that bears on the organization of different spectral classes of cones. The pigeon, *Columbia livia,* serves as a useful paradigm for discussing retinal organization in avians.

The cones of pigeons are either double or single and partition among themselves a variety of oil droplets: red, orange, yellow, or colorless. There have been numerous contemporary studies on the spectral properties of avian oil droplets (Strother, 1963; King-Smith, 1969; Bowmaker, 1977; Bowmaker and Knowles, 1977). Unfortunately there has been continuing disagreement in the literature as to the appropriate assignment of oil droplet types to morphological cones types. Almost all workers agree that there are single cones containing red oil droplets and single cones with small yellow oil droplets; such cones are easily found by light microscopy. When orange and yellow oil droplets are distinguished, large yellow oil droplets are usually assigned to LD cones and orange oil droplets to single cones (e.g., Bowmaker, 1977; Mariani and Leure-Dupree, 1978). There is substantial variability in the spectral properties of these kinds of droplets as a function of retinal locus (Bowmaker, 1977), the systematics of which have not been characterized. For the present it is sufficient to note that most workers assign some yellowish oil droplets to LD cones. The controversy arises when one attempts to place colorless oil droplets and assign a droplet, if any, to the SD member of the doubles. Bowmaker (1977) asserts that SD (accessory) cones lack an oil droplet, while Mariani and Leure-Dupree (1978) contend the opposite, further assigning the colorless droplet to SD cones. They are both at once correct and incorrect. Similar to the situation in turtles and many diurnal lizards, there is no oil droplet in the SD cones of pigeons but rather a pigment granule (Morris and Shorey, 1967; R. E. Marc, unpublished observations) that is cytochemically distinct in stainability from all the carotenoid-rich oil droplets. The granule is distinctly and unequivocally yellow in all fresh preparations, as are the SD granules of turtles and lizards. In no case have I ever seen a colorless oil droplet in any SD cone; they are in fact much too large. Mariani and Leure-Dupree (1978) claim that the clear oil droplets are the smallest such structures in the pigeon retina and thus assign them to SD cones. The granules are smaller by half, however, and more precisely fit SD cones. Figure 8 summarizes my view of the most probable oil droplet–cone pairings.

Assignment of cone pigments has also been a difficult task owing to the smallness of avian outer segments (1–2 μm in diameter), but an admirable effort has been made by Bowmaker (1977). While assignments were attempted, Bowmaker (1977) provided no photographic documentation of measured cones. The LD cones and single cones with red oil droplets contained the long-wave pigment, as expected, with an

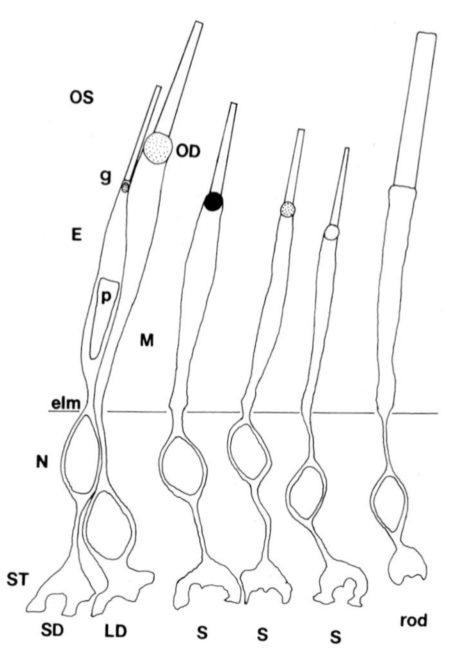

Fig. 8. General morphologies of photoreceptor types in the yellow field of the pigeon retina. OS, Outer segment; OD, oil droplets; g, colored granule (SD cone only); E, ellipsoid; p, paraboloid (SD cone only); M, myoid; elm, external limiting membrane; N, nucleus; ST, synaptic terminal. Shown are long (LD) and short (SD) members of the double cones, three single cones (S) and a rod. Oil droplet types indicated are red (solid), yellow or orange-yellow (dotted), and colorless (open). Adapted from camera lucida tracings of semithin sections. Notice the stratification of oil droplet levels, the tallest being LD cones and the shortest the single cones with colorless droplets.

absorbance maximum of 567 nm. One type of single cone with a yellowish oil droplet contained a middle-wave pigment absorbing maximally at 514 nm. One deviation from expected results came from assignment of the long-wave pigment to SD as well as LD cones: Not only is this finding at variance with the putative homology betwen reptilian and avian double cones, but it renders questionable the hope that structural homologies between turtle and pigeon horizontal cells will have a common functional basis (Mariani and Leure-Dupree, 1977; Leeper, 1978). At this juncture, one either has to ignore blithely such conflicts (and place less emphasis on homology) or presume that additional measurements may indeed find LD and SD cones to contain different pigments. On the basis of few measurements, Bowmaker tentatively assigned a long-wave pigment to single cones with colorless oil droplets, and a short-wave pigment (not measureably photolabile, however) to some single cones with yellow droplets. Again, this is different from the situation in turtles which have a colorless droplet–blue-absorbing pigment, yellow droplet–green-abosrbing pigment, red droplet–red-absorbing pigment arrangement for single cones. These discrepancies need to be resolved.

A violet- or ultraviolet-absorbing cone may also exist in the pigeon retina. Wright (1972) found that the presence of ultraviolet light in one half of a bipartite field was discriminable by pigeons engaged in a hue discrimination task. Subsequently, Kreithen and Eisner (1978) described an ultraviolet peak in the pigeon spectral sensitivity function obtained by heart rate conditioning; there were very few spectral points used in the study, however. A more compelling study was that of Graf and Norren (1974) using vector-voltmeter electroretinography to define the spectral sensitivity of the pigeon eye. They characterized a violet-sensitive mechanism (peak sensitivity 400 nm) that was relatively insensitive to very intense yellow-adapting lights. Adding a 6.5-log-troland yellow background to the stimulus resulted in a 3-log unit loss in sensitivity at 600 nm but less than a 0.25-log unit loss, with no change in the wavelength of maximal sensitivity, for a mechanism at 400 nm. This is the behavior expected of a true pigment, rather than an interaction between two color channels. The early-receptor potential recorded from the pigeon retina (Govardovskii and Zueva, 1977) lends credence to the presence of a violet-absorbing pigment. If such a system exists, it would make obvious sense to place the pigment in a cone with a colorless oil droplet.

The organizations of the cone types, regardless of their inconclusive pigment assignments, can be assessed on the basis of oil droplet types and general morphology. At first examination, avian retinas show only minor regularities; Engstrom (1958) thought he observed basic mosaic patterns in the retina of *Parus major,* the great tit, with four double cones surrounding a single cone, but the pattern is hardly as organized as those of fishes. Morris (1970) carefully analyzed the spatial organization of chick cones and found that each type of receptor had a roughly hexagonal organization, as if there were mechanisms that would lead to even spacing for each class of cones. This is the kind of organization I have called "critical spacing." Even though the relative proportions of the cone types change with retinal locus in the pigeon, each type apparently has its own critically spaced pattern. Figure 9 shows local patterning for LD cones with large yellow droplets, and single cones with small yellow droplets from nasal peripheral retina in the pigeon. The selective yellow oil droplet images were obtained by exciting the

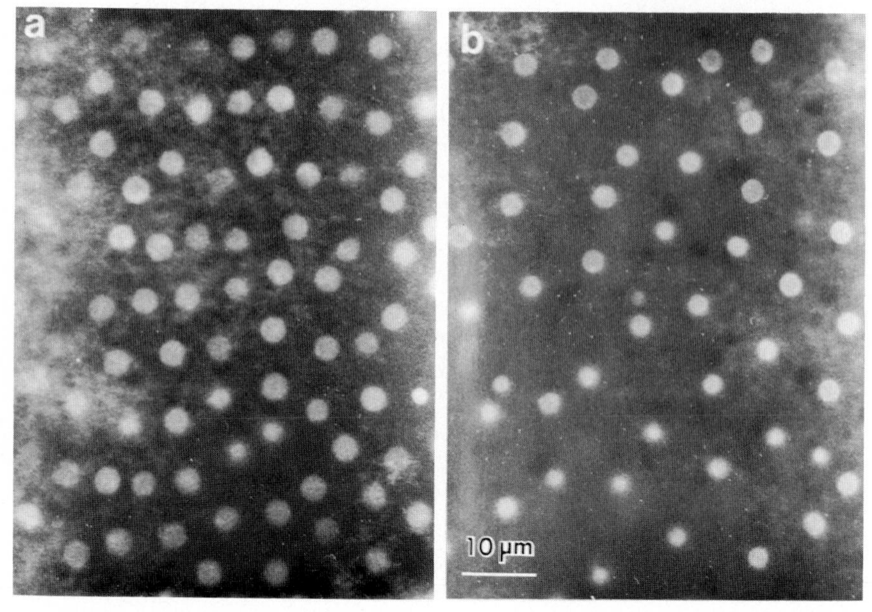

Fig. 9. Fluorescence micrographs of yellowish oil droplets excited by 365-nm light. (a) Distribution of LD cones in peripheral yellow field of the pigeon. (b) Focus about 10 μm below the level of the LD cones, revealing a second pattern of fluorescing yellow oil droplets in single cones.

yellow oil droplets to fluoresce with ultraviolet light (366 nm). The two yellow oil droplet images are of the same field, but since the double cones are taller than the single cones, the oil droplets are stratified in planes separated by about 10 μm. The degree of order represented in the yellow oil droplets is usually difficult to see in bright-field microscopy.

The avian eye represents the pinnacle of vertebrate intraretinal specialization. The pigeon has less dramatic changes in receptor frequency with retinal eccentricity than certain other avians but is typical in having a fovea, albeit shallow, in the approximate center of the eye (displaced slightly ventrotemporally) and a generalized increase in receptor frequency in a large portion of the dorsotemporal retina. Overall, receptor frequencies in the pigeon do not change by much more than a factor of 2 over most of the retina (Galifret, 1968). This dual system is a weak manifestation of a bifoveate condition that exists in many diurnal avians, especially hawks (Walls, 1942; Fite and Rosenfield-Wessels, 1975). There is usually one well-defined central fovea in most diurnal birds and a shallower but distinct fovea in temporal retina. Often the central fovea is very deep, with a cusplike valley; such foveas are termed "convexiclivate" (Walls, 1937). In diurnal birds the receptor frequencies in the central fovea can exceed 300,000 cones/mm^2 in the central fovea and 200,000 cones/mm^2 in the temporal retina (Fite and Rosenfield-Wessels, 1975). Both these frequencies exceed primate foveal cone frequencies by a factor of 2–5. Even the pigeon with its relatively homogeneous retina (by avian standards) has a receptor frequency of about 80,000 cones/mm^2 over much of its retina.

Another remarkable specialization of some avian eyes, well exemplified in the pigeon, is parcelation of the retina into zones possessing different proportions of color-specific cone classes. Most of the pigeon retina is called the yellow field because of its yellow coloration when observed fresh after removal from the eye. In the dorsotemporal retina there is a large red-orange zone, encompassing about a quarter of the retina, known as the red field. The terms "red" and "yellow" reflect the relative enrichments of long-wave-transmitting oil droplets in the former and middle-wave cutoff oil droplets in the latter. It has been shown that these two areas mediate spectral sensitivity functions with different shapes: The red field system generates a spectral sensitivity curve with a large short-wave deficit; the yellow field spectral sensitivity is a relatively continuous function from red to blue, peaking at about 560 nm (Martin and Muntz, 1978). Because of the potential for applying cone population analysis to spectral sensitivity data, as well as other psychophysical correlates of color, the definitive analysis of pigeon cone populations seems to be a critical undertaking.

E. Primates

The only type-3 retinas that have been analyzed by MSP in any detail are those of the primates. Human color vision is trichromatic; this is a statement of the psychophysical constraints on chromatic processing. MSP has provided explicit evidence for three classes of cones in the human and rhesus monkey retinas: a long-wave cone (commonly called a red-sensitive cone, though its absorption maximum is in the yellow), a middle-wave cone (green-sensitive), and a short-wave cone (blue-sensitive). Over the past 15 years, the presence of the red- and green-sensitive systems has been rather firmly established (Marks *et al.*, 1964; Brown and Wald, 1964; Liebman, 1972; Bowmaker *et al.* 1978), with the most recent measures placing the absorbance maxima at 565 and 536 nm for these two cone classes. Blue-sensitive cones have been and continue to be rare in MSP. Bowmaker *et al.* (1978) were not able to identify photolabile blue-absorbing cones in the rhesus retina but found two blue-absorbing cones in *Macaca fascicularis* (Bowmaker *et al.*, 1980) and three in the human (Bowmaker and Dartnall, 1980) with absorbance maxima between 415 and 420 nm.

Given that there are three spectral classes of primate cones and that they are monomorphic, methods that classify spectral type directly in the intact receptor array are obligatory. Harry Sperling and I attempted such measurements in 1975–1976 using NBT, much as we had in the goldfish. White and spectral lights were used to irradiate isolated baboon retinas which were then incubated in a saline-buffer-substrate-NBT mixture (Marc and Sperling, 1977). Because primate cones are monomorphic and we anticipated no order in the mosaic, MSP of the NBT-DF in responsive ellipsoids was used to characterize the strength of response and histograms of cone responses constructed for the different stimulus conditions. The basis of these methods is reviewed in Marc (1977). In summary, we found that white light effectively activated all the retinal cones (Fig. 10a) and spectral lights activated only fractions of the population. Blue light exposure in peripheral retina produced an astounding result: A regular array of strongly responsive cones representing about 12% of all the cones was encountered (Fig. 10b). Spatial analysis revealed that the cones were indeed critically spaced. Though we could not resolve any such order among the red-sensitive cones (Marc and Sperling, 1977; Marc, 1977), hence by subtraction no order among the green-sensitive cones, primate retinas possess a feature found in many if not most type-2 retinas: The short-wave system of the cones is the least frequent population but is usually highly regular in distribution. The few foveal and parafoveal experiments that were possible showed spatial changes in the frequency of blue-sensitive cones that were different from those in the other spectral classes (Fig. 11). Blue-sensitive cones are relatively rare near the foveola; in the baboon I measured a proportion of about 3% for blue-sensitive cones. I might add that this could be less in the human (cf. Williams *et al.* 1978) or that I actually missed the baboon foveola, which is not an unusual problem with NBT preparations. The frequency of blue-sensitive cones climbed rapidly with eccentricity, giving a peak value of near 6000 cones/mm^2 and a proportion of about 16% at 1–2 deg from which it gradually fell to a rather constant 11–13% by 10 deg and beyond.

F. M. de Monasterio *et al.* (1981) have described a similar pattern of cone organization revealed by selective staining of macaque (rhesus monkey) cones with the fluorescent dye Procion yellow M4RAN applied by intraviteral injection. The stained cones were arranged in

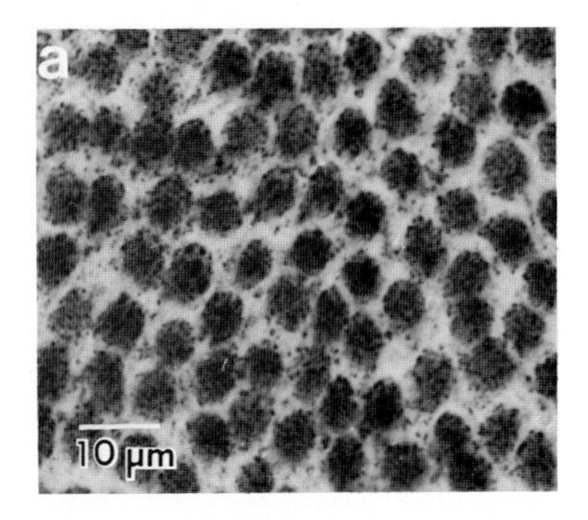

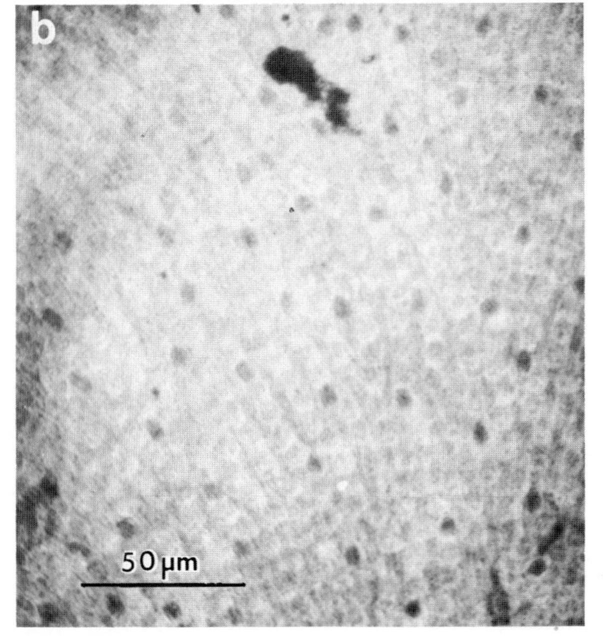

Fig. 10. Mosaic views of baboon cone responses to white and blue light stimulation using the NBT method. (a) White light, 10^7 quanta sec^{-1} μm^{-2} over 400–700 nm. All cones respond. (b) Blue 440-nm light, 4×10^4 quanta sec^{-1} μm^{-2}. A regular array of stained blue-absorbing cones is shown. From Marc (1977).

an orderly (general hexagonal) array and exhibited a spatial distribution similar to that of blue-sensitive baboon cones, with a foveal minimum, a density maximum at 1 degree eccentricity and a low density in the periphery. Differences exist in the absolute densities, however. For example, macaque blue-sensitive cones had a peak density of

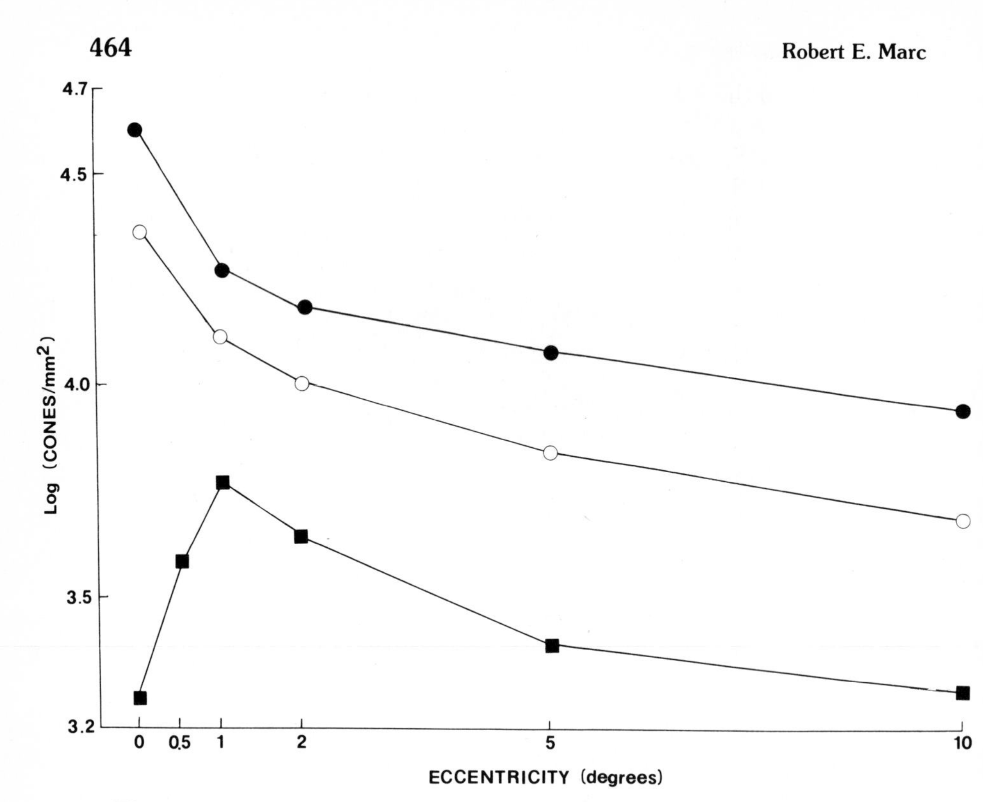

Fig. 11. Log_{10} cone frequency for blue-absorbing (solid squares), green-absorbing (solid circles), and red-absorbing (open circles) cones as a function of retinal eccentricity. Points indicate where actual measurements were made in individual experiments. Green-absorbing cone distribution calculated on the basis of the difference between the total cone distribution and the blue- and red-absorbing cone distributions.

about 1600 cones/mm^2, three to four times less than baboon peak blue-sensitive cone density; and rhesus blue-sensitive cones reached a peripheral asymptote at about 150–200 cones/mm^2, compared to 500–600 cones/mm^2 in baboons. Technical errors aside, which are unlikely, this discrepancy may reflect true species differences between the Asian macaques, dwelling predominantly in jungle canopy or otherwise heavily foliated regions, and the African baboons which roam semi-arid savannahs.

VI. CHROMATIC ORGANIZATION OF CONES AND VISUAL FUNCTION

There have been many discussions of the contributions of different spectral classes of cones to visual performance. A number of authors have discussed the roles of cone pigments in teleostean photopic vision in terms of the prevailing spectral environment (e.g., Lythgoe, 1972; Munz and MacFarland, 1977; Leow and Lythgoe, 1978; Levine and MacNichol, 1979). Similarly the issue of oil droplet distribution and spectral sensitivity in avian vision has been discussed (e.g., Bowmaker, 1977; Bowmaker and Knowles, 1977). I am going to concentrate on the distribution and patterning of cones rather than discuss summary functions of pigments. Three specific topics will be considered though they overlap substantially: (1) the organization of blue-absorbing cones, (2) mosaic patterns and their functions, (3) spectral acuity. Such discussions of ostensible function are usually speculative, and in some respects this one will be no different. However, I will attempt to delineate some hypotheses that give rise to explicit experimental tests.

A. The Organization of Blue-Absorbing Cones

The short-wave photoreceptors of teleostean fishes, frogs and toads, turtles, and mammals are the sparsest and usually the most regularly distributed cones of all. This may be true for birds as well, but the assignment of blue-absorbing pigments to cones containing small, yellow oil droplets (which is a rather frequent cell type) confuses the issue somewhat (Bowmaker, 1977). For the other species, however, the implication that blue-sensitive channels are not designed for maximal acuity remains clear. The reduction in frequency of blue-absorbing cones in the primate fovea also indicates that blue signals are of less importance for acuity than other functions. Why should this be so? An obvious fact is that there are few environmental structures possessing selectively blue-biased reflectance. There are few, if any, natural objects which when illuminated by sunlight from behind the observer are a spectral match for blue skylight. Most structures are long-wave-dominated and generate spectral contrast with skylight. There are many structures which, under the same conditions, could generate a close brightness match with the long-wave portion of skylight, but probably never a spectral match. It therefore makes some sense that blue-absorbing cones not occupy much space in a limited sensor surface if, as Morris suggests (1970), blue-absorbing cones literally monitor blue (sky) backgrounds. Logically, then, blue signals should contribute greatly to color opponency but would be of no great advantage in coding achromatic brightness. These are actually well-known features of blue-absorbing cone systems in humans; i.e., blue signals contribute more strongly to color opponency than luminosity channels. What is important about the spacing of blue-absorbing cones? Scholes (1975) may have provided an important clue to this question. He focused a small green bar of light on the retina of a rudd eye through the natural lens. By cutting a window in the sclera he could measure the spatial distribution of the image in the plane of the retina and after achieving optimal focus merely substituted a blue filter for the green filter and remeasured the spread. Chromatic aberration led to an increase in the half-width of the image spread that was proportional to the increase in interreceptor spacing for blue-absorbing SS versus double cones. This suggests that any increase in blue cone density would be "empty" as far as increased acuity is concerned. Optimal focus on detailed long-wave biased structures may place a limit on the useful frequency of blue-absorbing cones in primates as well.

Why should blue-absorbing cones be regularly spaced in fishes and primates? We do not know that such cones are not regularly spaced in turtles and pigeons; in fact it is likely that they are. One possible answer to this question is considered in the next section, but a homogeneous, critically spaced distribution may place cones at the practical minimum, so that blue acuity under long-wave focus conditions is as good as it can be without wasting mosaic positions.

Wald (1967) espoused the chromatic aberration hypothesis to explain the apparent absence or weakness of blue sensitivity in the foveola. If the foveola is responsible for fixation and accommodation signals, then one would not choose to have an ambiguous set of optimal planes of focus from which to choose (i.e., blue to red). Thus a small fixation area devoid of or poor in blue-absorbing cones would serve well, subtracting little from the encoding of blue signals.

B. Mosaic Patterns and Their Functions

The question of what mosaic order has to do with vision has existed since the discovery of teleostean cone patterns in the last century. There is a good correlation between the degree of order in teleostean retinas and ethological characteristics: More predatory fishes (e.g., pike) have more precise mosaics than vegetarian or scavenging fishes (e.g., carp). I suggest that mosaic order should be strongly associated with the ability of a fish to make velocity judgments, i.e., to track moving objects. In fact, the mosaic precision should probably correlate better (inversely) with the variability in performance than with the limits of performance. This hypothesis should be psychophysically testable. The rationale for the hypothesis is that the measurement of velocity may be a simple calculation of the displacement over time, which will be a reliable measure if the distance between sensors is constant (or at least orderly). Randomly placed or poorly organized sensors would yield reliable judgments of velocity only if the analyzing system had a spatial map of all sensor placements. This is clearly an oversimplification of complex tasks: Visual tracking involves far more than just cone patterns. I do suggest, however, that mosaic order will have a measurable influence on performance.

If an image moves on the retina, it will cross different spectral classes of cones at different intervals. If one supposes that encoding velocity involves summing analog versions of the outputs of cone-neuron chains, some interesting problems arise. With our summing constraint, the analog signals generated by different velocities must be unique values, and conversely the analog values generated by systems with different spacings must ultimately be the same for a given velocity. To achieve this with a system that sums the cone signals generated by brief exposure to the moving edge (i.e., spatiotemporal summation), I argue that more widely spaced cones should have slower times to peak. Put simply, blue-absorbing cones should be slower than other cones in proportion to their increased spacing. The blue system seems slower psychophysically (Wisowaty and Boynton, 1980) and electrophysiologically. In the pigeon, the phase lag of the short-wave mechanism is much greater than in the other cone systems (Graf and Norren, 1974). I raise these issues not to espouse a model for velocity coding but to suggest that many features of cone function may have their origins in the patterning of cones.

C. Spectral Acuity

Based on the previous section, the distribution of the spectral cone classes should influence spatial performance. With large, diffuse fields of light, Northmore and Muntz (1974) showed the behavioral spectral sensitivity of the rudd to be heavily blue-biased, whereas moving bars (4.75 deg in angular subtense) elicited a red-biased sensitivity function. This suggests, motion apart, that the spatial distribution of different spectral classes of cones is a significant determinant of chromatic performance. Unfortunately, detailed studies that carefully assess spatial features of color vision have yet to be performed in fishes or other nonhumans.

The human retina is the best known psychophysically, and many features of color vision may be considered in terms of possible spatial mechanisms. The distribution of cone types in the baboon predict rather well some aspects of human vision. Blue sensitivity, as measured by increment threshold, is minimal for small spots of light centered in the fovea and increases to a maximum at about 1–2 deg eccentricity (Brindley, 1954; Wald, 1967). This corresponds well with the form of the blue-absorbing cone frequency at these loci, but many other characteristics are possible influences as well (summation, cone size; see Marc, 1977). Based on the results of Campbell and Green (1965), it might be supposed that the grating acuity of the blue mechanism in humans would be poorer than that of the red and green mechanisms. The mean foveal intercone spacing in the baboon is 6–7′ of arc for blue-absorbing cones, 1–2′ of arc for red-absorbing cones, and 1′ of arc for green-absorbing cones (Marc and Sperling, 1977). Many psychophysical studies on spectral grating acuity have arrived at similar findings for the blue mechanism, i.e., 6.5–7.5′ of arc (Brindley, 1954; Porkorny *et al.*, 1968; Green, 1972; Daw and Enoch, 1973). Williams *et al.* (1978) suggest that the spacing is somewhat broader: 8–10′ of arc. A recent study on visually evoked cortical potential using spectral checkerboards of different sizes yielded an additional estimate of 7′ of arc for the blue system (Klingaman and Moskowitz-Cook, 1979). Overall, the evidence seems convincing that the spectral acuity of the blue mechanism is cone-limited.

VII. SUMMARY

A cone is indivisible: It occupies a position in the photoreceptor layer and thereby precludes the presence of another cone at that site. As we have seen, the placement of cones in the array is hardly haphazard, and one might assume that the forces that generated the patterns were evolutionary ones. Whether the forces are photic remain to be seen, but I have suggested some that might have influenced cone proportions and patterning, e.g., the paucity of selective blue reflectances in the natural world and chromatic aberration. There may be developmental forces as well.

Blue-absorbing cones are invariably the least frequent spectral variety and the most widely spaced of all. Psychophysical correlates of this phenomenon have been achieved only in humans so far, but certain animal models, such as avians and teleosts, would be a great boon to understanding the functional consequences of various types of receptor spacing.

The important issue of neuronal connectivity of chromatically characterized cones with retinal neurons has been purposefully neglected. The color-selective contacts of retinal neurons are not well known, though the beginnings are encouraging (Stell and Lightfoot,

1975; Scholes, 1975; Leeper, 1978). Human psychophysics, at least, indicates that the spatial limits of each spectral class of cone are, in many conditions, preserved through the neural transforms leading to behavioral performance. We are developing a structuralist perspective of color processing: The morphology of color-specific retinal neurons should reflect the spatial distributions of their inputs. Preliminary efforts in the goldfish, rudd, and turtle support this view, but there is much we need to know about the populations of retinal neurons and their organization. One might suspect that the neural retina should manifest a spatial chromatic organization similar to that of the cones, but will the brain? I think the answer is yes, as the concept of spatial adjacency demands. New technologies, such as radiolabeled 2-deoxy-glucose localization, will allow us probe the next question: How does the chromatic organization of the retina map itself onto the brain?

REFERENCES

Ali, M. A., Hárosi, F. I., and Wagner, H.-J. (1978). Photoreceptors and visual pigments in a cichlid fish, *Nannacara anomala. Sens. Process.* **2**, 130–145.

Baylor, D. A., and Fettiplace, R. (1975). Light path and photon capture in turtle photoreceptors. *J. Physiol.* **248**, 433–464.

Bowmaker, J. K. (1977). The visual pigments, oil droplets and spectral sensitivity of the pigeon. *Vision Res.* **17**, 1129–1138.

Bowmaker, J. K., and Dartnall, H. J. A. (1980). Visual pigments of rods and cones in a human retina. *J. Physiol.* **298**, 501–511.

Bowmaker, J. K., and Knowles, A. (1977). The visual pigments and oil droplets of the chicken retina. *Vision Res.* **17**, 755–764.

Bowmaker, J. K., Dartnall, H. J. A., Lythgoe, J. N., and Mollon, J. D. (1978). The visual pigments of rods and cones in the rhesus monkey, *Macaca mulatta. J. Physiol.* **274**, 329–348.

Bowmaker, J. K., Dartnall, H. J. A., and Mollon, J. D. (1980). Microspectrophotometric demonstration of four classes of photoreceptor in an Old World primate, *Macaca fascicularis. J. Physiol.* **298**, 131–143.

Brindley, G. S. (1954). The summation areas of human color receptive mechanisms at increment threshold. *J. Physiol.* **124**, 400–408.

Brown, K. T. (1969). A linear area centralis extending across the turtle retina and stabilized to the horizon by non-visual cues. *Vision Res.* **9**, 1053–1062.

Brown, P. K., and Wald, G. (1964). Visual pigments in single rods and cones of the human retina. *Science* **144**, 45–52.

Campbell, F. W., and Green, D. G. (1965). Optical and retinal factors affecting visual resolution. *J. Physiol.* **181**, 576–593.

Carey, R. G. (1976). The anuran "mystery cone": A new receptor type? *Assoc. Res. Vision Ophthalmol. Abstr.* 101.

Crescitelli, F. (1958). The natural history of visual pigments. *Ann N.Y. Acad. Sci.* **74**, 230–255.

Crescitelli, F. (1972). The visual cells and visual pigments of the vertebrate eye. *In*

"Photochemistry of Vision. Handbook of Sensory Physiology VII/1" (H. J. A. Dartnall, ed.), pp. 245-363. Springer-Verlag, Berlin and New York.

Crescitelli, F. (1977). The visual pigments of geckos and other vertebrates: An essay in comparative biology. *In* "The Visual System in Vertebrates. Handbook of Sensory Physiology VIII/5" (F. Crescitelli, ed.), pp. 391-449. Springer-Verlag, Berlin and New York.

Crescitelli, F., Dartnall, H. J. A., and Leow, E. R. (1977). The gecko visual pigments: A microspectrophotometric study. *J. Physiol.* **268**, 559-573.

Dartnall, H. J. A. (1967). The visual pigment of the green rods. *Vision Res.* **7**, 1-16.

Daw, N. W., and Enoch, J. M. (1973). Contrast sensitivity, Westheimer function, and Stiles-Crawford effect in a blue cone monochromat. *Vision Res.* **13**, 1669-1680.

de Monasterio, F. M., Schein, S. J. and McCrane, E. P. (1981). Staining of blue-sensitive cones of the macaque retina by a fluorescent dye. *Science 213:* 1278-1281.

Denton, E. J., and Wyllie, J. H. (1955). Study of the photosensitive pigments in the pink and green rods of the frog. *J. Physiol.* **127**, 81-89.

Dowling, J. E., and Ripps, H. (1970). Adaptation in skate photoreceptors. *J. Gen. Physiol.* **60**, 698-719.

Dunn, R. F. (1966). Studies on the retina of the gecko *Coleonyx variegatus*. II. The rectilinear visual cell mosaic. *J. Ultrastruct. Res.* **16**, 672-684.

Engstrom, K. (1958). On the cone mosaic in the retina of *Parus major*. *Acta Zool.* **38**, 65-69.

Engstrom, K. (1960). Cone types and cone arrangements in the retina of some cyprinids. *Acta Zool.* **41**, 277-295.

Engstrom, K. (1963a). Structure, organization and ultrastructure of the visual cells in the teleost family Labridae. *Acta Zool.* **44**, 1-41.

Engstrom, K. (1963b). Cone types and cone arrangement in teleost retinae. *Acta Zool.* **44**, 179-243.

Enoch, J. M. (1963). The use of tetrazolium to distinguish between retinal receptors exposed and not exposed to light. *Invest. Ophthalmol.* **2**, 16-23.

Enoch, J. M. (1964a). Validation of an indicator of mammalian retinal receptor response: Action spectrum. *J. Opt. Soc. Am.* **54**, 368-374.

Enoch, J. M. (1964b). Validation of an indicator of mammalian retinal receptor response: Absolute threshold. *J. Opt. Soc. Am.* **54**, 1027-1030.

Enoch, J. M. (1966). Validation of an indicator of mammalian retinal receptor response: Density of stain as a function of stimulus magnitude. *J. Opt. Soc. Am.* **56**, 116-123.

Fernald, R. D., and Liebman, P. A. (1980). Visual receptor pigments in the African cichlid fish, *Haplochromis burtoni*. *Vision Res.* **20**, 857-864.

Fineran, B. A., and Nicol, J. A. C. (1974). Studies on the eyes of New Zealand parrotfishes (Labridae). *Proc. R. Soc. London Ser. B* **186**, 217-247.

Fite, K. V., and Carey, R. (1973). The photoreceptors of the frog and toad: A quantitative study. *Assoc. Res. Vision Ophthalmol. Abstr.* 48.

Fite, K. V., and Rosenfield-Wessels, S. (1975). A comparative study of deep avian foveas. *Brain. Behav. Evol.* **12**, 97-115.

Galifret, Y. (1968). Les diverses aires functionelles de la rétine du pigeon. *Z. Zellforsch. Mikrosk. Anat.* **86**, 535-545.

Goldstein, E. B., and Wolf, B. M. (1973). Regeneration of the green rod pigment in the isolated frog retina. *Vision Res.* **13**, 527-534.

Govardovskii, V. I., and Zueva, L. V. (1977). Visual pigments of chicken and pigeon. *Vision Res.* **17**, 527-534.

Graf, V., and Norren, D. V. (1974). A blue sensitive mechanism in the pigeon retina: λ_{max}400 nm. *Vision Res.* **17**, 537-543.

Green D. G. (1972). Visual acuity in a blue cone monochromat. *J. Physiol.* **222,** 419–426.

Gruber, S. H., Hamasaki, D. H., and Bridges, C. D. B. (1963). Cones in the retina of the lemon shark, *Negaprion brevirostrus. Vision Res.* **3,** 397–399.

Hannover, A. (1844). "Recherches Microscopiques sur le System Nerveux." P. G. Philipsen, Copenhagen.

Hárosi, F. I. (1976). Spectral relations of cone pigments in goldfish. *J. Gen. Physiol.* **68,** 65–80.

Hárosi, F. I. (1981). Recent results obtained with single-cell microspectrophotometry: Cone pigments in frog, fish, and monkey. *Color Res. Appl.* (in press).

Hárosi, F. I., and MacNichol, E. F., Jr. (1974). Visual pigments of goldfish cones: Spectral properties and dichroism. *J. Gen. Physiol.* **63,** 279–304.

Hester, F. J. (1968). Visual contrast thresholds of the goldfish (*Carassius auratus*). *Vision Res.* **8,** 1315–1335.

Ishida, A. T., Stell, W. K., and Lightfoot, D. O. (1980). Rod and cone inputs to bipolar cells in goldfish retina. *J. Comp. Neurol.* **191,** 315–335.

King-Smith, P. E. (1969). Absorption spectra and function of the colored oil drops in the pigeon retina. *Vision Res.* **9,** 1391–1399.

Klingaman, R. L., and Moskowitz-Cook, A. (1979). Assessment of the visual acuity of human color mechanisms with the visually evoked cortical potential. *Invest. Ophthalmol.* **18,** 1273–1278.

Kreithen, M. L., and Eisner, T. (1978). Ultraviolet light detection by the homing pigeon. *Nature (London)* **272,** 347–348.

Lasansky, A. (1971). Synaptic organization of cone cells in the turtle retina. *Philos. Trans. R. Soc. London Ser. B* **262,** 365–381.

Leeper, H. F. (1978). Horizontal cells of the turtle retina. II. Analysis of interconnections between photoreceptor cells and horizontal cells by light microscopy. *J. Comp. Neurol.* **182,** 795–810.

Leow, E. R., and Dartnall, H. J. A. (1976). Vitamin A1/A2-based visual pigment mixtures in the cones of the rudd. *Vision Res.* **16,** 891–896.

Leow, E. R., and Lythgoe, J. N. (1978). The ecology of cone pigments in teleost fishes. *Vision Res.* **18,** 715–722.

Levine, J. S., and MacNichol, E. F., Jr. (1979). Visual pigments in teleost fishes: Effects of habitat, microhabitat and behavior on visual system evolution. *Sens. Process.* **3,** 95–131.

Liebman, P. (1972). Microspectrophotometry of photoreceptors. *In* "Photochemistry of Vision. Handbook of Sensory Physiology VII/1" (H. J. A. Dartnall, ed.), pp. 481–528. Springer-Verlag, Berlin and New York.

Liebman, P. A., and Entine, G. (1964). Sensitive low level microspectrophotometer detection of photosensitive pigments in retinal cones. *J. Opt. Soc. Am.* **54,** 451–459.

Liebman, P. A., and Entine, G. (1968). Visual pigments of frog and tadpole. *Vision Res.* **8,** 761–775.

Liebman, P. A., and Granda, A. M. (1971). Microspectrophotometric measurements of visual pigments in two species of turtle, *Pseudemys scripta* and *Chelonia mydas. Vision Res.* **11,** 105–114.

Liebman, P. A., and Liegh, R. A. (1969). Autofluorescence of visual receptors. *Nature (London)* **221,** 1249–1251.

Lythgoe, J. (1972). The adaptation of visual pigments to the photic environment. *In* "Photochemistry of Vision. Handbook of Sensory Physiology VII/1" (H. J. A. Dartnall, ed.), pp. 566–603. Springer-Verlag, Berlin and New York.

Marc, R. E. (1977). Chromatic patterns of cone photoreceptors. *Am. J. Optomet. Physiol. Opt.* **54,** 212–225.

Marc, R. E., and Sperling, H. G. (1976a). Color receptor identities of goldfish cones. *Science* **191,** 487-489.

Marc, R. E., and Sperling, H. G. (1976b). The chromatic organization of the goldfish cone mosaic. *Vision Res.* **16,** 1211-1224.

Marc, R. E., and Sperling, H. G. (1977). The chromatic organization of primate cones. *Science* **196,** 454-456.

Marks, W. B. (1965). Visual pigments of single goldfish cones. *J. Physiol.* **178,** 14-32.

Marks, W. B., Dobelle, W. H., and MacNichol, E. F., Jr. (1964). Visual pigments of single primate cones. *Science* **143,** 1181-1183.

Mariani, A. P., and Leure-Dupree, A. E. (1977). Horizontal cells of the pigeon retina. *J. Comp. Neurol.* **175,** 13-26.

Mariani, A. P., and Leure-Dupree, A. E. (1978). Photoreceptors and oil droplet colors in the red area of the pigeon retina. *J. Comp. Neurol.* **182,** 821-838.

Martin, G. R., and Muntz, W. R. A. (1978). Spectral sensitivity of the red and yellow oil droplet fields of the pigeon (*Columbia livia*). *Nature (London)* **274,** 620-621.

Miller, W. H., and Snyder, A. W. (1972). Optical function of myoids. *Vision Res.* **12,** 1841-1848.

Morris, V. B. (1970). Symmetry in a receptor mosaic demonstrated in the chick from the frequencies, spacing and arrangement of the types of retinal receptors. *J. Comp. Neurol.* **140,** 359-398.

Morris, V. B., and Shorey, C. D. (1967). An electron microscope study of the types of receptor in the chick retina. *J. Comp. Neurol.* **129,** 313-340.

Muntz, W. R. A. (1972). Inert absorbing and reflecting pigments. *In* "Photochemistry of Vision. Handbook of Sensory Physiology VII/1" (H. J. A. Dartnall, ed.), pp. 529-565. Springer-Verlag, Berlin and New York.

Munz, F. W., and McFarland, W. N. (1977). Evolutionary adaptations of fishes to the photic environment. *In* "The Visual System in Vertebrates. Handbook of Sensory Physiology VIII/5" (F. Crescitelli, ed.), pp. 193-275. Springer-Verlag, Berlin and New York.

Nilsson, S. E. G. (1964). An electron microscopic classification of retinal receptors of the leopard frog (*Rana pipiens*). *J. Ultrastruct. Res.* **10,** 390-416.

Niwa, H., and Tamura, T. (1975). Investigation of fish vison by means of S-potentials. III. Photoreceptors and spectral sensitivity in elasmobranchs' retinae. *Bull Jpn. Soc. Sci. Fish.* **41,** 393-401.

Northmore, D. P. M., and Muntz, W. R. A. (1974). Effects of stimulus size on spectral sensitivity in fish (*Scardinius erythrophthalmus*), measured with a classical conditioning paradigm. *Vision Res.* **14,** 503-515.

O'Connell, C. P. (1963). The structure of the eye of *Sardinops caerula, Engraulis mordax* and four other pelagic marine teleosts. *J. Morphol.* **13,** 287-329.

Pokorny, J., Graham, C. H., and Lanson, R. N. (1968). Effect of wavelength on grating acuity. *J. Opt. Soc. Am.* **58,** 1410-1414.

Reuter, T., White, R., and Wald, G. (1971). Rhodopsin and porphyropsin fields in the adult bullfrog retina. *J. Gen. Physiol.* **58,** 351-371.

Scholes, J. (1975). Color receptors, and their synaptic connections, in the retina of a cyprinid fish. *Philos. Trans. R. Soc. London Ser. B* **270,** 61-118.

Stell, W. K. (1972). The structure and morphologic relations of rods and cones in the retina of the spiny dogfish, *Squalus. Comp. Biochem. Physiol.* **42A,** 141-151.

Stell, W. K. (1978). Inputs to bipolar cell dendrites in goldfish retina. *Sens. Process.* **2,** 339-349.

Stell, W. K. (1980). Photoreceptor-specific pathways in goldfish retina: A world of color, a wealth of connections. *In* "Color Deficiencies V" (G. Verriest, ed.), pp. 1-14. Adam Hilger, London.

Stell, W. K., and Hárosi, F. I. (1976). Cone structure and visual pigment content in the retina of the goldfish, *Carassius auratus. Vision Res.* **16,** 647-657.

Stell, W. K., and Lightfoot, D. O. (1975). Color-specific interconnections of cones and horizontal cells in the retina of the goldfish. *J. Comp. Neurol.* **159,** 473-502.

Stell, W. K., and Witkovsky, P. (1973). Retinal structure in the smooth dogfish, *Mustelus canis:* Light microscopy of photoreceptor and horizontal cells. *J. Comp. Neurol.* **148,** 33-46.

Stell, W. K., Lightfoot, D. O., Wheeler, T. G., and Leeper, H. F. (1975). Goldfish retina: Functional polarization of cone horizontal cell dendrites and synapses. *Science* **190,** 989-990.

Stell, W. K., Ishida, A. T., and Lightfoot, D. O. (1977). Structural basis for on- and off-center responses in retinal bipolar cells. *Science* **198,** 1269-1271.

Strother, G. K. (1963). Absorption spectra of retinal oil globules in turkey, turtle and pigeon. *Exp. Cell Res.* **29,** 349-355.

Svaetichin, G., Negishi, K., and Fatechand, R. (1965). Cellular mechanisms of a Young-Hering visual system. *In* "Ciba Symposium on Color Vision" (A. V. S. de-Reuk and J. Knight, eds.), pp. 178-203. Little, Brown, Boston, Massachusetts.

Toyoda, J.-I., Saito, T., and Kondo, H. (1978). Three types of horizontal cells in the stingray retina: Their morphology and physiology. *J. Comp. Neurol.* **179,** 569-580.

Tsou, K. C., Cheng, C. S., Nachlas, M. M., and Seligman, A. M. (1956). Syntheses of some *p*-nitrophenyl substituted tetrazolium salts as electron acceptors for the demonstration of dehydrogenase. *J. Am. Chem. Soc.* **78,** 6139-6144.

Underwood, G. (1970). The eye. *In* "Biology of Reptilia" (C. Gans and T. S. Parsons, eds.), pp. 1-97. Academic Press, New York.

Wald, G. (1967). Blue-blindness in the normal fovea. *J. Opt. Soc. Am.* **57,** 1289-1300.

Walls, G. (1937). The significance of the foveal depression. *Arch. Ophthal. N.Y.* **18,** 912-919.

Walls, G. (1942). "The Vertebrate Eye and Its Adaptive Radiation." *Cranbrook Inst. Sci. Bull.* 19. Cranbrook Inst. Sci., Bloomfield Hills, Michigan.

Wheeler, T. G. (1978). Goldfish retina: Dorsal versus ventral areas. *Vision Res.* **18,** 1329-1336.

Williams, D. R., MacLeod, D. I. A., and Hayhoe, M. M. (1978). Distribution of blue-sensitive cones in the fovea. *Invest. Ophthalmol. Visual Sci. Suppl.* 177.

Wisowaty, J., and Boynton, R. M. (1980). Temporal modulation of the blue mechanism: Measurements made without chromatic adaptation. *Vision Res.* **11,** 895-909.

Wright, A. A. (1972). The influence of ultraviolet radiation on the pigeon's color discrimination. *J. Exp. Anal. Behav.* **17,** 325-337.

10

Biosynthesis and Morphogenesis of Outer Segment Membranes in Vertebrate Photoreceptor Cells

DAVID S. PAPERMASTER AND BARBARA G. SCHNEIDER

Light sensitivity in vertebrate retinas is a special response of a unique membranous organelle of rods and cones called the outer seg-

Cell Biology of the Eye

ISBN 0-12-483180-X

ment. The membranes of rod outer segments (ROS) are composed almost entirely of the visual pigment rhodopsin and highly unsaturated lipids and are organized in the form of flattened disks stacked within a plasma membrane from which they are separated (Fig. 1). Cones have homologous outer segments which are formed as highly folded lamellae still in bilayer continuity with the cone plasma membrane and open to the extracellular space (Cohen, 1968; Laties *et al.*, 1976). The disks and lamellae of ROS and cone outer segments (COS) are not stable—a circadian rhythm of membrane destruction leads to a shed-

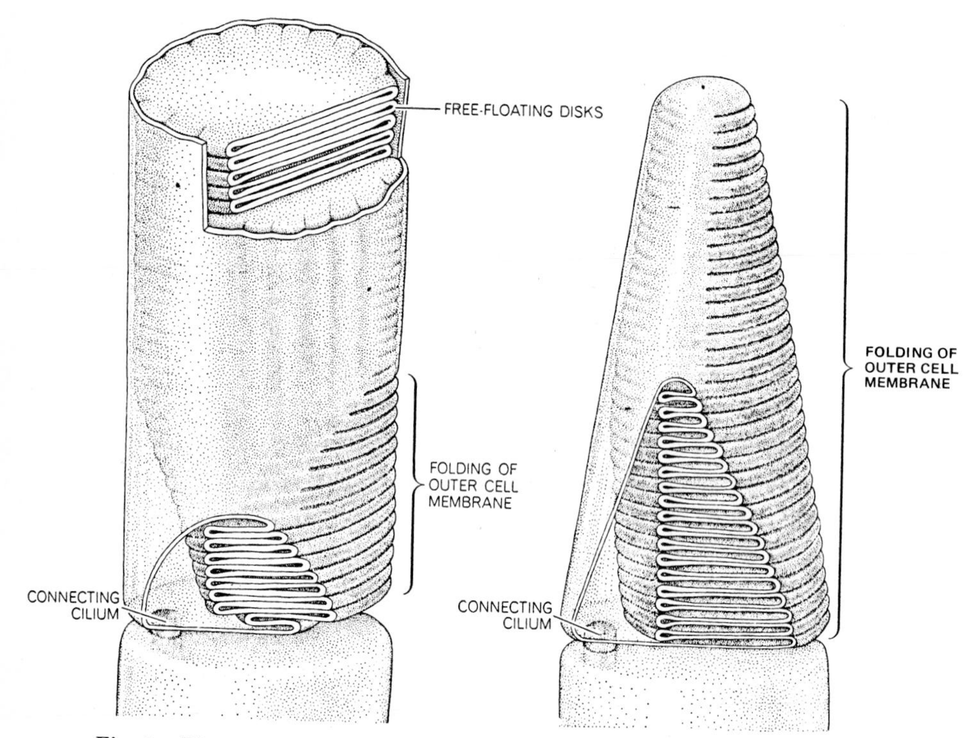

Fig. 1. Diagram of ROS and COS. ROS are composed of a stack of separated disks within an enveloping plasma membrane. At the base of the ROS, new disks are formed by evagination of the plasma membrane near the connecting cilium. Eventually, the disks become separated during maturation and morphogenesis. The borders of the disks are often scalloped by incisures. In this diagram the incisures are shallow. In some amphibian ROS, the incisures penetrate deeply into the disk interior (see Fig. 14). COS are formed by a continuous folding of the plasma membrane. The COS lamellae do not entirely separate from the plasma membrane in most species, except in the long, thin foveal cones of some mammalian and avian retinas. (From Young, 1970, with permission.)

ding of large portions of the tips of rod cells each morning and a shedding of cone tips in the evening—we eat our rod tips for breakfast and our cone tips for supper. To maintain a constant outer segment length, both rods and cones renew their outer segment membranes each day throughout adult life. Thus these cells are committed not only to the synthesis of a membrane containing a unique light-sensitive protein but also to its continuous renewal. Because of their unique turnover, structure, function, and ease of isolation, ROS membranes have become the object of intense investigation.

Membrane biosynthesis in rod cells can be divided into four stages which correspond to different synthetic and morphogenic steps: (1) membrane glycoprotein and lipid biosynthesis in the rough endoplasmic reticulum (RER) and movement of the glycoprotein opsin to the Golgi apparatus where it is further modified, (2) vectorial transport from the Golgi apparatus to the site of opsin insertion near the connecting cilium at the apex of the inner segment, (3) insertion and migration of opsin along the connecting cilium to the outer segment, and (4) disk morphogenesis.

The studies summarized here support the hypothesis that opsin is inserted into a membrane at its site of synthesis in the RER and leaves the Golgi apparatus bound to membranes. After vectorial transport through the inner segment the membranes bearing opsin fuse with the apical plasma membrane of the inner segment near the connecting cilium. Opsin then appears to move along the plasma membrane of the connecting cilium to a site of membrane expansion at the base of the stack of outer segment disks, where disk morphogenesis begins. This chapter concentrates on evidence gathered predominantly in the last 4 years. Earlier work in this area has been extensively reviewed and therefore will be summarized only briefly (Young, 1976; Bok *et al.*, 1977; O'Brien, 1978b; Olive, 1980; Holtzman and Mercurio, 1980).

I. BIOSYNTHESIS OF OUTER SEGMENT MEMBRANE COMPONENTS IN THE ROUGH ENDOPLASMIC RETICULUM

A. Biosynthesis of Opsin

Dynamic flow of newly synthesized protein from the RER to the outer segment was first demonstrated by light and electron microscopic autoradiographs of retinas which had incorporated radioactive amino acids into protein for successively longer time periods. Young and his colleagues observed the early accumulation of radioactivity over the RER. Radiolabeled protein(s) then gathered in the Golgi apparatus, passed alongside and in between the mitochondria in a region called the ellipsoid, and clustered in the apical portions of the inner segment near the base of the connecting cilium. Generation of a densely labeled band of radioactive disks at the bottom of the stack of disks in the ROS was followed by progressive displacement up the ROS as new disks were created each day. Synthesis of new disks continued throughout the adult life of all vertebrates studied (for review, see Young, 1976). Comparable studies on carbohydrate and lipid biosynthesis using radiolabeled precursors have demonstrated a similar pathway for these components of the ROS membrane as well (Bibb and Young, 1974; Bok *et al.,* 1977; O'Brien, 1977). The pathway of migration of newly synthesized proteins in rods is illustrated in Fig. 2.

At some point during this migration, probably as it enters the outer segment, opsin combines with 11-*cis*-retinaldehyde to become the red, light-sensitive protein rhodopsin (Hall and Bok, 1974, 1976; O'Brien

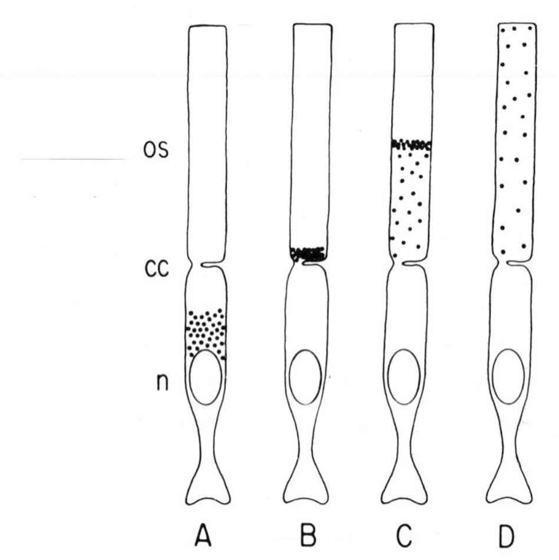

Fig. 2. Diagram of the migration of newly synthesized protein from sites of synthesis in the inner segment (myoid) to sites of disk assembly at the base of the outer segment (OS). (A) Shortly after injection of radiolabeled amino acids, the RER in the inner segment near the nucleus (n) is labeled. (B) Labeled protein subsequently migrates past the mitochondria clustered in the upper portion of the inner segment and along the connecting cilium (CC) to form labeled disks at the base of the outer segment. (C) Labeled disks are displaced apically by new disks from below. (D) Disks are finally shed into the adjacent pigment epithelium, and only residual label remains in the outer segment. The rate of renewal varies in different species. Cold-blooded amphibia complete renewal and shed new disks in about 40–50 days; this process is completed in 10 days in warm-blooded mammals. (Modified from Young and Bok, 1969.)

and Muellenberg, 1975). Opsin is fully destroyed as disks containing it are shed into the adjacent pigment epithelium (PE) cell layer (Hall *et al.*, 1969). It should be noted that aquatic vertebrates have visual pigments using the vitamin A_2 homolog of retinal and form porphyropsins (Wald, 1968). For the purposes of this chapter, this distinction will be ignored, since rhodopsins absorbing near 500 nm can be generated from porphyropsins by exchange of the chromophore from vitamin A_2 to vitamin A_1 retinaldehydes.

One of the outstanding questions to be answered involves identification of the means of transport of the newly synthesized opsin from the Golgi apparatus to the apex of the inner segment. At the time of the initial studies, it was considered possible that opsin was synthesized in a soluble precursor form which could migrate through the cilium interior and then assemble into a newly growing disk at the base of the outer segment (Young, 1968). During the early 1970s, as more membrane proteins were studied in detail, this hypothesis was largely abandoned because many membrane proteins appeared to be synthesized on elements of RER and then transported to other cellular sites on membrane vesicles (Hirano *et al.*, 1972; Palade, 1975; Katz *et al.*, 1977). Recently, the possible alternative pathways for transport of membrane proteins to their site of function have increased in number. Membrane proteins of some organelles are synthesized on free polysomes in the cytosol, transported to the organelle through the cytosol in soluble form, and then modified locally and inserted into the organelle membrane. This alternative has been especially evident in the biosynthesis of transiently soluble membrane proteins of the chloroplasts (Grossman *et al.*, 1980). Thus a dogmatic approach to the synthesis of membrane proteins is untenable. For these reasons, the details of the mechanism of transport of opsin to the ROS have continued to be investigated by a combination of biochemical methods and autoradiographic and immunocytochemical techniques at the ultrastructural level.

As a result of extensive studies on secreted glycoproteins such as immunoglobulins (Milstein *et al.*, 1972; Schechter, 1974) and pancreatic zymogens (Blobel and Sabatini, 1971; Blobel and Dobberstein, 1975), a hypothesis describing the mechanism of protein synthesis in RER was proposed, which proved for some time to be a consistent finding for nearly all proteins secreted from or inserted into the membranes of both eukaryotic and prokaryotic cells. It was observed that most proteins destined for secretion were synthesized initially from mRNA as precursor proteins with an additional 20 or so amino acids on the N-terminal. The extra peptides were not part of the mature protein because they were cleaved from it during its earliest stages of passage into the cisternae of the RER. The extra peptides on the N-terminals of myeloma proteins and pancreatic zymogens had markedly hydrophobic sequences which resembled each other. This led to the hypothesis that the transient extra peptide segment served as a signal for recognition by the RER. Membrane binding of polyribosomes to the RER membrane was thought to begin as the N-terminal signal peptide segment passed below the base of the large subunit of the ribosomes during mRNA translation. As more proteins were demonstrated to have transient N-terminal signal peptides, it was proposed that membrane proteins would have similar signal peptides for synthesis and insertion of the protein into or through the lipid bilayer. In order to arrest transmembrane insertion, the signal peptide would be followed by a "stop-transfer" hydrophobic peptide which would trap the protein within the lipid bilayer (Blobel, 1980). One consequence of this hypothesis was that membrane proteins would have an orientation such that the N-terminal would be extracellular if it passed through the bilayer, while the C-terminal would be on the cytoplasmic side of the membrane. Precisely this orientation was observed for some membrane glycoproteins such as human erythrocytic glycophorin A, although an N-terminal signal peptide has not yet been identified. Evidence for a comparable orientation of opsin included binding of concanavalin A (Con A)-ferritin conjugates or Con A-Sepharose to the interior of the disk and the outside of the ROS plasma membrane (Röhlich 1976; Adams *et al.*, 1978). If these conjugates bind exclusively to opsin in the outer segment, the results indicate that the N-terminal, which is glycosylated, had passed through the membrane. Proteolytic cleavage studies were also most consistent with this orientation (Fong, 1978; Fung and Hubbell, 1978a,b; Hargrave *et al.*, 1980). Despite the widespread existence of such extra peptides on the N-terminals of secreted proteins and bacterial membrane and viral membrane glycoproteins, the study of the translation products of isolated opsin mRNA provided an outstanding exception to this generalization.

B. *In Vitro* Translation of Opsin mRNA

In collaboration with Schechter and his colleagues at the Weizmann Institute of Science in Rehovot, Israel, we partially purified opsin mRNA by immunoprecipitation with antiopsin antibodies (Schechter *et al.*, 1979; Papermaster *et al.*, 1980). The mRNA, when translated in the wheat germ cell-free translation system, was capable of synthesizing opsin *in vitro*. After immunoprecipitation by antiopsin antibodies, the radiolabeled translation products were sequenced, and the position of the incorporated labeled amino acid was correlated with the known sequence of the N-terminal of opsin previously described by Hargrave (1977) and shown in Fig. 3. Synthesis with radiolabeled methionine, phenylalanine, asparagine, valine, proline, and tyrosine resulted in incorporation of label into 18 positions up to a proline at position 34. The translation system faithfully synthesized opsin, since the translation product had the same N-terminal sequence as mature opsin in the outer segment. In order to establish that the methionine incorporated into the first position was the initiator methionine, we employed eukaryotic tRNAs which bound exclusively to either the initiator or internal methionyl AUG codon. Only the initiator methionyl-tRNA donated methionine to the newly synthesized protein at the first position. Thus, the N-terminal methionine in the translation product of opsin mRNA is the initiator site *in vivo* as well as *in vitro,* since opsin isolated after *in vivo* synthesis has the same sequence as the *in vitro* translation product. Mature opsin isolated from ROS cannot be directly sequenced by Edman degradation, because the N-terminal is blocked by acetylation (Tsunawawa *et al.*, 1979). This blockage is probably the result of posttranslational acetylation. Similar acetylation of ovalbumin, a secreted protein of hen oviduct, was shown to be posttranslational (Palmiter *et al.*, 1978). Our results, which were confirmed by Goldman and Blobel (1981), clearly established that there was no transient N-terminal signal peptide at any stage in the synthesis of opsin.

Absence of a signal peptide on rhodopsin raised serious problems about the generalization of the signal hypothesis. Palmiter *et al.* (1978) had also demonstrated the absence of an N-terminal signal peptide on ovalbumin. However, Lingappa *et al.* (1980) proposed that an internal signal sequence in the middle of the molecule may lead ovalbumin through the RER bilayer in a hairpin fashion. Rhodopsin appears to cross the membrane of the disk bilayer several times (Pober and Stryer, 1975; Fung and Hubbel, 1978a,b) and has been termed a "polytopic" protein in the nomenclature described by Blobel (1980).

Ac–Met–Asn(CHO)–Gly–Thr–Glu–Gly–Pro–Asn–Phe–Tyr–Val–Pro–Phe–Ser–Asn(CHO)–

Lys–Thr–Gly–Val–Val–Arg

Fig. 3. The N-terminal amino acid sequence of opsin. This region of opsin bears two oligosaccharides bound to asparagines in positions 2 and 15. The methionine on the amino terminal is blocked by acetylation. Since there are no large clusters of hydrophobic residues, the N-terminal of rhodopsin is in the aqueous domain rather than embedded within the lipid bilayer. (From Hargrave, 1977, by permission of the author.)

Polytopic proteins have been hypothesized to enter lipid bilayers of RER by a sequentially repeated process of synthesis of a signal (or insertion) peptide followed by a stop-transfer peptide. Cycles of insertion and stop-transfer peptides, in combinations of even or odd numbers, would generate an N-terminal on the cytoplasmic or on the extracellular surface of the RER membrane depending on the position of the first signal peptide. Rhodopsin might have such an internal signal peptide beyond residue 38. An alternative trigger hypothesis, suggesting a spontaneous transfer of the protein into the bilayer as a result of hydrophobic and other forces, would allow the protein to adopt its preferred conformation during insertion into the bilayer after biosynthesis (Wickner, 1979). Goldman and Blobel (1981) have presented evidence against this mechanism of spontaneous transfer in the case of rhodopsin by showing that *in vitro* translation of opsin mRNA in the presence of dog pancreas microsome membranes involves cotranslational insertion into the membranes. If the microsomal membranes are added after translation has been completed, opsin does not enter the membranes and is susceptible to extensive proteolysis by trypsin. When synthesized in the presence of membranes, opsin is only slightly degraded by proteolysis in a pattern which parallels its cleavage when it is within the disk bilayer. They have proposed that opsin inserts with appropriate orientation into dog pancreas microsomal membranes only during cotranslational synthesis. Further amino acid sequence analysis and direct determination of the orientation of each opsin segment within the lipid bilayer will be helpful in testing some of these hypotheses about the earliest stages of rhodopsin biosynthesis.

C. Oligosaccharide Biosynthesis and Glycosylation of Opsin in the RER and Golgi Apparatus

Autoradiographic evidence for the coupling of glucosamine and mannose to opsin in the RER and additional conjugation of glucosamine in the Golgi apparatus was reviewed by Bok *et al.* (1977). Compositional studies by Heller and Lawrence (1970) and Plantner and Kean (1976), and the autoradiographic results, suggested that the structure resembled "complex" asparagine-linked oligosaccharides of other glycoproteins.

Fukuda *et al.* (1979) examined opsin's oligosaccharide structure in detail. They showed that the chitobiose linkage of the oligosaccharide to asparagine was specifcally cleaved with β-endoglucosaminidase D, leaving one N-acetylglucosamine bound to asparagine and the other linked to the oligosaccharide as a reducing terminal. After sodium borohydride reduction, the oligosaccharide alcohols were fractionated by gel chromatography. A major oligosaccharide was isolated, and its size was estimated to be 1250 daltons. Sequential α-mannosidase digestion released two mannose residues per mole of oligosaccharide. Further digestion by β-mannosidase released an additional mannose, resulting in a final product of *N*-acetylglucosaminitol. The anomeric linkages of the sugars were determined by methylation and gas chromatography, which demonstrated that two mannoses were linked by $\alpha 1 \rightarrow 6$ and $\alpha 1 \rightarrow 3$ linkages to the central mannose. Cleavage of these residues by α-mannosidases suggested that the nonreducing terminals were free. This was shown to be artificially generated as a result of a contaminating β-exoglucosaminidase which was present in the initially purified β-endoglucosaminidase D. When the action of the β-exoglucosaminidase was blocked by the addition of *N*-acetylglucosamino(1 → 5)lactone, only the mannose linked $\alpha 1 \rightarrow 6$ was cleaved, which indicated that at least one residue of *N*-acetylglucosamine was bound to the mannose linked $\alpha 1 \rightarrow 3$ and was the nonreducing terminal sugar. Support for this conclusion was obtained directly by high-resolution mass spectroscopy of the released oligosaccharide alcohol. Details of the interpretation of the mass spectroscopy are discussed fully in the report by Fukuda *et al.* (1979).

Rhodopsin's major oligosaccharide is diagrammed in Fig. 4. Both sites of attachment, Asn^2 and Asn^{15}, bear this chain. Liang *et al.* (1979) proposed the same structure, although they approached the structure determination slightly differently using hydrazinolysis. Rhodopsin's oligosaccharides are not homogeneous. A minor proportion of the oligosaccharides contain at least two additional mannose residues and constitute approximately 20% of the total oligosaccharide content of rhodopsin. This larger oligosaccharide is a possible precursor of the smaller major hexasaccharide. Current models for the generation of

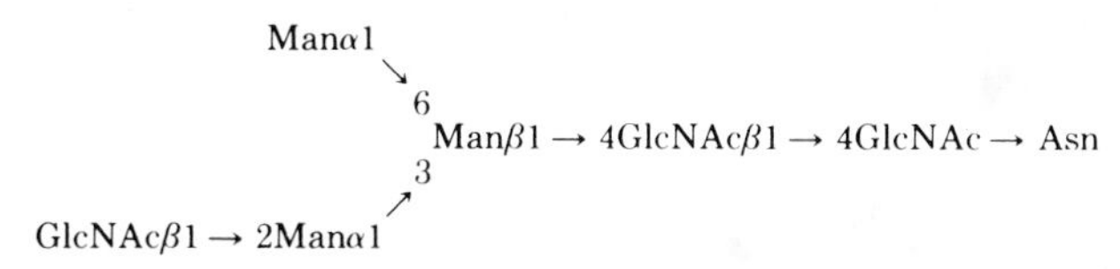

Fig. 4. Rhodopsin's major oligosaccharide. The sugars are linked to an asparagine in positions 2 and 15 (see Fig. 3). The innermost pair of *N*-acetylglucosamines constitutes the chitobiose linkage originally attached to a polyisoprenoid precursor and is probably transferred to asparagine during passage of the protein through the bilayer in the RER. The three mannose residues are the remnants of the larger high-mannose precursor. The *N*-acetylglucosamine on the nonreducing terminal is added in the Golgi apparatus. (From Fukuda *et al.*, 1979, with permission.)

complex oligosaccharides in glycoproteins (Robbins *et al.,* 1977; Tabas and Kornfeld, 1978; Chen and Lennarz, 1978) suggest that a larger oligosaccharide precursor rich in mannose is donated to asparagine from a polyisoprenoid donor. A small, fractional galactose content was also detected (Fukuda *et al.,* 1979), which may represent a further modification of opsin's oligosaccharides. O'Brien (1976) has postulated a role for terminal galactose as a mediator in the recognition of shed outer segments by PE cells.

Determination of the structure of the predominant hexasaccharide has many correlates with the earlier biosynthetic studies. Using radioactive glucosamine precursors, Bok *et al.* (1977) demonstrated the incorporation of glucosamine at two subcellular sites, the RER and the Golgi apparatus. Early attachment of glucosamine in the RER correlates with the position of the inner glucosamines in the chitobiose structure next to the asparagine. Involvement of a polyisoprenoid carrier in transfer of the mannose-rich oligosaccharide precursor to rhodopsin has been supported by biosynthetic studies and the isolation of such a precursor from retinal tissues by Kean (1977a,b,c,d, 1980a,b; see also Kean and Plantner, 1976), although direct evidence of their presence in photoreceptor cells has not yet been presented. Polyisoprenoid-linked oligosaccharide transfer to nascent polypeptide chains of glycoproteins is likely to occur in the RER (Chen and Lennarz, 1978). The position of *N*-acetylglucosamine on the nonreducing terminal is appropriate for the transfer of this sugar to complex, trimmed oligosaccharides which have lost most of their mannose as the result of sequential α-mannosidase digestion during passage from the RER to the Golgi apparatus. O'Brien and Muellenberg (1975) found that inhibition of protein synthesis blocked only a portion of total glucosamine incorporation. The blocked portion of glucosamine incorporation may represent the glucosamine residues added in RER, while unblocked residues may be added to completed peptide chains already in the Golgi zone. Addition of *N*-acetylglucosamine is thought to occur in the Golgi apparatus and to be catalyzed by an *N*-acetylglucosaminyltransferase. Furthermore, the studies on *in vitro* translation of opsin mRNA in the presence of dog pancreas microsomes by Goldman and Blobel (1981) showed that the translation product synthesized in the presence of membranes migrated more slowly on sodium dodecyl sulfate (SDS)-polyacrylamide gels than the translation product in the absence of membranes, which was free of oligosaccharides. The glycosylated translation product also migrated more slowly than opsin isolated from ROS which has the two short hexasaccharides described above. This result suggested that opsin was inserted appropriately into the dog pancreas microsomes to be glycosylated. Apparently lacking the processing enzymes and membrane continuity for transfer to the Golgi apparatus, the pancreatic microsomes generated opsin which was probably glycosylated with the larger mannose-rich precursor. Opsin glycosylation apparently was not required for its transport to the ROS. Plantner *et al.* (1980) observed that tunicamycin blocked the transfer of oligosaccharides to newly synthesized opsin, yet transport of opsin continued. Thus the combined study of protein and oligosaccharide synthesis has not only supported the role of the RER in opsin biosynthesis but has also demonstrated some of the later modifications of opsin's final structure as it is glycosylated in the RER and transferred to and modified within the Golgi apparatus.

D. Lipid Biosynthesis

Lipid composition and biosynthesis in rod cells have become subjects of recently intensified study. The ROS lipid composition is unusual in that the cholesterol content is low (5% or 6 mol/mol of rhodopsin, Hendriks *et al.*, 1976). Glycolipids of bovine ROS make up less than 1 mol/mol of rhodopsin (de Grip *et al.*, 1980), and much of the fatty acid content is highly unsaturated (Anderson and Maude, 1970). The major lipids are divided among five classes of phospholipids: phosphatidylethanolamine (PE, 35–47%); phosphatidylcholine (PC, 35–45%); phosphatidylserine (PS, 11–15%); phosphatidylinositol (PI, 2–5%), and sphingomyelin (about 3%). There is disagreement about the asymmetry of these lipid classes in rod disk membranes. Smith *et al.* (1977) found that the amino groups of PE and PS were modified by the reaction of disk membrane fragments with trinitrobenzenesulfonic acid (TNBS), a reagent thought not to penetrate to the disk interior. Crain *et al.* (1978) concluded that TNBS could penetrate the disk bilayer at pH 8.5 and may not accurately reflect the lipid orientation. Drenthe *et al.* (1980a,b, 1981) reexamined this problem and also concluded that TNBS labeled both sides of the disk membrane. By selective digestion with phospholipases, they observed no evidence of asymmetry of any of the major subclasses of phospholipids.

The renewal of ROS lipids is mandated by the prodigious shedding of old membranes at the tip of the ROS each day. Autoradiographic studies illustrated the difference between opsin sequestration during disk formation and lipid insertion. Opsin, once assembled within the disk, is unable to leave the disk bilayer and remains a stable but translationally mobile component until the disk is shed (Hall *et al.*, 1969). In contrast, newly synthesized lipid quickly becomes randomly dispersed throughout the ROS (Bibb and Young, 1974). Each class of phospholipid has a different half-life because of interconversion and catabolism (Anderson *et al.*, 1980a,b,c,d; Ilincheta de Boschero *et al.*, 1980). Anderson *et al.* (1980a,b,c) have shown that the half-life of the major phospholipids (PC, PE, PS) is 18–23 days in the frog compared to a lifetime of 35–50 days for opsin. Lipid turnover studies are complicated by the finding that conversion of PS to PE by decarboxylation is a major source of PE and that PE is a precursor of PC. PI turnover is much more rapid (half-life 3–5 days) and is also complex, since PI hydrolysis to diglycerides may be a response to fluctuations in calcium concentration in ROS cytosol. Diglyceride formation may also play a role in membrane fusion as new disks form or old disks are shed (Anderson *et al.*, 1980a).

Choline incorporation is used not only for PC formation from choline phosphate stores but also for biosynthesis of acetylcholine (Masland and Mills, 1980). While PC incorporation into disk membranes parallels opsin synthesis (Hall *et al.*, 1973; Basinger and Hoffman, 1976), it is not inhibited by short-term puromycin treatment which inhibits protein synthesis.

Polyunsaturated long-chain (20–24 carbons) fatty acids with four to six double bonds make up to 50% of the total acyl chains in ROS. Docosahexanoic acid (22:6) is a major fatty acid of PC, PE, and PS. Many phospholipids are unsaturated at both positions. The high fluidity of ROS membranes (Brown, 1972; Poo and Cone, 1974; Liebman and Entine, 1974) and the lack of a transition temperature above 0°C are consequences of this unsaturation.

The random insertion of lipids throughout the outer segment is probably a consequence of the rapid exchange of phospholipids by transfer proteins characterized by Dudley and Anderson (1978). It is unclear how specific lipids—especially diunsaturated lipids—are sequestered in the ROS when a potential for mixing throughout the continuous plasma membrane along the connecting cilium is theoretically not excluded. Moreover, phospholipid transfer proteins must be partially sequestered within the ROS to avoid redistribution into the inner segment through the cytosol of the connecting cilium. Thus the mechanism for attaining polarity of cell lipids in rods remains an important challenge for understanding the biosynthesis of the outer segment.

II. VECTORIAL TRANSPORT OF OPSIN FROM THE GOLGI APPARATUS TO THE APEX OF THE INNER SEGMENT

When opsin leaves the Golgi apparatus, it travels toward the outer segment. The absence of immunocytochemical labeling with antiopsin antibodies in the lower portion of the cell between the nucleus and the synaptic terminal supports this conclusion. Autoradiographic studies alone could not evaluate this question, since other radiolabeled proteins are distributed to the synaptic terminal in the rod cell, while opsin moves toward the outer segment (Young, 1976). Membrane vesicles appear to be involved in transporting opsin to the outer segment in amphibian retinas. (Preliminary studies on rat retinas support the identification of a continuous cisternal system running from the myoid along the ciliary rootlet to the ciliary base as the analogous pathway for vectorial transport of opsin.) In collaboration with Besharse, we observed that immunocytochemical labeling was restricted to intermitochondrial spaces in the ellipsoid, the region of the inner segment containing the clustered mitochondria (Papermaster *et al.*, 1979). Numerous small vesicles ranging in size from 0.05 to 0.1 μm were distributed throughout the intermitochondrial spaces. These vesicles were readily seen in plastic-embedded retinas (Fig. 5; Kinney and Fisher, 1978b; Besharse and Pfenninger, 1980). *Xenopus laevis* tadpoles injected with radioactive leucine were sacrificed at times varying from 1 hr before the onset of light to 2 hr after light exposure. The tadpoles had been entrained to a 12-hr-light–12-hr-dark cycle. A second group was kept in the dark for 5 days prior to the experiment. Our results showed that radioactive leucine was incorporated into protein to a greater extent in tadpoles exposed to light after 5 days of darkness. Slightly less radioactivity was incorporated in animals exposed to light after 12 hr of darkness. Movement of radiolabeled protein from the RER to the Golgi apparatus and up to the apex of the inner segment followed a time course identical to that described earlier by Hall *et al.* (1969) in adult *Rana pipiens*. Possible subcellular sites of radiolabeled protein were analyzed by the technique of Salpeter *et al.* (1969). When all components of the ellipsoid were considered, the highest label density was associated with vesicles lying in between the mitochondria and clustered at the base of the connecting cilium (Fig. 6). Cytosol was relatively unlabeled, and mitochondria were nearly unlabeled. This autoradiographic evidence strongly supported an earlier suggestion by Papermaster *et al.* (1976) that membranes were the vectors for transport of newly synthesized frog opsin to the outer segment.

Parallel analysis of the paired eye was conducted using immunocytochemical techniques capable of high-resolution localization at the ultrastructural level. Our approach to immunocytochemistry uses a rather unconventional embedding technique which will be described in more detail so that the illustrations in this chapter will be more understandable (McLean and Singer, 1970; Kraehenbuhl and Jamieson, 1976). After fixation with either formaldehyde or glutaral-

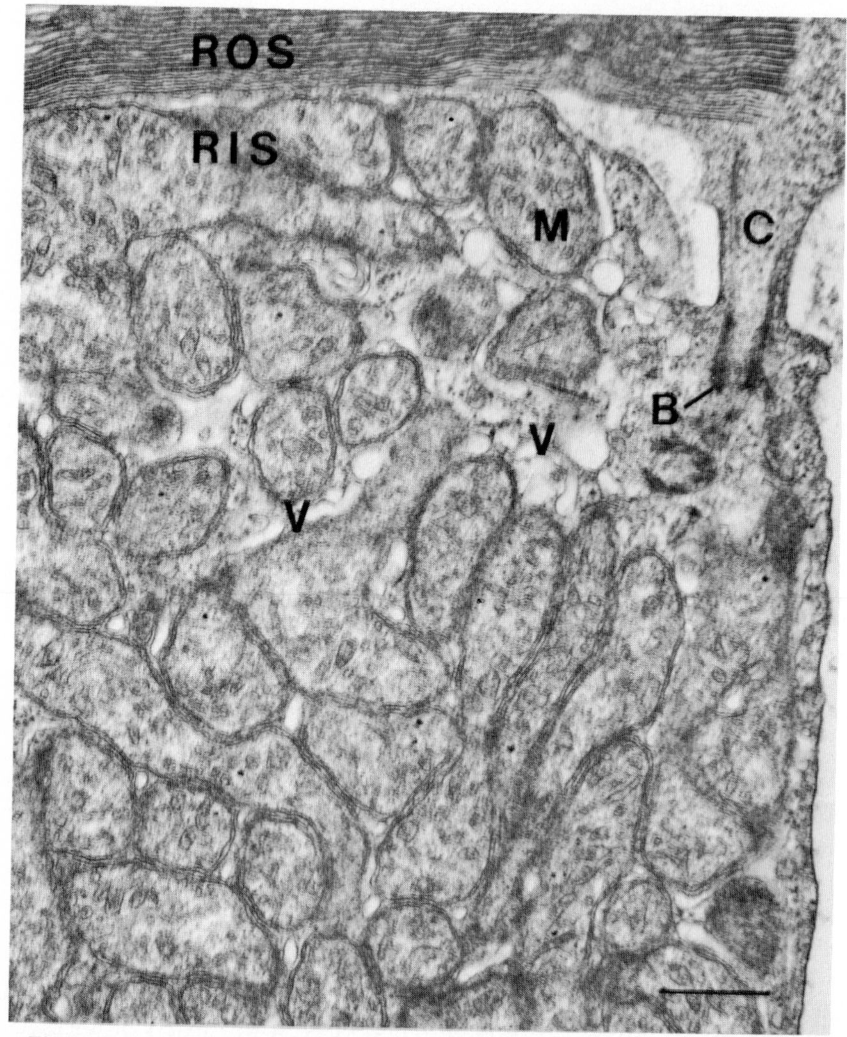

Fig. 5. Electron micrograph of the junctional region of the *Xenopus* tadpole rod inner segment (RIS) and outer segment (ROS). The tissue was osmicated and embedded in Epon, and sections were stained with lead and uranyl salts. The connecting cilium (C) is the only apparent site of contact between these two compartments in the cell. The basal body (B) and centriole are at the base of the cilium. The cytoplasm beneath the cilium is filled with mitochondria (M) and vesicles (V). In most species these mitochondria are elongated and appear oriented toward the cilium. Between the tightly packed mitochondria, numerous vesicles (V) are clustered in the periciliary cytoplasm which is otherwise largely free of mitochondria. Glycogen particles and ribosomes are also present in the cytoplasm of this region. (Bar, 0.5 μm; $\times$ 30,000.) (From Papermaster, Schneider, and Besharse, 1979.)

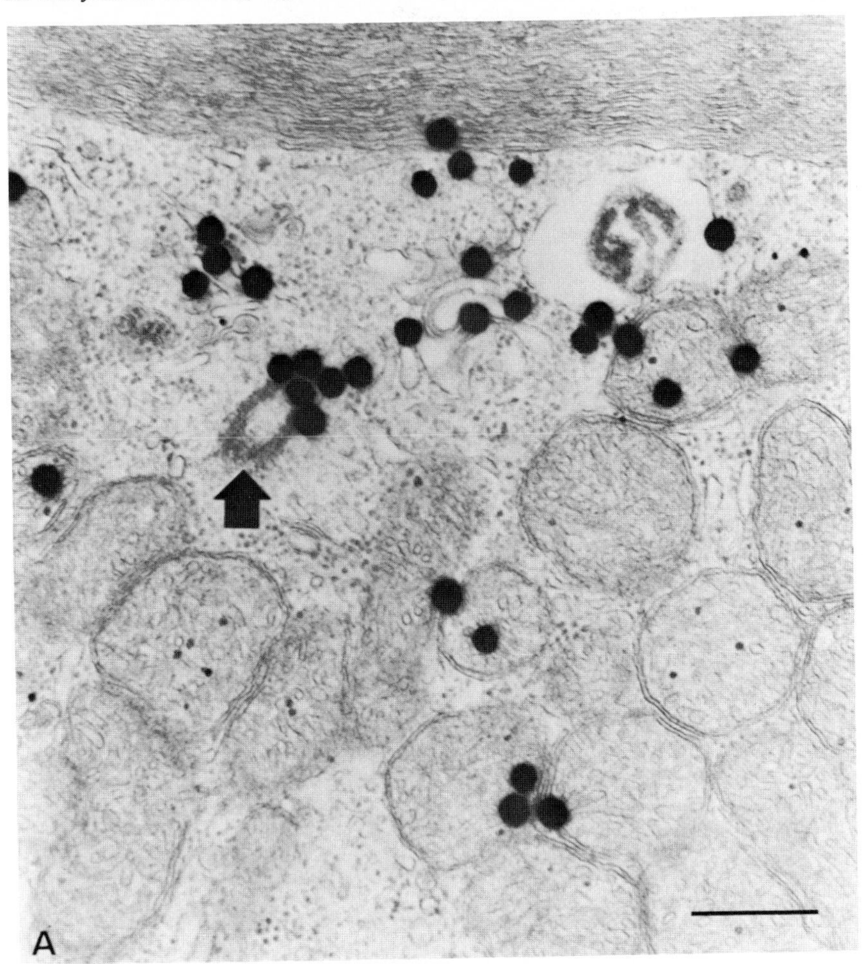

Fig. 6A. Autoradiograph of a *Xenopus* tadpole rod inner segment. (A) Juxtaciliary region after 2 hr of incorporation of radiolabeled amino acids. The cross-sectioned centriole (arrow) indicates that this section is immediately adjacent to the connecting cilium. The periciliary cytoplasm is rich in vesicles. Numerous large black silver grains indicate the presence of radiolabeled protein on these vesicles. Morphometric analysis of the distribution of this label indicates that the vesicle population is the most intensely labeled subcellular component in this region and that many of the silver grains over mitochondria have originated from vesicles that lie between them. (Bar, 0.5 μm; $\times$ 35,000.) (From Papermaster, Schneider, and Besharse, 1979.)

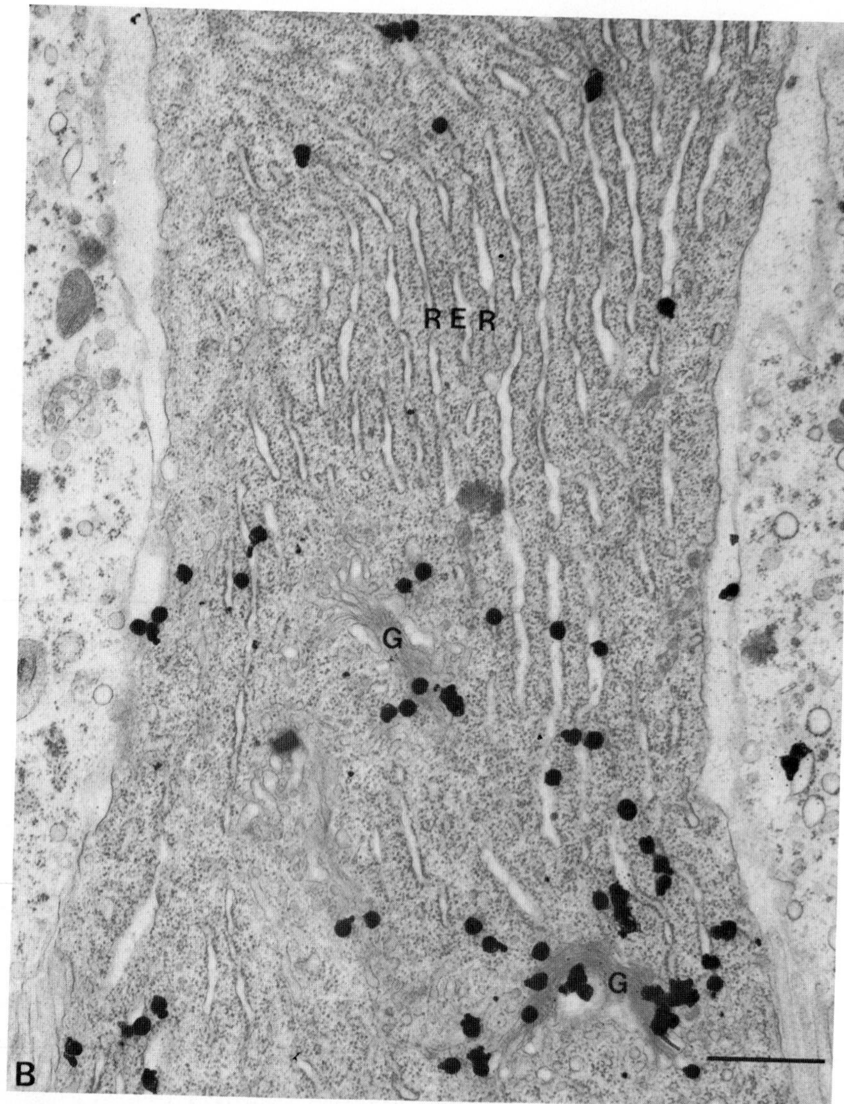

Fig. 6B. Autoradiograph of a *Xenopus* tadpole rod inner segment. (B) The myoid region contains the Golgi apparatus (G) and RER which are also labeled after 1 hr of incorporation. (Bar, 1 μm; $\times$ 16,000.) (From Papermaster, Schneider, and Besharse, 1979.)

dehyde retinas are infiltrated by a solution of 30% bovine serum albumin (BSA), which is cross-linked by a second glutaraldehyde or formaldehyde fixation. The retina embedded in BSA is dried, which renders the tissue hard and suitable for thin-sectioning with a diamond knife. Thin sections are obtained which are comparable in thickness to sections of tissues traditionally embedded in plastics such as Epon. However, the only chemical perturbation to which the tissue has been subjected consists of fixation and drying. No osmication, heating, organic chemical extraction, or exposure to a heavy metal occurs prior to labeling with antibodies.

Thin sections of retina are then successively exposed to a series of preparative washes, BSA, and the first-stage antibody, e.g., rabbit antiopsin. After washing, the bound first-stage antibody can be detected by either a two-stage or three-stage technique. In the two-stage technique, the second-stage reagent can be an antibody–ferritin conjugate reactive with the species of the first-stage antibody (e.g., ferritin-conjugated sheep anti-rabbit immunoglobulin). An alternative two-stage label involves a reaction of the first-stage antibody with the biotin (vitamin H), a small organic compound. Avidin, an egg-white protein, has an enormous affinity for biotin (10^{-15} M). It can be conjugated to ferritin and used to localize the binding of the biotinyl first-stage antibody. Two-stage labeling couples an average of one or two molecules of electron-dense second-stage conjugate to each bound first-stage antibody. Both these techniques have given comparable labeling densities, high resolution, and low backgrounds and were first used for quantitative analysis of the labeling densities of ROS, COS, and Golgi zones (Papermaster *et al.*, 1978a).

The detection of bound first-stage antibody can be enhanced by use of the three-stage labeling technique (Roll *et al.*, 1980). Bound first-stage antibody was detected by reaction successively with biotinyl anti-antibody (second stage) and with avidin–ferritin conjugates (third stage), which results in binding of about five to seven ferritins to each first-stage antibody. Three-stage labeling was especially useful in analyzing the distribution of opsin in regions of lower antigen density such as the ellipsoid. It was also applied to localization of the large protein in ROS (see below).

Labeling densities were estimated by stereological techniques developed by Weibel (1979). Quantitation of labeling density is quite straightforward. A grid of fine lines is placed over the micrograph in the region of interest, and the ferritin grains per square determined by point counting. The surface density on the thin section is calculated according to the magnification of the electron micrograph and the size of the grid square (Kraehenbuhl *et al.,* 1978). BSA-embedded retinas are relatively impermeable to immunocytochemical reagents. When stereo electron micrographs are obtained from paired images taken on a goniometer stage at angles of $+6°$ and $-6°$; the ferritin grains are seen exclusively on the upper surface of the thin section. Thus ferritin grains are not superimposed upon one another to generate a false surface density. Labeling densities of 600–2000/μm^2 are obtained. When either two-stage technique is used, the opsin labeling densities are usually closer to the lower value, while the triple-stage technique generates densities closer to the upper value because of the greater multiplicity of ferritins bound to each first-stage antibody by the "sandwich" of biotinyl anti-antibody and avidin–ferritin. Regardless of the technique employed, quantitative comparisons of one region of a cell with another or of one cell to another on the same section are readily obtained. Moreover, the technique has been so reproducible that sections can be compared from one grid to another from experiments done over several months. Thus labeling densities can be determined from many cells with a very low variation (standard error of the mean $\pm 5\%$), which not only increases the validity of the qualitative impressions of localization but also provides micrographs which illustrate the mean of the overall results. Tables of labeling density describe the level of antibody binding quantitatively and the variability of binding for the entire population of cells studied. The need for extensive sampling of many cells and many tissue orientations cannot be overemphasized. Electron micrographs serve to illustrate the data, but the quantitative results in the tables are the most important test of the validity of the illustrations and the conclusions drawn from them. A similar rationale for quantitation of electron microscope (EM) autoradiographs has been emphasized by Salpeter *et al.* (1969), however, quantitative autoradiographic analysis is primarily devoted to localization of the source of the decaying particle, since the silver grains can be deposited several hundred angstroms away from the source.

Because of the reproducibility of the immunocytochemical technique and its ability to yield repeated sections for several years from the same tissue blocks, we are encouraged to study opsin density not only in regions of high concentration such as the ROS and Golgi apparatus but also in the inner segment over the RER and in apical portions of rod inner segments. We sought localization of opsin on vesicle membranes that were highly radiolabeled and which traversed between mitochondria in that region. While autoradiography provided quantitative evidence that the most heavily labeled cell compartment in upper portions of the inner segment was composed of vesicles, it could not establish that the newly synthesized protein on these vesicle membranes was opsin. Biosynthetic studies had demonstrated that the major protein continuously synthesized in bovine and frog retinas was opsin (Hall *et al.*, 1969; O'Brien *et al.*, 1972; Basinger *et al.*, 1976a; Papermaster *et al.*, 1975, 1976). We also showed that newly synthesized opsin was easily recovered from sedimentable membranous fractions obtained by homogenization of frog retinas after short periods of radiolabeled amino acid incorporation (Papermaster *et al.*, 1975). Immunocytochemical approaches appeared to us to represent the most direct way to localize opsin as a component of the highly labeled vesicles.

In order to interpret the electron micrographs of BSA-embedded retinas it is important to recall that the tissues had not been osmicated. Membranes do not have an affinity for lead or uranyl stains unless they have been osmicated. Rather, the membranes appear as pale, negatively stained structures against a more darkly stained cytosol or extracellular matrix.

In many retinas the RER is relatively unordered. Longitudinally oriented parallel stacks of RER cisternae are often seen, however, in *X. laevis* tadpoles and *Rana pipiens* adults. Retinas of these tadpoles rapidly synthesize membranes (Besharse *et al.*, 1977a,b). Immunocytochemical analysis of the RER was easily evaluated in these retinas. Ferritin grains representing bound antiopsin antibodies were distributed along the membranes of the RER almost exclusively on the cytoplasmic side (Fig. 7). Although this observation could be interpreted to indicate that opsin at this site had not yet passed into the bilayer or across to the cisternal interior, alternatively, it could represent a mechanical barrier to the binding of antiopsin antibodies to opsin determinants in the tightly collapsed cisternae. Antibodies with specificity exclusively against smaller peptide domains at the N- and C-terminals will be of value for testing these observations. In no other region of the inner segment was it possible to obtain so predominant a label on one side of a membrane.

The Golgi apparatus was prominently labeled by antiopsin antibodies (Papermaster *et al.*, 1978a). This supported suggestions that the Golgi zone was involved in the addition of *N*-acetylglucosamine to opsin (Bok *et al.*, 1974, 1977) and that newly synthesized protein, most probably opsin, was transferred there from RER prior to being distributed to the ROS (Hall *et al.*, 1969). Since photoreceptors also synthesized other proteins, however, immunocytochemical localization of opsin at a high density in the Golgi membranes established that opsin was at least a major component of this compartment in rod cells.

Once opsin leaves the Golgi apparatus, it becomes more difficult to

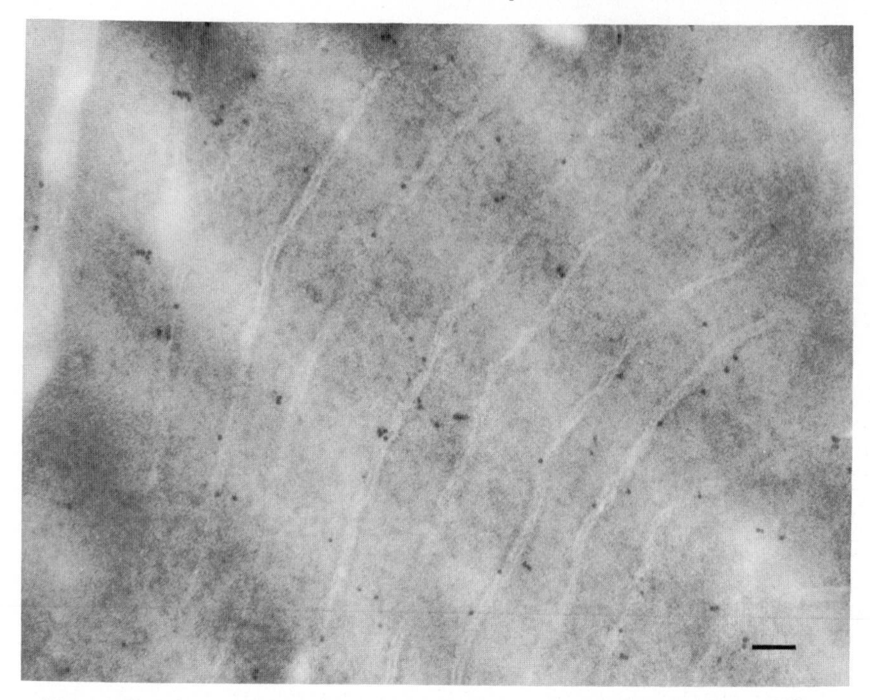

Fig. 7. Immunocytochemical localization of opsin in the RER of the *Xenopus* tadpole retina. The tissue in this figure and in Figs. 7–9, 12, 14, and 15 was embedded in BSA cross-linked by gluteraldehyde. The section in this figure was exposed to antiopsin (first stage) and stained with ferritin conjugated to sheep anti-rabbit immunoglobulin (second stage). This double-label technique results in the binding of one or two ferritin grains wherever the first-stage antiopsin is bound. The small, black dots of ferritin are distributed almost exclusively along the oriented cisternae of the RER which appear as pale-gray membranes negatively stained with uranyl salts. The adjacent intercisternal cytoplasm is not significantly labeled. In this region, over 90% of the ferritin grains are within one ferritin diameter (10 nm) of the RER membranes. This indicates that the radiolabeled protein noted in the RER in Fig. 6B is newly synthesized opsin. The Golgi apparatus is also labeled by antiopsin antibody (see Papermaster *et al.*, 1978a). (Bar, 0.1 μm; $\times$ 60,000.) (From Papermaster, Schneider, and Besharse, 1979.)

establish which of the smooth membranes of the inner segment are involved in its transport. Holtzman and Mercurio (1980) have claimed that a smooth "agranular reticulum" is prominent in this region. While the smooth membranes in the inner segment vary in distribution and their association with microtubules and ciliary rootlets in various species, in the *Xenopus* tadpole and adult and in *Rana* adults, small vesicles are abundant and readily observed in plastic-embedded retinas (Fig. 5). In the owl monkey, vesicles are surrounded by tubules and thin (actin?) filaments (B. Burnside, personal communication), while in mice and rats, longitudinally oriented membrane cisternae are closely apposed to the ciliary rootlet deep in the inner segment and vesicles are rarely seen (Cohen, 1960; Schneider and Papermaster, 1981). Preliminary results (Schneider and Papermaster, 1981) indicate that these cisternae in the rat may be labeled by antiopsin antibodies in BSA thin sections, however, only *Xenopus* tadpoles have been subjected to quantitative analysis at this time.

Within the ellipsoid, quantitative analysis of the distribution of subcellular sources of radiolabeled protein demonstrated that the vesicles in transit between the closely packed mitochondria were the most highly labeled components. The vesicles are often seen as thin, flattened sacules (Fig. 5). In this region, antiopsin is bound almost entirely in the intermitochondrial space (Fig. 8). The labeling density over mitochondria approaches the density of background labeling. Cross sections demonstrate a nonhomogeneous distribution of the vesicles, since most are between mitochondria which lie beneath the basal body. The mitochondria are excluded from the cytoplasm below the basal body and are also tapered and oriented toward this structure. Here vesicles cluster at a high density and are extremely radioactive after 1.5–2 hr of incorporation of radiolabeled precursors (Fig. 6A; Young, 1976; Papermaster *et al.*, 1979). Here, too, immunocytochemical labeling densities are at the highest level observed in the ellipsoid (Fig. 9). However, the small diameter of the vesicles makes tangential sections of the vesicle surface a frequent result. Consequently, a comparison of the labeling densities of these vesicles and of the other inner segment structures whose shapes make tangential sections less probable is not justifiable. Cross-sectioned vesicles are noted, however, and appear as rings of ferritin label. Labeling density over some of the tangentially sectioned vesicles approaches the density of antiopsin labeling of outer segments, indicating a high opsin concentration in the vesicle membrane. A high density of intramembranous particles (IMPs) in vesicles beneath the cilium was observed in freeze-fracture electron micrographs by Besharse and Pfenninger (1980). The IMPs in these vesicles (Fig. 10) had diameters comparable in size to those observed in ROS, which are most likely aggregates of three or four rhodopsin molecules (Chen and Hubbell, 1973; Corless *et al.*, 1976). Since other membranous areas of the inner segment and the plasma membrane contained IMPs of a different size and density, Besharse and Pfenninger (1980) postulated that the vesicles transported opsin to this region. Of particular note, however, was the observation that the IMP density of the

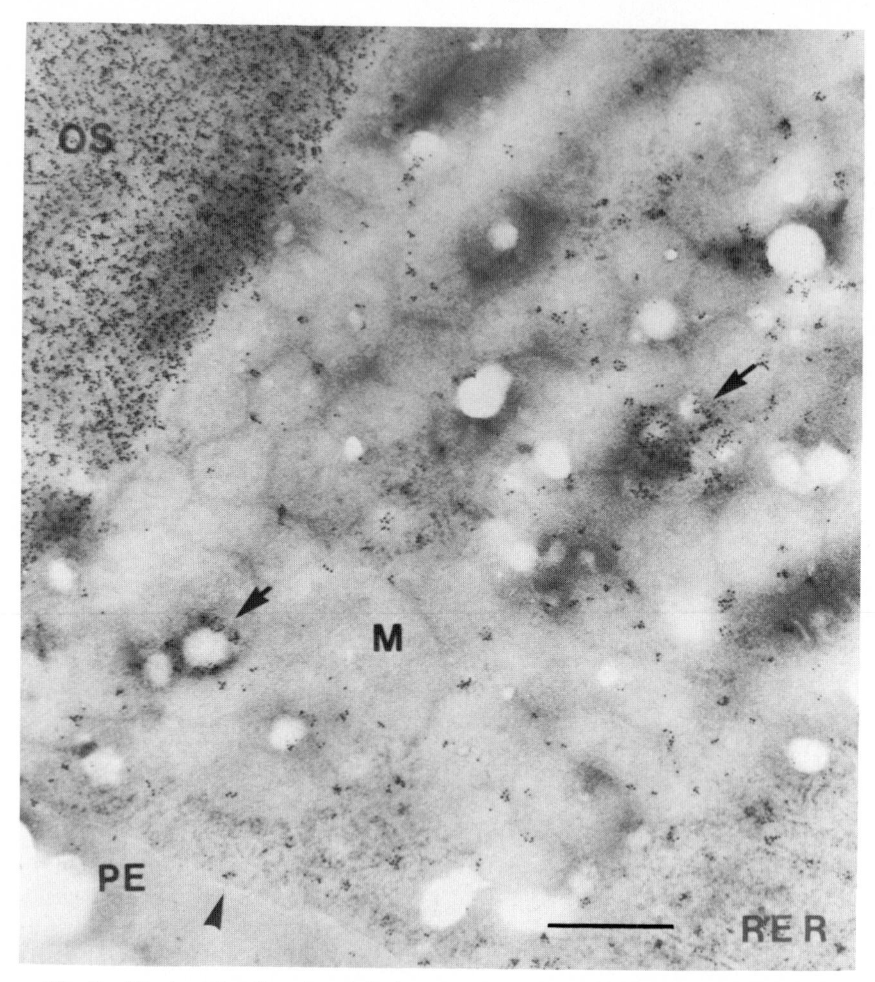

Fig. 8. The junctional region of the inner and outer segments of the *Xenopus* tadpole rod. This section of BSA-embedded retina is labeled by antiopsin, especially in the ROS. The bound antibody (first stage) is detected by a three-stage label of biotinyl sheep anti-rabbit immunoglobulin (second stage) and avidin ferritin (third stage) which results in the deposit of five to seven ferritin grains over each bound first-stage antibody. The majority of the ferritin grains in the inner segment are clustered over pale vesicles lying between the mitochondria (arrow) which lie between the mitochondria (M) (cf. vesicles in Fig. 5). The adjacent PE is unlabeled as is the lateral plasma membrane (arrowhead). The area beneath the ellipsoid contains elements of RER cisternae which are labeled by antibody. In the ellipsoid, 80% of the ferritin grains lie in the darker staining cytoplasm between the pale, negatively stained mitochondria. Only rare ferritin grains are seen over the interior of the mitochondria. This indicates that opsin is transported between the mitochondria. (Bar, 0.5 μm; $\times$ 35,000.) (From Papermaster, Schneider and Besharse, 1979.)

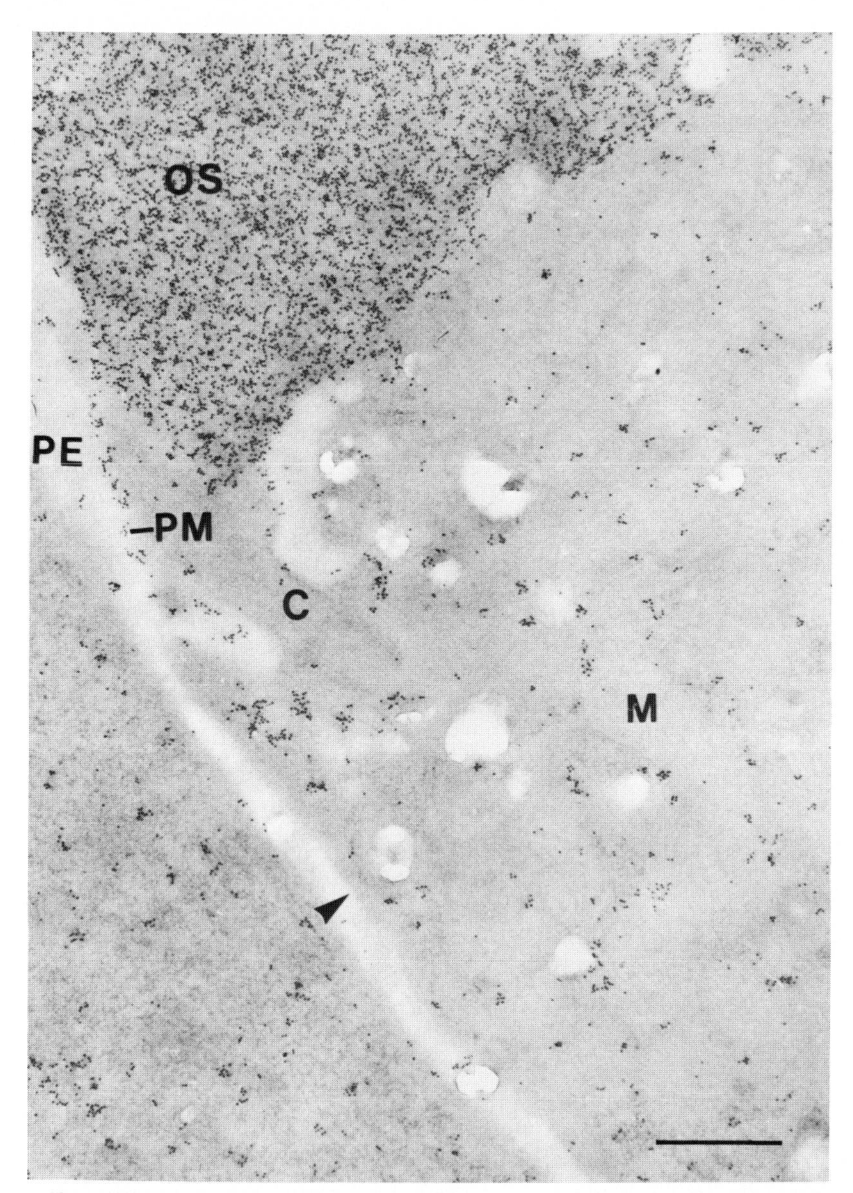

Fig. 9. Junctional region of *Xenopus* tadpole rod. This BSA section passes through the center of the connecting cilium (C). The mitochondria (M) are unlabeled, and the borders of many of the vesicles are labeled with antiopsin. The plasma membrane (PM) of the upper portions of the cilium and ROS is intensely labeled with antiopsin, whereas the lateral plasma membrane of the inner segment (arrowhead) and the base of the cilium are unlabeled. The base of the cilium may act as a one-way gate permitting passage of opsin and other ROS proteins to the ROS but blocking back diffusion to the inner segment. (Bar, 0.5 μm; × 35,000.) (From Schneider and Papermaster, 1981, with permission.)

TABLE I

Comparison of Mean Density and Size of IMPs in the PF-Leaflet of Membranes of Rod Photoreceptors[a,b]

Membrane	IMP per μm_2	Mean IMP size (nm)
Inner segment		
plasma membrane	2989 ± 81	7.6 ± 0.12
Calycal processes		
Disks	3523 ± 161	7.5 ± 0.12
Periciliary	4629 ± 63	10.2 ± 0.14
vesicles	1921 ± 121	9.9 ± 0.16

[a] Modified from Besharse and Pfenninger (1980).

[b] Note that the IMP density is greater in calycal processes and disks than in the inner segment and that the mean IMP size is greater in disks and vesicles than in the inner segment plasma membrane and calycal processes.

vesicles was half that of the ROS (Table I). One interpretation of this lower IMP density in vesicles is that the rod transports twice as much membrane surface area to the base of the connecting cilium as it eventually uses to form the disks in the ROS. A possible consequence of this observation is that a major retrieval of the unused membrane must occur or significant local destruction will be necessary. Attempts to resolve these alternatives are underway and include incubation of retinas *in vitro* with horseradish peroxidase (HRP) to determine if any HRP is engulfed by endocytosis during membrane retrieval (Besharse and Forestner, 1980). It will be important to account quantitatively for the magnitude of membrane turnover predicted as a consequence of the lower IMP density of inner segment vesicles. Immunocytochemical localization of high concentrations of opsin on this vesicle cluster beneath the cilium (Fig. 9) strongly supports the suggestion of Besharse and Pfenninger (1980) that the IMPs in freeze-fractured vesicles represent newly synthesized opsin. Antibody binding to the vesicles also indicates that the autoradiographic studies on protein synthesis in rods localize newly synthesized opsin in transit through the inner segment (Fig. 8). Thus, while each ultrastructural approach to analysis of the vesicles beneath the connecting cilium is intrinsically limited in its ability to characterize the molecules on the vesicle membrane, in concert they provide strong support for the concept that membrane vesicles are the vehicles of opsin transport from the Golgi apparatus in amphibian retinas. More extensive study of mammalian retinas will be required to establish if a homologous process transports opsin in their

Fig. 10. Freeze-fracture replica of *Xenopus* tadpole rod junctional region. The disks of the outer segment are filled with intramembranous particles (large arrow). Within the inner segment, numerous vesicles are seen, which contain intramembranous particles of the same size as the outer segment (small arrows). The direction of shadowing is indicated by the open arrow. (Modified from Besharse, 1980, with permission.) (Bar, 0.2 μm; × 65,000.)

rods. Either a vesicular (amphibian) or cisternal (mammalian?) system would involve comparable topological constraints on the orientation of opsin during its transport and incorporation into the plasma membrane near the base of the cilium (Hirano *et al.*, 1972). The two alternatives would differ to the extent that concentration of opsin from the molar proportions within the Golgi apparatus to those in the vesicles or ellipsoidal cisternae would involve either separated or continuous membranes in the respective systems.

III. TRANSPORT OF OPSIN TO THE ROS FROM THE BASE OF THE CONNECTING CILIUM

The terminal site of opsin's insertion after its transport to the base of the connecting cilium is yet to be established. As shown above, the vesicles of amphibian retinas cluster near the base of the cilium and are highly radiolabeled after 1.5–2 hr of incorporation. They are also labeled by antiopsin antibodies at high density and contain IMPs of appropriate size. For this reason attention has turned to the apical plasma membrane of the inner segment near the connecting cilium as a possible site for delivery of opsin by an exocytic fusion of vesicles and the plasma membrane at this site (Papermaster *et al.*, 1976; Besharse and Pfenninger, 1980; Kinney and Fisher, 1978b).

In nearly all vertebrate photoreceptors the plasma membrane of the inner segment immediately adjacent to the cilium is highly convoluted (Fig. 5). Freeze-fracture studies by Andrews (1981) demonstrated parallel arrays of IMPs radiating from the cilium along the apical plasma membrane of the inner segment. Scanning electron microscopy (SEM) of this region at high resolution presents a striking image of this new cell structure (Peters *et al.*, 1981). The apical plasma membrane is thrown into a folded highly ordered complex of ridges and grooves termed the periciliary ridge complex (Fig. 11). There appear to be nine ridges around each cilium, usually in parallel groups of three separated from each other by an angle of 120 deg. It is not yet known whether the nine ridges are a numerical correlate in some structural way to the 9 + 0 array of microtubules that form the axoneme in the core of the connecting cilium. The same invagination seen in fixed and critical point-dried retinas studied with SEM is also seen in living frog rods viewed by video-enhanced differential interference-contrast light microscopy (Papermaster and Allen, 1981).

The top of the ridges of the percilliary ridge complex may represent a site of opsin insertion into the apical plasma membrane of the inner

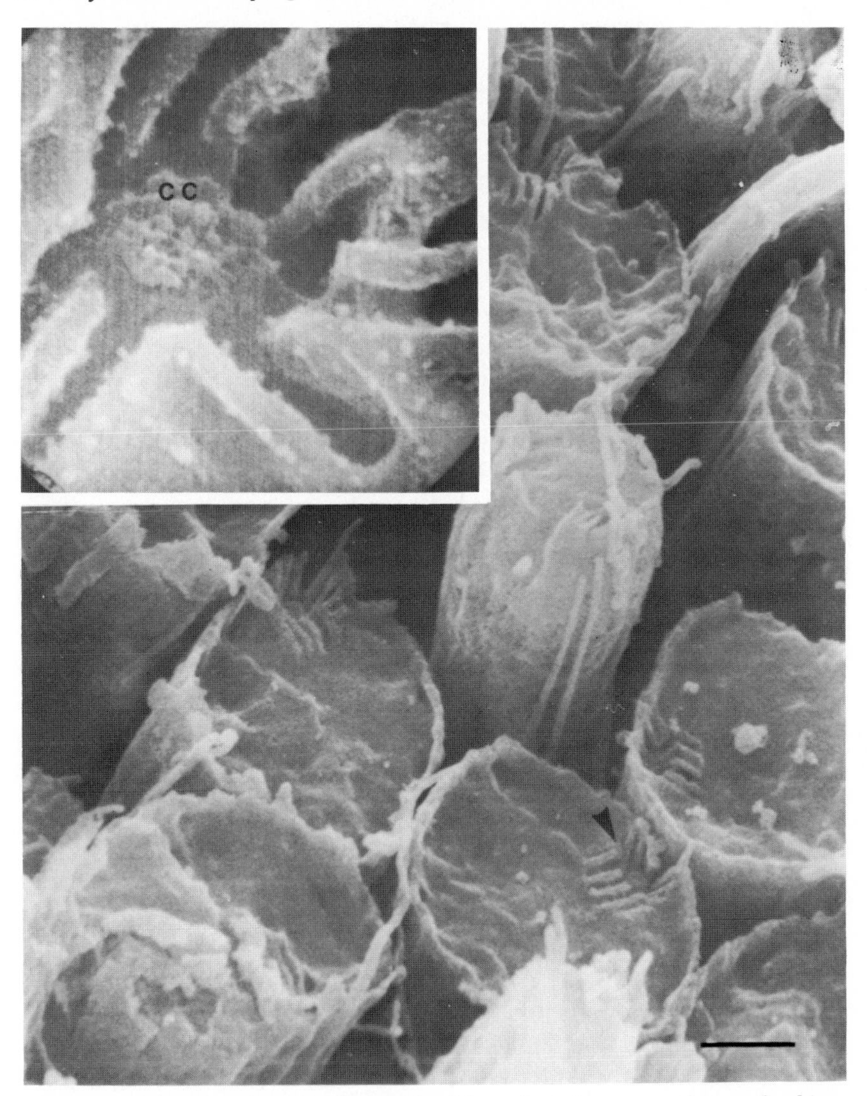

Fig. 11. Scanning electron micrograph of the apical plasma membrane of rod inner segment. The outer segments were broken off to reveal this surface. Each cell contains an array of nine ridges and grooves (arrowhead) surrounding the broken connecting cilium (CC). (From Peters *et al.*, 1981, with permission.) (Bar, 1 μm; $\times$ 13000; inset $\times$ 81,000.)

segment. Immunocytochemical studies should be able to localize this important site. It is clear from both light and electron microscopic studies, however, that the periciliary ridge complex is a stable morphological unit and that the deep invagination of the plasma membrane around the base of the cilium is not simply a dynamic response to local exocytosis. If opsin is inserted in or near this complex, it must traverse the apical plasma membrane to the adjacent plasma membrane of the connecting cilium to reach the ROS.

A preliminary study on the ontogeny of the cilium of several mammalian rods also indicates the presence of "foot processes" radiating from the centriole as it approaches the plasma membrane and participates in the organization of the cilium. The initially flat plasma membrane invaginates around the centriole and then protrudes above the edge of the cell as ciliary elongation proceeds (Bodley *et al.*, 1981). In earlier ontogenetic studies, Nilsson (1964) and Kinney and Fisher (1978a) also showed that the plasma membrane of the inner segment of amphibian retinas invaginated around the cilium, however, the invagination was not emphasized as a stable unit until it was visualized by recent SEM and freeze-fracture reports and shown to be a much more complex structure.

The structure of the connecting cilium has become the subject of increasing attention, as its possible role in opsin transport has been emphasized. Freeze-fracture studies on rat rod connecting cilia indicate that the IMP arrays in the ciliary plasma membrane may form concentric rings along the length of the cilium (Matsusaka, 1974), although this appearance is not always observed (Röhlich, 1975; Besharse and Pfenninger, 1980). Besharse and Pfenninger (1980) noted that the IMP density of *X. laevis* cilia was comparable to the density of particles in ROS disks and was about twice the density observed on the vesicles clustered below and basal body (Table II). This suggested that a dramatic concentration of opsin occurred over a very small domain of the rod cell's surface.

Autoradiographic studies on the connecting cilium have presented images which contain abundant silver grains within the cilium interior (Young, 1976). Above the cilium, EM autoradiographs of retinas synthesizing opsin in organ culture showed early arrival of newly synthesized protein, presumably opsin, in the plasma membrane of the outer segment (Basinger *et al.*, 1976a). To establish this conclusion, however, an analysis of the distribution of possible sites of origin of radioactive decay by the procedure of Salpeter *et al.* (1969) was valuable. EM autoradiographic resolution is too low to assign the site of origin to the ROS plasma membrane by direct observation of the posi-

TABLE II

Ferritin-Labeling Density of Ciliary Plasma Membrane[a]

Rat (14)	
Lowest fifth	2.9 ± 3.8
Next fifth	11.5 ± 3.7
Upper fifth	16.8 ± 3.5
p	<0.001
Length	1.1 μm
Xenopus (9)	
Lower half	2.2 ± 0.5
Upper half	11.2 ± 1.3
p	<0.001
Length	0.5 μm

[a] Values are numbers of ferritin grains per 0.25 μm ± SE. The number of cilia counted is shown in parentheses. P values calculated by Student's t test compare the density over the lower cilium with the density over the upper cilium.

tion of the silver deposits. Because of this low resolution, it is also possible that a significant fraction of the silver grains seen over the ciliary interior also arise from sites on the ciliary plasma membrane. In addition, since cytosol proteins of the outer segment must also be renewed. some of these may contribute to the grain density of the ciliary interior. This inherent limitation of autoradiographic resolution and its inability to identify the radiolabeled proteins has served as another stimulus for our application of high-resolution immunocytochemical methods for analysis of cell surface and intracellular antigens.

Although Jan and Revel (1974, 1975) observed binding of anti-cattle opsin antibodies to the surface of rat rod cilia after immersion in antiserum, no quantitative analysis of the labeling distribution was reported. In order to evaluate the distribution of opsin quantitatively, we studied the binding of anti-frog opsin antibody to *X. laevis* and rat retinas embedded in cross-linked BSA. A common feature of these rod cilia was observed. The innermost 0.2 μm of the cilium was relatively unlabeled when compared to the plasma membrane of the upper cilium and the outer segment (Figs. 9 and 12 and Table II). The only alternative pathway available for passage of opsin to the ROS is the unlikely possibility of formation of a transient cytoplasmic bridge between the apical plasma membrane of the inner segment and the lowermost disk of the outer segment. Although one report of such a bridge supports this possibility (Richardson, 1969), such bridges are usually not seen in

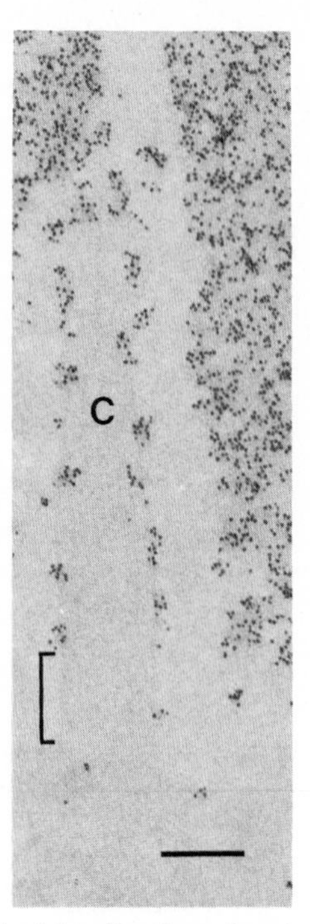

Fig. 12. Immunocytochemical localization of opsin in the connecting cilium of rat rod cells. This BSA-embedded retina is heavily labeled by antiopsin over outer segments and along the plasma membrane of the long connecting cilium (C). The lowest 0.2 μm of the cilium (bracket) is nearly unlabeled. Compare to *Xenopus* cilium (Fig. 9) which is also unlabeled at the base of the cilium. (See Table II.) (Bar 0.2 μm; $\times$ 43,000.) (From Schneider and Papermaster, 1981, with permission.)

undamaged retinas when studied by conventional EM, freeze-fracture, or SEM.

The lateral plasma membranes of the inner segments of *X. laevis* rods are not labeled by antiopsin antibodies (Fig. 12). Thus the relatively unlabeled base of the cilium may constitute a locus of stable ciliary membrane components between which opsin passes at low density in the ciliary membrane on its way to the upper portion of the connecting cilium and the outer segment. If this interpretation is correct, the membranous base of the cilium may serve as a "one-way" gate allowing free passage of opsin to the ROS and as a barrier to back diffusion along the plasma membrane of the inner segment. Such membrane "fences" have been postulated in other cells to account for the local specializations of cell surfaces. Junctional complexes alongside apical brush borders of epithelia and synaptic contacts are just two examples of such components. Rarely, however, has the behavior of an intrinsic membrane protein in two regions of a cell been contrasted so dramatically over such a small region of the cell's surface. Above the cilium, opsin translates and rotates rapidly in a membrane of relatively low viscosity (Brown, 1972; Poo and Cone, 1974; Liebman and Entine, 1974) and diffuses throughout the plasma membrane (Basinger *et al.*, 1976a). Yet beneath this turbulent zone is the cell membrane of the inner segment with little or no opsin in it, despite its continuity with the ROS. Lectin-binding sites also show a differential distribution in inner and outer segments of rods and cones (Bridges and Fong, 1979; Nir and Hall, 1979; Bridges, 1981). It is still too early to know if the distribution of opsin in amphibian retinas can be the basis for a generalization about opsin distribution in all rod photoreceptor cells, since studies on mammelian rods are at an early stage. Nonetheless, the existence of a barrier to back diffusion even in a few species demonstrates that such fences can be detected by the new cell biological techniques developed during the past few years.

IV. OUTER SEGMENT DISK STRUCTURE AND MORPHOGENESIS

A. Rhodopsin Structure

Most recent studies have converged around an average molecular weight of approximately 35,000. This includes at least 12 sugar residues, leaving a molecular weight for the protein of 32,000–33,000. The amino acid sequence of opsin has resisted analysis because of the presence of peptides with hydrophobic sequences that are difficult to separate after proteolytic cleavage. Heller and Lawrence (1970) isolated and sequenced a small, unblocked glycopeptide containing one oligosaccharide unit linked to asparagine. The peptide had been isolated by column chromatography in acid and had been stored for some time in acid prior to Edman degradation. Subsequent studies demonstrated that this peptide was blocked at its N-terminal (Hargrave, 1977) and was the N-terminal region of mature rhodopsin. The N-terminal methionine is blocked by acetylation (Tsunasawa *et al.*, 1979). Hargrave (1977) further showed that there were two sites of oligosaccharide attachment on asparagines in the second and fifteenth positions. Fukuda *et al.* (1979) and Liang *et al.* (1979) determined the exact structure of these oligosaccharides as described above. Of interest was the complete absence of leucine until the thirtieth residue. Thus, this portion of rhodopsin did not resemble in any structural way the signal peptides of secreted or membrane glycoproteins, which had short hydrophobic leucine-rich peptides preceding the N-terminal of the mature protein (Fig. 3). The next 150–200 residues are still unsequenced at the time of this review, although partial sequences from many peptides in this region have been obtained which indicate that this region contains several hydrophobic peptide sequences with blocks of uncharged amino acids separating charged residues (P. A. Hargrave, personal communication).

Pober and Stryer (1975) demonstrated that rhodopsin could be cleaved by a variety of proteolytic enzymes into two large fragments, F1, the N-terminal two-thirds, and F2, the C-terminal one-third. Most laboratories involved in studies on rhodopsin's proteolytic fragments have agreed that the retinaldehyde is linked to a lysine as a protonated Schiff base, most probably in F2. Wang *et al.* (1980) reported that the sequence of this site was Ala-Lys.

Hargrave *et al.* (1977, 1980) have sequenced the C-terminal 40 amino acids. Threonine and serine residues phosphorylated by rhodopsin kinase are located in this region (Fong, 1978). Virmaux and co-workers (Pellicone *et al.*, 1981a,b) extended the available sequence information in an abstract describing a peptide segment 80 amino acids long. A portion of the C-teminal of their sequence overlapped with the N-terminal sequence of the C-terminal peptide described by Hargrave *et al.* (1980), indicating that these two long peptide segments constituted a significant portion of F2. Recent studies by Hargrave (1982) have extended this sequence to the thermolysin cleavage site at an alanine which generates F2 and F1. Their work indicates that the N-terminal of F2 is 108 residues from the C-terminal of opsin. Nearly all current studies on proteolytic cleavage suggest that the site of thermolysin cleavage which generates F1 and F2 is on the cytoplasmic side of the disk membrane. Within F2 are several long segments containing only uncharged hydrophobic amino acids. Circular dichroism studies on rhodopsin in disk membranes indicate a significant helical content (Schichi and Shelton, 1974), and neutron diffraction indicates that much of the opsin is inserted in the lipid bilayer (Saibil *et al.*, 1976). One proposal for displaying the sequence which incorporates these observations is illustrated in Fig. 13. This model also maximizes the structural homology of bovine rhodopsin and bacteriorhodopsin of *Halobacterium*. It is clear that additional sequence data will make a major contribution to our understanding of rhodopsin structure and function. Several labora-

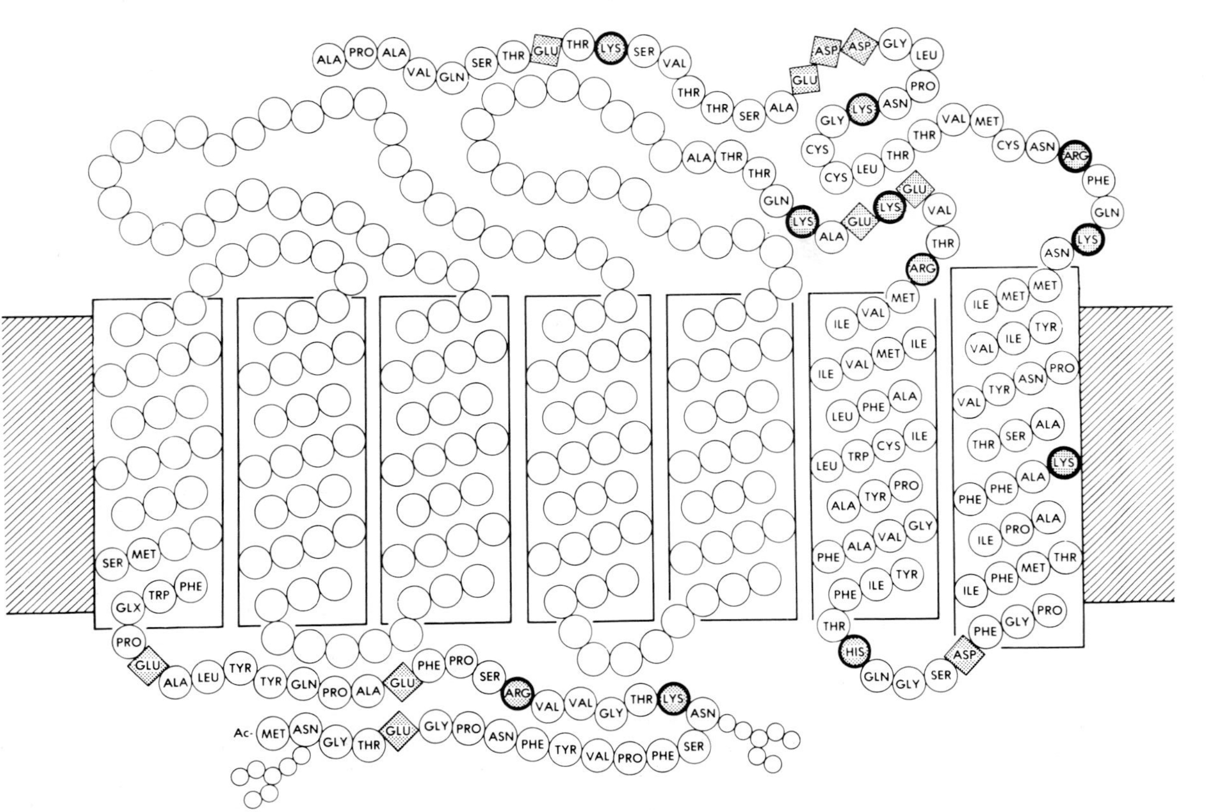

Fig. 13. Model of bovine opsin in a lipid bilayer. The amino acid sequence of the C-terminal region is from Hargrave *et al*. (1982) and Pellicone *et al*. (1981a,b). The hydrophobic portions have been arrayed in helices oriented perpendicular to the plane of the membrane. (From P. A. Hargrave, personal communication.)

tories are now actively pursuing the sequence of the rhodopsin gene in order to establish its structure. These combined amino acid sequence and nucleotide sequence studies should give us an exact analysis of rhodopsin's linear structure. Assignments of portions of the molecule to the lipid domain (helical?) or to hydrophilic surfaces will be a fascinating challenge.

An extensive comparison of opsin and the visual pigments of cones (iodopsins) would still be premature at this point. Few studies have been conducted, since cones are the predominant photoreceptor only in chickens, reptiles, some rodents, and the foveal region of diurnal animals (see Chapter 9, this volume). One difference between rhodopsin and iodopsin of rabbit and goldfish cones is the observation by Bunt and Klock (1980a,b) that some cones are intensely labeled by radioactive fucose incorporation. Cones of the goldfish retina differ in their extent of fucosylation; those sensitive to green light seem to be labeled the least, while red- and blue-sensitive cones are heavily labeled. SDS–polyacrylamide gels of fucosylated goldfish proteins indicate a migration more rapid than that of opsin, which is consistent with either a lower molecular weight or a less extensive total glycosylation (Bunt *et al.*, 1981). Opsin's oligosaccharide resembles the core structure seen in other cells which usuallly add further sugars to fashion a more elaborate chain. Fucosylation may occur on the *N*-acetylglucosamine of the chitobiose linkage next to asparagine if iodopsin has a similar oligosaccharide.

Little is known about the homology of the polypeptide chains of iodopsins and rhodopsins. If the fucosylated proteins are the iodopsins of rabbits and goldfish, their molecular weights are similar or slightly smaller than that of opsin (Saari and Bunt, 1980; Bunt *et al.*, 1981). Chicken iodopsins are also approximately the same size as opsin (Fager and Fager, 1978). Antibodies to frog opsin which were affinity-purified on cattle opsin immunoadsorbants were shown to react immunocytochemically with frog COS and green ROS (Papermaster *et al*, 1978a). While these results all support the homology of rhodopsins and iodopsins, the isolation and sequencing of their genes would provide invaluable evidence for determining the sources of this variability in spectral sensitivity. Especially fascinating is the evidence that the stability of some portion of the opsin structure has been so great that opsins of the lowest and highest vertebrates cross-react immunocytochemically. Al-Mahdawi *et al.* (1981) have also shown homology in peptide maps of proteolytic fragments of bovine and frog opsin. At the same time variability in the sequence within the retinaldehyde-binding pocket can be presumed to be the structural basis of the different opsins in fish and the various iodopsins in many species.

B. Stoichiometry of ROS Membrane and Cytosol Proteins: Consequences for Theories of Excitation by Light

1. *Enzymes Mediating Cyclic Nucleotide Concentration and Protein Phosphorylation in Outer Segments*

Since the early pioneering spectrophotometric studies by Wald and his colleagues it has been known that rhodopsin is present in the ROS in high concentration (Wald, 1968). These studies demonstrated that at least 50% of ROS protein was rhodopsin. As techniques of subcellular fractionation improved, modifications of the original isolation procedures were developed, all of which resulted in isolation of the organelle relatively free of contamination by membranes from other regions of photoreceptor cells or from other cells in the retina. As early as 1938, Saito demonstrated that ROS could be purified on sucrose gradients. McConnell (1965) emphasized the need for extensive homogenization to eliminate contamination of ROS by membranes of the adjacent ellipsoid. A reduction in the shear forces in the initial stages of homogenization and the addition of divalent cations to stabilize nuclear membranes and to shift membranes of endoplasmic reticulum to more dense regions of the gradient further reduced contamination (Bownds *et al.*, 1971; Papermaster and Dreyer, 1974). Once relatively pure ROS membranes were obtained, all laboratories were in agreement that rhodopsin constituted 85–90% of the outer segment membrane protein. In fact, only one other intrinsic membrane protein has consistently appeared in outer segments along with rhodopsin. This protein, one of the largest known, averages 290,000 daltons in amphibia and varies from 230,000 to 260,000 daltons in mammals (Papermaster and Dreyer, 1974; Papermaster *et al.*, 1976, 1978b). The large intrinsic membrane protein of outer segments will be discussed in more detail below.

Given the enormous molar ratio of rhodopsin to other proteins in outer segments, it became perplexing to explain the capacity of rod cells to transduce photon capture into a change in sodium permeability of the plasma membrane surrounding the ROS disks (Tomita, 1970). Hagins (1972) and Cone (1973) discussed the role of calcium ions and other transmitters as mediators. A transmitter hypothesis postulated that an ionic transmitter, concentrated in the disk interior, was released through a pore in the disk membrane induced by a conformational change in rhodopsin upon bleaching. When bound to the plasma membrane, this putative transmitter might block sodium permeability and act as the initiation site for hyperpolarization of the cell.

Light-activated phosphodiesterase in the outer segment rapidly lowers cyclic GMP (cGMP) levels (Miki *et al.,* 1975). In an elegant physiological demonstration of the effects of cGMP, Miller and Nicol (1979) injected cGMP by microiontophoresis into living rod photoreceptors and demonstrated a marked inability of the injected cells to be hyperpolarized by light until destruction of the additional cGMP, presumably by activated phosphodiesterase, had lowered the concentration sufficiently. Earlier studies in this field were reviewed by Pober and Bitensky (1979).

Once light is captured by rhodopsin, the photopigment initiates a series of interactions with an enzymatic cascade which was discovered recently and has been the subject of lively debate and recent reviews (see Miller, 1981). Godchaux and Zimmerman (1979) identified a set of cytosol proteins which bound GDP and hydrolyzed GTP. Kühn (1980) also observed this group of proteins and showed that they were reversibly bound to light-exposed membranes—a property he exploited in their purification. That this complex of three peptide chains and guanyl nucleotides may serve as the nexus between the activated rhodopsin molecule and phosphodiesterase has been demonstrated by Fung and Stryer (1980) and Fung *et al.* (1981) who coined the term "transducin" to emphasize its possible function. Control of amplification of the input of energy by a single photon may occur by sequential collisions of an activated rhodopsin molecule with these enzyme complexes as the light-exposed rhodopsin translationally diffuses across the disk membrane (Liebman and Pugh, 1979; Fung and Stryer, 1980; Fung *et al.,* 1981). If activation of an enzymatic cascade is the mechanism of transduction, it remains to be demonstrated which molecules in the plasma membrane are affected by the fall in cyclic nucleotide levels and how the sensitivity of rods can vary over such a wide range of light intensity. Rhodopsin is phosphorylated upon light exposure by a protein kinase which is not regulated by cyclic nucleotides (Kuhn *et al.,* 1973) but may be regulated by free calcium (Hermolin, 1981). Two low-molecular-weight peptides in the ROS are phosphorylated in the dark and become dephosphorylated in the light and may be the site affected by cGMP in the cytosol (Polans *et al.,* 1979). A cGMP-dependent protein kinase may phosphorylate these two small ROS proteins in the dark (high cGMP) and be deactivated as phosphodiesterase lowers cGMP levels in the light (Farber *et al.,* 1979). In this review of membrane biosynthesis, it is not possible to do justice to this important aspect of photoreceptor function. For a recent compendium of reports from many of the groups active in this field see Miller (1981).

In addition, there is, one important biosynthetic problem associated with this group of cytosol proteins of the ROS, namely, the mechanism by which cells sequester cytosol proteins. Some of these proteins should theoretically be capable of diffusing freely through the interior of the connecting cilium if they are not membrane-bound. How the cell achieves high concentrations of these special proteins in the cytosol of the outer segment is not clear, since no membrane barriers divide the cytoplasm of the inner segment from the outer segment. One possibility is that most of these soluble proteins are transiently associated with the cytoplasmic side of the vesicle membrane and may be carried to the outer segment by weak associations with the opsin molecules of the vesicle or other components in the vesicle membrane. Rhodopsin kinase, transducin, and phosphodiesterase are associated with the cytoplasmic surface of ROS (see Miller, 1981; Kuhn and Hargrave, 1981). It is clear that the cytosol of the outer segment is by no means as complex as the cytosol released from retinal homogenates (Kuhn *et al.*, 1973; Papermaster *et al.*, 1975; Godchaux and Zimmerman, 1979). Concentrated samples of cytosol released from ROS contain only a few proteins detectable by SDS-gel electrophoresis, most of which are now identified and associated with the enzymes discussed above. Thus, the rod cell provides an extraordinary opportunity to determine not only the conditions required to obtain membrane polarity in differentiated neurons but also cytosol polarity.

The striking simplicity of ROS membranes contrasts markedly with the complexity of the adjacent membranes of the inner segment of photoreceptor cells and other neuronal elements of the retina. It illustrates most dramatically the problem of membrane compartmentalization in contiguous regions of the same cell. The fluid-mosaic model of membrane structure proposed by Singer and Nicolson (1972) emphasized the capacity of membrane proteins to redistribute throughout lipid bilayer membranes. Randomization of cell structures was thought to be avoided by stabilization of the position of some components of cell membranes by cytoskeletal proteins attached to the cytoplasmic side of the membrane protein. Indeed ROS presented one of the most dramatic examples of the fluidity of membranes, since rhodopsin could be shown both to rotate and translate across the disk membrane. Brown (1972), using polarized light, demonstrated the rotation of rhodopsin in the plane of the disk. Poo and Cone (1974) and Liebman and Entine (1974) demonstrated lateral migration of bleached rhodopsin from one side of a half-bleached rod to the other. Diffusion time was calculated by all these techniques and averaged approximately 5×10^{-9} cm^2/sec. Given this high mobility, the mechanism for restraint of rhodopsin's diffusion over the entire surface of the photoreceptor cell is still to be determined. There is direct membrane continuity of the plasma membrane surrounding the outer segment, the plasma membrane of the connecting cilium, the lateral plasma membrane, and the apical plasma membrane of the inner segment. The lower portion of the connecting cilium may serve as a barrier to the free diffusion of opsin once it has passed through to the outer segment (Schneider and Papermaster, 1981).

2. *The Large Intrinsic Membrane Protein of Outer Segments*

In addition to rhodopsin, vertebrate ROS contain an extremely large intrinsic membrane protein (Bownds *et al.,* 1971; Dreyer *et al.,* 1972; Papermaster and Dreyer, 1974). Despite the use of reducing agents and powerful denaturants, it has been impossible to show any subunit structure of this very large protein. In order to ascertain whether this protein was synthesized in concert with rhodopsin a study of radiolabeled amino acid incorporation into this protein was conducted. Subcellular fractions demonstrated only membrane transport of the newly synthesized large protein. The cytosol fraction did not contain the protein. The relative molar proportions of this large protein and opsin were estimated in three ways. At the end of several days of biosynthesis, the proportion of radioactivity incorporated into the large protein and into opsin was determined in isolated ROS. Their molar ratios were also estimated by staining of the SDS–polyacrylaminde gels and by amino acid analysis of the respective proteins. The amino acid composition of the large protein was not unusual, it showed no predominance of proline as seen in some other large proteins such as collagen (Papermaster *et al.*, 1976). Immunocytochemical studies on its localization demonstrated that the large protein was uniquely localized to the incisures and margins of ROS disks in red rods of frogs (Papermaster *et al.,* 1978b). With a more sensitive three-stage immunocytochemical technique, more recent studies demonstrated that this protein was also distributed on the incisures and margins of green ROS and on the margins of COS (Fig. 14). In addition, some cones and most green ROS

Fig. 14. Cross sections of red (RROS) and green (GROS) outer segments of *R. pipiens.* (A) Epon section illustrates RROS divided by deeply penetrating incisures (IN). GROS are scalloped by shorter incisures. PE processes containing melanosomes surround each outer segment. (Bar, 2 μm; $\times$7000). (B) BSA sections are labeled exclusively on incisures and margins of RROS and GROS by antibody to the large intrinsic membrane protein. Bound antibody was detected by the three-stage technique. (Bar, 2 μm; $\times$ 6400; inset $\times$ 10,000.)

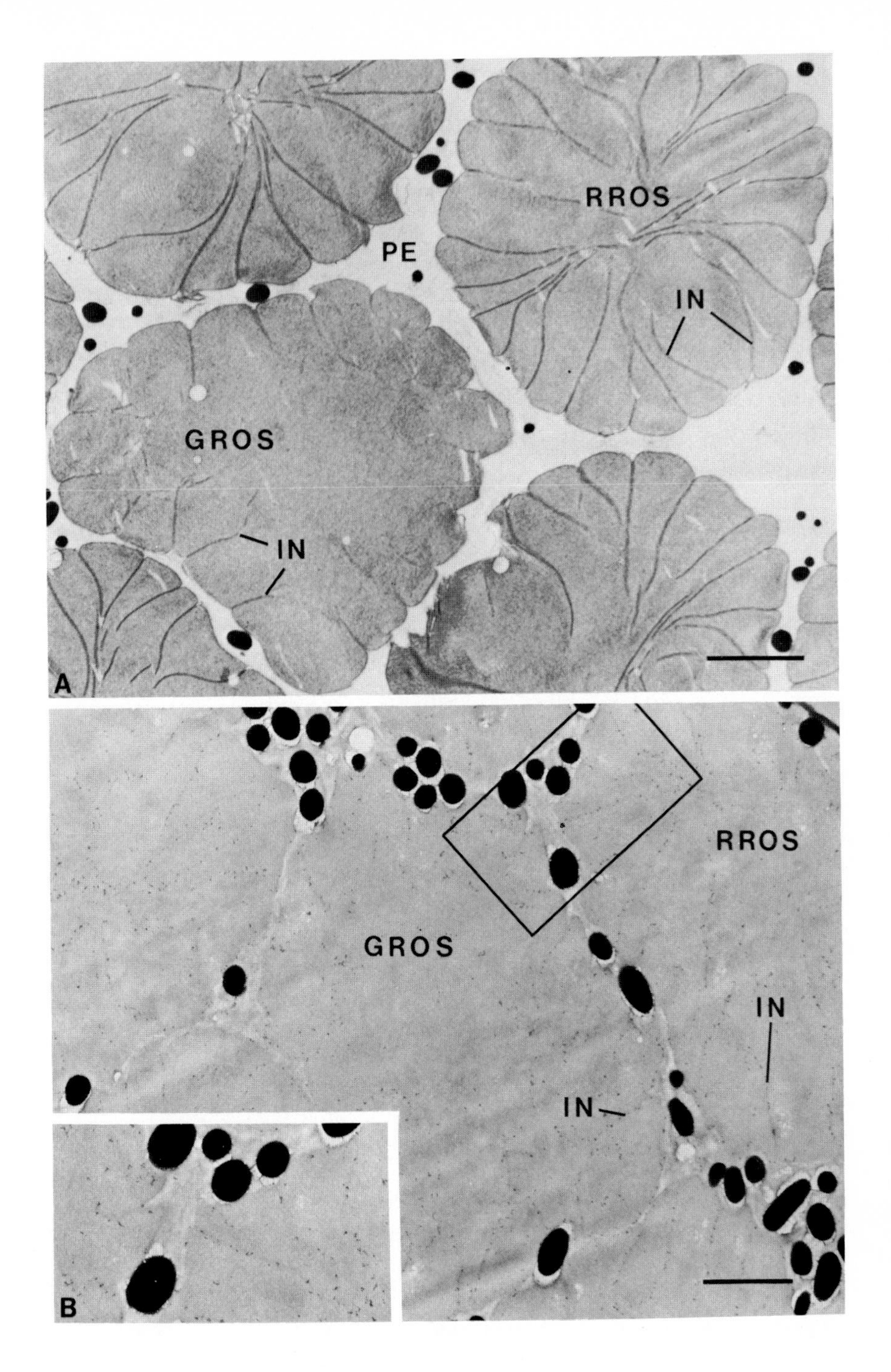

RROS
PE
IN
GROS
IN
A
RROS
GROS
IN
IN
B

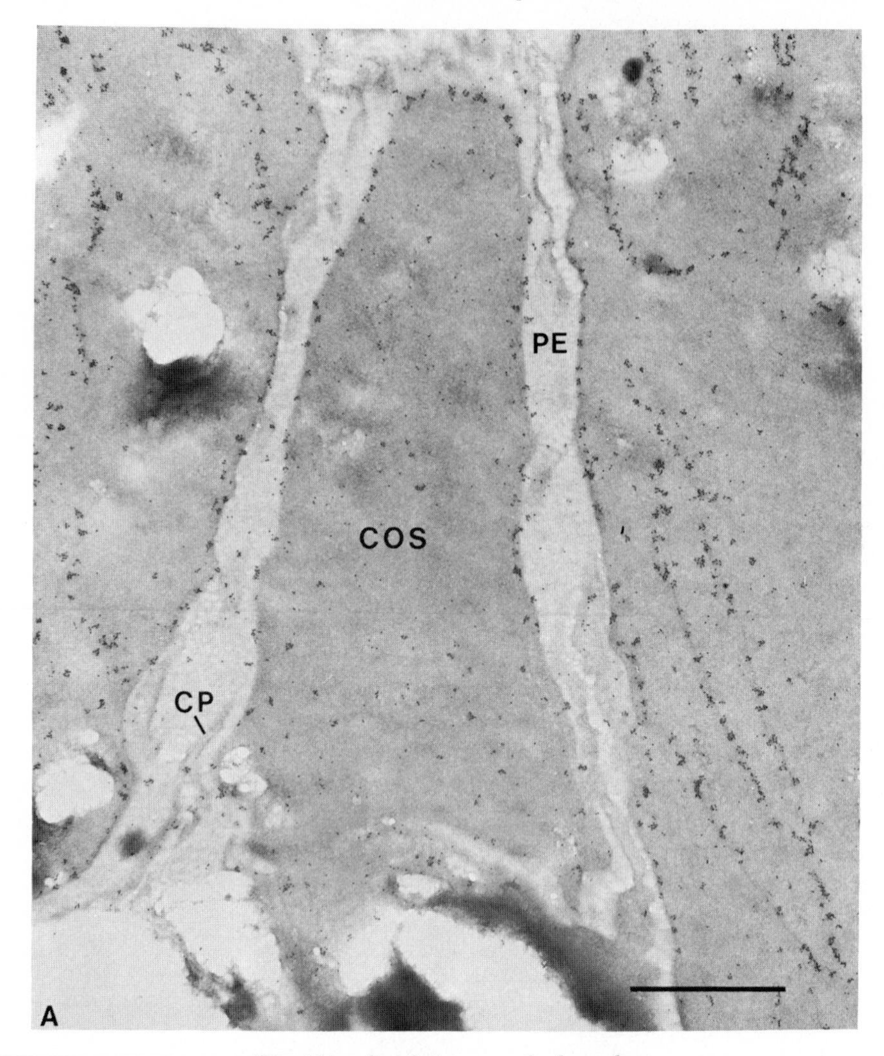

Fig. 15A. See facing page for legend.

contained label in the disk interior, which was probably a reflection of binding by a low-affinity antibody (Papermaster *et al.*, 1982).

Longitudinally sectioned and cross-sectioned COS were also labeled by the antibody to the large protein (Fig. 15). The interior of the COS lamella is analogous structurally to the margin of ROS disks. The portion of the lamella next to the outer surface, however, is folded back into the interior in a reverse fold when compared to the interior fold (Fig. 1). This difference between outside and inside folds in the cone

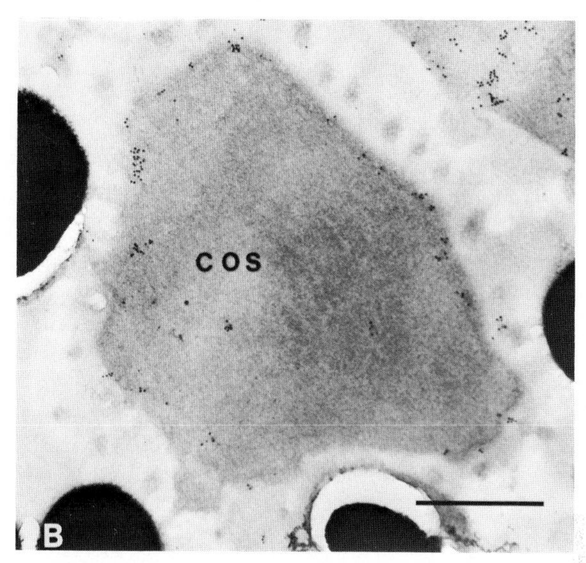

Fig. 15. Immunocytochemical localizaton of the large intrinsic membrane protein to the margins of COS. Longitudinal (A) and cross-sectioned (B) COS are labeled at a density comparable to that of adjacent RROS disk margins. The PE and calyceal processes (CP) are unlabeled. Thus this protein is present on the edges of disks and lamellae of all three photoreceptor classes of frog retinas. (A) Bar, 1 μm; × 21,000. (B) Bar, 0.5 μm; × 34,000.)

lamellae may explain why many cross-sectioned COS showed noncircumferential labeling around the entire cone. These results illustrate the presence of this protein in all three classes of photoreceptors in the frog and emphasize its role in the organization of the edges of these three photoreceptor outer segments.

Since nearly all vertebrates have incisures in their rods (Steinberg and Wood, 1975; Papermaster *et al.,* 1978b), it is likely that the incisures play an important role in the structure and function of these organelles. It is, however, not at all obvious what the function of these incisures or the large protein on them might be. Roof *et al.* (1980) have observed filaments spanning the gap between ROS disks and the disk margin. If these images correspond to fibrils or other molecules at the disk margin, the large protein may serve a structural role in maintaining space between the disks and stabilize the rod against torsional stress when the eye changes shape during mechanical compression and physiological accommodation. Other possible functions of incisures include slight expansion of the cytosol space within the ROS, which might permit more rapid equilibration of molecular events within the cytosol occurring after transduction of photon capture. The incisures may also serve to divide rod disks into smaller units of diffusion. Thus a rhodopsin molecule undergoing a random walk from one edge of the disk to the other cannot traverse laterally beyond the lobe within which it is confined. Since the average lobe of a frog red ROS disk is 1 μm in diameter, this restricts the diffusion distance to one-seventh the diameter of the average frog ROS. This reduces by a factor of 49 the average diffusion time between the collision of rhodopsin with a margin or incisure compared to the diffusion time if the 7-μm-wide disk contained no incisures. As a result, rhodopsin should collide with an incisure or margin in less than 100 msec during a random walk (Papermaster *et al.*, 1978b). Whether this reduction in diffusion time plays a role in restoring excitability to rhodopsin (dark adaptation), or in the transduction of photon capture energy to the plasma membrane, is still extremely speculative. Yau *et al.* (1977) showed that the generation of a graded potential by low levels of light exposure of single rods followed a constant delay in onset of potential change. A diffusion–collision model would predict a variation in time of onset of rise in potential if rhodopsins nearer the margin had been hit and collision were an initiating event of excitation. Thus diffusion of light-exposed rhodopsin to the disk margin or incisure is not likely to be the source of the time delay separating the onset of light exposure and the beginning of hyperpolarization. Steinberg *et al.* (1980) have presented a coherent model for the generation of disk margins and incisures and disk separation. A line of closure spreading laterally from the cilium to the other side of the outer segment could enclose the disks.

V. CIRCADIAN CONTROL OF DISK SHEDDING AND OF RHODOPSIN BIOSYNTHESIS

A. Light-Activated Shedding of Rod Tips

After traversing the inner segment and inserting into the disk membrane, opsin becomes the photopigment rhodopsin by the addition of 11-*cis*-retinaldehyde to its active site (O'Brien *et al.*, 1972; Hall and Bok, 1974). The molecular intermediates of the light cycle determined in the pioneering studies of Wald, Hubbard, and their colleagues have been summarized (Wald, 1968). Recent spectroscopic studies have emphasized the rapidity of the early steps in the bleaching of rhodopsin (Busch *et al.*, 1972). The enzymatic cascade initiated by bleaching rhodopsin not only has profound effects on the levels of cyclic nucleotides but also leads to extensive phosphorylation of serine and threonine side chains of opsin (Kuhn *et al.*, 1973).

Aside from these extraordinary cyclic modifications of opsin's structure and conformation, the protein is biosynthetically stable in the disk, as it resides within the center of the ROS. No evidence of protein turnover of labeled disks is observed in EM autoradiographs or biochemically (Hall *et al.*, 1969). Rather the entire disk as a unit is destroyed by shedding of the rod tip into the adjacent pigment epithelium where it is broken down in phagolysosomes (Young and Bok, 1969).

The shedding of outer segment membranes and rhodopsin biosynthesis do not proceed at a constant rate throughout the day. In a landmark study, LaVail (1976) showed that rat rods shed their tips in the morning and that constant darkness did not prevent the shedding the next day at about the same time. Thus shedding is a circadian event. Frog rod shedding is also triggered by light (Basinger *et al.*, 1976b). In contrast to the situation in rats, however, darkness inhibits shedding so that rods elongate. In all vertebrates studied, constant light also inhibits shedding and elongates rods at a more rapid rate (Currie *et al.*, 1978; Goldman *et al.*, 1980; Teirstein *et al.*, 1980). PE phagosomes usually contain about 50–100 disks, which constitutes most of the product of 1 day's synthesis. Additional shedding of terminal disks in small packets is seen in *X. laevis* (Besharse *et al.*, 1977b; Kinney and Fisher, 1978c) and probably occurs in other species.

Light not only stimulates rod disk shedding but also may accelerate both opsin biosynthesis and disk assembly. In *X. laevis* tadpoles maintained on a 12-hr-light–12-hr-dark cycle, rates of disk formation were determined by measuring the displacement of radiolabeled disks from the base of the outer segment. Tadpoles kept in the dark for 5 days displaced the labeled disks about half the distance compared to controls. When restored to light, a burst of disk formation led to displacement of the labeled disks to a position comparable to the distance seen in tadpoles maintained in cyclic light. Thus during the first day of light exposure, the dark-reared tadpoles generated nearly 4 days worth of membrane (Besharse *et al.*, 1977a). Still to be resolved is the locus of the light effect on disk generation. At least two loci can be considered: (1) At the level of the polysome, acceleration of opsin biosynthesis may occur either by increasing the content or rate of translation of opsin mRNA or the cell may prolong the duration of maximal synthesis beyond the first few hours in the morning, or (2) at the insertion site near the connecting cilium, continued synthesis at the normal rate each day may be followed by inhibition of disk membrane insertion. The latter alternative would predict a gathering of unassembled vesicles beneath the cilium in dark-reared tadpoles. Once the opsin gene has been isolated, it should become relatively straightforward to determine the levels of opsin mRNA quantitatively throughout the light cycle to evaluate any circadian changes.

One obstacle to *in vitro* study of the shedding process was overcome when Besharse *et al.* (1980) showed that high levels of bicarbonate ion (35 m*M*) were required in organ cultures of retina to induce shedding of tadpole outer segments. This advance should accelerate evaluation of the metabolic control of disk shedding and opsin biosynthesis. Once shed, the disks are rapidly ingested by the processes on the apical surface of the pigment epithelium and incorporated into phagolysosomes (Young and Bok, 1969; Hall *et al.*, 1969). Degradation is essentially completed in 1 day, so that the PE is prepared to resume this function the next day.

B. Shedding of Cones in Darkness

COS membranes are also continuously renewed. In contrast to the situation in rods, however, the radiolabeled proteins inserted into COS lamellae do not form a discrete band but distribute throughout the organelle. In view of the evidence for translation of rhodopsin in rod disks, it has been assumed that comparable diffusion of iodopsin throughout the connected lamellae in cones accounts for the difference. Shedding of the tips of cones was not consistently observed until Young (1977) correctly reasoned that both rods and cones may shed their tips at the end of their functional half-day. When he sought cone tip shedding in animals sacrificed in the evening, a maximum number of shed cone membranes was reproducibly obtained.

C. Retinal Dystrophies Associated with Disordered Photoreceptor Function or Pigment Epithelial Cell Function

Disorders of phagocytosis and degradation have been observed in humans and experimental animals. The RCS rat retina develops normally shortly after birth but gradually is destroyed by a relentless pileup of uningested ROS membranes which progressively fill the subretinal space (Dowling and Sidman, 1962). That the defect resides in the PE cell was established in tetraparental allophenic rats created by the fusion of blastulas of normal and affected rats (Mullen and LaVail, 1976). In regions of retina containing the PE from normal strains, ROS from both strains were shed normally, ingested, and destroyed, while beneath the retinal regions covered by RCS PE cells, uningested disks accumulated. *In vitro* evidence supporting this conclusion was obtained by Edwards and Szamier (1977) and by Hall (1978) who showed that primary cultures of normal and RCS rat PE cells could be used to evaluate the ingestion of ROS fragments. Both cell lines could ingest inert latex spheres comparably. In an innovative quantitative study, Hall and Chaitin (1981) determined the degree of ingestion by labeled ROS fragments with anti-ROS antiserum. Those on the surface could be detected with rhodamine-conjugated anti-rabbit immunoglobulin without disruption of cell membranes. Those inside were then revealed by disruption of the PE cell membrane with acetone and reaction with the same anti-ROS antiserum and a fluorescein conjugate. Comparison of micrographs before and after disruption simplified the analysis of ingestion considerably. In this way, RCS PE cultures were shown to bind ROS to a comparable degree but to ingest them at a markedly slower rate and to ingest only about 10% of the bound fragments when compared to the rapid and nearly complete ingestion by normal PE cells. Battelle and LaVail (1980) observed no defects in the rates of opsin synthesis in RCS rats until degeneration had begun to affect cell vaiability in older rats.

The RCS rat lesion differs from most human retinal dystrophies, since the onset is relatively early. A new defect in Wag/Rij rats was described by Lai *et al.* (1975), which appears to consist of a slow degeneration of individual photoreceptors throughout adult life. This model deserves to be the object of further study.

Disorders of cyclic nucleotide regulation have been identified in mice and humans. The *rd* allele in mice inhibits full development of photoreceptors. Outer segments begin to develop, but by the eighth postnatal day a progressive deterioration of the structure of subcellular organelles begins, which completely destroys retinal photosensitivity by 1 month. Lolley and Farber (1975, 1980) have shown that this disorder may be the result of an accumulation of high levels of cGMP, possibly as a result of a labile phosphodiesterase. Chader *et al.* (1980) have also observed a defect in development or regulation of phosphodiesterase in some strains of Irish setters.

Both humans and English setters have a defect in phagolysosomal function that leads to the gradual accumulation of incompletely digested debris generically called lipofuscin and ceroid. The disorder, termed ceroid lipofuscinosis, is characterized by the presence of multilamellar electron-dense membrane-bound deposits in the retina and PE cytoplasm as well as the nervous system and viscera. As aging continues, the accumulating mass is accompanied by degeneration of both PE and rod cells (Armstrong *et al.*, 1980). In the miniature poodle, a general progressive retinal atrophy of slower onset mimics the clinical features of human retinitis pigmentosa (Aguirre and Rubin, 1972).

VI. GENERAL PRINCIPLES OF MEMBRANE BIOSYNTHESIS SHARED BY PHOTORECEPTORS AND OTHER CELLS

Since the earliest studies on membrane biosynthesis by ultrastructural techniques, the dictum, "All membranes come from membranes," has dominated concepts of biogenesis. Now, more than two decades since extrapolation of the findings in pancreatic zymogen granule biosynthesis to other cell systems (Palade, 1959) the photoreceptor has shown rather faithful adherence to this prediction. Little about the early stages of opsin synthesis and transport through the Golgi apparatus is unique except for the absence of an N-terminal signal peptide and the small size of opsin's oligosaccharide chains. Its orderly vectorial transport to the apex of the inner segment shares with other polarized cells the avoidance of a random, haphazard distribution of membrane proteins throughout the cell. Not only do other neurons and epithelial cells confront similar tasks to become and remain polarized, but viruses which selectively mature on one surface rather than around the entire surface of the infected cell also are transported and shed in a polarized fashion (Boulin and Sabatini, 1978). However, when cell contacts are disrupted, virus shedding can become depolarized and circumferential. Thus, intercellular contacts can markedly affect the phenotypic expression of the cell's potential for polarized organization. Experimental manipulation of the intercellular contacts of developing photoreceptor cells is a potential probe for evaluating this factor in the polarized biosynthesis of outer segment membranes.

Photoreceptors differ from other cells in their elaboration of a unique membranous organelle at the end of an apical cilium. Until the outer segment begins to organize, the photoreceptor cell resembles other sensory receptors in the olfactory bulb and ear. The transcription and translation of the rhodopsin and iodopsin genes are unique functions of the photoreceptor cells (with the possible exception of pinealocytes and iris epithelium). The vectorial transport of opsin to the outer segment is accompanied by a parallel concentration of specific cytosol proteins in the outer segment. The photoreceptor therefore remains an extraordinary model for developing new strategies for the study of cell polarization of both membranes and cytosol.

Finally one is tempted to speculate on the origins of the opsin gene, especially since little direct information is known. As the gene is isolated and complementary DNA probes are used to search for homologous genes, it should be possible to conduct this search in an exact way. Although bacteria and invertebrates have also evolved retinaldehyde-linked photopigments, no evidence of immunochemical homology of the proteins has been observed. Thus these proteins seem to be examples of convergent evolution in their retinaldehyde-binding sites,probably as a result of the unique properties of retinaldehyde as a chromophore.

The diversity of rhodopsins and iodopsins in different species begs a molecular analysis. Fish adapted for various depths in water have rhodopsins with quite variable wavelengths of maximal absorption (Lythgoe, 1972). The rhodopsin absorption spectra obtained from deep-diving fish retinas may be displaced 80 nm to the blue of the 500 nm peak of nearly all rhodopsins from land-based vertebrates. Despite the diversity of absorption maxima in fish opsins, there is nearly no diversity in the opsins of land vertebrates. Moreover, antibodies to frog opsin cross-react with fish and mammalian opsins, which indicates a considerable evolutionary conservation of some peptide segments of opsin structure. Both aquatic and land vertebrates have also diversified their visual pigments to construct their families of iodopsins. In the frog many of the cone photopigments cross-react with antiopsin antibodies despite their manipulation of the polypeptide sequence of the retinaldehyde-binding site in order to shift the spectral sensitivity. One possible mechanism for creating polypeptide domains which are conserved (the sites cross-reactive with antisera) and domains which are variable (the retinaldehyde-binding sites of rhodopsins and iodopsins) has been shown to operate in the generation of diversity in immunoglobulins. The gene regions coding for both constant and variable portions of heavy and light chains are separated by long, untranslated DNA segments. During ontogeny and specific expansion of lymphocyte clones the gene segments translocate to become more closely apposed. The variable regions are created during the linking of one of many variable gene (V-gene)segments to the constant gene (C-gene) segment (Maki *et al.*, 1980; Early *et al.*, 1980). Could there be an analogous set of V-gene segments coding for the retinaldehyde-binding site which move and link to C-gene segments coding for the constant regions of opsins and iodopsins which account for their cross-reactivity? It is a most exciting prospect to anticipate the analysis of such hypotheses by direct genetic sequence determination and detection of homologous gene systems.

ACKNOWLEDGMENTS

The research reported from the authors' laboratory was supported in part by USPHS grants EY-3239, EY-845, GM-21714, the American Cancer Society grant BC-129, and Josiah Macy Jr. Foundation, and the Veterans Administration.

REFERENCES

Adams, A. J., Tanaka, M., and Shichi, H. (1978). Concanavalin A binding to rod outer segment membranes: Usefulness for preparation of intact disks. *Exp. Eye Res.* **27,** 595–605.

Aguirre, G., and Rubin, L. (1972). Progressive retinal atrophy in the miniature poodle: An electrophysiologic study. *J. Am. Vet. Med. Assoc.* **160,** 191–201.

Al-Mahdawi, S. A. H., Converse, C. A., and Cooper, A. (1981). Partial proteolysis of cattle and frog rhodopsins. *Biochem. Soc. Trans.* **9,** 319–320.

Anderson, R. E., and Maude, M. D. (1970). Phospholipids of bovine rod outer segments. *Biochemistry* **9,** 3624–3628.

Anderson, R. E., Kelleher, P. A., and Maude, M. B. (1980a). Metabolism of phosphatidylethanolamine in the frog retina. *Biochim. Biophys. Acta* **620,** 227–235.

Anderson, R. E., Kelleher, P. A., Maude, M. B., and Maida, T. M. (1980b). Synthesis and turnover of lipid and protein components of frog retinal rod outer segments. *Neurochemistry* **1,** 29–42.

Anderson, R. E., Maude, M. B., and Kelleher, P. A. (1980c). Metabolism of phosphatidylinositol in the frog retina. *Biochim. Biophys. Acta* **620,** 236–246.

Anderson, R. E., Maude, M. B., Kelleher, P. A., Maida, T. M., and Basinger, S. F. (1980d). Metabolism of phosphatidylcholine in the frog retina. *Biochim. Biophys. Acta* **620,** 212–226.

Andrews, L. (1981). Freeze fracture studies of vertebrate photoreceptor membranes. *In* "Structure of the Eye. IV. Proceedings of the Fourth International Symposium on the Structure of the Eye" (J. G. Hollyfield and E. A. Vidrio, eds.), Elsevier, Amsterdam (in press).

Armstrong, D, Neville, H. Siaktos, A., Wilson, B. Wehlins, C., and Koppans, N. (1980). Morphological and biochemical abnormalities in a model of retinal degeneration: Canine ceroid-lipofuscinosis (CCL). *Neurochemistry* **1,** 405–426.

Basinger, S., and Hoffman, R. (1976). Phosphatidyl choline metabolism in the frog rod photoreceptor. *Exp. Eye Res.* **23,** 117–126.

Basinger, S., Bok, D., and Hall, M. O. (1976a). Rhodopsin in the rod outer segment plasma membrane. *J. Cell Biol.* **69,** 29–42.

Basinger, S., Hoffman, R., and Matthes, M. (1976b). Photoreceptor shedding is initiated by light in the frog retina. *Science* **194,** 1074–1076.

Battelle, B.-A. and LeVail, M. M. (1980). Protein synthesis in retinas of rats with inherited retinal dystrophy. *Exp. Eye Res.* **31,** 251–269.

Besharse, J. C. (1980). Light and membrane biogenesis in rod photoreceptors of vertebrates. *In* "The Effects of Constant Light on Visual Processes" (T. P. Wiliiams and B. N. Baker, eds.), pp. 409–431. Plenum, New York.

Besharse, J. C., and Forestner, D. M. (1981). Horseradish peroxidase uptake by rod photoreceptor inner segments accompanies outer segment disk assembly. *39th Ann. Proc. Electron Microsc. Soc. Am., Atlanta,* pp. 486–487.

Besharse, J. C., and Pfenninger, K. H. (1980). Membrane assembly in retinal photoreceptors. I. Freeze-fracture analysis of cytoplasmic vesicles in relationship to disc assembly. *J. Cell Biol.* **87,** 451–463.

Besharse, J. C., Hollyfield, J. G., and Rayborn, M. E. (1977a). Photoreceptor outer segments: Accelerated membrane renewal in rods after exposure to light. *Science* **196,** 536–538.

Besharse, J. C., Hollyfield, J. G., and Rayborn, M. E. (1977b). Turnover of rod photoreceptor outer segments. II. Membrane addition and loss in relationship to light. *J. Cell Biol.* **75,** 507–527.

Besharse, J. C., Terrill, R. O., and Dunis, D. A. (1980). Light-evoked disk shedding by rod photoreceptors *in vitro:* Relationship to medium, bicarbonate concentration. *Invest. Ophthalmol. Visual Sci.* **19,** 1512–1517.

Bibb, C., and Young, R. W. (1974). Renewal of fatty acids in the membranes of visual cell outer segments. *J. Cell Biol.* **61,** 327–343.

Blobel, G. (1980). Intracellular protein topogenesis. *Proc. Natl. Acad. Sci. U.S.A.* **77,** 1496–1500.

Blobel, G., and Dobberstein, B. (1975). Transfer of proteins across membranes. I. Presence of proteolytically processed and unprocessed nascent immunoglobulin light chains on membrane bound polysomes of murine myeloma. *J. Cell Biol.* **67,** 835–851.

Blobel, G., and Sabatini, D. (1971). Ribosome-membrane interaction in eukaryotic cells. *Biomembranes* **2,** 193–195.

Bodley, H. D., Greiner, J. V., Weldman, T. A., and Greiner, C. A. M. (1981). Ciliogenesis in retinal photoreceptors. *Invest. Ophthalmol. Visual Sci. Suppl. 1,* **20,** 148.

Bok, D., and Hall, M. (1971). The role of the pigment epithelium in the etiology of inherited retinal dystrophy in the rat. *J. Cell Biol.* **49,** 664–682.

Bok, D., Basinger, S., and Hall, M. (1974). Autoradiographic and radiobiochemical studies on the incorporation of [6-^{3}H]glucosamine into frog rhodopsin. *Exp. Eye Res.* **18,** 225–240.

Bok, D., Hall, M. O., and O'Brien, P. (1977). The biosynthesis of rhodopsin as studied by membrane renewal in rod outer segments. *Int. Cell Biol. Pap. Int. Congr., 1st, 1976* pp. 608–617.

Boulan, E. R., and Sabatini, D. D. (1978). Asymmetric budding of viruses in epithelial monolayers: A model system for study of epithelial polarity. *Proc. Natl. Acad. Sci. U.S.A.* **75,** 5071–5075.

Bownds, D., Gordon-Walker, A., Guide-Huguenin, A. C., and Robinson, W. (1971). Characterization and analysis of frog photoreceptor membranes. *J. Gen. Physiol.* **58,** 225–237.

Bridges, C. D. B. (1981). Lectin receptors of rods and cones visualization by fluorescent label. *Invest. Ophthalmol. Visual Sci.* **20,** 8–16.

Bridges, C. D. B., and Fong, S. L. (1979). Different distribution of receptors of peanut and ricin agglutinins between inner and outer segments of rod cells. *Nature (London)* **282,** 513–515.

Brown, P. K. (1972), Rhodopsin rotates in the visual receptor membrane. *Nature (London) New Biol.* **236,** 35–38.

Bunt, A. H., and Klock, I. B. (1980a). Fine structure and autoradiography of retinal cone outer segments in goldfish and carp. *Invest. Ophthalmol. Visual Sci.* **19,** 707–719.

Bunt, A. H., and Klock, I. B. (1980b). Comparative study of ^{3}H-fucose incorporation into vertebrate photoreceptor outer segments. *Vision Res.* **20,** 739–747.

Bunt, A. H., Saari, J. C., and Barth, R. (1981). Characterization of cone fucosylated protein. *Invest. Ophthalmol. Visual Sci. Suppl. 1,* **20,** 151.

Busch, G. E., Applebury, M. L., Lamola, A., and Rentzepis, P. M. (1972). Formation and decay of prelumirhodopsin at room temperature. *Proc. Natl. Acad. Sci. U.S.A.* **69,** 2802–2806.

Chader, G., Liu, Y., O'Brien, P., Fletcher, R., Krishna, G., Asuirre, G., Farber, D., and Lolley, R. (1980). Cyclic GMP phosphodiesterase activator: Involvement in hereditary retinal degeneration. *Neurochemistry* **1,** 441–458.

Chen, Y. S., and Hubbell, W. L. (1973). Temperature- and light-dependent structural changes in rhodopsin-lipid membranes. *Exp. Eye Res.* **17,** 517–532.

Chen, W. W., and Lennarz, W. J. (1978). Enzymatic excision of glucosyl units linked to the oligosaccharide chains of glycoproteins. *J. Biol. Chem.* **253,** 5780–5785.

Cohen, A. I. (1960). The ultrastructure of the rods in the mouse retina. *Am. J. Anat.* **107,** 23–48.

Cohen, A. I. (1968). New evidence supporting the linkage to extracellular space of outer segment saccules of frog cones but not rods. *J. Cell Biol.* **37,** 424–444.

Cone, R. A. (1973). The internal transmitter model for visual excitation: Some quantitative implications. *In* "Biochemistry and Physiology of Visual Pigments" (H. Langer, ed.), pp. 275–284. Springer-Verlag, Berlin and New York.

Corless, J. M., Cobbs, W. H., III, Costello, M. J., and Robertson, J. D. (1976). On the assymetry of frog retinal rod outer segment membranes. *Exp. Eye Res.* **23,** 295–324.

Crain, R. C., Marinetti, G. V., O'Brien, D. F. (1978). Topology of aminophospholipids in bovine retinal rod outer segment disk membranes. *Biochemistry* **17,** 4186–4192.

Currie, J. R., Hollyfield, J. G., and Rayborn, M. E. (1978). Rod outer segments elongate in constant light: Darkness is required for normal shedding. *Vision Res.* **18,** 995–1003.

deGrip, W. J., Daemen, F. J. M., and Bonting, S. L. (1980). Isolation and purification of bovine rhodopsin. *Methods Enzymol.* **67,** 301–320.

Dowling, J. E., and Sidman, R. L. (1962). Inherited retinal dystrophy in the rat. *J. Cell Biol.* **14,** 73–109.

Drenthe, E. H., Bonting, S. L., and Daemen, F. J. M. (1980a). Transbilayer distribution of phospholipids in photoreceptor membrane studied with various phospholipases. *Biochim. Biophys. Acta* **603,** 117–129.

Drenthe, E. H., Klompmakers, A. A., Bonting, S. L., and Daemen, F. J. M. (1980b). Transbilayer distribution of phospholipids in photoreceptor membrane studied by trinitrobenzene sulfonate alone and in combination with phospholipase D. *Biochim. Biophys. Acta* **603,** 130–141.

Drenthe, E. H., Klompmakers, A. A., Bonting, S. L., and Daemen, F. J. M. (1981). Transbilayer distribution of phospholipid fatty acyl chains in photoreceptor membrane. *Biochim. Biophys. Acta.* **641,** 377–385.

Dreyer, W. J., Papermaster, D. S., and Kuhn, H. (1972). On the absence of ubiquitous structural protein subunits in biological membranes. *Ann. N.Y. Acad. Sci.* **195,** 61–74.

Dudley, P. A., and Anderson, R. E. (1978). Phospholipid transfer protein from bovine retina with high affinity toward retinal rod disc membranes. *FEBS Lett.* **95,** 57–60.

Early, P., Rogers, J., Davis, M., Calame, K., Bond, M., Wall, R., and Hoods, L. (1980). Two mRNAs can be produced from a single immunoglobulin mu gene by alternative RNA processing pathways. *Cell* **20,** 313–319.

Edwards, R. B., and Szamier, R. B. (1977). Defective phagocytosis of isolated rod outer segments by RCS rat pigment epithelium in culture. *Science* **197,** 1001–1003.

Fager, L. Y., and Fager, R. S. (1978). Separation of rod and cone pigments from the chicken retina. *Invest. Ophthalmol. Visual Sci. Suppl. 1,* **17,** 126.

Farber, D. B., Brown, R. M., and Lolley, R. N. (1979). Cyclic nucleotide dependent protein kinase and the phosphorylation of endogenous proteins of retinal rod outer segments. *Biochemistry* **18,** 370–378.

Fong, S.-L. (1978). Partial structural characterization of bovine rhodopsin: A photoreceptor protein. PhD Thesis, Southern Illinois Univ., Carbondale, Illinois.

Fukuda, M. N., Papermaster, D. S., and Hargrave, P. A. (1979). Rhodopsin carbohydrate:

Structure of small oligosaccharides attached at two sites near the NH_2 terminus. *J. Biol. Chem.* **254,** 8201–8207.

Fung, B. K. K., and Hubbell, W. L. (1978a). Organization of rhodopsin in photoreceptor membranes. I. Proteolysis of bovine rhodopsin in native membranes and the distribution of sulfhydryl groups in the fragments. *Biochemistry* **17,** 4396–4402.

Fung, B. K. K., and Hubbell, W. L. (1978b). Organization of rhodopsin in photoreceptor membranes. II. Transmembrane organization of bovine rhodopsin: Evidence from proteolysis and lactoperoxidase-catalyzed iodination of native and reconstituted membranes. *Biochemistry* **17,** 4403–4410.

Fung, B. K. K., and Stryer, L. (1980). Photolyzed rhodopsin catalyzes the exchange of GTP for bound GDP in retinal rod outer segments. *Proc. Natl. Acad. Sci. U.S.A.* **77,** 2500–2504.

Fung, B. K. K., Hurley, J. B., and Stryer, L. (1981). Flow of information in the light-triggered cyclic nucleotide cascade of vision. *Proc. Natl. Acad. Sci. U.S.A.* **78,** 152–156.

Godchaux, W., III, and Zimmerman, W. F. (1979). Membrane-dependent guanine nucleotide binding and GTPase activities of soluble protein from bovine rod cell outer segments. *J. Biol. Chem.* **16,** 7874–7884.

Goldman, A. I., Teirstein, P. S., and O'Brien, P. J. (1980). The role of ambient lighting in circadian disc shedding in the rod outer segment of the rat retina. *Invest. Ophthalmol. Visual Sci.* **19,** 1257–1267.

Goldman, B. M., and Blobel, G. (1981). *In vitro* biosynthesis, core-glycosylation and membrane integration of opsin. *J. Cell Biol.* **90,** 236–242.

Grossman, A., Bartlett, S., and Chua, N.-H. (1980). Energy-dependent uptake of cytoplasmically synthesized polypeptides by chloroplasts. *Nature (London)* **285,** 625–628.

Hagins, W. A. (1972). The visual process: Excitatory mechanisms in the primary receptor cells. *Annu. Rev. Biophys. Bioeng.* **1,** 131–158.

Hall, M. O. (1978). Phagocytosis of light and dark-adapted rod outer segments by cultured pigment epithelium. *Science* **202,** 526–528.

Hall, M. O., and Bok, D. (1974). Incorporation of [^{3}H]vitamin A into rhodopsin in light- and dark-adapted frogs. *Exp. Eye Res.* **18,** 105–117.

Hall, M. O., and Bok, D. (1976). Reduction of the retinal-opsin linkage in isolated frog retinas. *Exp. Eye Res.* **22,** 595–609.

Hall, M. O., and Chaitin, M. H. (1981). The engulfment phase of phagocytosis is deficient in the dystrophic (RCS) rat pigment epithelium. *Invest. Ophthalmol. Visual Sci. Suppl. 1* **20,** 96.

Hall, M. O., Bok, D., and Bacharach, A. D. E. (1969). Biosynthesis and assembly of the rod segment membrane system: Formation and fate of visual pigment in the frog retina. *J. Mol. Biol.* **45,** 397–406.

Hall, M. O. Basinger, S. F., and Bok, D. (1973). Studies on the assembly of rod outer segment disc membranes. *Biochem. Physiol. Visual Pigm. Symp.,* 1972 pp. 319–326.

Hargrave, P. A. (1977). The amino-terminal tryptic peptide of bovine rhodopsin. *Biochim. Biophys. Acta* **492,** 83–94.

Hargrave, P. A. (1982). Rhodopsin chemistry, structure and topography. *In* "Progress in Retinal Research" (N. Osborne and G. Chader, eds.). Pergamon, New York (in press).

Hargrave, P. A., and Fong, S.-L. (1977). The amino- and carboxy-terminal sequence of bovine rhodopsin. *J. Supramol. Struct.* **6,** 559–570.

Hargrave, P. A., Fong, S. L., McDowell, J. H., Mas, M. T., Curtis, D. R., Wang, J. K., Jiuszczak, E., and Smith, D. P. (1980). The partial primary structure of bovine rhodopsin and its topography in the retinal rod cell disc membrane. *Neurochemistry* **1,** 231–244.

Heller, J., and Lawrence, M. A. (1970). Structure of the glycopeptide from bovine visual pigment 500. *Biochemistry* **9,** 864–869.

Hendriks, Th., Klompmakers, A. A., Daemen, F. J. M., and Bonting, S. L. (1976). Biochemical aspects of the visual process. XXXII. Movement of sodium ions through bilayers composed of retinal and rod outer segment lipids. *Biochim. Biophys. Acta* **433,** 271–281.

Hermolin, J. (1981). Light-influenced outer segment protein phophorylation in frog retina: Effects of cyclic nucleotides and Ca. *In* "Cold Spring Harbor Conference on Cell Proliferation" (O. M. Rosen and E. G. Krebs, eds.), Vol. 8. (in press).

Hirano, H. Parkhouse, B., Nicolson, G. L., Lennox, E. S., and Singer, S. J. (1972). Distribution of saccharide residues on membrane fragments from a myeloma cell homogenate: Its implications for membrane biogenesis. *Proc. Natl. Acad. Sci. U.S.A.* **69,** 2945–2949.

Hollyfield, J. G., Besharse, J. C., and Rayborn, M. E. (1977). Turnover of rod photoreceptor outer segments. I. Membrane addition and loss in relationship to temperature. *J. Cell Biol.* **75,** 490–506.

Holtzman E., and Mercurio, A. M. (1980). Membrane circulation in neurons and photoreceptors: Some unresolved issues. *Int. Rev. Cytol.* **67,** 1–67.

Ilincheta de Boschero, M. G., Guisto, N. M., and Bazan, N. G. (1980). Biosynthesis of membrane lipids in the retina: Subcellular distribution and propanalol action on phosphatidic acid, phosphatidyl serine and phosphatidyl ethanolamine. *Neurochemistry* **1,** 17–28.

Jan, L. Y., and Revel, J.-P. (1974). Ultrastructural localization of rhodopsin in the vertebrate retina. *J. Cell Biol.* **62,** 257–273.

Jan, L. Y., and Revel J.-P. (1975). Hemocyanin antibody labeling of rhodopsin in the vertebrate retina. *J. Supramol. Struct.* **3,** 61–66.

Katz, F. N., Rothman, J. E., Lingappa, V. R., Blobel, G., and Lodish, H. F. (1977). Membrane assembly *in vitro:* Synthesis glycosylation and assymetric insertion of a transmembrane protein. *Proc. Natl. Acad. Sci. U.S.A.* **74,** 3278–3282.

Kean, E. L. (1977a). Mannosyl transferases of the retina: Mannolipid and complex glycan biosynthesis. I. Kinetic properties: Product identification. *Exp. Eye Res.* **25,** 405–417.

Kean, E. L. (1977b). Concerning the catalytic hydrogenation of polyprenyl phosphate-mannose synthesized by the retina. *J. Biol. Chem.* **252,** 5619–5621.

Kean, E. L. (1977c). GDP-mannose-polyprenyl phosphate mannosyltransferases of the retina. *J. Biol Chem.* **252,** 5622–5629.

Kean, E. L. (1977d). The biosynthesis of mannolipids and mannose-containing complex glycans by the retina. *J. Supramol. Struct.* **7,** 381–385.

Kean, E. L. (1980a). Stimulation by GDP-mannose of the biosynthesis of *N*-acetylglucosaminylpyrophosphoryl polyprenols by the retina. *J. Biol. Chem.* **255,** 1921–1927.

Kean, E. L. (1980b). The lipid intermediate pathway in the retina for the activation of carbohydrates involved in the glycosylation of rhodopsin. *Neurochemistry* **1,** 59–68.

Kean, E. L., and Plantner, J. J. (1976). Biosynthesis of mannose-containing heteropolymers by cell-free preparations of bovine retina. *Exp. Eye Res.* **23,** 89–104.

Kinney, M. S., and Fisher, S. K. (1978a). The photoreceptors and pigment epithelium of the adult *Xenopus* retina: Morphology and outer segment renewal. *Proc. R. Soc. London Ser. B* **201,** 131–147.

Kinney, M. S., and Fisher, S. K. (1978b). The photoreceptors and pigment epithelium of the larval *Xenopus* retina: Morphogenesis and outer segment renewal. *Proc. R. Soc. London Ser. B* **201,** 149–167.

Kinney, M. S., and Fisher, S. K. (1978c). Changes in length and disk shedding rate of *Xenopus* rod outer segments associated with metamorphosis. *Proc. R. Soc. London Ser. B.* **201,** 169–177.

Kraehenbuhl, J. P., and Jamieson, J. D. (1976). Enzyme labeled antibody markers for electron microscopy. *Methods Immunol. Immunocytochem.* **5,** 482–495.

Kraehenbuhl, J. P., Weibel, E. R., and Papermaster, D. S. (1978). "Quantitative Immunocytochemistry at the Electron Microscope" (W. Knapp, K. Holubar, and G. Wick, eds.), pp. 245–253. Elsevier/North Holland, Amsterdam.

Kühn, H. (1980). Light-induced reversible binding of proteins to bovine photoreceptor membranes. Influence of nucleotides. *Neurochemistry* **1,** 269–285.

Kühn, H., and Hargrave, P. A. (1981). Light-induced binding of guanosinetriphosphatase to bovine photoreceptor membranes: Effect of limited proteolysis of the membranes. *Biochemistry* **20,** 2410–2417.

Kühn, H., Cook, J. H., and Dreyer, W. J. (1973). Phosphorylation of rhodopsin in bovine photoreceptor membranes: A dark reaction after illumination. *Biochemistry* **12,** 2495–2499.

Lai, Y. L., Jacoby, R. O., Jonas, A. M., and Papermaster, D. S. (1975). A new form of hereditary retinal degeneration in Wag/Rij rats. *Invest. Ophthalmol.* **14,** 62–67.

Laties, A. M., Bok, D., and Liebman, P. (1976). Procion yellow: A marker dye for outer segment disc patency and for rod renewal. *Exp. Eye Res.* **23,** 139–148.

LaVail, M. M. (1976). Disc shedding in rat retina: Relationship to cyclic lighting. *Science* **194,** 1071–1074.

Liang, C.-J., Yamashita, K., Schichi, H., Muellenberg, C. G., and Kobata, A. (1979). Structure of the carbohydrate moiety of bovine rhodopsin. *J. Biol. Chem.* **254,** 6414–6418.

Liebman, P. A., and Entine, G. (1974). Lateral diffusion of visual pigment in photoreceptor disk membranes. *Science* **185,** 457–459.

Liebman, P. A., and Pugh, E. N., Jr. (1979). The control of phosphodiesterase in rod disk membranes: Kinetics, possible mechanisms and significance for vision. *Vision Res.* **19,** 375–380.

Lingappa, V. R., Lingappa, J. R., and Blobel, G. (1980). Signal sequences for early events in protein secretion and membrane assembly. *Ann. N.Y. Acad. Sci.* **343,** 3656–361.

Lolley, R. N., and Farber, D. B. (1975). Cyclic nucleotide phosphodiesterases in dystrophic rat retinas: Guanosine 3′,5′-cyclic monophosphate anomalies during photoreceptor cell degeneration. *Exp. Eye Res.* **20,** 585–597.

Lolley, R. N., and Farber, D. B. (1980). Cyclic GMP metabolic defects in inherited disorders of rd mice and RCS rats. *Neurochemistry* **1,** 427–440.

Lythgoe, J. N. (1972). List of vertebrate visual pigments. *In* "Handbook of Sensory Physiology." VII/1. Photochemistry of Vision (H. J. A. Dartnall, ed.), pp. 604–624. Springer-Verlag, Berlin and New York.

McConnell, D. G. (1965). The isolation of retinal rod outer segment fragments *J. Cell Biol.* **27,** 459–473.

McLean, J. D., and Singer, S. J. (1970). A general method for the specific staining of intracellular antigens with ferritin-antibody conjugates. *Proc. Natl. Acad. Sci. U.S.A.* **65**, 122–128.

Maki, R., Traunecker, A., Sakano, H., Roeder, W., and Tonegawa, S. (1980). Exon shuffling generates an immunoglobulin heavy chain gene. *Proc. Natl. Acad. Sci. U.S.A.* **77**, 2138–2142.

Masland, R. H., and Mills, J. W. (1980). Choline accumulation by photoreceptor cells of the rabbit retina. *Proc. Natl. Acad. Sci. U.S.A.* **77**, 1671–1675.

Matsusaka, T. (1974). Membrane particles of the connecting cilium. *J. Ultrastruct. Res.* **48**, 305–312.

Miki, N., Baraban, J. M., Keirns, J. J., Boyce, J. J., and Bitensky, M. W. (1975). Purification and properties of the light-activated cyclic nucleotide phosphodiesterase of rod outer segments. *J. Biol. Chem.* **250**, 6320–6327.

Miller, W. H., ed. (1981). "Molecular Mechanisms of Photoreceptor Transduction" Academic Press, New York (in press).

Miller, W. H., and Nicol, G. D. (1979). Evidence that cyclic GMP regulates membrane potential in rod photoreceptors. *Nature (London)* **280**, 64–66.

Milstein, C., Brownlee, G. G., Harrison, T. M., and Matthews, M. B. (1972). A possible procursor of immunoglobulin light chains. *Nature (London) New Biol.* **239**, 117–120.

Mullen, R. J., and LaVail, M. M. (1976). Inherited retinal dystrophy: A primary defect in pigment epithelium determined with experimental rat chimeras. *Science* **192**, 799–801.

Nilsson, S. E. G. (1964). Receptor cell outer segment development and ultrastructure of the disk membranes in the retina of the tadpole (*Rana pipiens*). *J. Ultrastr. Res.* **11**, 581–620.

Nir, I., and Hall, M. O. (1979). Ultrastructural localization of lectin binding sites on the surface of retinal photoreceptors and pigment epithelium. *Exp. Eye Res.* **28**, 191.

O'Brien, P. J. (1976). Rhodopsin as a glycoprotein: A possible role for the oligosaccharide in phagocytosis. *Exp. Eye Res.* **23**, 127–137.

O'Brien, P. J. (1977). Incorporation of mannose into rhodopsin in isolated bovine retina. *Exp. Eye Res.* **24**, 449–458.

O'Brien, P. J. (1978a). Characteristics of galactosyl and fucosyl transfer to bovine rhodopsin. *Exp. Eye Res.* **26**, 197–206.

O'Brien, P. J. (1978b). Rhodopsin: A light-sensitive membrane glycoprotein. *Recept. Recognition Ser. A* **6**, 107–150.

O'Brien, P. J., and Muellenberg, C. G. (1973). Incorporation of glucosamine into rhodopsin in isolated bovine retina. *Arch. Biochem. Biophys.* **158**, 36–42.

O'Brien, P. J., and Muellenberg, C. G. (1975). Synthesis of rhodopsin and opsin *in vitro*. *Biochemistry* **14**, 1695–1699.

O'Brien, P. J., Muellenberg, C. G., and Bungenberg de Jong, J. J. (1972). Incorporation of leucine into rhodopsin in isolated bovine retina. *Biochemistry* **11**, 64–69.

Olive, J. (1980). The structural organization of mammalian retinal disc membrane. *Int. Rev. Cytol.* **64**, 107–169.

Palade, G. (1959). Functional changes in the structure of cell components. *In* "Subcellular Particles" (T. Hayashi, ed.), p. 64. Ronald, New York.

Palade, G. E. (1975). Intracellular aspects of the process of protein synthesis. *Science* **189**, 347–358.

Palmiter, R. D., Gagnon, J., and Walsh, K. A. (1978). Ovalbumin: A secreted protein without a transient hydrophobic leader sequence. *Proc. Natl. Acad. Sci. U.S.A.* **75**, 94–98.

Papermaster, D. S., and Allen, R. D. (1981). Clustered vesicles in the juxtaciliary cytoplasm of rod inner segments revealed in living cells by high resolution Allen video-enhanced-differential interference contrast microscopy. *Invest. Ophthalmol. Visual Sci. Suppl. 1,* **20,** 152.

Papermaster, D. S., and Dreyer, W. J. (1974). Rhodopsin content in the outer segment membranes of bovine and frog retinal rods. *Biochemistry* **13,** 2438–2444.

Papermaster, D. S., Converse, C. A., and Siu, J. (1975). Membrane biosynthesis in the frog retina: Opsin transport in the photoreceptor cell. *Biochemistry* **14,** 2438–2442.

Papermaster, D. S., Converse, C. A., and Zorn, M. (1976). Biosynthetic and immunochemical characterization of a large protein in frog and cattle rod outer segment membranes. *Exp. Eye Res.* **23,** 105–115.

Papermaster, D. S., Schneider, B. G., Zorn, M. A., and Kraehenbuhl, J. P. (1978a). Immunocytochemical localization of opsin in the outer segments and Golgi zones of frog photoreceptor cells. *J. Cell Biol.* **77,** 196–210.

Papermaster, D. S., Schneider, B. G., Zorn, M. A., and Kraehenbuhl, J. P. (1978b). Immunocytochemical localization of a large intrinsic membrane protein to the incisures and margins of frog rod outer segment disks. *J. Cell Biol.* **78,** 415–425.

Papermaster, D. S., Schneider, B. G., and Besharse, J. C. (1979). Assembly of rod photoreceptor membranes: Immunocytochemical and autoradiographic localization of opsin in smooth vesicles of the inner segment. *J. Cell Biol.* **83,** 275a.

Papermaster, D. S., Reilly, P., and Schneider, B. G. (1982). Cone lamellae and red and green rod outer segment disks contain a large intrinsic membrane protein on their margins: An ultrastructural immunocytochemical study of frog retinas. *Vision Res.* (in press).

Papermaster, D. S., Burstein, Y., and Schechter, I. (1980). Opsin mRNA isolation from bovine retina and partial sequence of the *in vitro* translation product. *Ann. N.Y. Acad. Sci.* **343,** 347–355.

Pellicone, C., Bouillon, P., Virmaux, N., and Vincendon, G. (1981a). Amino-acid sequence determination of a hydrophobic region of bovine rhodopsin. *Biochimie* **63,** 671–676.

Pellicone, C., Virmaux, N., Nullans, G., and Mandel, P. (1981b). Chemical cleavage of bovine rhodopsin at tryptophanyl bonds: Characterization of the polypeptide fragments at the phosphorylated site. *Biochimie* **63,** 197–209.

Peters, K.-R., Schneider, B. G., and Papermaster, D. S. (1981). Ultrahigh resolution scanning electron microscopy of a periciliary ridge complex of frog retinal rod cells. *J. Cell. Biol.* **89,** 273a.

Plantner, J. J., and Kean, E. L. (1976). Carbohydrate composition of bovine rhodopsin. *J. Biol. Chem.* **251,** 1548–1552.

Plantner, J. J., Poncz, L., and Kean, E. L. (1980). Effect of tunicamycin on the glycosylation of rhodopsin. *Arch. Biochem. Biophys.* **201,** 527–532.

Pober, J. S., and Bitensky, M. W. (1979). Light-regulated enzymes of vertebrate retinal rods. *Adv. Cyclic Nucleotide Res.* **11,** 265–301.

Pober, S., and Stryer, L. (1975). Light dissociates enzymatically-cleaved rhodopsin into two different fragments *J. Mol. Biol.* **95,** 477–481.

Polans, A. S., Hermolin, J., and Bownds, M. D. (1979). Light induced dephosphorylation of two proteins in frog rod outer segments: Influence of cyclic nucleotides and calcium. *J. Gen. Physiol.* **74,** 595–613.

Poo, M. M., and Cone, R. A. (1974). Lateral diffusion of rhodopsin in the photoreceptor membrane. *Nature (London)* **247,** 438–441.

Richardson, T. M. (1969). Cytoplasmic and ciliary connections between the inner and outer segments of mammalian visual receptors. *Vision Res.* **9,** 727–731.

Robbins, P. W., Hubbard, S. C., Turco, D. F., and Wirth, D. F. (1977). Proposal for a common oligosaccharide intermediate in the synthesis of membrane glycoproteins. *Cell* **12**, 895–900.

Röhlich, P. (1975). The sensory cilium of retinal rods is analogous to the transitional zone of motile cilia. *Cell Tissue Res.* **161**, 421–430.

Röhlich, P. (1976). Photoreceptor membrane carbohydrate on the intradiscal surface of retinal rod disks. *Nature (London)* **263**, 789–791.

Roll, F. J., Madri, J. A., Albert, J., and Furthmayr, H. (1980). Codistribution of collagen types IV and AB2 in basement membranes and mesangium of the kidney. *J. Cell Biol.* **85**, 597–616.

Roof, D. J., Korenbrot, J. I., and Heuser, J. (1980). Surfaces of rod photoreceptor disc membranes, exposed by deep etching. *J. Cell Biol.* **87**, 201a.

Saari, J. C., and Bunt, A. H. (1980). Fucosylation of rabbit photoreceptor outer segments: Properties of the labeled components. *Exp. Eye Res.* **30**, 231–244.

Saibil, H., Chabre, M., and Worcester, D. (1976). Neutron diffraction studies of retinal rod outer segment membranes. *Nature (London)* **262**, 266–270.

Saito, Z. (1938). Isolierung der Stabchenaussenglieder und spektrale Untersuchung des daraus hergestellten Sehpurpurextraktes. *Tohoku J. Exp. Med.* **32**, 432–446.

Salpeter, M. M., Bachmann, L., and Salpeter, E. E. (1969). Resolution in electron microscope radioautography. *J. Cell Biol.* **41**, 1–20.

Schechter, I. (1974). Use of antibodies for the isolation of biologically pure messenger ribonucleic acid from fully functional eukaryotic cells. *Biochemistry* **13**, 1875–1885.

Schechter, I, Burstein, Y., Zemell, R., Ziv, E., Kantor, F., and Papermaster, D. S. (1979). Messenger RNA of opsin from bovine retina: Isolation and partial sequence of the *in vitro* translation product *Proc. Natl. Acad. Sci. U.S.A.* **76**, 2654–2658.

Schneider, B. G., and Papermaster, D. S. (1981). Immunocytochemical localization of opsin in the connecting cilium of rat and *Xenopus* rods. *Invest. Ophthalmol. Visual Sci. Suppl. 1,* **20**, 76.

Shichi, H., and Shelton, E. (1974). Assessment of physiological integrity of sonicated retinal rod membranes. *J. Supramol. Struct.* **2**, 7–17.

Singer, S. J., and Nicolson, G. (1972). The fluid mosaic model of the structure of cell membranes. *Science* **175**, 720–723.

Steinberg, R. H., and Wood, I. (1975). Clefts and microtubules of photoreceptor outer segments in the retina of the domestic cat. *J. Ultrastruct. Res.* **51**, 397–403.

Steinberg, R. H., Fisher, S. K., and Anderson, D. H. (1980). Disc morphogenesis in vertebrate photoreceptors. *J. Comp. Neurol.* **190**, 501–518.

Tabas, I., and Kornfeld, S. (1978). The synthesis of complex-type oligosaccharides III: Identification of an alpha-D-mannosidase activity involved in a late stage of processing of complex-type oligosaccharides. *J. Biol. Chem.* **253**, 7779–7786.

Teirstein, P. S., Goldman, A. I., and O'Brien, P. J. (1980). Evidence for both local and central regulation of rat rod outer segment disc shedding. *Invest. Ophthalmol. Visual Sci.* **19**, 1268–1273.

Tomita, T. (1970). Electric activity of vertebrate photoreceptors. *Q. Rev. Biophys.* **3**, 179–185.

Tsunasawa, S., Narita, K., and Shichi, H. (1979). The blocking group of the N-terminal residue of opsin is acetyl group. *Am. Soc. Photobiol. Abstr. 7th, Annu. Meet. Asilomar,* p. 71.

Wald, G. (1968). Molecular basis of visual excitation. *Science* **162**, 230–239.

Wang, J. K., McDowell, J. H., and Hargrave, P. A. (1980). Site of attachment of 11-*cis*-retinal in bovine rhodopsin. *Biochemistry* **19**, 5111–5117.

Weibel, E. R. (1979). "Stereological Methods for Biological Morphometry," Vol. I. Academic Press, New York.

Wickner, W. (1979). The assembly of proteins into biological membranes: The membrane trigger hypothesis *Annu. Rev. Biochem.* **48**, 23–45.

Yau, K.-W., Lamb, T. D., and Baylor, D. A. (1977). Light-induced fluctuations in membrane current of single toad rod outer segments. *Nature (London)* **269**, 78–80.

Yee, R., and Liebman, P. A. (1978). Light-activated phosphodiesterase of the rod outer segment. *J. Biol. Chem.* **253**, 8902–8909.

Young, R. W. (1968). Passage of newly formed protein through the connecting cilium of retinal rods in the frog. *J. Ultrastruct. Res.* **23**, 462–473.

Young, R. W. (1970). Visual cells. *Sci. Am.* **223**, 81–91.

Young, R. W. (1976). Visual cells and the concept of renewal. *Invest. Ophthalmol.* **15**, 700–725.

Young, R. W. (1977). The daily rhythm of shedding and degradation of cone outer segment membranes in the lizard retina. *J. Ultrastruct. Res.* **61**, 172–185.

Young, R. W., and Bok, D. (1969). Participation of the retinal pigment epithelium in the rod outer segment renewal process. *J. Cell Biol.* **42**, 392–403.

11

Dopamine Neurons in the Retina: A New Pharmacological Model

WILLIAM W. MORGAN

I. INTRODUCTION

The vertebrate retina possesses many features which make it a unique tissue in which to investigate the physiology as well as the pharmacology of central neurotransmitter systems. Many of the putative neurotransmitters of the brain, including acetylcholine, γ-aminobutyric acid (GABA), dopamine, and glycine, are also found in subpopulations of neurons, particularly amacrine cells, of the retina. Compared to the brain the retina is a highly organized, layered structure which appears to contain a relatively small number of cell types. Although the structure of the retina can hardly be considered simple, the basic synaptic interconnections of the retinal cells are better understood than similar circuits in the brain. The vitreal cavity provides

Cell Biology of the Eye

ISBN 0-12-483180-X

an excellent site for the application of drugs, and this route of administration virtually eliminates the confounding peripheral effects which might be produced following the systemic administration of these agents. Indeed, the vitreal cavity presents many of the appealing characteristics of a perfusion chamber, and the retina many of the characteristics of an *in vivo* brain slice. This latter comparison seems justified considering the ontogenic origin of the retina as an outpocketing of the diencephalon.

The thrust of this chapter deals with a small population of amacrine cells in the vertebrate retina, which have been shown to contain the putative neurotransmitter dopamine (DA). These neuronal cells have been shown to be highly responsive to light, and we have recently observed that they receive an inhibitory GABA-like input. As I shall show in the subsequent discussion, these dopaminergic neurons and their GABA-like input not only provide an excellent new model for studying the interactions between GABA and DA-containing neurons in the central nervous system but also provide, in combination with other techniques, a new model for studying GABA receptor physiology and pharmacology.

II. BACKGROUND

Catecholamine-containing neurons were first demonstrated in the rabbit retina by histofluorescence techniques (Häggendal and Malmfors, 1965). Fine cell processes and intensely fluorescent varicosities or terminals of these neurons were observed within the inner plexiform layer. With the use of spectrofluorometric procedures, Nichols *et al.* (1967) subsequently showed that DA was the predominant if not the exclusive catecholamine in these cells. Autoradiographic studies demonstrated the selective uptake of [^{3}H]DA by a group of cells at the junction of the inner nuclear and the inner plexiform layers (Kramer, 1971; Ehinger and Falck, 1971). The localization of these cells was strikingly similar to that of the catecholamine-containing cells described by Häggendal and Malmfors. Morphologically these cells appeared to be amacrine cells, but they constituted only about 10% of the total amacrine population.

A group of DA-containing interplexiform cells have also been described in the retinas of cebus monkeys and goldfish (Ehinger, 1977). These particular cells have not been described in the rat retina and thus are not believed to be a confounding factor in our experiments which will be discussed in subsequent sections of this chapter.

Nichols and co-workers (1967) reported that exposure to light increased the level of DA in the retinas of animals that had been previously dark-adapted. Their observation was the first to suggest that the DA-containing cells in the retina were responsive to light.

Kramer (1971, 1976) first observed that the DA-containing neurons or amacrine cells of the intact, dark-adapted retina of cats released DA at a more rapid rate during periods of photic stimulation. He observed that a flashing light stimulus increased the release of radioactively labeled DA from the dark-adapted retina at a rate which was proportional to the stimulus frequency. A flashing light stimulus also augmented labeled DA release from the light-adapted retina, but the rate was no longer proportional to that of the stimulus. Bilateral section of the optic nerves did not affect the light-induced increase in DA release. This latter observation provided strong evidence that the enhanced release of DA was mediated at the level of the retina and was not secondary to efferent activation from the brain.

Exposure to light produces a more than twofold enhancement of tyrosine hydroxylase activity in previously dark-adapted retinas. This effect is obvious whether tyrosine hydroxylase is measured *in vitro* (Iuvone *et al.*, 1978) or *in vivo* by measuring the accumulation of dihydroxyphenylalanine (DOPA) following the inhibition of L-aromatic amino acid decarboxylase (DaPrada, 1977). This light-mediated increase in DOPA accumulation in the dark-adapted rat retina is clearly demonstrated within 15 min after exposure to room light (Table I). The turnover of DA as measured by the decline in this amine following the inhibition of DA synthesis with α-methyl-*p*-tyrosine (αMPT) is also

TABLE I

The Effect of Exposure to Light on the Accumulation of DOPA in the Retinas of Dark-Adapted Rats Following Treatment with a L-Aromatic Amino Acid Decarboxylase Inhibitor (NSD 1015)[a]

Treatment	DOPA (ng/retina)
Dark-maintained	1.6 ± 0.1 (6)
Light-exposed (15 minutes)	2.4 ± 0.4 (6)

[a] The rats were maintained in darkness for 15 hr before sacrifice or before exposure to light for 15 min and subsequent sacrifice. *m*-Hydroxybenzylhydrazine dihydrochloride (NSD 1015) (100 mg/kg) was given 15 min before killing. The levels of DOPA were analyzed in duplicate by the method of Demarest and Moore (1980), and the statistical significance of the observed difference ($A = p < 0.001$) was determined by the analysis of variance (Sokal and Rohlf, 1969). The means plus or minus the standard error of the means is shown. The numbers in parentheses indicate the number of animals.

increased by three- to fourfold following the stimulation of dark-adapted retinas of rats *in vivo* with light (Iuvone *et al.*, 1978; Morgan and Kamp, 1980a). At least in the nigrostriatal pathway the turnover of DA has been shown to be correlated with the impulse activity of the dopaminergic neurons (Aghajanian *et al.*, 1973). Thus, these data all provide evidence for a dramatic light-mediated increase in the activity of the DA neurons in the retina.

The increase in DA turnover in the retina in response to light is not an all-or-none phenomenon. In a recent study conducted in our laboratory,rats were housed in the dark for 15 hr. At the end of this time period they were either sacrificed without exposure to light or were exposed to 2–4 or 240–260 ft-c of light for 30 min or 1 hr before sacrifice. Two-thirds of the animals in each group were given αMPT (250 mg/kg intraperitoneally) just before exposure to light. The sacrifice of animals in the dark was conducted with illumination provided by a dim red, filtered (Wratten 1A) photographic safelight. Dim red light does not appear to activate retinal DA neurons. However, the validation of this particular assumption requires further investigation (De-Vries *et al.*, 1978). The retinas of each animal were collected, and the level of DA per retina was quantified by a radioisotope–enzyme procedure (Palkovits *et al.*, 1974). The method of least squares (Sokal and Rohlf, 1969) was used to determine the best fit regression line for the αMPT-induced decline in DA in the retinas of rats maintained in the dark and rats exposed to 2–4 or 240–260 ft-c of light, separately. The data in Fig. 1 show that the light-mediated increase in the rates of decline in DA after tyrosine hydroxylase inhibition in dark-adapted rat retinas was proportional to the intensity of the light stimulus. These data suggest that the activity of retinal DA neurons is proportional to the intensity of light reaching the retina.

The functional role of DA-containing neurons in the retina has not been extensively studied. Based on the observation that reserpine-treated rats are photophobic, Ehinger (1976) suggested that dopaminergic neurons may be involved in adaptation of the retina to light. If DA neurons do serve an important role in light adaptation of the retina, it is possible that these neurons are more active in the retinas of albino rats, which have no pigment in the epithelial layer of their retinas partially to screen the photoreceptors from light. To test this hypothesis, we recently dark-adapted both Long–Evans rats (pigmented retinas) and albino Sprague–Dawley rats (nonpigmented retinas) and then exposed these animals to room light for 30 or 60 min before sacrifice. As before, one-half of the animals in each group received αMPT just before exposure to light. DA levels were quantified

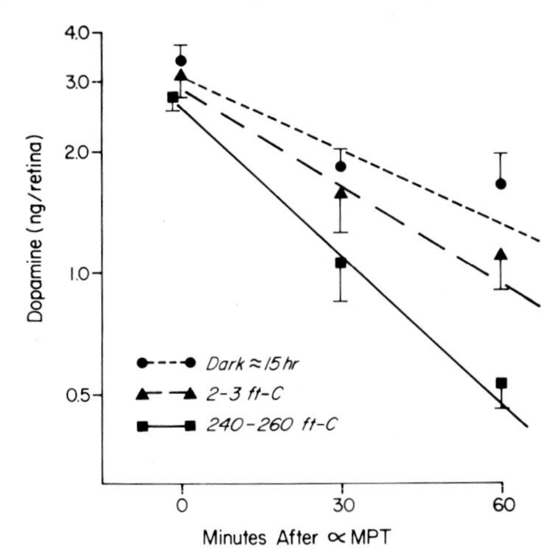

Fig. 1. The effect of light intensity on the αMPT-induced decline in DA in the retinas of dark-adapted, albino male Sprague–Dawley rats. αMPT (250 mg/kg) was administered 30 or 60 min before sacrifice. The regression lines were calculated by the method of least squares, and the statistical significance of the differences in the slopes among the three regression lines was determined by the method of Steel and Torrie (1960). The slope of the DA decline in rats exposed to 240–260 ft-c of light was statistically greater than that of dark-maintained rats ($p < 0.01$) or that of rats exposed to 2–3 ft-c of light ($p < 0.05$). The points plotted on the regression lines indicate the mean retinal DA content plus or minus the standard error of the mean for eight animals.

as described before, and the αMPT-induced decline in this amine provided an indirect measure of DA turnover. As shown in Fig. 2, the decline in DA after αMPT treatment was not significantly different in rats with pigmented retinas as compared to rats with nonpigmented retinas. The calculated turnover rate of DA in the retinas of the dark-adapted Long–Evans rats and the Sprague–Dawley rats was 0.04 and 0.03 ng/retina/min, respectively. Following exposure to light, the turnover rate of DA in the Long–Evans rats increased 2.8-fold to 0.10 ng/retina/min, and in the Sprague–Dawley rats DA turnover increased by 3.3-fold to 0.09 ng/retina/min. These results show that retinal DA neurons do not compensate for the absence of pigment by being more active in the retinas of albino rats. On the other hand, these neurons may still serve an important but more subtle role in light adaptation.

Some electrophysiological evidence suggests that the DA neurons in the retina may supply an inhibitory input to retinal ganglion cells. Straschill and Perwein (1969, 1975) observed that electrophoretically applied DA suppressed the spontaneous activity of 90% of the ganglion cells tested in the retinas of cats. Further, DA was found to be inhib-

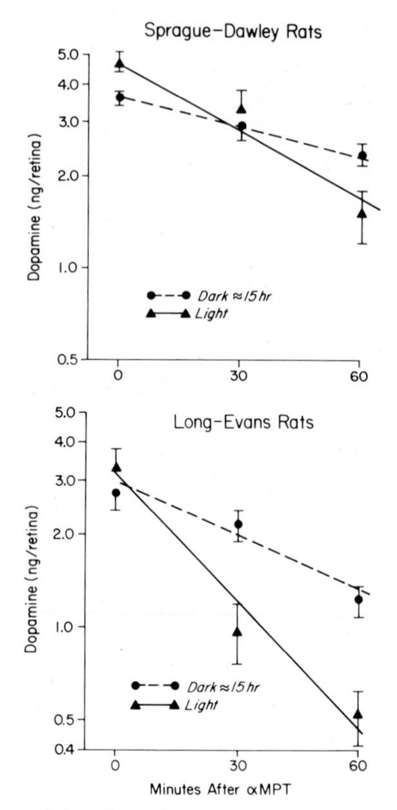

Fig. 2. A comparison of the effect of exposure to light on the α-MPT-induced decline in DA in the retinas of dark-adapted Long-Evans versus Sprague-Dawley rats. The experimental procedure was similar to that outlined in the caption to Fig. 1, except that the statistical significance of differences was determined by the analysis of variance and subsequently by the Student-Newman Keuls test (Sokal and Rohlf, 1969).

itory to the excitatory reaction of "on" and "off" ganglion cells and was antagonistic to glutamate-induced excitation of retinal ganglion cells.

Other cell populations in the retina also seem to have an inhibitory dopaminergic input. Starr (1975) showed that the amplitudes of both the a and the b waves of the electroretinogram (ERG) were reduced by DA superfusion with a potency greater and an effect more long-lasting than that produced by glycinelike inhibitory amino acids.

The action of a retinal DA system on higher visual pathways was recently shown by McCulloch *et al.* (1980). These investigators observed that the intravenous administration of apomorphine enhanced metabolic activity, as evidenced by 2-deoxyglucose incorporation, in the superficial layer of the superior colliculus. This effect of apomorphine was completely lost in enucleated animals, suggesting that it was mediated via the DA pathways in the retina. Since dopaminergic fibers are not believed to be located in the optic nerve, this effect on the superior colliculus is probably mediated via a multineuronal pathway. No effects of apomorphine on metabolic activity were observed in the lateral geniculate body or the visual cortex. These latter data suggest that retinal DA neurons are involved primarily in retinal information conveyed to the superficial layer of the superior colliculus and not to other visual centers.

The transjunctional effects of DA in the retina are mediated, at least in part if not entirely, via an adenylate cyclase-coupled mechanism (Brown and Makman, 1972; Watling *et al.,* 1979). In the retina, haloperidol, a DA antogonist, almost totally blocks the effect of DA on adenylate cyclase (Spano *et al.,* 1976). DA-sensitive adenylate cyclase activity is predominantly localized in the inner layers of the retina, with little if any activity in the photoreceptor layer (Lolley *et al.,* 1974; Thomas *et al.,* 1978). DA receptors, as identified by high-affinity spiroperidol binding, have also been characterized in the rat, bovine, and goldfish retina (Magistretti and Schorderet, 1979; Schaeffer, 1980; Redburn *et al.,* 1980). Since *d*-lysergic acid diethylamide (*d*-LSD) also appears to stimulate the activity of retinal adenylate cyclase via the DA receptor, DA receptors in the retina may be of importance in the production of LSD-induced visual hallucinations (Spano *et al.,* 1977).

Trabucchi *et al.* (1976) observed that the exposure of rats to continuous darkness for 96 hr produced a supersensitivity of the DA-sensitive adenylate cyclase in the retina. Similar results have been obtained in our laboratory, although in our hands 12 days of continuous darkness appears to be the minimum required to produce an enhancement of DA-sensitive adenylate cyclase in the adult albino rat retina. These data provide further evidence that dopaminergic neurons of the retina are relatively inactive in darkness and that prolonged exposure to darkness is probably equivalent to functional denervation of adenylate cyclase-related DA receptors in this tissue.

III. REGULATION OF THE LIGHT-MEDIATED ACTIVATION OF RETINAL DA NEURONS

Since the DA neurons of the retina have not been shown to contain photosensitive pigments, it is likely that the light-induced activation of these cells is indirect and that the information about the presence of light is conveyed to DA neurons by other cells in the retina. Interestingly, however, the response of these neurons to light is still clearly maintained in rat retinas which have been very severely depleted of photoreceptors (Morgan and Kamp, 1980b). The animals utilized in these studies were albino Sprague–Dawley rats which had been exposed to continuous light for 4 months before being returned to a normal 14:10-hr alternating light–dark regimen. As reported by others (O'Steen and Anderson, 1971; O'Steen *et al.*, 1972; Bennett *et al.*, 1972), exposure to continuous light of normal intensity destroyed all but a very few scattered photoreceptor cell bodies in the outer nuclear layer, and no photoreceptor segments remained intact. The continued response of DA neurons to light following destruction of the photoreceptors is consistent with the observations of others (Bennett *et al.*, 1972, 1973; Anderson and O'Steen, 1972, 1974; Lemmon and Anderson, 1979) who concluded that similarly treated rats were still able to perceive light. These results suggest that the very few remaining photoreceptor cell bodies may contain sufficient visual pigment to maintain light sensitivity. The remaining photoreceptor cells may also be cone cells which appear to be more resistant to destruction during exposure to continuous light (LaVail, 1976; Cicerone, 1976). Alternatively, the continued ability of retinal DA neurons to respond to light stimulation may be dependent on an as yet unidentified photoreceptor in the retina. Although it is unlikely that DA neurons themselves are this photoreceptor, these cells may serve an important function in the continued ability of rats with severely depleted photoreceptors to respond to light.

Electron microscopic studies have shown that DA-containing amacrine cells probably receive no direct input from retinal cells outside the inner plexiform layer and certainly have no synaptic contacts with the classical photoreceptors (Ehinger, 1976). Within the inner plexiform layer these cells receive input from other amacrine cells but reportedly receive no direct input from either bipolar neurons or from ganglion cells (Ehinger, 1977). Logically, an investigation of the mechanisms which regulate the activity of retinal DA neurons should therefore begin with an investigation of the effects of various putative transmitters of the amacrine cells on the activity of these neurons.

IV. GABA AS A TRANSMITTER IN THE INNER NUCLEAR LAYER

The presence of GABA in the eye was first demonstrated by Kojema *et al.* (1958) who found quantities of this amino acid in the retina and the choroid layer but not in the iris or the ciliary body of the dog or the ox. The highest levels of GABA and glutamate decarboxylase (GAD), the biosynthetic enzyme for GABA, were subsequently localized to the inner plexiform layer (Kuriyama *et al.*, 1968; Graham, 1972). Based on these data, these investigators proposed that GABA was probably localized in amacrine cells. Subsequent autoradiographic studies have linked this amino acid with a small population of amacrine cells, although in mammalian species the Müller or glial cells of the retina also take up substantial quantities of labeled GABA (Voaden, 1976). In the goldfish retina, radioactive GABA is also incorporated into a specific subpopulation of horizontal cells, where it may also function as a transmitter (Marc *et al.*, 1978). Based on immunochemical techniques Vaughan *et al.* (1978) has concluded that "some classes of amacrine cells are the only GABAergic neurons in the rat retina."

The best evidence that the release of GABA from the mammalian retina is affected by exposure to light has been provided by Bauer and Ehinger (1977). These investigators observed that a flashing light produced an increase in [^{3}H]GABA release from isolated rabbit retinas following a 30-min period of dark adaptation. It is pertinent that in this study the retinas were preloaded with labeled GABA *in vivo* 4 hr before the isolation and superfusion of these tissues. In mammalian retinas large amounts of [^{3}H]GABA are taken up by Müller or glial cells shortly after infusion of this labeled amino acid. Apparently, the [^{3}H]GABA in the Müller cells is rapidly metabolized, and by 4 hr the majority of the remaining labeled GABA is associated with neural elements of the inner nuclear layer (Voaden, 1976). Bauer and Ehinger were unable to show a significant effect of light on [^{3}H]GABA release during the early time periods when [^{3}H]GABA was localized predominantly in the Müller cells.

Because the release of GABA is enhanced by photic stimulation and since GABA is found in the amacrine cell population of the inner nuclear layer, this amino acid is a logical candidate for a neurotransmitter involved in regulation of the light-evoked activation of retinal DA neurons.

V. EVIDENCE THAT A GABA RECEPTOR MECHANISM AFFECTS THE ACTIVITY OF RETINAL DA NEURONS

We recently showed that the light-induced increase in DA turnover in the dark-adapted rat retina was blocked in a dosage-related fashion by muscimol, a potent GABA agonist (Figs. 3 and 4; Morgan and

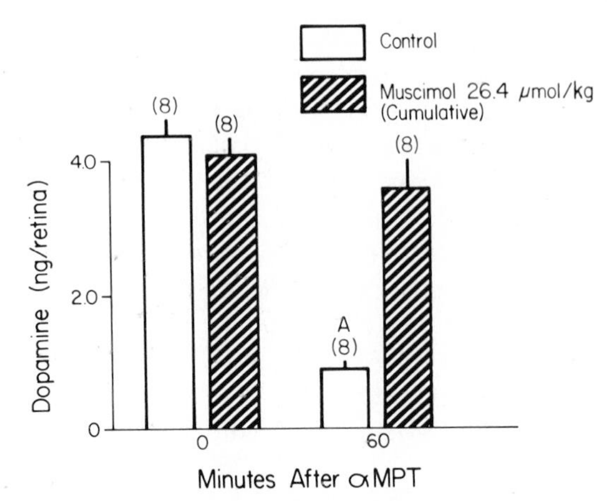

Fig. 3. The effect of muscimol on the light-mediated increase in the αMPT-induced decline in DA in the retinas of dark-adapted rats. The rats were adapted to darkness for 15 min and killed 60 min after exposure to light. Muscimol was administered in three equal dosages 15 min before, immediately before, and 30 min after exposure to light. The cumulative dosage of muscimol is shown in this figure. Control rats received saline injections by the same schedule. One half of the control and one half of the muscimol-treated rats received αMPT 60 min before killing. The statistical significance of the differences was determined by the analysis of variance. ($A = p < 0.001$.)

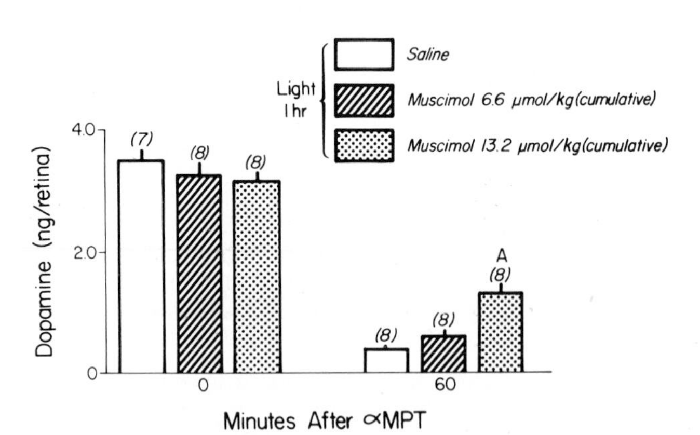

Fig. 4. A dosage-response study on the effects of two lower dosages of muscimol on the light-mediated increase in the αMPT-induced decline in DA in the retinas of dark-adapted rats. This experiment was designed as that outlined in Fig. 3. The cumulative dosages of muscimol are shown in the figure. ($A = p < 0.001$.)

Kamp, 1980a). Cumulative intravenous dosages of 13.2 or 26.4 μmol muscimol/kg effectively antagonized the light-mediated enhancement of DA turnover in the rat retina. A lower cumulative dosage of 6.6 μmol muscimol/kg was ineffective. The inhibitory effects of muscimol were blocked by picrotoxinin, a GABA antagonist (Fig. 5). When administered intravenously at this same dosage, picrotoxinin alone had no significant effects on the light-evoked response of retinal DA neurons. Picrotoxinin rather than picrotoxin was used in these studies because it is the active metabolite of picrotoxin and because it is more water-soluble (Jarboe and Porter, 1965). In contrast, strychnine had no antagonistic action on the inhibitory effects of muscimol in the retina (Fig. 6). Strychnine alone also had no effect on the response of dopaminergic neurons to light. The intravitreal administration of much lower dosages of muscimol was also effective in inhibiting the light-induced increase in retinal DA turnover (Fig. 7). Collectively, these data provide strong evidence for an inhibitory GABA-like input to retinal DA neurons. It is not evident from these data whether the GABA-like input is directly to retinal DA neurons or whether it is mediated via one or more interneurons. Recent studies with dissociated retina preparations, however, suggest that GABA receptors are present directly on retinal DA neurons (Iuvone *et al.*, 1980).

The data just presented provide evidence for a potential GABA in-

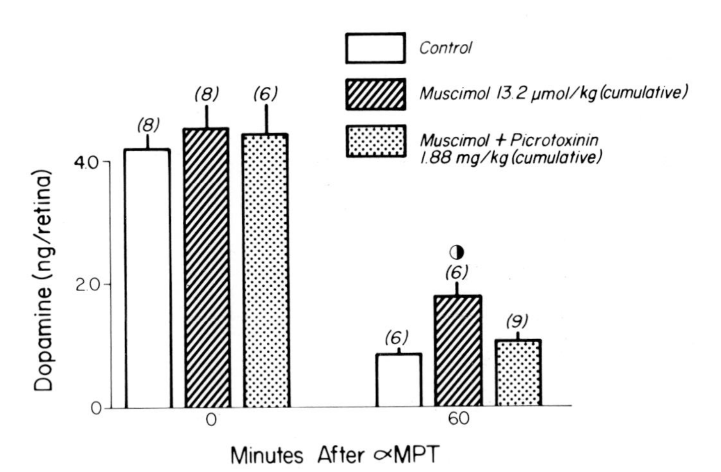

Fig. 5. The antagonism by picrotoxinin of the inhibitory effect of muscimol on DA turnover in the rat retina. The experiment was designed like that outlined in Fig. 3 except that picrotoxinin was administered intravenously in two equal dosages, the first 30 min before exposure to light and the second 15 min after exposure to light. The cumulative dosage is shown in the figure. ($B = p < 0.005$.)

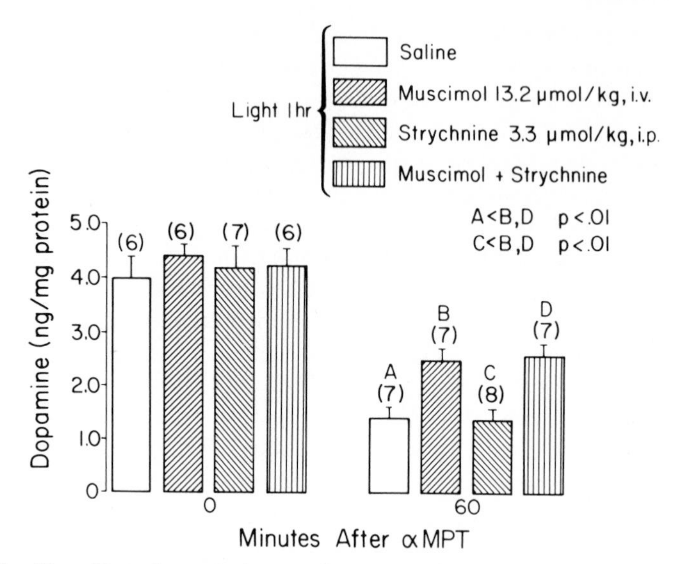

Fig. 6. The effect of strychnine on the action of muscimol in blocking the light-mediated increase in DA turnover in the retinas of dark-adapted rats. This experiment was designed like that outlined in Figs. 3 and 5, except that strychnine was administered intravenously only once 30 min before exposure to light.

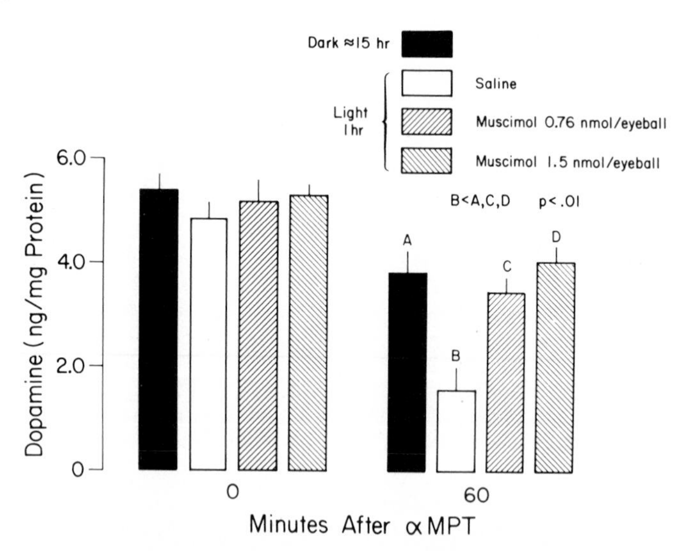

Fig. 7. The inhibitory effect of intravitreally administered muscimol on the light-mediated increase in DA turnover in the retinas of dark-adapated rats. The experiment was designed exactly like that outline in Fig. 3, except that muscimol was administered intravitreally to ether-anesthetized rats in equal dosages 10 min before and 30 min following exposure to light.

fluence on retinal dopaminergic neurons but provide no information on the physiological significance of this input. In the course of studies ascertaining the physiological significance of these observations, we found that the intravitreal administration of low dosages of picrotoxinin or bicuculline, selective GABA antagonists, markedly enhanced DA turnover in the retinas of rats maintained in darkness (Kamp and Morgan, 1980b). The increase in DA turnover induced by the higher dosages of these drugs was equivalent to that observed in response to stimulation with room light (90–150 ft-c). These data suggest that in the dark there is a tonic GABA receptor-mediated, inhibitory influence on retinal DA neurons. When comparable intravitreal dosages of bicuculline or picrotoxinin were given to rats also exposed to room light, neither of these drugs produced a much greater enhancement of DA turnover over that induced by exposure to light alone. Thus, there appears to be relatively little inhibitory GABA tone on retinal DA neurons following exposure to room light. Further, these latter observations show that picrotoxinin and bicuculline do not nonspecifically activate retinal DA neurons. Collectively, these data suggest that an inhibitory GABA influence on retinal dopaminergic neurons may have a predominant if not exclusive role in regulating the change in the activity of these DA neurons in response to light.

It could be argued that GABA antagonists indirectly enhance the activity of retinal DA neurons in dark-adapted retinas by releasing the inhibitory effects of GABA on cholinergic neurons in the retina, which subsequently activate DA neurons. A potential GABA-like input to cholinergic neurons in the retina has been recently shown (Massey and Neal, 1979). However, this possibility is refuted by our observation that atropine is without effect on the enhancement of DA turnover induced by picrotoxinin (Kamp and Morgan, 1980b).

Interestingly, Bauer and Ehinger (1977) provided data suggesting that the release of GABA from some retinal cells was enhanced by light. In contrast, our data show that at least one population of GABAergic cells in the retina may be more active in darkness than in light. It appears then that more than one population of GABA neurons may exist in the rat retina and that the role of GABA in the regulation of retinal function may be more complex than originally supposed.

VI. THE EFFECT OF BARBITURATES ON CENTRAL GABA PATHWAYS

The neurochemical basis for the pharmacological action of central nervous system depressants has been controversial for many years. In no case has the confusion been greater than with barbiturates. However, new insight into the mode of action of barbiturates has been recently obtained by a series of electrophysiological studies which provide abundant evidence that depressant barbiturates exert a significant portion of their pertinent pharmacological actions by augmenting both pre- and postsynaptic inhibition conveyed through GABAergic pathways (Ransom and Barker, 1975, 1976; MacDonald and Barker, 1978). The potentiating effects on GABA appear to be mediated by depressant but not by excitatory barbiturates. Further, the specificity of action of barbiturates for GABA-related as compared to glycine-related pathways argues that these agents exert their effects at the level of C1 conductance (Barker and Ransom, 1978). The available data suggest that barbiturates exert their augmenting effect at a modulator site associated with the GABA receptor, although pentobarbital also appears to exert some direct GABA-mimetic effects which are blocked by GABA antagonists (MacDonald and Barker, 1978).

VII. INFLUENCE OF BARBITURATES ON THE INHIBITORY ACTION OF MUSCIMOL ON LIGHT-ACTIVATED RETINAL DA NEURONS

The data obtained by others linking the pharmacological manifestation of barbiturates to effects on GABA led us to investigate the effects of these drugs on the action of muscimol in depressing the response of retinal DA neurons to light (Kamp and Morgan, 1980a). An anesthetic dosage of pentobarbital alone (40 mg/kg) did not affect the light-stimulated increase in DA turnover observed in dark-adapted retinas of rats. As in previous studies, a low cumulative dosage of muscimol (6.6 μmol/kg) also had no significant effect on this response. However, when given in combination with an anesthetic dosage of pentobarbital, the same low dosage of muscimol exerted an inhibitory effect on the light-mediated increase in retinal DA turnover comparable to that shown previously to be exerted by a fourfold higher dosage of muscimol (Fig. 8; Kamp and Morgan, 1980a). Since both drugs were ineffective alone, the very magnitude of the response induced by the combination of drugs argues that pentobarbital produced a true potentiation of the inhibitory action of muscimol. The inhibitory effect of the low dosage of muscimol in combination with pentobarbital was reversed by picrotoxinin (Kamp and Morgan, 1980a). These latter data show that the inhibitory action of the combination of muscimol and pentobarbital is mediated via a GABA receptor mechanism.

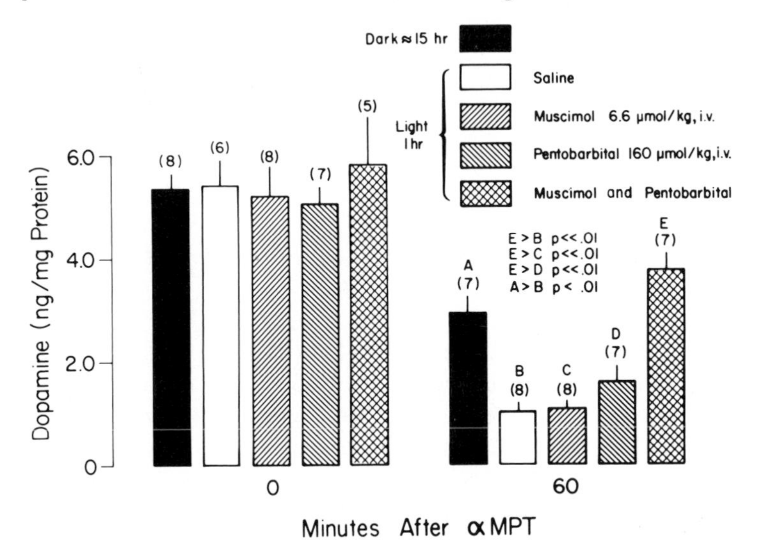

Fig. 8. The effect of muscimol and pentobarbital on the light-induced increase in retinal DA turnover. Dark-adapted rats were maintained in darkness for 15 hr before sacrifice or were exposed to light for 60 min before sacrifice. Muscimol was administered intravenously following the schedule in the caption to Fig. 3. Pentobarbital was given intravenously 30 min before (five-eighths of total dosage) and 15 min (three-eighths total dosage) after light exposure. αMPT was given to some rats 60 min before sacrifice. Cumulative dosages for the drugs are shown in the figure. Statistical analyses were outlined in Fig. 2. This figure is from Kamp and Morgan (1980), *J. Pharmacol. Exp. Ther.* **213**, 333; copyright 1980, American Society for Pharmacology and Experimental Therapeutics.

Subsequent studies showed that phenobarbital (sodium salt) also significantly augmented the inhibitory effect of muscimol on the light-induced increase in retinal DA turnover. Like pentobarbital, phenobarbital alone had no inhibitory effect on the response of DA neurons to light. Sodium barbital when given alone was unusual in that it significantly decreased the response of retinal DA neurons to light, but it was unable to augment significantly the inhibitory effect of muscimol on this same parameter (Kamp and Morgan, 1980a). The relative inability of barbital to augment the inhibitory effect of muscimol may be related to the low therapeutic potency of this compound. The suppressive effect of barbital alone on the light-mediated increase in retinal DA turnover is not blocked by picrotoxinin and thus does not appear to be conveyed via a GABA receptor.

The convulsant 5-(2-cyclohexylidene-ethyl)-5-ethyl barbiturate (CHEB) when given in the maximally tolerable intravenous dosage did not significantly affect the light-mediated increase in DA turnover when given alone or in combination with muscimol (Kamp and Morgan, 1980a).

To my knowledge these data were the first biochemical evidence obtained *in vivo* for barbiturate-induced enhancement of a GABA-mediated response.

VIII. FURTHER USE OF THE RETINA AND RETINAL DA NEURONS AS A PHARMACOLOGICAL MODEL

Although it is generally agreed that GABA is a widely distributed inhibitory neurotransmitter in the mammalian nervous system, the role of this amino acid in the regulation of central neuronal function or in the expression of various pharmacological agents is only very poorly understood. This lack of adequate understanding relates in part to the paucity of good neurochemical tools with which to study the role of GABA as a neurotransmitter.

One relatively new tool for use in studying the transmitter role of GABA is the measurement of high-affinity GABA binding. The sodium-independent, high-affinity binding site at which GABA binds to neuronal membranes possesses many of the characteristics expected of the physiological GABA receptor (Enna and Snyder, 1975). However, it has been difficult to relate changes in sodium-independent GABA binding to changes in physiological GABA receptors unambiguously. In order to better understand the functional significance of changes in high-affinity GABA binding it is essential to relate these to changes in a functioning GABA receptor-mediated response which can be measured *in vivo*. Based on our own results and those of others presented in the preceding discussion, I propose use of the inhibitory GABAergic input to retinal DA neurons as such an *in vivo* model.

By combining studies on high-affinity GABA or muscimol binding in the retina with studies on the effects of muscimol or other GABA agonists on the light-mediated increase in DA turnover in the retina, it may be possible to gain a clearer understanding of the relationship of high-affinity GABA-binding sites to functional GABA receptors. This combined approach may be especially useful in studying the effects of chronic drug treatments on the number and function of GABA receptors in the central nervous system. In our laboratory we are presently investigating the effects of chronic barbiturate treatment on the ability of muscimol to block the light-induced increase in retinal DA turnover. Since depressant barbiturates appear to potentiate the inhibitory effects of GABA, long-term treatment with these drugs may lead to a compensatory decrease in the numbers or in the affinity of central GABA receptors. Such a compensation might account at least in part for the abstinence syndrome (hyperactivity) associated with abrupt withdrawal of barbiturates following long-term administration. If such a compensation occurs, then a shift to the right in the dosage response curve for the effect of muscimol on DA turnover may be observed in the retinas of chronic barbiturate-treated rats.

Obviously, the effects of other depressant drugs on the GABA system could be tested with the use of the retina model. Benzodiazepines, whose actions have been clearly shown to be intimately associated with the GABA system (Costa *et al.*, 1975; Haefely, 1977; Geller *et al.*, 1978), and ethanol are logical examples which are presently under investigation in our laboratory. Binding sites for benzodiazepines have been clearly demonstrated in mammalian retinas (Borbe *et al.*, 1980; Osborne, 1980; Regan *et al.*, 1980). Preliminary results from our laboratory clearly show that diazepam and flurazepam block the light-mediated enhancement of DA turnover in the dark-adapted retina (Kamp and Morgan, 1980c). However, we have not conclusively established whether this effect is mediated via a GABA receptor mechanism. Nevertheless our results are the first to demonstrate a pharmacological effect of benzodiazepines in the retina and the first to suggest a physiological role for the endogenous benzodiazepines in this tissue. Studies with ethanol would be particularly exciting in that it has only recently been shown that ethanol in low dosages can potentiate the effects of iontophoretically applied GABA (Nestoros, 1980).

Of course, the value of retinal DA neurons as a pharmacological model is hardly limited to the study of interrelationships with the GABA system. As briefly noted in Section I, numerous other putative neurotransmitter substances have been localized to cell populations in the inner nuclear layer of the retina. Radioactively labeled glycine is selectively accumulated by cells in the inner nuclear layer of the rabbit, as well as in the human retina (Bruun and Ehinger, 1972; Ehinger 1972). Further, a flashing light stimulus has been shown to increase the efflux of tritiated glycine from the dark-adapted retinas of cats (*in vivo* eye cup preparation) (Ehinger and Lindberg, 1974). Electrophysiological studies have shown that strychnine, a selective glycine antagonist, increases the spontaneous activity of ganglion cells (Straschill, 1968). This observation implies that some of these neurons receive an endogenous glycinergic input.

There is also good evidence that acetylcholine is an excitatory neurotransmitter in a population of amacrine cells in the retina. Compared to other brain areas, the retina has the richest source of choline acetyltransferase, and this enzyme is localized predominantly in the inner plexiform layer with smaller amounts in the amacrine and ganglion cell layers (Graham, 1974). The acetylcholine-containing cells in the latter population are apparently displaced amacrine cells (Hayden *et al.*, 1980). Masland and Livingstone (1976) observed a fourfold increase in the release of [^{14}C]acetylcholine following the stimulation of isolated, dark-adapted rabbit retina with light. Further, a number of electrophysiological studies have provided evidence for a physiological influence of acetylcholine on retinal ganglion cells (Ames and Pollen, 1969; Straschill and Perwein, 1973; Masland and Ames, 1976). It appears from these results that ganglion cells also receive a cholinergic input.

Immunohistochemical procedures have also recently localized populations of specific enkephalin-like, substance P-like, neurotensin-like, and somatostatin-like immunoreactive amacrine cells in the retina of the pigeon and chicken (Brecha *et al.*, 1980). It is likely that many of these substances are also found in amacrine cell populations in mammalian retinas.

Given the localization of the peptide putative transmitters as well as glycine and acetylcholine in the amacrine cell population, it is likely that some if not all of these substances have an influence on the activity of dopaminergic amacrine cells. Future studies on how these substances affect the activity of retinal DA neurons will provide insight not only into the role of these putative transmitters in the retina but also into the role of these substances in regulation of the activities of the central nervous system as well.

The discussion presented in this chapter clearly shows that the retina in general and retinal DA neurons in particular provide exceptionally promising material for study of the function and interaction of numerous neurotransmitter candidates in the central nervous system.

ACKNOWLEDGMENT

Supported by NIDA Grant No. R01 DA 00755, NINCDS R01 NS 14855, and Research Scientist Development Award 5K02 MH 0028.

REFERENCES

Aghajanian, G. K., Bunney, B. S., and Kuhar, M. J. (1973). Use of single unit recording in correlating transmitter turnover with impulse flow in monoamine neurons. *In*

"New Concepts in Neurotransmitter Regulation" (A. J. Mandell, ed.), pp. 115–134. Plenum, New York.

Ames, A., III, and Pollen, D. A. (1969). Neurotransmission in central nervous tissue: A study of isolated rabbit retina. *J. Neurophysiol.* **32,** 424–442.

Anderson, K. V., and O'Steen, W. K. (1972). Black-white and pattern discrimination in rats without photoreceptors. *Exp. Neurol.* **34,** 446–454.

Anderson, K. V., and O'Steen, W. K. (1974). Altered response latencies in visual discrimination tasks in rats with damaged retinas. *Physiol. Behav.* **12,** 633–637.

Barker, J. L., and Ransom, B. R. (1978). Pentobarbitone pharmacology of mammalian central neurones grown in tissue culture. *J. Physiol.* **280,** 355–372.

Bauer, B., and Ehinger, B. (1977). Light evoked release of radioactivity from rabbit retinas preloaded with ^{3}H-GABA. *Experientia 33,* 470–471.

Bennett, M. H., Dyer, R. F., and Dunn, J. D. (1972). Light induced retinal degeneration: Effect upon light-dark discrimination. *Exp. Neurol. 34,* 434–445.

Bennett, M. H., Dyer, R. F., and Dunn, J. D. (1973). Visual-deficit following long-term continuous light exposure. *Exp. Neurol.* **38,** 80–89.

Borbe, H. O., Müller, W. E., and Wollert, U. (1980). The identification of benzodiazepine receptors with brain-like specificity in bovine retina. *Brain Res.* **182,** 466–469.

Brecha, N., Karten, H. J., and Davis, B. (1980). Localization of neuropeptides, including vasoactive intestinal polypeptide and glucagon, within adult and developing retina. *Soc. Neurosci. 10th Annu. Meet.* **6,** 346.

Brown, J. H., and Makman, M. H. (1972). Stimulation by dopamine of adenylate cyclase in retinal homogenates and of adenosine 3′:5′-cyclic monophosphate formation in intact retina. *Proc. Natl. Acad. Sci. U.S.A.* **69,** 539–543.

Bruun, A., and Ehinger, B. (1972). Uptake of the putative neurotransmitter, glycine, into the rabbit retina. *Invest. Ophthalmol. 11,* 191–198.

Cicerone, C. M. (1976). Cones survive rods in the light-damaged eye of the albino rat. *Science 19,* 1183–1185.

Costa, E., Guidotti, A., Mao, C. C., and Suvia, A. (1975). New concepts on the mechanism of action of benzodiazepines. *Life Sci.* **17,** 167–186.

Demarest, K. T., and Moore, K. E. (1980). Accumulation of L-DOPA in the median eminence: An index of tuberoinfundibular dopaminergic nerve activity. *Endocrinology* **106,** 463–468.

DePrada, M. (1977). Dopamine content and synthesis in retina and n. accumbens septi: Pharmacological and light-induced modifications. *Adv. Biochem. Psychopharmacol.* **16,** 311–319.

DeVries, G. W., Cohen, A. I., Hall, I. A., and Ferrendelli, J. A. (1978). Cyclic nucleotide levels in normal and biologically fractionated mouse retina: Effects of light and dark adaptation. *J. Neurochem.* **31,** 1345–1351.

Ehinger, B. (1972). Uptake of tritiated glycine into neurons of the human retina. *Experientia 28,* 1042–1043.

Ehinger, B. (1976). Biogenic monoamines as transmitters in the retina. *In* "Transmitters in the Visual Process" (S. L. Bonting, ed.), pp. 145–163. Pergamon, Oxford.

Ehinger, B. (1977). Synaptic connections of the dopaminergic retinal neurons. *Adv. Biochem. Psychopharmacol.* **16,** 299–306.

Ehinger, B., and Falck, B. (1971). Autoradiography of some suspected neurotransmitter substances: GABA, glycine, glutamic acid, histamine, dopamine and L-DOPA. Brain Res. **33,** 157–172.

Ehinger, B., and Lindberg, B. (1974). Light-evoked release of glycine from the retina. *Nature (London)* **251,** 727–728.

Enna, S. J., and Snyder, S. H. (1975). Properties of γ-aminobutyric acid (GABA) receptor binding in rat brain synaptic membrane fractions. *Brain Res. 100,* 81-97.

Geller, H. M., Taylor, D. A., and Hoffer, B. J. (1978). Benzodiazepines and central inhibitory mechanisms. *Naunyn-Schmiedeberg's Arch. Pharmacol.* **304,** 81-88.

Graham, L. T. (1972). Intraretinal distribution of GABA content and GAD activity. *Brain Res.* **36,** 476-479.

Graham, L. T. (1974). Comparative aspects of neurotransmitters in the retina. *In* "The Eye" (H. Davson and L. T. Graham, eds.), Vol. 5 pp. 283-342. Academic Press, New York.

Haefely, W. E. (1977). Synaptic pharmacology of barbiturates and benzodiazepines. *Agents Actions* **7,** 353-359.

Häggendal, J., and Malmfors, T. (1965). Identification and cellular localization in the retina and the choroid of the rabbit. *Acta Physiol. Scand.* **64,** 58-66.

Hayden, S. A., Mills, J. W., and Masland, R. M. (1980). Acetylcholine synthesis by displaced amacrine cells. *Science* **210,** 435-437.

Iuvone, P. M., Galli, C. L., Garrison-Gund, C. K., and Neff, N. H. (1978). Light stimulates tyrosine hydroxylase and dopamine synthesis in retinal amacrine neurons. *Science* **200,** 901-902.

Iuvone, P. M., Marshburn, P. B., and Neff, N. H. (1980). Regulation of tyrosine hydroxylase activity in whole retina retinal cell suspensions and retinal homogenates. *Soc. Neurosci. 10th Annu. Meet.* **6,** 143.

Jarboe, C. H., and Porter, L. A. (1965). The preparative column chromatographic separation of picrotoxinin. *J. Chromatog.* **19,** 427-428.

Kamp, C. W., and Morgan, W. W. (1980a). Some barbiturates enhance the effect of muscimol on dopamine turnover in the rat retina. *J. Pharmacol. Exp. Ther.* **213,** 332-336.

Kamp, C. W., and Morgan, W. W. (1980b). GABA antagonists enhance dopamine turnover in the rat retina *in vivo. Eur. J. Pharmacol.* **69,** 273-279.

Kamp, C. W., and Morgan, W. W. (1980c). Diazepam inhibits the light-induced increase in retinal dopamine turnover. *Soc. Neurosci. 10th Annu. Meet.* **6,** 142.

Kojema, K., Mizuno, K., and Miyazaki, M. (1958). Gamma-aminobutyric acid in ocular tissue. *Nature (London)* **181,** 1200-1201.

Kramer, S. G. (1971). Dopamine: A retinal neurotransmitter. I. Retinal uptake, storage and light-stimulated release of ^{3}H dopamine *in vivo. Invest. Ophthalmol.* **10,** 438-452.

Kramer, S. G. (1976). Dopamine in retinal neurotransmission. *In* "Transmitters in the Visual Process" (S. P. Bonting, ed.), pp. 165-198. Pergamon, Oxford.

Kuriyama, K., Sisken, B., Haber, B., and Roberts, E. (1968). The γ-aminobutyric acid system in rabbit retina. *Brain Res.* **9,** 165-168.

LaVail, M. M. (1976). Survival of some photoreceptor cells in albino rats following long-term exposure to continuous light. *Invest. Ophthalmol.* **15,** 64-70.

Lemmon, V., and Anderson, K. V. (1979). Central neurophysiological correlates of constant light-induced retinal degeneration. *Exp. Neurol.* **63,** 50-75.

Lolley, R. N., Schmidt, S. Y., and Farber, D. B. (1974). Alterations in cyclic AMP metabolism associated with photoreceptor cell degeneration in the C3H mouse. *J. Neurochem.* **22,** 701-707.

McCulloch, J., Savaki, H. E., McCulloch, M. C., and Sokoloff, L. (1980). Retina-dependent activation by apomorphine of metabolic activity in the superficial layer of the superior colliculus. *Science* **207,** 313-315.

MacDonald, R. L., and Baker, J. L. (1978). Differential actions of anticonvulsant and

anesthetic barbiturates revealed by use of cultured mammalian neurons. *Science* **200,** 775–777.
Magistretti, P. J., and Schorderet, M. (1979). Dopamine receptors in bovine retina: Characterization of the ^{3}H-spiroperidol binding and its use for screening dopamine receptor affinity of drugs. *Life Sci.* **25,** 1675–1686.
Marc, R. E., Stell, W. K., Bok, D., and Lam, D. M. K. (1978). GABAergic pathways in the goldfish retina. *J. Comp. Neurol.* **182,** 221–246.
Masland, R. H., and Ames, A., III (1976). Responses to acetylcholine of ganglion cells is an isolated mammalian retina. *J. Neurophysiol.* **39,** 1220–1225.
Masland, R. H., and Livingston, C. J. (1976). Effect of stimulation with light on synthesis and release of acetylcholine by isolated mammalian retina. *J. Neurophysiol.* **39,** 1210–1219.
Massey, S. C., and Neal, M. J. (1979). The light-evoked release of acetylcholine from the rabbit retina *in vivo* and its inhibition by γ-aminobutyric acid. *J. Neurochem.* **32,** 1327–1329.
Morgan, W. W., and Kamp, C. W. (1980a). A GABAergic influence on the light-induced increase in dopamine turnover in the dark-adapted rat retina *in vivo. J. Neurochem.* **34,** 1082–1086.
Morgan, W. W., and Kamp, C. W. (1980b). Dopaminergic amacrine neurons of rat retinas with photoreceptor degeneration continue to respond to light. *Life Sci.* **26,** 1619–1626.
Nestoros, J. N. (1980). Ethanol specifically potentiates GABA-mediated neurotransmission in feline cerebral cortex. *Science* **209,** 708–710.
Nichols, C. W., Jacobowitz, D., and Hottenstein, M. (1967). The influence of light and dark on the catecholamine content of the retina and choroid. *Invest. Ophthalmol.* **6,** 642–646.
Osborne, N. N. (1980). Benzodiazepine binding to bovine retina. *Neurosci. Lett.* **16,** 167–170.
O'Steen, W. K., and Anderson, K. V. (1971). Photically evoked responses in the visual system of rats exposed to continuous light. *Exp. Neurol.* **30,** 525–534.
O'Steen W. K., Shear, C. R., and Anderson, K. V. (1972). Retinal damage after prolonged exposure to visible light: A light and electron microscopic study. *Am. J. Anat. 134,* 5–22.
Palkovits, M., Brownstein, M., Saavedra, J. M., and Axelrod, J. (1974). Norepinephrine and dopamine content of hypothalamic nuclei of the rat. *Brain Res.* **77,** 137–149.
Ransom, B. R., and Barker, J. L. (1975). Pentobarbital modulates transmitter effects on mouse spinal neurons grown in tissue culture. *Nature (London)* **254,** 703–705.
Ransom, B. R., and Barker, J. L. (1976). Pentobarbital selectively enhances GABA-mediated post synaptic inhibition in tissue cultured mouse spinal neurons. *Brain Res.* **114,** 530–535.
Redburn, D. A., Clement-Cormier, Y., and Lam, D. M. K. (1980). Dopamine receptors in the goldfish retina: ^{3}H spiroperidol and ^{3}H domperidone binding; and dopamine-stimulated adenylate cyclase activity. *Life Sci.* **27,** 23–31.
Regan, J. W., Roeske, W. R., and Yamamura, H. I. (1980). ^{3}H-flunitrazepam binding to bovine retina and the effect of GABA thereon. *Neuropharmacology* **19,** 413–414.
Schaeffer, J. M. (1980). Identification of dopamine receptors in the rat brain. *Exp. Eye Res.* **30,** 431–437.
Sokal, R. R., and Rohlf, F. J. (1969). "Biometry," pp. 242–245, p. 226, and pp. 208–209. Freeman, San Francisco, California.
Spano, P. F., Kumakura, K., and Trabucchi, M. (1976). Dopamine-sensitive adenylate

cyclase in the retina: A point of action for D-LSD. *Adv. Biochem. Psychopharmacol.* **15**, 357-365.

Spano, P. F., Govoni, S., Hofmann, M., Kumakura, K., and Trabucchi, M. (1977). Physiological and pharmacological influences on dopaminergic receptors in the retina. *Adv. Biochem. Psychopharmacol.* **16**, 307-310.

Starr, M. S. (1975). The effects of various amino acids, dopamine, and some convulsants on the electroretinogram of the rabbit. *Exp. Eye Res.* **21**, 79-87.

Steel, R. G. D., and Torrie, J. H. (1960). "Principles and Procedures of Statistics," p. 173 McGraw-Hill, New York.

Straschill, M. (1968). Actions of drugs on single neurons in the cat's retina. *Vision Res.* **8**, 35-47.

Straschill, M., and Perwein, J. (1969). The inhibition of retinal ganglion cells by catecholamines and γ-aminobutyric acid. *Pfluegers Arch.* **312**, 45-54.

Straschill, M., and Perwein, J. (1973). The effect of iontophoretically applied acetylcholine upon the cat's retinal ganglion cells. *Pfluegers Arch.* **339**, 289-298.

Straschill, M., and Perwein, J. (1975). Effects of biogenic amines and amino acids on the cat's retinal ganglion cells. "Golgi Centennial Symposium Proceedings" (M. Santini, ed.), pp. 583-591. Raven, New York.

Thomas, T. N., Clement-Cormier, Y. C., and Redburn, D. A. (1978). Uptake and release of ^{3}H dopamine and dopamine-sensitive adenylate cyclase activity in retinal synaptosomal fractions. *Brain Res.* **155**, 391-396.

Trabucchi, M., Govoni, S. Tonon, G. C., and Spano, P. F. (1976). Dopamine receptor supersensitivity in rat retina after light deprivation. *In* "Catecholamines and Stress" (E. Usdin, R. Kvetnansky, and I. J. Kopin, eds.), pp. 225-230. Pergamon, Oxford.

Vaughan, J. E., Barber, R. P., Saito, K., Roberts, E., and Famiglietti, E. V., Jr. (1978). Immunological identification of GABAergic neurons in rat retina. *Anat. Rec.* **190**, 571-572.

Voaden, M. J. (1976). Gamma-aminobutyric acid and glycine as retinal neurotransmitters. *In* "Transmitters in the Visual Process" (S. L. Bonting, ed.), pp. 107-125. Pergamon, Oxford.

Watling, K. J., Dowling, J. E., and Iversen, L. L. (1979). Dopamine receptors in the retina may all be linked to adenylate cyclase. *Nature (London)* **281**, 578-580.

Index

B

D

E

F

N

O

P

S

W

X

Z

CELL BIOLOGY: A Series of Monographs

EDITORS

G. M. Padilla, G. L. Whitson, and I. L. Cameron (editors). THE CELL CYCLE: Gene-Enzyme Interactions, 1969

A. M. Zimmerman (editor). HIGH PRESSURE EFFECTS ON CELLULAR PROCESSES, 1970

I. L. Cameron and J. D. Thrasher (editors). CELLULAR AND MOLECULAR RENEWAL IN THE MAMMALIAN BODY, 1971

I. L. Cameron, G. M. Padilla, and A. M. Zimmerman (editors). DEVELOPMENTAL ASPECTS OF THE CELL CYCLE, 1971

P. F. Smith. The BIOLOGY OF MYCOPLASMAS, 1971

Gary L. Whitson (editor). CONCEPTS IN RADIATION CELL BIOLOGY, 1972

Donald L. Hill. THE BIOCHEMISTRY AND PHYSIOLOGY OF *TETRAHYMENA*, 1972

Kwang W. Jeon (editor). THE BIOLOGY OF AMOEBA, 1973

Dean F. Martin and George M. Padilla (editors). MARINE PHARMACOGNOSY: Action of Marine Biotoxins at the Cellular Level, 1973

Joseph A. Erwin (editor). LIPIDS AND BIOMEMBRANES OF EUKARYOTIC MICROORGANISMS, 1973

A. M. Zimmerman, G. M. Padilla, and I. L. Cameron (editors). DRUGS AND THE CELL CYCLE, 1973